DEDICATION

To my wife JoAnn—JWS

To my immediate and extended family—RBJ

To Dee, Amelia, and Alex—SDB

BRIEF CONTENTS

CONTENTS

FEATURES

Systems Analysis and Design in a Changing World, Sixth Edition, was written and developed with instructor and student needs in mind. Here is just a sample of the unique and exciting features that help bring the field of systems analysis and design to life.

The innovative and **entirely new text organization** starts with a complete beginning-to-end system development example, moves immediately to systems analysis models and techniques, and then to system design concepts emphasizing system architecture, user interfaces, and system interfaces. Analysis and much of design is covered in the first seven chapters. Next, the text focuses on managing system development projects, including contemporary approaches such as Agile development, the Unified Process, Extreme Programming, and Scrum, after the student has a chance to learn what is involved in system development. Finally, the text covers detailed design topics, deployment topics, and current trends.

The text uses a **completely updated** integrated case study of moderate complexity—**Ridgeline Mountain Outfitters (RMO)**—to illustrate key concepts and techniques. In addition, a smaller RMO application—the Tradeshow System—is used in Chapter 1 to introduce the entire system development process.

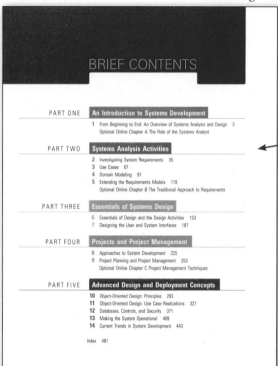

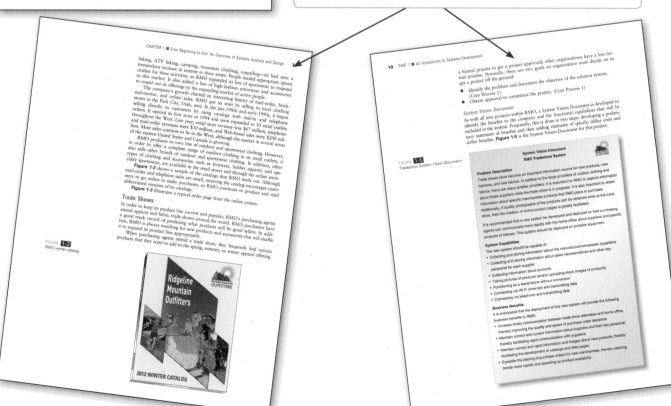

FEATURES

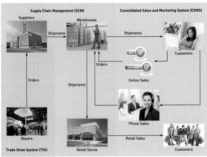

An overview of the RMO Consolidated Sales and Marketing System (CSMS) is presented in Chapters 1 and 2 to place the projects in context. The **planned system architecture** provides for rich examples—a Web-based component, a wireless smartphone/tablet application, and a client/server Windows-based component.

The new **Consolidated Sales and Marketing System (CSMS)** is the system development project used throughout the text for examples and explanations. It is strategically important to RMO, and the company must integrate the new system with legacy systems and other planned systems. There are four subsystems, and the requirements and design models are all new in this edition. UML diagrams are used throughout for examples and exercises.

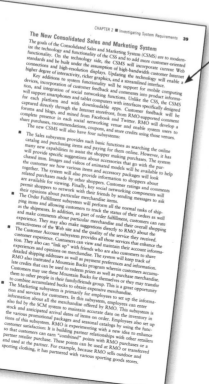

FEATURES

The text describes **predictive and adaptive approaches to the SDLC** and recommends **Agile, iterative development** for most projects. The SDLC used in the text features a generic, condensed version of the Unified Process SDLC, which emphasizes iterations and core development processes found in most current development methodologies. Core processes and iterations are emphasized over phases to reduce the confusion that ordinarily occurs when students are taught "phases" and then told to use iterations. Project planning and project management are emphasized throughout, and the book focuses more on systems analysis and systems design as development disciplines rather than phases.

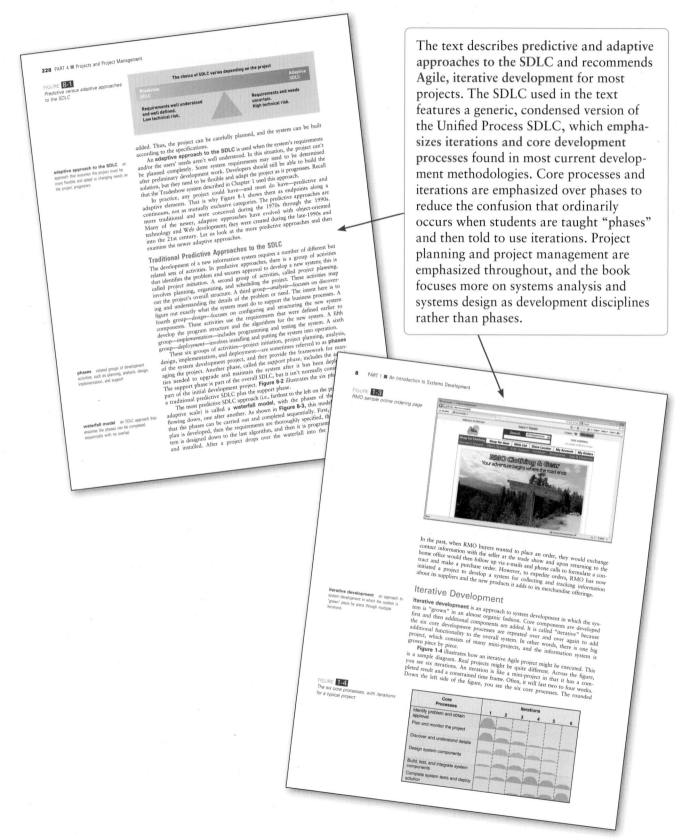

228 PART 4 ▪ Projects and Project Management

FIGURE 8-1
Predictive versus adaptive approaches to the SDLC

The choice of SDLC varies depending on the project

Predictive SDLC — Requirements well understood and well defined. Low technical risk.

Adaptive SDLC — Requirements and needs uncertain. High technical risk.

added. Thus, the project can be carefully planned, and the system can be built according to the specifications.

An **adaptive approach to the SDLC** is used when the system's requirements and/or the users' needs aren't well understood. In this situation, the project can't be planned completely. Some system requirements may need to be determined after preliminary development work. Developers should still be able to build the solution, but they need to be flexible and adapt the project as it progresses. Recall that the Tradeshow system described in Chapter 1 used this approach.

In practice, any project could have—and most do have—predictive and adaptive elements. That is why Figure 8-1 shows them as endpoints along a continuum, not as mutually exclusive categories. The predictive approaches are more traditional and were conceived during the 1970s through the 1990s. Many of the newer, adaptive approaches have evolved with object-oriented technology and Web development; they were created during the late-1990s and into the 21st century. Let us look at the more predictive approaches and then examine the newer adaptive approaches.

adaptive approach to the SDLC an approach that assumes the project must be more flexible and adapt to changing needs as the project progresses

Traditional Predictive Approaches to the SDLC

The development of a new information system requires a number of different but related sets of activities. In predictive approaches, there is a group of activities that identifies the problem and secures approval to develop a new system; this is called *project initiation*. A second group of activities, called *project planning*, involves planning, organizing, and scheduling the project. These activities map out the project's overall structure. A third group—*analysis*—focuses on discovering and understanding the details of the problem or need. The intent here is to figure out exactly what the system must do to support the business processes. A fourth group—*design*—focuses on configuring and structuring the new system components. These activities use the requirements that were defined earlier to develop the program structure and the algorithms for the new system. A fifth group—*implementation*—includes programming and testing the system. A sixth group—*deployment*—involves installing and putting the system into operation.

These six groups of activities—are sometimes referred to as **phases** of the system development project, and they provide the framework for managing the project. Another phase, called the *support phase*, includes the activities needed to upgrade and maintain the system after it has been deployed. The support phase is part of the overall SDLC, but it isn't normally considered part of the initial development project. **Figure 8-2** illustrates the six phases of a traditional predictive SDLC plus the support phase.

The most predictive SDLC approach (i.e., farthest to the left on the predictive-adaptive scale) is called a **waterfall model**, with the phases of the SDLC flowing down, one after another. As shown in **Figure 8-3**, this model assumes that the phases can be carried out and completed sequentially. First, the project plan is developed, then the requirements are thoroughly specified, then the system is designed down to the last algorithm, and then it is programmed, tested, and installed. After a project drops over the waterfall into the next

phases related groups of development activities, such as planning, analysis, design, implementation, and support

waterfall model an SDLC approach that assumes the phases can be completed sequentially with no overlap

---

8 PART 1 ▪ An Introduction to Systems Development

FIGURE 1-3
RMO sample online ordering page

In the past, when RMO buyers wanted to place an order, they would exchange contact information with the seller at the trade show and upon returning to the home office would then follow up via e-mails and phone calls to formulate a contract and make a purchase order. However, to expedite orders, RMO has now initiated a project to develop a system for collecting and tracking information about its suppliers and the new products it adds to its merchandise offerings.

Iterative Development

Iterative development is an approach to system development in which the system is "grown" in an almost organic fashion. Core components are developed first and then additional components are added. It is called "iterative" because the six core development processes are repeated over and over again to add additional functionality to the overall system. In other words, there is one big project, which consists of many mini-projects, and the information system is grown piece by piece.

Figure 1-4 illustrates how an iterative Agile project might be executed. This is a sample diagram. Real projects might be quite different. Across the figure, you see six iterations. An iteration is like a mini-project in that it has a completed result and a constrained time frame. Often, it will last two to four weeks. Down the left side of the figure, you see the six core processes. The rounded

iterative development an approach to system development in which the system is "grown" piece by piece through multiple iterations

FIGURE 1-4
The six core processes, with iterations for a typical project

Core Processes	Iterations					
	1	2	3	4	5	6
Identify problem and obtain approval						
Plan and monitor the project						
Discover and understand details						
Design system components						
Build, test, and integrate system components						
Complete system tests and deploy solution						

FEATURES

Each chapter provides a chapter outline, states clear learning objectives, and includes an opening case study.

7

Designing the User and System Interfaces

Chapter Outline

- User and System Interfaces
- Understanding the User Interface
- User-Interface Design Concepts
- The Transition from Analysis to User-Interface Design
- User-Interface Design
- Identifying System Interfaces
- Designing System Inputs
- Designing System Outputs

Learning Objectives

After reading this chapter, you should be able to:

- Describe the difference between user interfaces and system interfaces
- Describe the historical development of the field of human-computer interaction (HCI)
- Discuss how visibility and affordance affect usability
- Describe user-interface guidelines that apply to all types of user-interface types and additional guidelines specific to Web pages and mobile applications
- Create storyboards to show the sequence of forms used in a dialog
- Discuss examples of system interfaces found in information systems
- Define system inputs and outputs based on the requirements of the application program
- Design printed and on-screen reports appropriate for recipients

---

OPENING CASE

Blue Sky Mutual Funds: A New Development Approach

Jim Williams, vice president of finance for Blue Sky Mutual Funds, spoke first. "There are some things I like about this new approach, but other things worry me," he told Gary Johnson, the company's director of information technology.

"This idea of 'growing' the system through several iterations makes a lot of sense to me. It is always hard for my people to know exactly what they need a new information system to do and what will work best for the company. So, if they can get their hands on the system early, they can begin acceptance testing and try it out to see whether it addresses their needs in the best way.

"Let me see if I understand the big picture, though. Your development team and my investment advisors will decide on a few core processes that the system needs to support and then your team will design and build a system to support those core processes. You will do that in a mini-project that will last about six weeks. Then, you will continue adding more functionality through several other mini-projects until the system is complete and functioning well. Is that right?"

Jim was becoming more enthusiastic about this new approach to system development.

"Yes, that's the basic idea," Gary said. "Your users need to understand that the first few versions of the system won't be complete and may not be completely robust either. But these early versions will give them something to work with and try out. We also need good feedback from their acceptance testing so the system will be thoroughly tested by the time we are through."

"I realize that," Jim said. "My people will like not having to think from the very beginning about everything they need the system to do. They'll like being able to try things out. As I said earlier, I like this approach. However, the part I don't like about this approach is that it will be more difficult for you to give me a firm time schedule and project cost. That worries me. In the past, those have been two of the major tools we used to monitor a project's progress. Are you saying that now we won't have a schedule at all? And you want an open budget?" Jim frowned.

"It's not as bad as it first sounds," Gary said. "This approach is an 'adaptive' approach, by which I mean that because the system is growing, the project is more open ended. The project manager will still create a schedule and estimate the project costs, but she won't even try to identify and lock in all the required functionality for several of the iterations. Because the system's scope is going to continually be refined over the first few iterations, there is the risk of 'scope creep.' That is one of the biggest risks with adaptive approaches. You and I should meet with the project manager fairly frequently to ensure that the scope is controlled and the project doesn't get out of control."

"Okay," Jim said. "You have convinced me to try this new approach. However, let's treat this project as a pilot and see how it works. If it's successful, we will consider using this iterative approach on our other projects." Jim and Gary agreed that a pilot was the best way to get started. Gary then headed off to meet with the project manager and get the project started.

Overview

Chapter 8 introduced you to the SDLC and the various alternatives for organizing software development activities. By now, you may be asking yourself such questions as:

- "How are all these activities coordinated?"
- "How do I know which tasks to do first?"
- "How is the work assigned to the different teams and team members?"
- "How do I know which parts of the new system should be developed first?"

The purpose of project planning and project management is to bring some order to all these (sometimes seemingly unrelated) tasks. As you will learn in this chapter, the success of any given project highly depends on the skills and abilities of those managing the project. You will also learn that project management skills aren't only for project managers—that all the project team members contribute to the management of the project and thus to its success.

Margin definitions of key terms are placed in the text when the term is first used. Each chapter includes extensive **figures** and **illustrations** designed to clarify and summarize key points and to provide examples of UML models and other deliverables produced by an analyst.

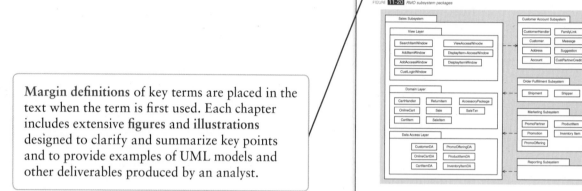

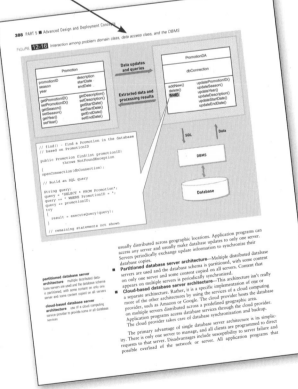

FEATURES

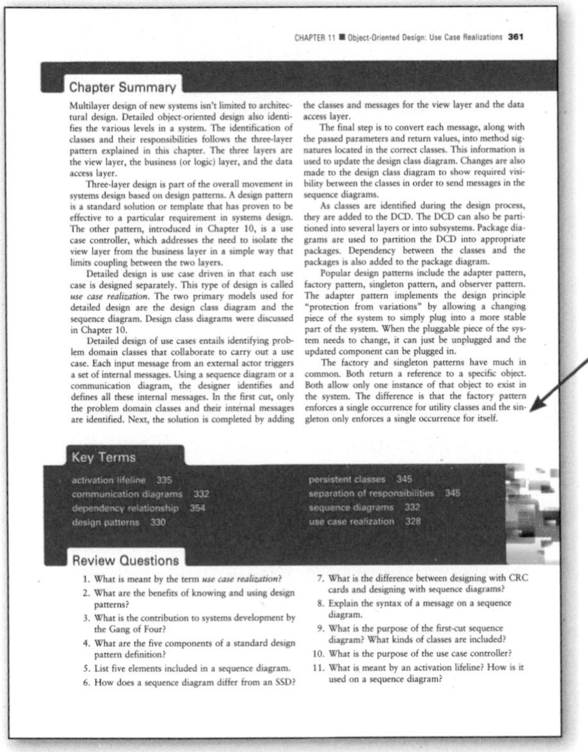
CHAPTER 11 ■ Object-Oriented Design: Use Case Realizations 361

Chapter Summary

Multilayer design of new systems isn't limited to architectural design. Detailed object-oriented design also identifies the various levels in a system. The identification of classes and their responsibilities follows the three-layer pattern explained in this chapter. The three layers are the view layer, the business (or logic) layer, and the data access layer.

Three-layer design is part of the overall movement in systems design based on design patterns. A design pattern is a standard solution or template that has proven to be effective to a particular requirement in systems design. The other pattern, introduced in Chapter 10, is a use case controller, which addresses the need to isolate the view layer from the business layer in a simple way that limits coupling between the two layers.

Detailed design is use case driven in that each use case is designed separately. This type of design is called *use case realization*. The two primary models used for detailed design are the design class diagram and the sequence diagram. Design class diagrams were discussed in Chapter 10.

Detailed design of use cases entails identifying problem domain classes that collaborate to carry out a use case. Each input message from an external actor triggers a set of internal messages. Using a sequence diagram or a communication diagram, the designer identifies and defines all these internal messages. In the first cut, only the problem domain classes and their internal messages are identified. Next, the solution is completed by adding the classes and messages for the view layer and the data access layer.

The final step is to convert each message, along with the passed parameters and return values, into method signatures located in the correct classes. This information is used to update the design class diagram. Changes are also made to the design class diagram to show required visibility between the classes in order to send messages in the sequence diagrams.

As classes are identified during the design process, they are added to the DCD. The DCD can also be partitioned into several layers or into subsystems. Package diagrams are used to partition the DCD into appropriate packages. Dependency between the classes and the packages is also added to the package diagram.

Popular design patterns include the adapter pattern, factory pattern, singleton pattern, and observer pattern. The adapter pattern implements the design principle "protection from variations" by allowing a changing piece of the system to simply plug into a more stable part of the system. When the pluggable piece of the system needs to change, it can just be unplugged and the updated component can be plugged in.

The factory and singleton patterns have much in common. Both return a reference to a specific object. Both allow only one instance of that object to exist in the system. The difference is that the factory pattern enforces a single occurrence for utility classes and the singleton only enforces a single occurrence for itself.

Key Terms

activation lifeline 335
communication diagrams 332
dependency relationship 354
design patterns 330
persistent classes 345
separation of responsibilities 345
sequence diagrams 332
use case realization 328

Review Questions

1. What is meant by the term *use case realization*?
2. What are the benefits of knowing and using design patterns?
3. What is the contribution to systems development by the Gang of Four?
4. What are the five components of a standard design pattern definition?
5. List five elements included in a sequence diagram.
6. How does a sequence diagram differ from an SSD?
7. What is the difference between designing with CRC cards and designing with sequence diagrams?
8. Explain the syntax of a message on a sequence diagram.
9. What is the purpose of the first-cut sequence diagram? What kinds of classes are included?
10. What is the purpose of the use case controller?
11. What is meant by an activation lifeline? How is it used on a sequence diagram?

End-of-chapter material includes a **detailed summary**, an **indexed list of key terms**, and ample review questions.

Each chapter also includes a collection of **problems and exercises** that involves additional research or problem solving, an end-of-chapter **case study** that invites students to complete analysis and design tasks appropriate to the chapter, **four running cases** that create challenging and integrated course assignments, and a list of **further resources**.

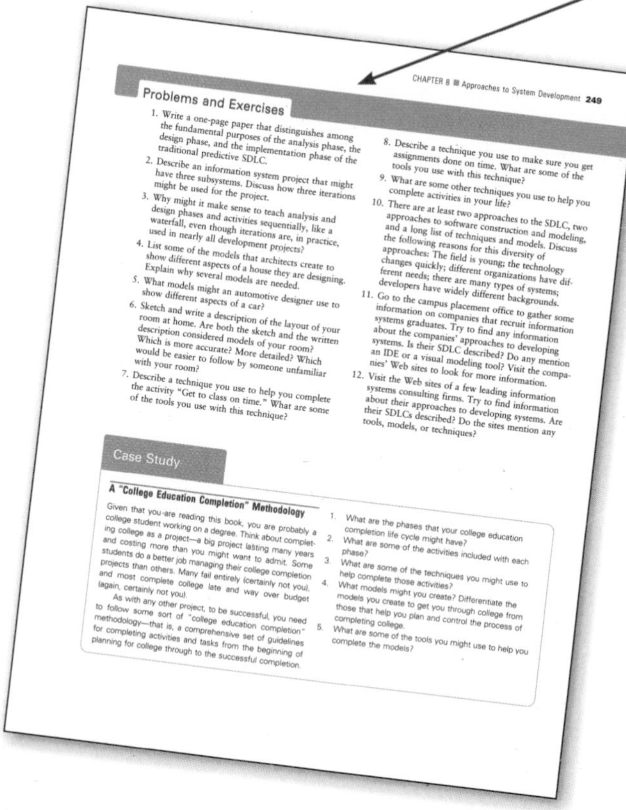
CHAPTER 8 ■ Approaches to System Development 249

Problems and Exercises

1. Write a one-page paper that distinguishes among the fundamental purposes of the analysis phase, the design phase, and the implementation phase of the traditional predictive SDLC.
2. Describe an information system project that might have three subsystems. Discuss how three iterations might be used for the project.
3. Why might it make sense to teach analysis and design phases and activities sequentially, like a waterfall, even though iterations are, in practice, used in nearly all development projects?
4. List some of the models that architects create to show different aspects of a house they are designing. Explain why several models are needed.
5. What models might an automotive designer use to show different aspects of a car?
6. Sketch and write a description of the layout of your room at home. Are both the sketch and the written description considered models of your room? Which is more accurate? More detailed? Which would be easier to follow by someone unfamiliar with your room?
7. Describe a technique you use to help you complete the activity "Get to class on time." What are some of the tools you use with this technique?
8. Describe a technique you use to make sure you get assignments done on time. What are some of the tools you use with this technique?
9. What are some other techniques you use to help you complete activities in your life?
10. There are at least two approaches to the SDLC, two approaches to software construction and modeling, and a long list of techniques and models. Discuss the following reasons for this diversity of approaches: The field is young; the technology changes quickly; different organizations have different needs; there are many types of systems; developers have widely different backgrounds.
11. Go to the campus placement office to gather some information on companies that recruit information systems graduates. Try to find any information about the companies' approaches to developing systems. Is their SDLC described? Do any mention an IDE or a visual modeling tool? Visit the companies' Web sites to look for more information.
12. Visit the Web sites of a few leading information systems consulting firms. Try to find information about their approaches to developing systems. Are their SDLCs described? Do the sites mention any tools, models, or techniques?

Case Study

A "College Education Completion" Methodology

Given that you are reading this book, you are probably a college student working on a degree. Think about completing college as a project—a big project lasting many years and costing more than you might want to admit. Some students do a better job managing their college completion projects than others. Many fail entirely (certainly not you), and most complete college late and way over budget (again, certainly not you).

As with any other project, to be successful, you need to follow some sort of "college education completion" methodology—that is, a comprehensive set of guidelines for completing activities and tasks from the beginning of planning for college through to the successful completion.

1. What are the phases that your college education completion life cycle might have?
2. What are some of the activities included with each phase?
3. What are some of the techniques you might use to help complete those activities?
4. What models might you create? Differentiate the models you create to get you through college from those that help you plan and control the process of completing college.
5. What are some of the tools you might use to help you complete the models?

250 PART 4 ■ Projects and Project Management

RUNNING CASE STUDIES

Community Board of Realtors

The Board of Realtors Multiple Listing Service (MLS) system isn't very large in terms of use cases and domain classes. In that respect, the functional requirements are simple and well understood. MLS needs a Web site with public access to the listings, and it also needs to allow agents and brokers to log in to the system to add and update listings. There is very little back-end administrative data maintenance required, except to add or update a real estate office or agent.

1. Compared to the Tradeshow application described in Chapter 1, how long might this project take, and which approach to the SDLC would be most appropriate?
2. If you use a predictive SDLC, how much time might each phase of the project take? How much overlap of phases might you plan for? Be specific about how you would overlap the phases.
3. If you use an adaptive SDLC, how many iterations might you plan to include? What use cases would you analyze, design, and implement in the first iteration? What use cases would you work on in the second iteration? In additional iterations? Think in terms of getting the core functionality implemented early and then building the supporting functionality.
4. Let us say this project focused on Web access to the MLS. If you also plan to deploy a smartphone application for use by the public and by the agents and brokers, how might this affect your choice of the approach to the SDLC? What are the implications for including the smartphone application in the initial project versus having a separate project for wireless later?
5. Consider using incremental development to include the Web application and the wireless support. Describe what would be included in the first and second deployments of the project. Take into consideration that you might want to work on some initial problem solving for requirements, design, and implementation of the wireless support at the same time you are working on the Web application.

The Spring Breaks 'R' Us Travel Service

Recall from Chapter 2 that SBRU's initial system included four major subsystems: Resort relations, Student booking, Accounting and finance, and Social networking. The project calls for an adaptive approach to the SDLC for several reasons. One, it is relatively large in scope. Two, there is a diverse set of users in several functional areas, internal and external to the company and in several foreign countries. Three, the project needs to use an assortment of newer technologies that can communicate anytime and anywhere.

1. The SBRU information system includes four major subsystems: Resort relations, Student booking, Accounting and finance, and Social networking. Although you have only worked with the domain model class diagram for the Social networking subsystem, list as many of the domain classes that would probably be involved in each of the subsystems. Note which classes are used by more than one subsystem.
2. Based on the overlapping classes, what domain classes seem to be part of the core functionality for SBRU? Draw a domain model class diagram that shows these classes and their associations.
3. Let us say you plan to implement the basic use cases that create and maintain the classes that are part of the core functionally you just modeled. Describe what domain classes you would focus on in each iteration if you assumed that you would need two iterations for the initial core functionality and two additional iterations to complete each of the subsystems.
4. How might you use incremental development to get some core functionality or some subsystems deployed and put into use before the project is completed?

(continued on page 251)

PREFACE

When we wrote the first edition of this textbook, the world of system development was in a major transition period—from structured methodologies to object-oriented methodologies. We were among the first to introduce a comprehensive treatment of object-oriented methodologies, and *Systems Analysis and Design in a Changing World, Sixth Edition*, continues to be the leader in teaching object-oriented techniques.

However, change continues. Today, many new initiatives and trends have become firmly embedded in the world of system development. First and foremost is the ubiquitous access to the Internet throughout the global economy. The resulting explosion of connectivity means that project teams are now distributed around the world. In addition, large providers (such as Microsoft) and a proliferation of small providers now contribute to a wonderfully rich and varied software development environment.

In order to manage system development teams in today's distributed, fast-paced, connected, ever-changing environment, the techniques for software development and the approach to project management have expanded. Along with the foundational project management principles, additional approaches and philosophies provide new, success-oriented methodologies, such as iterative, incremental development approaches. These are thoroughly covered in this edition.

Even though *Systems Analysis and Design in a Changing World, Sixth Edition*, continues to be the leader in its field, with thorough treatment of such topics as use cases, object-oriented modeling, comprehensive project management, the unified modeling language, and Agile techniques, it was time to take another step forward in textbook design. This edition uses an innovative approach to teaching systems analysis and design, taking advantage of the new teaching tools and techniques that are now available. As a result, not only is systems analysis and design easier to learn by using this approach, it is also easier to teach. It brings together the best approaches for teachers *and* students.

In this edition, we accomplish three major new objectives. First, we teach all the essential principles of system development—principles that must be followed in today's connected environment. Second, we teach and explain the new methodologies and techniques that are now available because of widespread connectivity. Third, we have organized and revamped the textbook so that it teaches these new concepts in a new way.

For example, Chapter 1 presents a complete iteration in the development of a new system. Students get to see that complete iteration—from beginning to end (through implementation and testing)—before having to learn abstract principles or memorize terms. Also, the newly written running cases throughout the book focus on current issues of communication and connectedness and take the students through all aspects of system development. Along with the textbook itself, there are teaching tools, such as video explanations of complex models and topics. We have also expanded the *Instructor's Materials* and enhanced the aids available through CourseMate, our online resource. Additional online chapters are also available to enhance and extend the learning experience.

We are excited about this new approach. The time is right for new materials and new tools for teaching systems analysis and design. Instructors will find this textbook intuitive, powerful, and easy to use. Students will find it engaging and empowering. Many concepts are presented so the students can teach themselves, with coaching and direction provided by the professor. It will be an incredible experience to teach and learn with this textbook.

Innovations

This edition is innovative in many respects. It includes key concepts from traditional and object-oriented approaches, covers the use case–driven object-oriented approach (with UML modeling being detailed in depth), emphasizes Agile and iterative development, and incorporates the latest concepts in project management. Also, the material is completely reorganized to better support learning systems analysis and design.

Coverage of Object Orientation and Traditional Analysis and Design

This textbook is unique in its integration of key systems-modeling concepts that apply to the traditional structured approach and the object-oriented approach— user goals and events that trigger system use cases, plus objects/entities that are part of the system's problem domain. We devote one chapter to identifying use cases and another chapter to modeling key objects/entities, including coverage of entity-relationship diagrams, while emphasizing UML domain model class diagrams. After completing these chapters, instructors can cover structured analysis and design by including an online chapter, or they can focus on object-oriented analysis and design by using the chapters in this textbook. It is assumed from the beginning that everyone should understand the key object-oriented concepts. The traditional approach isn't discarded; key structured concepts are included. But these days, most instructors are emphasizing the object-oriented approach.

Full Coverage of UML and the Object-Oriented Approach

The object-oriented approach presented in this textbook is based on the Unified Modeling Language (UML 2.0) from the Object Management Group, as originated by Grady Booch, James Rumbaugh, and Ivar Jacobson. A model-driven approach to analysis starts with use cases and scenarios and then defines problem domain classes involved in the users' work. We include requirements modeling with use case diagrams, domain modeling, use case descriptions, activity diagrams, and system sequence diagrams. The FURPS+ model is used to emphasis functional and nonfunctional requirements.

Design principles and design patterns are discussed in depth, and system architecture is modeled by using UML component diagrams and package diagrams. Detailed design models are also discussed in detail, with particular attention given to use case realization with CRC cards, sequence diagrams, and design class diagrams.

Project Management Coverage

Many undergraduate programs depend on their systems analysis and design course to teach project management principles. To satisfy this need, we cover project management by taking a four-pronged approach. First, specific project management techniques, skills, and tasks are included and highlighted throughout this book. This integration teaches students how to apply specific project management tasks to the various activities of the systems development life cycle, including iterative development. Second, complete coverage of project planning and project management is included in a separate chapter. Third, we include a 120-day trial version of Microsoft Project 2010 Professional in the back of this book so students can obtain hands-on experience with this important tool. Fourth, a more in-depth treatment of project management techniques and principles is provided in an online chapter on this book's Web site. This information is based on the Project Management Body of Knowledge (PMBOK), as developed by the Project Management Institute—the primary professional organization for project managers in the United States.

Organized for More Effective Learning

This edition's innovative and entirely new organization starts with a complete beginning-to-end example of system development, moves immediately to systems analysis models and techniques, and then proceeds to system design concepts, emphasizing system architecture, user interfaces, and system interfaces. The student sees analysis and much of design covered in the first seven chapters. Next, the text focuses on managing system development projects, including Agile development, after the student has had a chance to understand what is actually involved in system development. Finally, the text covers detailed design topics, deployment topics, and current trends, going into more depth about such contemporary approaches as the Unified Process, Extreme Programming, and Scrum.

Coursemate Companion Web Site

Cengage Learning's *Systems Analysis and Design in a Changing World, Sixth Edition*, CourseMate brings course concepts to life with interactive learning, study, and exam preparation tools that support the printed textbook. Watch student comprehension soar as your class works with the printed textbook and the textbook-specific Web site. CourseMate goes beyond the book to deliver what you need! Learn more at **cengage.com/coursemate**.

Engagement Tracker

How do you assess your students' engagement in your course? How do you know your students have read the material or viewed the resources you have assigned? How can you tell if your students are struggling with a concept? With CourseMate, you can use the included Engagement Tracker to assess student preparation and engagement. Use the tracking tools to see progress for the class as a whole or for individual students. Identify students at risk early in the course. Uncover which concepts are most difficult for your class. Monitor time on task. Keep your students engaged.

Interactive Teaching and Learning Tools

CourseMate includes interactive teaching and learning tools:

- Quizzes
- Case projects
- Flashcards
- Videos
- PowerPoint presentations

These assets enable students to review for tests, prepare for class, and address the needs of students' varied learning styles.

Interactive eBook

In addition to interactive teaching and learning tools, CourseMate includes an interactive eBook. Students can take notes, highlight, search for, and interact with embedded media specific to their book. Use it as a supplement to the printed text or as a substitute—the choice is your students' with CourseMate.

Organization and Use

Systems Analysis and Design in a Changing World, Sixth Edition, includes this printed textbook, electronic editions, and supporting online chapters. The current printed textbook provides a compact, streamlined, and focused presentation of those topics that are essential and most important for information systems developers. The online chapters extend those concepts and provide a broader presentation of several topics. The online chapters may be integrated into the course or simply used as additional readings as prescribed by the instructor.

There are three major subject areas discussed in this book: systems analysis, systems design, and project management. There are additional subject areas, which are no less important but aren't discussed in as much depth. These include systems implementation, testing, and deployment. In addition, we have taken an approach that is quite different from other texts. Because students already have a basic understanding of systems analysis and design from Chapter 1, we immediately present in-depth concepts related to systems analysis and design. We present project management topics later in the text. This allows students to learn those project management concepts after understanding the elements of systems analysis and design. We think it will be more meaningful for students at that point in the course.

Part 1: Introduction to Systems Development

Part 1, comprising Chapter 1 and Online Chapter A, presents an overview of systems development. The first chapter begins by briefly explaining the objectives of systems analysis and systems design. Then, it provides a detailed, concrete example of what is required in a typical software development project. Many students who take a programming class think that programming is all you need to develop software and deploy a system. This chapter and the rest of the book should dispel that myth.

Online Chapter A, "The Role of the Systems Analyst," describes the many skills required of a systems analyst. It also discusses the various career options available to information systems majors. For students who are new to the discipline of information systems, this chapter will provide interesting and helpful knowledge about information systems careers.

Part 2: Systems Analysis Tasks

Chapters 2 through 5 cover systems analysis in detail. Chapter 2 discusses system requirements, analysis activities, and techniques for gathering information about the business problem. Developing the right system solution is possible only if the problem is accurately understood. Chapter 2 also explains how to identify and involve the stakeholders and introduces the concept of models and modeling. Chapters 3 and 4 teach modeling techniques for capturing the detailed requirements for the system in a useful form. When discussing an information system, two key concepts are particularly useful: the use cases that define what the end users need the system to do and the data entities/domain classes that users work with while carrying out their work tasks. These two concepts—use cases and data entities/domain classes—are important no matter what approach to system development is being used. Chapter 5 presents more in-depth requirements models, such as use case descriptions, use case diagrams, system sequence diagrams, and state machine diagrams.

Online Chapter B, "The Traditional Approach to Requirements," presents the traditional, structured approach to developing systems. To those instructors and students who desire to learn about data flow diagrams and structured English, this chapter provides an in-depth presentation.

All these modeling techniques provide in-depth analysis of user needs and allow the analyst to develop requirements and specifications. Again, the purpose of systems analysis is to thoroughly understand and specify the user's needs and requirements.

Part 3: Essentials of Systems Design

Chapters 6 and 7 provide the fundamental concepts related to systems design and designing the user experience. Chapter 6 provides broad and comprehensive coverage of important principles of systems design and architectural design. It serves not only as a broad overview of design principles but also as a foundation for later chapters that explain the detailed techniques, tasks, skills, and models used to carry out design.

Chapter 7 presents additional design principles related to designing the user interface and the system interfaces. Designing the user interface is a combination of analysis and design. It is related to analysis because it requires heavy user involvement and includes specifying user activities and desires. On the other hand, it is a design activity because it is creating specific final components that are used to drive the programming effort. The screens and reports and other user interaction components must be precisely designed so they can be programmed as part of the final system. Systems interfaces occur when one information system communicates or interacts with another information system without human intervention. System interfaces are becoming increasingly important because of Web services and cloud computing.

Part 4: Projects and Project Management

By this point, students will have a basic understanding of all the elements of systems development. Part 4 brings together all these concepts by explaining more about the process of organizing and managing development projects. Chapter 8 describes different approaches to systems development in today's environment, including several important system development life cycle models and Agile development. It is an important chapter to help you understand how projects actually get executed.

Chapter 9 extends these concepts by teaching foundation principles of project planning and project management. Every systems analyst is involved in helping organize, coordinate, and manage software development projects. In addition, most good students will eventually become team leaders and project managers. The principles presented in Chapter 9 are essential to a successful career.

Online Chapter C, "Project Management Techniques," goes into more detail regarding the tools and techniques used by systems analysts and project managers to plan and monitor development projects. For those instructors and students who would like to learn specific project management skills, this is an important chapter.

Part 5: Advanced Design and Deployment Concepts

Part 5 goes into more depth with respect to systems design, database design, and other important issues related to effective and successful system development and deployment.

Chapters 10 and 11 explain in detail the models, skills, and techniques used to design software systems. As mentioned earlier, systems design is a fairly complex activity, especially if it is done correctly. The objective of these two chapters is to teach the student the various techniques—from simple to complex—that can be used to effectively design software systems.

Chapter 12 explains how to design the database from the information gleaned during analysis and the identification of the object classes. Other related concepts, such as controls and security, are also presented in this chapter. Chapter 13 describes the final elements in systems development: final testing, deployment, maintenance, and version control. Chapter 14 wraps things up by looking at current trends and the future of software development.

Designing Your Analysis and Design Course

There are many approaches to teaching analysis and design courses, and the objectives of the course differ considerably from college to college. In some academic information systems departments, the analysis and design course is a capstone course in which students apply the material learned in prior database, telecommunications, and programming courses to a real analysis and design project. In other information systems departments, analysis and design is used as an introduction to the field of system development and is taken prior to more specialized courses. Some information systems departments offer a two-course sequence emphasizing analysis in the first semester and design and implementation in the second semester. Some information systems departments have only one course that covers analysis and design.

The design of the analysis and design course is complicated even more by the choice of emphasizing either traditional content or object-oriented content—again, depending on local curriculum priorities. Additionally, the more iterative approach to development in general has made choices about sequencing the analysis and design topics more difficult. For example, with iterative development, a two-course sequence can't be divided into analysis and then design as easily.

The objectives, course content, assignments, and projects have many variations. What we offer below are some suggestions for using this textbook in various approaches to the course.

Object-Oriented Analysis and Design Course

This is the course we designed the printed textbook to support, so all the printed chapters but none of the online chapters are included. Note that object-oriented design is included in detail. The course covers object-oriented analysis and design, user and system interface design, database design, controls and security, and implementation and testing. It is usually assumed that the projects will use custom development, including Web development. The course emphasizes iterative development with three-layer architecture, project management, information gathering, and management reporting. One-semester courses are usually limited to completing some prototypes of the user interface to give students closure. Sometimes, this course is spread over two semesters, with some implementation of an actual system in the second semester for a more complete development experience. Iterative development is emphasized.

A suggested outline for a course emphasizing object-oriented development is:

Chapter 1: From Beginning to End: An Overview of Systems Analysis and Design

Chapter 2: Investigating System Requirements

Chapter 3: Use Cases

Chapter 4: Domain Modeling

Chapter 5: Extending the Requirements Models

Chapter 6: Essentials of Design and the Design Activities

Chapter 7: Designing the User and System Interfaces

Chapter 8: Approaches to System Development

Chapter 9: Project Planning and Project Management

Chapter 10: Object-Oriented Design: Principles

Chapter 11: Object-Oriented Design: Use Case Realizations

Chapter 12: Databases, Controls, and Security (combine)

Chapter 13: Making the System Operational

Chapter 14: Current Trends in System Development

Traditional Analysis and Design Course

A traditional systems analysis and design course provides coverage of activities and tasks by using structured analysis, user and system interface design, database design, controls and security, and implementation and testing. It is usually assumed that the project will use custom development, including Web development. The course emphasizes the SDLC, project management, information gathering, and management reporting. One-semester courses are usually limited to completing some prototypes of the user interface to give students closure. Sometimes, this course is spread over two semesters, with some implementation of an actual system in the second semester for a more complete development experience.

For this approach to the analysis and design course, a reasonable outline would omit chapters and sections detailing object orientation and possibly current trends but include the online chapters on the role of the systems analyst and on traditional structured analysis. However, object-oriented concepts are introduced throughout the text, so students will still be familiar with them. Additionally, because of the amount of material to cover, the online chapter detailing project management, financial feasibility, and scheduling might be omitted.

A suggested outline for a course emphasizing the traditional structured approach is:

Chapter 1: From Beginning to End: An Overview of Systems Analysis and Design

Online Chapter A: The Role of the Systems Analyst

Chapter 2: Investigating System Requirements

Chapter 3: Use Cases

Chapter 4: Domain Modeling

Online Chapter B: The Traditional Approach to Requirements

Chapter 6: Essentials of Design and the Design Activities

Chapter 7: Designing the User and System Interfaces

Chapter 8: Approaches to System Development

Chapter 9: Project Planning and Project Management

Chapter 12: Databases, Controls, and Security (combine)

Chapter 13: Making the System Operational

In-Depth Analysis and Project Management

Some courses cover object-oriented systems analysis methods in more depth and briefly survey structured analysis—with not much about object-oriented design—while emphasizing project management. Sometimes, these courses are graduate courses; sometimes, they assume design and implementation are covered in more technical courses. In some cases, it might be assumed that packages are likely solutions rather than custom development, so defining requirements and managing the process are more important than design activities. The online chapters covering the role of the systems analyst, the traditional approach to structured analysis, and project management would be included.

A suggested outline for a course emphasizing object-oriented analysis, with in-depth coverage of project management, is:

Chapter 1: From Beginning to End: An Overview of Systems Analysis and Design

Online Chapter A: The Role of the Systems Analyst

Chapter 2: Investigating System Requirements

Chapter 3: Use Cases

Available Support

Systems Analysis and Design in a Changing World, Sixth Edition, includes teaching tools to support instructors in the classroom. The ancillary materials that accompany the textbook include an *Instructor's Manual,* solutions, test banks and test engine, PowerPoint presentations, and figure files. Please contact your Cengage Course Technology sales representative to request the Teaching Tools CD-ROM if you haven't already received it. Or go to the Web page for this book at **login.cengage.com** to download all these items.

The Instructor's Manual

The *Instructor's Manual* includes suggestions and strategies for using the text, including course outlines for instructors that emphasize the traditional structured approach or the object-oriented approach. The manual is also helpful for those teaching graduate courses on analysis and design.

Solutions

We provide instructors with answers to review questions and suggested solutions to chapter exercises and cases. Detailed traditional and UML object-oriented models are included for all exercises and cases that ask for modeling solutions.

ExamView

This objective-based test generator lets the instructor create paper, LAN, or Web-based tests from test banks designed specifically for this Course Technology text. Instructors can use the QuickTest Wizard to create tests in fewer than five minutes by taking advantage of Course Technology's question banks or instructors can create customized exams.

WebTUTOR Plug and Play!

Jumpstart your course with customizable, text-specific content within your Course Management System!

- **Jumpstart**—Instructors simply load a WebTutor cartridge or e-Pack into their Course Management System.
- **Content**—Text-specific content, media assets, quizzing, Web links, discussion topics, interactive games and exercises, and more.
- **Customizable**—Instructors can easily blend, add, edit, reorganize, or delete content.

Whether you want to Web-enable your class or put an entire course online, WebTutor delivers! Visit **academic.cengage.com/webtutor** to learn more.

Product Description

WebTutor and WebTutor Toolbox products are Course Cartridges and e-Packs that provide content natively on a Course Management System (WebCT, BlackBoard, Angel, D2L, and eCollege). The purpose of the product is to provide electronic solutions in an easy-to-use format with little upfront costs to instructors.

For more information on how to bring WebTutor to your course, instructors should contact their Cengage Learning sales representative.

PowerPoint Presentations

Microsoft PowerPoint slides are included for each chapter. Instructors might use the slides in a variety of ways, such as teaching aids during classroom presentations or as printed handouts for classroom distribution. Instructors can add their own slides for additional topics they introduce to the class.

Figure Files

Figure files allow instructors to create their own presentations by using figures taken directly from this text.

Credits and Acknowledgments

We have been very gratified as authors to receive so many supportive and enthusiastic comments about *Systems Analysis and Design in a Changing World*. Students and instructors in the United States and Canada have found our text to be the most up-to-date and flexible book available. The book has also been translated into many languages and is now used productively in Europe, Australia, New Zealand, India, China, and elsewhere. We truly thank everyone who has been involved in all the editions of our textbook.

This sixth edition was managed by Kate Mason, who was charged with recruiting a new developmental editor, negotiating with the production department for an accelerated writing and editing schedule, and dealing with numerous author uncertainties and scheduling conflicts. Our developmental editor, Kent Williams, was charged with the daunting task of pulling together material that had been completely reorganized and submitted in what probably seemed like a random order by what probably seemed like mad professors.

Last but certainly not least, we want to thank all the reviewers who worked so hard for us—beginning with an initial proposal and continuing throughout the completion of all six editions of this text. We were lucky enough to have reviewers with broad perspectives, in-depth knowledge, and diverse preferences. We listened very carefully, and the text is much better as a result of their input. Reviewers for the various editions include:

Rob Anson, *Boise State University*
Marsha Baddeley, *Niagara College*
Teri Barnes, *DeVry Institute—Phoenix*
Robert Beatty, *University of Wisconsin—Milwaukee*
James Buck, *Gateway Technical College*
Anthony Cameron, *Fayetteville Technical Community College*
Genard Catalano, *Columbia College*
Paul H. Cheney, *University of Central Florida*
Kim Church, *Oklahoma State University*
Jung Choi, *Wright State University*
Jon D. Clark, *Colorado State University*
Mohammad Dadashzadeh, *Oakland University*
Lawrence E. Domine, *Milwaukee Area Technical College*
Gary Garrison, *Belmont University*
Cheryl Grimmett, *Wallace State Community College*

Jeff Hedrington, *University of Phoenix*
Janet Helwig, *Dominican University*
Susantha Herath, *St. Cloud State University*
Barbara Hewitt, *Texas A&M University*
Ellen D. Hoadley, *Loyola College in Maryland*
Jon Jasperson, *Texas A&M University*
Norman Jobes, *Conestoga College—Waterloo, Ontario*
Gerald Karush, *Southern New Hampshire University*
Robert Keim, *Arizona State University*
Michael Kelly, *Community College of Rhode Island*
Rajiv Kishore, *The State University of New York—Buffalo*
Rebecca Koop, *Wright State University*
Hsiang-Jui Kung, *Georgia Southern University*
James E. LaBarre, *University of Wisconsin—Eau Claire*
Ingyu Lee, *Troy University*
Terrence Linkletter, *Central Washington University*
Tsun-Yin Law, *Seneca College*
David Little, *High Point University*
George M. Marakas, *Indiana University*
Roger McHaney, *Kansas State University*
Cindi A. Nadelman, *New England College*
Bruce Neubauer, *Pittsburgh State University*
Michael Nicholas, *Davenport University—Grand Rapids*
Mary Prescott, *University of South Florida*
Alex Ramirez, *Carleton University*
Eliot Rich, *The State University of New York—Albany*
Robert Saldarini, *Bergen Community College*
Laurie Schatzberg, *University of New Mexico*
Deborah Stockbridge, *Quincy College*
Jean Smith, *Technical College of the Lowcountry*
Peter Tarasewich, *Northeastern University*
Craig VanLengen, *Northern Arizona University*
Bruce Vanstone, *Bond University*
Haibo Wang, *Texas A&M University*
Terence M. Waterman, *Golden Gate University*

All of us involved in the development of this text wish you all the best as you take on the challenges of analysis and design in a changing world.

—John Satzinger
—Robert Jackson
—Steve Burd

PART 1

An Introduction to Systems Development

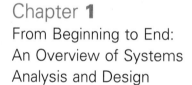

Chapter 1
From Beginning to End:
An Overview of Systems
Analysis and Design

Optional Online Chapter A
The Role of the Systems
Analyst

1

From Beginning to End: An Overview of Systems Analysis and Design

Chapter Outline

- Software Development and Systems Analysis and Design
- Systems Development Life Cycle
- Introduction to Ridgeline Mountain Outfitters
- Iterative Development
- Developing RMO's Tradeshow System
- Where You Are Headed—The Rest of This Book

Learning Objectives

After reading this chapter, you should be able to:

- Describe the purpose of systems analysis and design in the development of information systems
- Describe the characteristics of iterative systems development
- Explain the six core processes of the Systems Development Life Cycle
- Identify key documents that are used in planning a project
- Identify key diagrams used in systems analysis and systems design
- Explain the utility of identifying use cases in systems development
- Explain the utility of identifying object classes in systems development

Software Development and Systems Analysis and Design

Computers are everywhere today, and microchips impact every part of our lives. We live in a world not only of ubiquitous computing but of pervasive communication and connectivity. An incredibly large part of our everyday lives depends on computer chips, connection links, and application software.

You have grown up in this world of high technology. You use smartphones, laptops, iPads, notepads, electronic game equipment, and so on. Your mobile devices provide daily (if not hourly) text messages, tweets, videos, snapshots, Internet access, games, and much more. Many of you have already developed your own application software or you have a friend who has written applications for laptops, smartphones, iPads, or Facebook. Some of you have taken programming classes; others of you have taught yourself how to write computer application programs. Given that we live in this world of high-tech gadgets, we might ask, "What is systems analysis and design, and why is it important?" How does the development of new technology and new application software utilize systems analysis and design? In other words, what role does systems analysis and design play in the development of high-tech solutions and applications?

computer application (app)   a computer software program that executes on a computing device to carry out a specific function or set of related functions

information system   a set of interrelated computer components that collects, processes, stores, and provides as output the information needed to complete business tasks

First, let us provide two important definitions. A **computer application** is a computer software program that executes on a computing device to carry out a specific function or set of related functions. Sometimes, *computer application* is shortened to **app** (such as an iPhone app or a Facebook app). An **information system** is a set of interrelated computer components that collects, processes, stores (usually in a database), and provides as output the information needed to complete business tasks. Although these terms are sometimes used synonymously, an application usually refers to only the computer software involved, whereas an information system may include the software, the database, and even the related manual processes. Examples of computer applications include browsers that access the Internet to play games or calendaring apps. **Figure 1-1** shows a typical mobile digital device.

Why is systems analysis and design important in the development of information systems? To answer that question, let us consider an analogous

situation: the art and science of creating a beautiful building. In this scenario, there is the buyer who has the vision, the builder who will construct the building, and the architect who serves as the bridge between the buyer and the builder. The architect helps the buyer develop the vision but must also communicate the building's specifications to the builder. In doing so, the architect uses various tools to first capture the vision from the buyer and then provide the builder with instructions—including such tools as line drawings, blueprints, to-scale models, detail specifications, and even on-site inspections.

Just as a builder doesn't start construction without plans, programmers don't just sit down and start writing program code. They need someone (maybe themselves) to function like an architect—planning, capturing the vision, understanding details, specifying needs—before writing the code and verifying that it satisfies the vision. The software architect has to be able to understand and capture the vision of the persons funding the project. Usually, we call this person a *systems analyst*. In situations where you are the programmer as well as the analyst, it might be easy to keep track of the details without writing them down. However, in today's world, with some system development teams distributed worldwide, you may only be responsible for part of the programming, with the rest handled by team members around the world. In a distributed team situation, it is much more important to have written documents to assist in understanding, capturing, explaining, and specifying the software application.

In a nutshell, systems analysis and design (SA&D) is about providing the tools and techniques to you, the developer, so you can understand the need (business need), capture the vision, define a solution, communicate the vision and the solution, build the solution and direct others in building the solution, confirm that the solution meets the need, and launch the solution application.

Included in SA&D are all the skills, steps, guidelines, and tools that support and lead up to the actual programming of the system. SA&D includes such "soft" skills as interviewing and talking to users as well as such "hard" (more technical) skills as detailing specifications and designing solutions. Many of the technical skills are associated with creating models that capture specifications or define solutions. In this book, you will learn all these skills as well as how they work together to develop an information system.

systems analysis those activities that enable a person to understand and specify what the new system should accomplish

Let us conclude this section with a few more definitions. **Systems analysis** consists of those activities that enable a person to understand and specify what the new system should accomplish. The operative words here are "understand" and "specify." Systems analysis is much more than simply a brief statement of the problem. For example, a customer management system must keep track of customers, register products, monitor warranties, and keep track of service levels, among many other functions—all of which have myriad details. Systems analysis describes in detail the "what" that a system must do to satisfy the need or to solve the problem.

systems design those activities that enable a person to define and describe in detail the system that solves the need

Systems design consists of those activities that enable a person to describe in detail the system that solves the need. The operative word in this case is "solves." In other words, systems design describes "how" the system will work. It specifies in detail all the components of the solution system and how they work together to provide the desired solution.

Systems Development Life Cycle

project a planned undertaking that has a beginning and an end, and that produces some definite result

Initial development of a new system is usually done as a project. What this means is that the activities required to develop a new system are identified, planned, organized, and monitored. We can think of a **project** as a planned undertaking that has a beginning and an end and produces some definite result. Some projects are very formal, whereas others are so informal they can barely be recognized as projects.

To manage a project with analysis, design, and other development activities, you need a project management framework to guide and coordinate the work of the project team. The **Systems Development Life Cycle (SDLC)** identifies all the activities required to build, launch, and maintain an information system. Normally, the SDLC includes all the activities that are part of systems analysis, systems design, programming, testing, and maintaining the system as well as other project management processes that are required to successfully launch and deploy the new information system.

There are many approaches to the SDLC and many variations for projects that have various needs. However, there is a core set of processes that is always required, even though there is also an incredible number of variations of these core processes—in how each process is planned and executed and in how the processes are combined into a project. Here are six core processes required in the development of any new application:

1. Identify the problem or need and obtain approval to proceed.
2. Plan and monitor the project—what to do, how to do it, and who does it.
3. Discover and understand the details of the problem or the need.
4. Design the system components that solve the problem or satisfy the need.
5. Build, test, and integrate system components.
6. Complete system tests and then deploy the solution.

There are many ways to implement these six core processes of the SDLC. An **information systems development process** is the actual approach used to develop a particular information system. Most information systems you will develop are conceived and built to solve organizational problems, which are usually very complex, thus making it difficult to plan and execute a development project. In fact, many projects end up being much larger than expected—often resulting in late deliveries that are over budget. During the last 10 years, several new information systems development processes have been developed to enhance project success. One of the newer and more effective ones is called **Agile Development**. The basic philosophy of Agile Development is that neither team members nor the user completely understands the problems and complexities of a new system, so the project plan and the execution of the project must be responsive to unanticipated issues. It must be agile and flexible. It must have procedures in place to allow for, anticipate, and even embrace changes and new requirements during the development process.

Maybe the best way to understand these concepts is to see the way they play out in a complete example. That is the objective of this chapter, "From Beginning to End: An Overview of Systems Analysis and Design." Here, we will use a fairly small information system application to show you all six core processes (as much as is feasible in a textbook, anyway). We will illustrate one way to organize the various activities into an actual working project; in other words, we will show you one version of an information system development process. By going all the way through a very simple project, you will more easily learn and understand the complex concepts provided in the rest of the text. Our project involves Ridgeline Mountain Outfitters, a retailer and manufacturer of clothing for all types of outdoor activities.

Introduction to Ridgeline Mountain Outfitters

Ridgeline Mountain Outfitters (RMO) is a large retail company that specializes in clothing and related accessories for all types of outdoor and sporting activities. By the 2010s, the Rocky Mountain and Western states had seen tremendous growth in recreation activities, and with the increased interest in outdoor sports, the market for winter and summer sports clothes had exploded. Skiing, snowboarding, mountain biking, water skiing, jet skiing, river running, jogging, hiking, ATV biking, camping, mountain climbing, rappelling—all had seen a tremendous increase in interest in these states. People needed appropriate sports clothes for these activities, so RMO expanded its line of sportswear to respond to this market. It also added a line of high-fashion activewear and accessories to round out its offerings to the expanding market of active people.

The company's growth charted an interesting history of mail-order, brick-and-mortar, and online sales. RMO got its start by selling to local clothing stores in the Park City, Utah, area. In the late-1980s and early-1990s, it began selling directly to customers by using catalogs with mail-in and telephone orders. It opened its first store in 1994 and soon expanded to 10 retail outlets throughout the West. Last year, retail store revenue was $67 million, telephone and mail-order revenues were $10 million, and Web-based sales were $200 million. Most sales continue to be in the West, although the market in several areas of the eastern United States and Canada is growing.

RMO produces its own line of outdoor and sportswear clothing. However, in order to offer a complete range of outdoor clothing in its retail outlets, it also sells other brands of outdoor and sportswear clothing. In addition, other types of clothing and accessories, such as footwear, leather apparel, and specialty sportswear, are available in the retail stores and through the online store.

Figure 1-2 shows a sample of the catalogs that RMO mails out. Although mail-order and telephone sales are small, receiving the catalog encourages customers to go online to make purchases, so RMO continues to produce and mail abbreviated versions of its catalogs.

Figure 1-3 illustrates a typical order page from the online system.

Trade Shows

In order to keep its product line current and popular, RMO's purchasing agents attend apparel and fabric trade shows around the world. RMO purchasers have a good track record of predicting what products will be good sellers. In addition, RMO is always watching for new products and accessories that will enable it to expand its product line appropriately.

When purchasing agents attend a trade show, they frequently find various products that they want to add to the spring, summer, or winter apparel offering.

FIGURE
RMO winter catalog

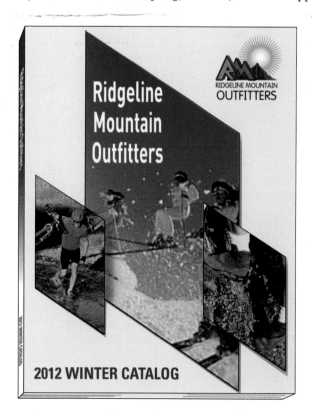

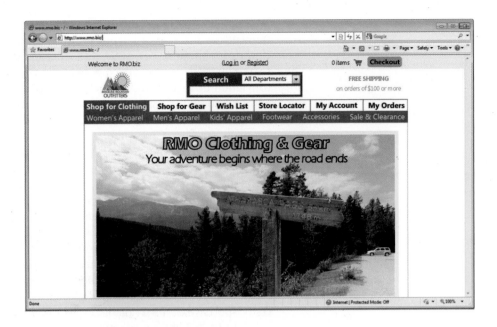

In the past, when RMO buyers wanted to place an order, they would exchange contact information with the seller at the trade show and upon returning to the home office would then follow up via e-mails and phone calls to formulate a contract and make a purchase order. However, to expedite orders, RMO has now initiated a project to develop a system for collecting and tracking information about its suppliers and the new products it adds to its merchandise offerings.

Iterative Development

iterative development   an approach to system development in which the system is "grown" piece by piece through multiple iterations

Iterative development is an approach to system development in which the system is "grown" in an almost organic fashion. Core components are developed first and then additional components are added. It is called "iterative" because the six core development processes are repeated over and over again to add additional functionality to the overall system. In other words, there is one big project, which consists of many mini-projects, and the information system is grown piece by piece.

Figure 1-4 illustrates how an iterative Agile project might be executed. This is a sample diagram. Real projects might be quite different. Across the figure, you see six iterations. An iteration is like a mini-project in that it has a completed result and a constrained time frame. Often, it will last two to four weeks. Down the left side of the figure, you see the six core processes. The rounded

FIGURE **1-4**
The six core processes, with iterations for a typical project

Core Processes	Iterations					
	1	2	3	4	5	6
Identify problem and obtain approval						
Plan and monitor the project						
Discover and understand details						
Design system components						
Build, test, and integrate system components						
Complete system tests and deploy solution						

mounds inside the graph represent the amount of effort for that core process during that iteration. The amount of area under a curve is an approximate indication of the amount of effort expended within an iteration with regard to a particular core process. For example, in Figure 1-4, Iteration 1 appears to primarily focus on identifying the problem and planning the project. Lesser amounts of discovery, design, and build and test may also be done. For this iteration, nothing is done with regard to deploying the system.

There are several benefits to iterative development. For one, portions of the often system can sometimes be deployed sooner. If there are core functions that provide basic support, these can be deployed in an early iteration. A second benefit is that by taking a small portion and developing it first, many tough problems can be addressed early in the project. Many of today's systems are so large and complex that it is impossible to remember and understand everything. By focusing on only a small portion, the requirements are fewer and easier to grasp and solve. Finally, developing a system in iterations makes the entire development process much more flexible and able to address new requirements throughout the project.

A key element of iterative development is choosing a part of the solution system that can be done in two to four weeks. During one iteration, often all the core development processes are involved, including programming and system-wide testing, so the result is a part of the working system, even though it may only have a portion of the functionality that is ultimately required.

Developing RMO's Tradeshow System

We will organize our sample project—the RMO Tradeshow System—for the first iteration, and our goal is to have the iteration last six days. However, our primary objective is to introduce you to the concepts and techniques of the six core processes. Therefore, in some instances, we may go a little deeper into a core process than we might do on the first iteration of a real project. It is a little unrealistic to complete an entire iteration with all the necessary details in only six days, but it should be a good learning experience. There will not be a one-to-one correspondence with the six SDLC core processes and the six days of the project, but we will include all the SDLC core processes within the project.

At the end of this chapter, we have provided a small case project that has also been divided into a six-day project. Each day of the project has several possible assignments that can be completed to help you solidify your understanding of systems analysis and design and the six core processes.

Most new applications require a project with several iterations. In the first iteration, there are usually three major objectives. The first objective is to get project approval. The second objective is to get a clear picture of the system's overall vision—all the major functions and data requirements. The third objective is to determine the detail specifications and develop a solution for one portion of the system (i.e., actually analyze, design, build, and test one part of the system).

In our project, we will touch on all these objectives. We will show an example of a System Vision Document and then develop one portion of the overall system. We have constrained the scope of the new system so we can complete it in one iteration. It should be noted that the division of this project into days and daily activities is somewhat arbitrary. There are numerous ways to partition and organize the work. The following organization is quite workable, but it is not the only way to organize the project.

Pre-Project Activities

Before the project actually begins, the head of RMO's purchasing department works with a systems analyst to identify and document the specific business need as well as define a specific project objective. RMO's management reviews the primary project objective and provides budget approval. Every organization has to give budget approval before a project can start. Some organizations have a formal process to get a project approved; other organizations have a less formal process. Normally, there are two goals an organization must decide on to get a project off the ground:

■ Identify the problem and document the objective of the solution system. (Core Process 1)
■ Obtain approval to commence the project. (Core Process 1)

System Vision Document

As with all new projects within RMO, a System Vision Document is developed to identify the benefits to the company and the functional capabilities that will be included in the system. Frequently, this is done in two steps: developing a preliminary statement of benefits and then adding estimates of specific dollar costs and dollar benefits. **Figure 1-5** is the System Vision Document for this project.

FIGURE **1-5**

Tradeshow System Vision Document

System Vision Document
RMO Tradeshow System

Problem Description

Trade shows have become an important information source for new products, new fashions, and new fabrics. In addition to the large providers of outdoor clothing and fabrics, there are many smaller providers. It is important for RMO to capture information about these suppliers while the trade show is in progress. It is also important to obtain information about specific merchandise products that RMO plans to purchase. Additionally, if quality photographs of the products can be obtained while at the trade show, then the creation of online product pages is greatly facilitated.

It is recommended that a new system be developed and deployed so field purchasing agents can communicate more rapidly with the home office about suppliers and specific products of interest. This system should be deployed on portable equipment.

System Capabilities

The new system should be capable of:
• Collecting and storing information about the manufacturer/wholesaler (suppliers)
• Collecting and storing information about sales representatives and other key personnel for each supplier
• Collecting information about products
• Taking pictures of products (and/or uploading stock images of products)
• Functioning as a stand-alone without connection
• Connecting via Wi-Fi (Internet) and transmitting data
• Connecting via telephone and transmitting data

Business Benefits

It is anticipated that the deployment of this new system will provide the following business benefits to RMO:
• Increase timely communication between trade show attendees and home office, thereby improving the quality and speed of purchase order decisions
• Maintain correct and current information about suppliers and their key personnel, thereby facilitating rapid communication with suppliers
• Maintain correct and rapid information and images about new products, thereby facilitating the development of catalogs and Web pages
• Expedite the placing of purchase orders for new merchandise, thereby catching trends more rapidly and speeding up product availability

As described earlier, RMO needs a portable system that can be used by its purchasing agents as they attend various product and clothing fabric trade shows. The system needs to fulfill two major requirements. First, it has to have the functionality to capture information about suppliers and products. Second, it needs to be able to communicate with the home office systems, and because these trade shows are held in various venues around the world, various methods of connectivity are needed.

Preliminary investigation considered various equipment options, including laptop computers, iPad computing devices, and smartphones. Even though smartphones appeared to have the best connection options, the small size made viewing the details of photographs somewhat difficult; the iPad and other similar portable devices with advanced technology also appear to be viable options. However, due to the similarity of the smartphones and tablets, it seems feasible to develop an application that will execute on either device. Each purchasing agent could use his or her preferred device.

Toward the end of the pre-project activities, a meeting is held involving all the key persons, including a representative of executive management. The decision is made to move ahead with the project and budget the necessary funds.

Day 1 Activities

RMO—Supplier Information Subsystem

The project actually begins with Day 1, which is essentially a planning day. Usually, the first activity is the project team reviewing the System Vision Document and verifying that the preliminary work is still valid. It reviews the scope of the project to become familiar with the problem to be solved, then it plans the iterations and activities for the remainder of the project. The second SDLC core process—planning the project—includes business analysis and project management activities. All these topics will be treated in depth in later chapters. These activities are completed on Day 1:

- Determine the major components (functional areas) that are needed. (Core Process 2)
- Define the iterations and assigning each functional area to an iteration. (Core Process 2)
- Determine team members and responsibilities. (Core Process 2)

Planning the Overall Project and the Project Iterations

Myriad details need to be considered in a project plan. For our project, we will only focus on the bare essentials. We will describe project planning more elaborately in later chapters.

The project team meets with the users to review the overall business need and the objectives of the new system. The System Vision Document serves as the starting point for these discussions. As is often the case, the list of System Capabilities provides the foundation information for determining the overall project plan. The first step is to divide the system into several subsystems or components. A **subsystem** is simply a portion of the overall system. Based on the list of System Capabilities, the project team identifies these functional subsystems:

subsystem an identifiable and partitioned portion of an overall system

- Supplier Information subsystem
- Product Information subsystem

The Supplier Information subsystem will collect and maintain information about the manufacturers or wholesalers and the contract people that work for them. The Product Information subsystem will capture information about the various products, including detailed descriptions and photographs.

The next step is to identify which subsystems will be developed in which order. Many issues are considered, such as dependencies between the various tasks, sequential versus parallel development, project team availability, and project urgency. In our case, the team decides that the project will proceed in a serial fashion, with the Supplier Information subsystem scheduled as the first iteration.

Planning the First Iteration

Each iteration is like a systems development mini-project. The core processes described earlier can all be applied, with the scope limited to the component that is to be developed during the iteration. The planning process for an iteration consists of these three steps:

- Identify the tasks required for the iteration.
- Organize and sequence these tasks into a schedule.
- Identify required resources (especially people) and assign people to tasks.

The first step is to identify—or attempt to identify—all the individual tasks that need to be done. As these tasks are identified, they are compiled and organized. Sometimes, this organized list of tasks is called a Work Breakdown Structure. **Figure 1-6** shows the Work Breakdown Structure for this iteration.

Part of the effort is trying to estimate how long each task will take. Because this project has a very limited scope (only six days), all the estimates will be in hours. These estimates do not include the time expended by those who are not on the team. However, of those on the team, the estimates include the time for the original work, the time for discussion, and the time for reviewing and checking the Work Breakdown Structure for accuracy and correctness.

The next step is to get these tasks organized into a schedule. Again, we can be very formal and use a sophisticated project-scheduling tool or we can just list the tasks in the order we think they need to be done. An important part of building the schedule is identifying any dependencies between the tasks. For

FIGURE **1-6**

Sample handwritten Work Breakdown Structure

Work Breakdown Structure

I. Discover and understand the details of all aspects of the problem.

    1. Meet with the Purchasing Department manager. ~ 3 hours

    2. Meet with several purchasing agents. ~ 4 hours

    3. Identify and define use cases. ~ 3 hours

    4. Identify and define information requirements. ~ 2 hours

    5. Develop workflows and descriptions for the use cases. ~ 6 hours

II. Design the components of the solution to the problem.

    1. Design (lay out) input screens, output screens, and reports. ~ 8 hours

    2. Design and build database (attributes, keys, indexes). ~ 4 hours

    3. Design overall architecture. ~ 4 hours

    4. Design program details. ~ 6 hours

III. Build the components and integrate everything into the solution.

    1. Code and unit test GUI layer programs. ~ 14 hours

    2. Code and unit test Logic layer programs. ~ 8 hours

IV. Perform all system-level tests and then deploy the solution.

    1. Perform system functionality tests. ~ 5 hours

    2. Perform user acceptance test. ~ 8 hours

example, it does not make sense to try to design the database before we have identified the information requirements. But many tasks can be done in parallel.

Again, the great benefit of a single iteration is that we can make the schedule informal, and we will be able to adjust the work day by day to respond to specific complexities that occur.

For our project, we will not build a complete schedule. You will learn how to do that in a later chapter. However, in order to organize our six-day project, we have taken the tasks from the Work Breakdown Structure and placed them on a day-by-day sequence that we call a *work sequence draft*, as shown in **Figure 1-7**. To develop a formal schedule, the project leader will use this diagram to assign people to the tasks as well as put the tasks on a specific schedule chart with calendar dates.

You should be aware that the sequence of activities and the dependencies of those activities are represented in this diagram with only partial accuracy. For example, we show that programming does not start until design has finished. However, in reality, there may be some overlap between the two activities.

FIGURE 1-7
Work sequence draft

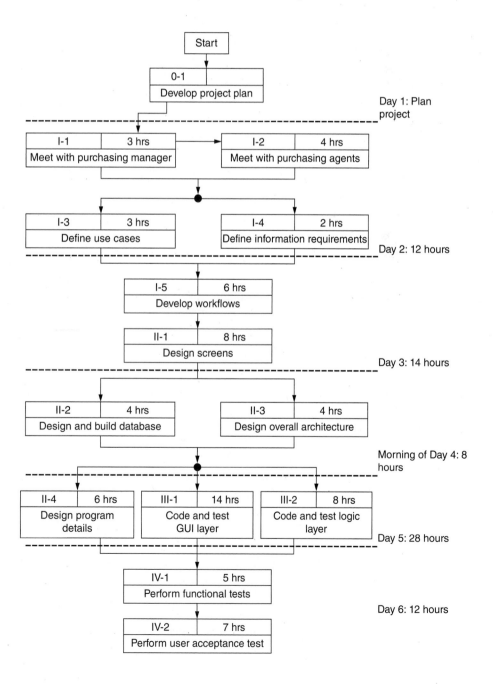

The benefit of a work sequence draft is threefold. First, it helps the team organize its work so there is enough time set aside to think through the critical design issues before programming begins. Second, it provides a measuring rod to see if the iteration is on schedule. For example, if meetings with the purchasing agents take all day or more than a day, the team will know early on that this iteration will take longer than expected. Third, the project leader can see that programming may require more resources if the project is going to stay on this schedule. Hence, the project leader can begin lining up resources early on to help with that part of the project. It should be obvious that even this simple dependency diagram can help a project manager plan and organize the work.

Day 2 Activities

Day 1 involved planning and organizing the project. Day 2 involves systems analysis activities that help us understand and document requirements. On Day 2, we will specify the functions in more detail. These activities are included:

- Do preliminary fact-finding tasks to understand the requirements. (Core Process 3)
- Develop a preliminary list of use cases and a use case diagram. (Core Process 3)
- Develop a preliminary list of classes and a class diagram. (Core Process 3)

Fact Finding and User Involvement

Before the project commenced, a preliminary, broad definition of functions was developed. It is now time to examine the specifics of those functions and determine exactly what the user needs the system to do. The first step is to identify the key users that will help define these details. Obviously, the manager of the purchasing department will be one of the first ones to meet with. She will probably designate one or two knowledgeable purchasing agents who can work with the team on an ongoing basis to develop the specifications and to verify that the system performs as required. All successful projects depend on heavy user involvement. In Chapter 2, you will learn more about identifying key stakeholders.

There are various techniques to ensure that the fact finding is complete and thorough. These include interviewing the key users, observing existing work processes, reviewing existing documentation and existing systems, and even researching other companies and other systems.

Identifying Use Cases

A use case documents a single user-triggered business event and the system's response to that event. For example, let us say a purchasing agent goes to a trade show and finds some new lightweight sports jackets that will work well for RMO's fall merchandise offerings. Maybe the first task the purchasing agent needs to do is find out if this supplier has worked with RMO before. Thus, the business event that requires the Tradeshow System might be "Look up a supplier." Activities leading up to the event of using the system are important, but we do not identify them as business events until the Tradeshow System is used—hence, the term *use case*—a case or situation where the system is used. One good way to help you identify use cases is to say, "The purchasing agent 'uses' the system to 'Look up a supplier.'"

There are multiple methods used to identify use cases, which you will learn about later in this book. **Figure 1-8** is a preliminary list of use cases for the entire Tradeshow System. When the project team meets with the purchasing agents in brainstorming sessions, they together identify every business event in which a purchasing agent might use the system. However, because this first iteration is focusing only on the Supplier Information subsystem, the project team will also focus its attention on only the first four use cases on the list.

Use Case	Description
Look up supplier	Using supplier name, find supplier information and contacts
Enter/update supplier information	Enter (new) or update (existing) supplier information
Look up contact	Using contact name, find contact information
Enter/update contact information	Enter (new) or update (existing) contact information
Look up product information	Using description or supplier name, look up product information
Enter/update product information	Enter (new) or update (existing) product information
Upload product image	Upload images of the merchandise product

Identifying Object Classes

Object classes identify those things in the real world that the system needs to know about and keep track of. In order to find object classes, we look for all the objects, or things, that the system uses or captures. Objects come in all types and variations, from tangible items (such as merchandise products that you can see and touch) to more abstract concepts that you cannot touch (such as an order), which, though intangible, definitely exist.

Object classes are identified in the discussions with purchasing agents by looking for the nouns that describe categories of things. For example, the agents will often talk about suppliers, merchandise products, or inventory items. More details about how to identify object classes and their attributes are provided later in this book.

Figure 1-9 illustrates which nouns have been determined to be fundamental object classes for the Tradeshow System. The attributes are descriptors that help define and describe an object class.

In addition to just providing a list of object classes, systems analysts often develop a visual diagram of the classes, their attributes, and their relationships to other classes. This diagram is called a class diagram. **Figure 1-10** illustrates a class diagram for the Tradeshow System.

Each box is a class and can be thought of as a particular set of objects that are important to the system. Important attributes of each class are also included in each box. These represent the detailed information about each object that will be maintained by the system. Note that some classes have lines between them. These represent relationships between the classes that need to be captured in the system. For example, a contact is a person who works for a particular supplier. A specific example might be that Bill Williams is the contact person for

Object Classes	Attributes
Supplier	supplier name, address, description, comments
Contact	name, address, phone(s), e-mail address(es), position, comments
Product	category, name, description, gender, comments
ProductPicture	ID, image

FIGURE **1-10**

*Preliminary class diagram for the
Tradeshow System*

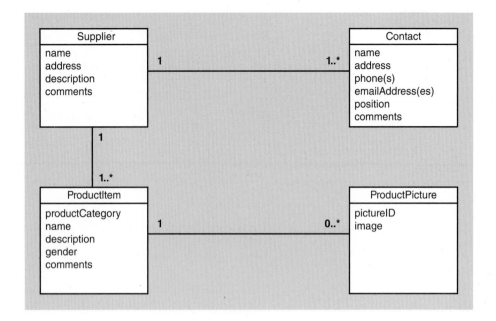

FIGURE **1-10**

*Preliminary class diagram for the
Tradeshow System*

the South Pacific Sportswear Company. Thus, the system needs to associate Bill Williams and the South Pacific Sportswear Company. The relationship line documents that requirement.

Class diagrams are a powerful and frequently used way to understand and document the information requirements of a system. The Tradeshow System is extremely simple, with only four classes identified—two of which belong to the Supplier Information subsystem. Most real-life systems are much larger and have dozens of classes.

Day 3 Activities

The purpose of Day 3 activities is to analyze in detail those use cases and classes that were selected to be implemented in this first iteration. During Day 3, we are still performing processes that are considered systems analysis. We are still trying to understand the requirements at a detailed level for the system. Included are these activities:

■ Perform in-depth fact finding to understand details. (Core Process 3)
■ Understand and document the detailed workflow of each use case. (Core Process 3)
■ Define the user experience with screens and reports. (Core Processes 3 and 4)

It is important to note that the use cases help the project team organize its work. These drill-down activities are done for each use case. As mentioned earlier, these use cases pertain to the Supplier Information subsystem:

■ *Look up supplier.*
■ *Enter/update supplier information.*
■ *Look up contact information.*
■ *Enter/update contact information.*

The project team will develop a workflow for each use case to better understand how it works and to identify what screens and possibly what reports will need to be developed. As the team gets more into the details, it may discover that some of the initial analysis is incomplete, if not incorrect. This is a good time to make such discoveries—much better than after the programs have been written.

Figure 1-11 illustrates a simple use case diagram. It shows the four aforementioned use cases identified and the user who will be the primary person

FIGURE **1-11**
Use case diagram

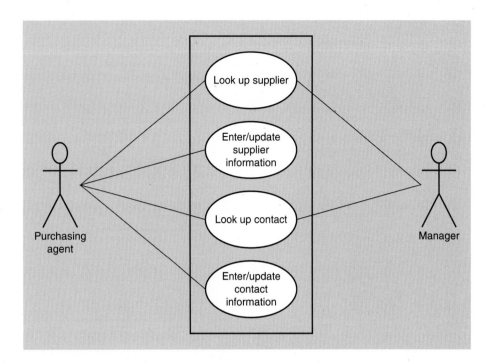

performing that function. What the diagram means in detail and how to develop one will be discussed in a later chapter.

Developing Use Case Descriptions and Workflow Diagrams

There are various methods for documenting the details of a use case. One that you will learn later in this text is called a *use case description*. Another method is developing a workflow diagram, which shows all the steps within the use case. The purpose with either method is to document the interactions between the user and the system (i.e., how the user interacts and uses the system to carry out a specific work task for a single use case).

Let us develop a workflow for one use case. To develop a workflow, we will use a simple type of diagram called an activity diagram. **Figure 1-12** illustrates the workflow for the *Look up supplier* use case. The ovals in the diagram represent the tasks, the diamonds represent decision points, and the arrows represent the sequence of the flow. The columns represent who performs which tasks. Usually, workflow diagrams are quite easy to understand.

The arrows that cross the center line represent the interactions between the system and the user. These are critically important because the developers must provide a screen or Web page that either captures or displays information. The arrows that cross the center line identify the data elements that become part of the user interface.

Looking at Figure 1-12, we see that the top arrow indicates that the supplier name enters into the system. Thus, we infer that the user must have an online form in order to enter the supplier name for the initial lookup. The next arrow indicates that there must be a form that displays all the details for an individual supplier, including a list of existing contacts. The user may also want to see more details about a specific contact person for this supplier, so the user may request detailed information for a particular person. Because the user can select one of the displayed results, it appears that we must design the form so each entry on the list is either a hotlink or has some mechanism to select it.

Defining Screen Layout

User-interface design includes all those tasks that describe the look and feel of the system to the user. Because the user interface is the window that the users

FIGURE **1-12**

Workflow diagram for the Look up supplier *use case*

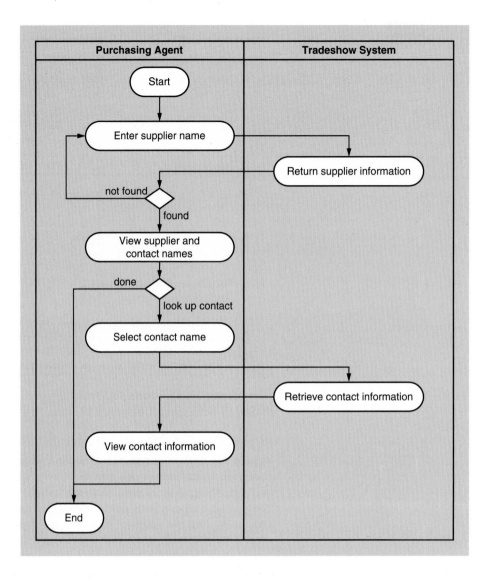

work with to utilize the functionality of the system, the user interface is essentially the system. If the interface is poorly designed, users will not be able to take full advantage of the system; they may even consider the system to be less than optimal. On the other hand, a well-designed user interface—one that is intuitive and easy to use, with a full range of features to facilitate navigation, and that provides good information—will enhance the utility of the system tremendously.

Figure 1-13 illustrates the layout of the first screen used for the workflow in the use case *Look up supplier*. The top portion of the screen provides the locations for the user to enter the supplier information, and the bottom portion of the screen shows the results. When results are provided, the search box for data entry will remain visible to allow the user to enter another search. Each entry in the results will be built as a hotlink, so the user can click on any particular supplier to retrieve more detailed information. This drill-down technique is a common method used in today's systems and will be intuitively easy for the users.

Searches are conducted on the RMO database, resulting in such RMO information as name, address, and contact information. An Internet-wide search is also possible. This allows the purchasing agents to look for and view the suppliers' own Web sites, which can be helpful, as can looking at forums and discussions about the supplier. Note that this tangential activity is not captured in the workflow of Figure 1-12. It is intended to assist in screen design, not document everything the user can or will do.

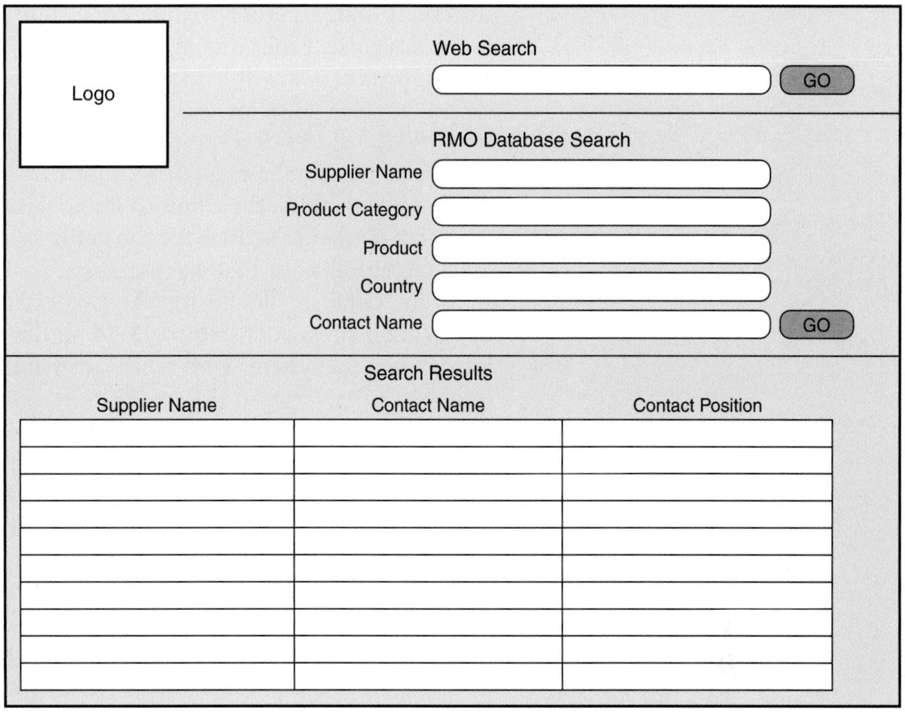

FIGURE **1-13**

Draft of screen layout for the Look up supplier *use case*

Day 4 Activities

The primary focus of Day 4 activities is to design the various components of the solution system. Up to now, we have mostly been trying to understand the user requirements. On Day 4, we carry out design activities that direct programming efforts. In that sense, design activities can be considered somewhat of a bridge. During analysis activities, the project team's objective is to understand user needs. During programming, the objective is to produce the solution. Thus, design is the bridge between understanding and construction. It provides the outline for how the solution will be structured and how it is to be programmed. System design also tends to involve the technical people, with less need for user participation.

Design can be a complex process. In our small project, we will limit our design examples to only a few models and techniques. Later in this text-book, you will learn additional design techniques. Day 4 Activities include the following:

■ Design the database structure (schema). (Core Process 4)
■ Design the system's high-level structure. (Core Process 4)

Database design is a fairly straightforward activity that uses the class diagram as input and develops the detailed database schema that can be directly implemented by a database management system. Such elements as table design, key and index identification, attribute types, and other efficiency decisions are made during this activity.

Designing the high-level system structure and the individual programs can be an intricate and complex process. First, the overall structure of the system is designed, including identifying the subsystems and connections to other systems. Within each subsystem, decisions are made about individual programs, such as user-interface programs, business logic programs, and database access programs. Then, at the lowest level, the login within each program is defined, including what program functions are required and what variables are used.

It is not uncommon for developers to begin writing program code as they develop portions of the design. It is a good idea to complete most of the structural design before writing code. But as the lower levels of the system are being designed, programming often begins. However, in the RMO Tradeshow System project, we will list them as separate activities.

Designing the Database

Designing the database uses the information provided by the class diagram to determine the tables, the columns in the tables, and other components. Sometimes, the database design is done for the entire system or subsystem. At other times, it is built piecemeal—use case by use case. To keep our project simple, we will just show the database design for the two classes that are required for the Supplier Information subsystem. **Figure 1-14** shows the database schema for the Supplier Information subsystem. Two tables are defined: Supplier and Contact.

Approaching High-Level Systems Design

There are fundamental design principles that will guide you through systems and program design; they will be explained in detail later in this book. For now, we will describe the general approach to design.

One of the first questions encountered in systems design is how and where to start. So far, we have three types of documents that can provide specifications to help answer that question. We have use cases, with their accompanying documentation, such as use case workflow diagrams. We have a class diagram that will help us identify some of the object-oriented classes that will be needed in the system. (In the previous section, we used the class diagram as the basis for the database design. Those same classes are important in developing object-oriented program classes.) Finally, we have screens and reports that also provide specifications for program logic and display logic.

Before we jump into design, let us briefly discuss the objective of systems design and what we expect to have as the output or result. Object-oriented programs are structured as a set of interacting classes. Therefore, in order to program, we need to know what those programming classes are, what the logic is within each class (i.e., the functions), and which programming classes must interact together. That is the final objective of systems design: to define the classes, the methods within those classes, and the interactions between classes.

FIGURE **1-14**

Database schema for Supplier Information subsystem

Table Name	Attributes
Supplier	SupplierID: integer {key} Name: string {index} Address1: string Address1: string City: string State-province: string Postal-code: string Country: string SupplierWebURL: string Comments: string
Contact	ContactID: integer {key} SupplierID: integer {foreign key} Name: string {index} Title: string WorkAddress1: string WorkAddress2: string WorkCity: string WorkState: string WorkPostal-code: string WorkCountry: string WorkPhone: string MobilePhone: string EmailAddress1: string EmailAddress2: string Comments: string

We perform this design by starting at the very highest level and then drilling down to the lowest level until we have defined all the functions within each class. Detailed design is the thought process of how to program each use case. Later in the text, you will learn techniques to carry out detailed design. For Day 4, we will focus only on the overall design.

Designing the Overall Architecture

Figure 1-15 shows the overall architecture or structure of the new system. Although the figure itself appears rather simple, some important decisions have been involved in the development of this design. First, note that the decision was made to build this application as a browser-based system. A different and very popular approach would have been to build smartphone or tablet applications. Browser-based systems sometimes do not provide the same connectivity speed and control as smartphone or tablet applications, but they are more versatile in that they can be more easily deployed on different equipment, such as laptops, without modification.

These high-level design decisions will determine the detailed structure of the system. A browser-based system is structured and constructed differently than an application system that runs on a smartphone or a tablet computer.

Defining the Preliminary Design Class Diagram

Given that the Tradeshow System will be built by using object-oriented programming (OOP) techniques, an important component of the design is developing the set of object classes and functions that will be needed for the system. This process can become quite detailed, and we will not try to explain it for this project. You will learn those techniques later in this book.

Figure 1-16 is a preliminary design class diagram for the Tradeshow System. A design class diagram (DCD) identifies the OOP classes that will be needed for the system. The set of design classes includes problem domain classes, view layer classes, sometimes separate data access classes, and utility classes. In Figure 1-16, we show only the problem domain classes and the view layer classes. Problem domain classes are usually derived from those classes that were identified during analysis activities—hence, the name: problem (user need) domain classes. You will also notice that they very closely correspond to the

FIGURE **1-15**

Tradeshow System architectural configuration diagram

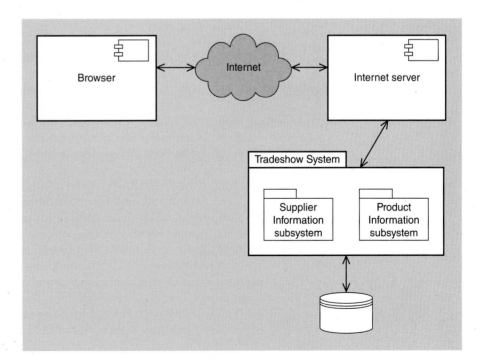

FIGURE **1-16**

Preliminary design class diagram

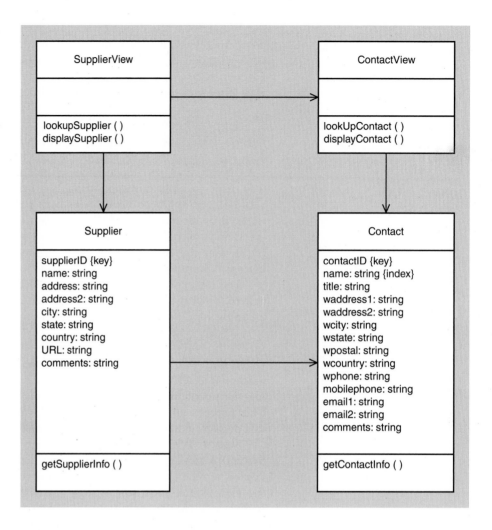

database tables; in fact, in this simple project, they are almost exactly the same as the database tables. On more complex systems, they will be similar but not exactly the same. However, remember that programming classes are distinct from database tables.

Other classes are required for the graphical user interface (GUI). In a dynamic Web system, such as the Tradeshow System, they are the classes that receive the input from the browser and format the output HTML files to be displayed by the browser.

The design classes in Figure 1-16 include the class-level variables that are needed for the class. These classes also show method names of the important methods within each class. These methods are identified and specified during high-level design and detailed design. One final element in the design class diagram are arrows that show which classes can access the methods of which other classes.

Designing Subsystem Architecture

Once we have an overall structure and an overall approach for implementing the new system, we begin to drill down to the subsystem design. **Figure 1-17** illustrates the architectural design of the Supplier Information subsystem. Notice that this subsystem is further divided into layers: a view layer and a model layer. You will learn much more about multilevel design later in this textbook. One of the advantages of partitioning the system into layers is that the system is much easier to build and maintain with this kind of structure. For example, the system will be browser based, but different browsers require different techniques. It is better not to get these complexities mixed in with the basic program functions. Hence, they are separated out into a distinct layer.

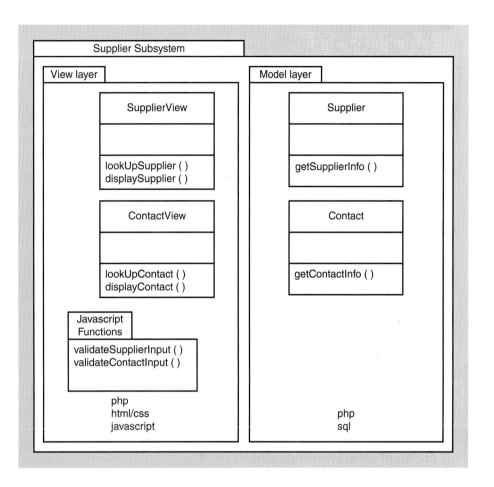

Figure 1-17 shows that the view layer has two PHP classes that process user inputs from the browser and format output HTML files. It also contains various JavaScript functions that will execute within the browser itself. The model layer classes are those classes that perform the business logic as well as access the database. Sometimes, the data layer and the business logic layer are further divided.

Managing the Project

Design is a complex activity with multiple levels—from high-level structural design to low-level detailed program design. In our project, we have separated the tasks for designing the overall system structure from detailed design of the programs themselves. However, these activities are often done concurrently. The basic high-level architectural structure is defined first, but mid-level and low-level design are often done concurrently with programming.

In Figure 1-17, we can see that detailed design and programming are quite time-consuming activities. A project manager must decide whether to extend the project or bring on additional programmers to help write the code. In our project, we have elected to insert a half-day of free time to bring in two additional programmers and train them. Of course, we could go ahead and begin Day 5's activities to ensure that we keep the project on schedule.

Day 5 Activities

Even though detailed design and programming may frequently begin earlier in the project, we have identified it as a separate day's activities. We do this for a couple of very important reasons. First, we want to emphasize that it is not a good practice to begin programming before critical information is obtained and decisions are made. Often, novice programmers will begin to program before the users' needs are adequately understood or even before the structure of the overall system has been determined. But a much better approach is to understand, design, and build small chunks of the system at a time. Agile Development anticipates and plans for the expected changes and refinements to the problem requirements that happen during detailed design and programming.

As the programmers write the code, they also perform individual testing on the classes and functions they program. This textbook does not focus on programming activities. However, we include an example of program code so you can see how systems design relates to the final program code. **Figure 1-18** is an example of a class that receives and processes the request for supplier information.

Day 6 Activities

The focus of Day 6 activities is to do the final testing that is required before the system is ready to be deployed. There are many types of testing that are required. In this example, we mention only two types of testing: overall system functional testing and user acceptance testing. Functional testing is usually a system-level test of all user functions and is often done by a quality assurance team. User acceptance tests are similar in nature, but they are done by the users, who test not only the correctness of the system but its "fitness" to accomplish the business requirements.

Each of the various testing activities in Day 6 has a somewhat similar sequence of tasks to perform. The tasks themselves highly depend on the test data and on the method for testing a particular test case. In some instances, the testing may be automated. In others, individuals may need to manually conduct the tests. Many new systems are interactive systems with user activity. There are some testing tools that automate that process somewhat, but it tends to be a rather complex task.

FIGURE **1-18**

Code for the SupplierView class

```php
<?php
    class SupplierView
    {
        private Supplier $theSupplier;

        function __construct()
        {
            $this->theSupplier = new Supplier();
        }

        function lookupSupplier()
        {
            include('lookupSupplier.inc.html');
        }

        function displaySupplier()
        {
            include('displaySupplierTop.inc.html');
            extract($_REQUEST);    // get Form data
        //Call Supplier class to retrieve the data
            $results = $theSupplier->getSupplierInfo($supplier, $category,
                                            $product, $country, $contact);

            foreach ($results as $resultItem){
            ?>
                <tr>
                    <td style="border:1px solid black">
                        <?php echo $resultItem->supplierName?></td>
                    <td style="border:1px solid black">
                        <?php echo $resultItem->contactName?></td>
                    <td style="border:1px solid black">
                        <?php echo $resultItem->contactPosition?></td>
                </tr>
            <?php }
            include('displaySupplierFoot.inc.html');
        }
    }
?>
```

FIGURE **1-19**
Generalized workflow of testing tasks

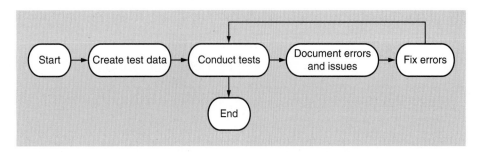

Figure 1-19 is a generalized workflow for testing the new system. In this workflow, we have shown the different testing tasks as separate steps. In reality, they all tend to be carried out together. However, any given test case will follow this flow.

First Iteration Recap

Figure 1-20 is a screenshot of the browser page that is used in the Tradeshow System to enter and view suppliers.

As stated previously, this is the first (six-day) iteration of a longer project. Using Agile techniques and iterations within an overall project allows flexibility in defining and building a new system. One of the Agile mandates is that the user should be heavily involved in the development of the new system. In this six-day project, the users have had major involvement during all days except Day 4 and Day 5.

A primary problem in developing a new system is that as the project progresses, new requirements are often identified. This happens because the users and the project team learn more about how to solve the business need. Agile, iterative projects are structured to handle these new requirements—often by adding another iteration to the overall project.

As a final step in a current iteration, or perhaps as part of the planning process for the next iteration, there should be a review of the processes and success of the current iteration. The lessons learned and issues to be carried forward create an environment of continual improvement and refinement. Iterative projects tend to improve and become more efficient during the life of the project.

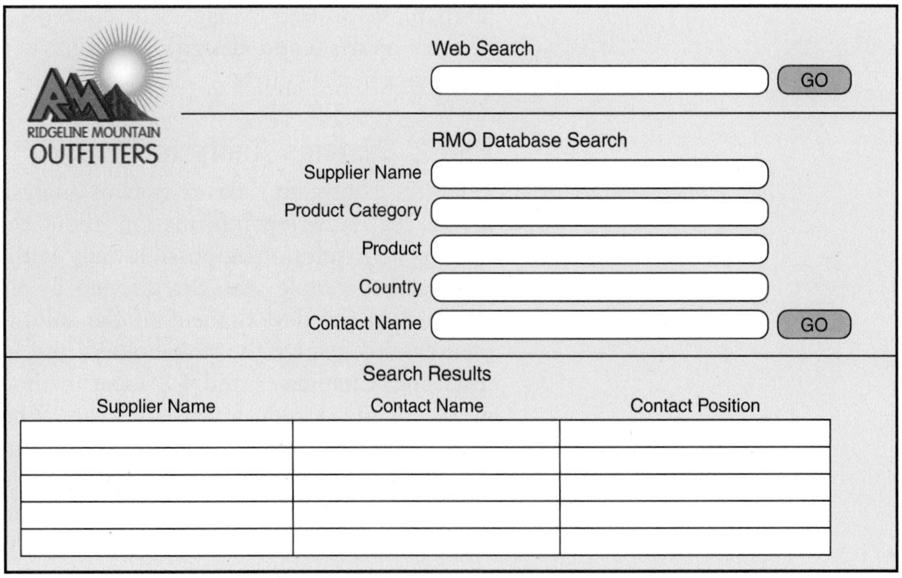

FIGURE **1-20**

Screen capture for Look up supplier *use case*

Where You Are Headed—The Rest of This Book

This sixth edition of *Systems Analysis and Design in a Changing World* includes the printed textbook and supporting online chapters. The current printed textbook provides a compact, streamlined, and focused presentation of those topics that are essential for information systems developers. The online chapters extend those concepts and provide a broader presentation of several topics. The online chapters may be integrated into the course or simply used as additional reading as prescribed by the instructor.

Part 1: Introduction to Systems Development

Part 1—comprising Chapter 1 and Online Chapter A—presents an overview of systems development. The introductory chapter provides a detailed, concrete example of what is required in a typical software development project. Of course, many details had to be left out to keep this chapter at a reasonable length. However, this chapter does present many processes, techniques, and diagrams. You are not expected to understand all the elements from this brief introduction. However, you should have a general idea of the approach to developing systems. You may want to refer back to this chapter from time to time to help understand the big picture.

This first chapter begins by briefly explaining the objectives of systems analysis and systems design. Many students who take a programming class think that programming is all you need to develop software and deploy a system. This chapter and the rest of this book should dispel that myth.

Online Chapter A, "The Role of the Systems Analyst," describes the many skills required of a systems analyst. It also discusses the various career options available to information systems majors. For those of you who are new to the discipline of information systems, this chapter will provide interesting and helpful knowledge about information systems careers.

There are three major subject areas discussed in this book: systems analysis, systems design, and project management. There is also one minor subject area, which is no less important but not discussed in as much depth, and that is systems implementation, testing, and deployment. In addition, we have taken an approach that is quite different from other texts. Because you already have a basic understanding of systems analysis and design from Chapter 1, we can immediately present in-depth concepts about systems analysis and design. We present project management topics later in this text. This will allow you to learn those project management concepts after understanding the elements of systems analysis and design. We think it will be more meaningful for you at that point in the course.

Part 2: Systems Analysis Tasks

Chapters 2 through 5 cover systems analysis in detail. Chapter 2 discusses techniques for gathering information about the business problem. Developing the right system solution is possible only if the problem is accurately understood. The various people who are affected by the system (the stakeholders) are also included in the development of the solution. Chapter 2 also explains how to identify and involve the stakeholders and introduces the concept of models and modeling. Chapters 3 and 4 present methods for capturing the detailed requirements for the system in a useful form. When discussing an information system, two key concepts are particularly useful: use cases, which define what the end users need the system to do, and the things, called data entities or classes, that users work with while carrying out their tasks. These two concepts—use cases and data entities/classes—are important no matter what approach to system development is being used. Chapter 5 presents more in-depth models, such as use case descriptions, use case diagrams, and system sequence diagrams.

Online Chapter B, "The Traditional Approach to Requirements," presents the traditional, structured approach to developing systems. To those instructors and students who desire to learn about data flow diagrams, this chapter provides an in-depth presentation.

All these modeling techniques provide in-depth analysis of the user's needs and allow the analyst to develop requirements and specifications. Again, the purpose of systems analysis is to thoroughly understand and specify the user's needs and requirements.

Part 3: Essentials of Systems Design

Chapters 6 and 7 provide the fundamental concepts related to systems design and to defining and designing the user experience. Chapter 6 provides broad and comprehensive coverage of important principles of systems design. It serves not only as a broad overview of design principles but as a foundation for later chapters that explain the detailed techniques, tasks, skills, and models used to carry out design.

Chapter 7 presents additional design principles related to designing the user interface and the system interfaces. Designing the user interface is a combination of analysis and design. It is related to analysis because it requires heavy user involvement and includes specifying user activities and desires. On the other hand, it is a design activity because it is creating specific final components that are used to drive the programming effort. The screens and reports and other user interaction components must be precisely designed so they can be programmed as part of the final system. System interfaces occur when one information system communicates or interacts with another information system without human intervention. System interfaces are becoming increasingly important because of Web services and cloud computing.

Part 4: Projects and Project Management

By this point, you will have a basic understanding of all the elements of systems development.

Part 4 brings together all these concepts by explaining more about the process of organizing and managing development projects. Chapter 8 describes different approaches to systems development in today's environment, including several important System Development Life Cycle models. It is an important chapter to help you understand how projects actually get executed.

Chapter 9 extends these concepts by teaching foundation principles of project management. Every systems analyst is involved in helping organize, coordinate, and manage software development projects. In addition, almost all of you will become team leaders and project managers. The principles presented in Chapter 9 are essential to a successful career.

Online Chapter C, "Project Management Techniques," goes into more detail regarding the tools and techniques used by systems analysts and project managers to plan and monitor development projects. For those instructors and students who would like to learn specific project management skills, this is an important chapter.

Part 5: Advanced Design and Deployment Concepts

Part 5 goes into more depth with respect to systems design, database design, and other important issues related to effective and successful system development and deployment.

Chapters 10 and 11 explain in detail the models, skills, and techniques used to design software systems. As mentioned earlier in this chapter, systems design is a fairly complex activity, especially if it is done correctly. The objective of these two chapters is to teach you the various techniques—from simple to complex—that can be used to effectively design software systems.

Chapter 12 explains how to design the database from the information gleaned during analysis and the identification of the object classes. Other related concepts, such as controls and security, are also presented in this chapter. Chapter 13 describes the final elements in systems development: final testing, deployment, maintenance, and version control. Chapter 14 concludes this book by looking toward the future of software development and assessing the current trends that may eventually enhance and improve the approaches to software development.

Chapter Summary

This chapter provided a quick overview of a particular software system development project called the Tradeshow System. The six core processes that control software development were presented. Then, the various activities that support the execution of these six core processes were explained as we went through an implementation of the Tradeshow System. The six core processes are:

1. Identify the problem or need and obtain approval to proceed.
2. Plan and monitor the project—what to do, how to do it, and who does it.
3. Discover and understand the details of the problem or the need.
4. Design the system components that solve the problem or satisfy the need.
5. Build, test, and integrate system components.
6. Complete system tests and then deploy the solution.

In order to facilitate learning and to help you remember the core processes and related activities, we divided the project into pre-project activities and then six other groupings, which we called *project days*. It should be noted that there is nothing magical or mandatory about this way of organizing a project. It was done in this manner simply to help you understand the various activities related to the core processes.

Key Terms

Agile Development   6

computer application (app)   4

information system   4

information systems development
   process   6

iterative development   8

project   5

subsystem   11

systems analysis   5

systems design   5

Systems Development Life
   Cycle (SDLC)   6

Review Questions

1. What is the difference between an information system and a computer application?
2. What is the purpose of systems analysis? Why is it important?
3. What is the difference between systems analysis and systems design?
4. What is a project?
5. What are the six core processes for software systems development?
6. What is meant by Agile Development?
7. What is the purpose of a System Vision Document?
8. What is the difference between a system and a subsystem?
9. What is the purpose of a Work Breakdown Structure?
10. What are the components of a Work Breakdown Structure? What does it show?
11. What information is provided by use cases or a use case diagram?
12. What information is provided by a class diagram?
13. How do a use case diagram and a class diagram drive the system development process?

14. What is another way to describe an activity diagram? What does it show?

15. How does an activity diagram help in user-interface design?

16. What is the purpose of architectural design?

17. What new information is provided in a design class diagram (more than a class diagram)?

18. What are the steps of system testing?

19. What is the purpose of user acceptance testing?

20. Why is it a good practice to divide a project into separate iterations?

21. What should be the objective or result of an iteration?

CHAPTER CASE

Keeping Track of Your Geocaching Outings

When Wayne Johansen turned 16, his dad bought him a new Garmin handheld GPS system. His family had always enjoyed camping and hiking, and Wayne was usually the member of the family who monitored their hikes with his dad's GPS system. He always liked to carry the GPS because he really enjoyed monitoring the routes, distances, and altitudes of their hikes. More recently, though, he had found a new hobby by using his GPS system: geocaching.

Geocaching is akin to the treasure hunts that most of us did when we were kids. The difference is that geocaching is a high-tech version of a treasure hunt that uses GPS but also calls on one's basic treasure-hunting skills.

As Wayne became more involved with his hobby, he discovered that there are many different kinds of activities for geocaching enthusiasts. The simplest ones are those that involve caches that can be found by using GPS coordinates, although even some of these can be difficult if the caches are well hidden and well camouflaged. Some of the activities involve multipoint drops in which there is a set of clues at multiple locations that must be followed in order to arrive at the final cache point. Some activities involve puzzles that must be solved in order to determine the coordinates and location of the final cache.

Before long, Wayne wanted to make his own caches and post them for people to find. He discovered that there were several Web sites, including individuals' blogs, with geocaching information, caches, and memberships. He joined one of the geocaching Web sites and used it to log his finds. It was fun to log and publish his finds and to post the geocaches that he created. But he decided he would like his own little system for keeping track of all the information he wanted to maintain about his caches. Conveniently, Wayne's older brother Nick, a college student majoring in information systems, was looking for a semester project for one of his programming classes. The two of them decided to

work together and develop a system to help Wayne keep track of all his geocaching activities.

In this end-of-chapter case, you will go through the various core processes of an SDLC and perform some of the activities of a development project. Of course, this is a very small project with very limited requirements. The project and various assignments are divided into days, as was our Tradeshow project. You have not learned all the skills required to effectively produce all the documents illustrated in the chapter. Hence, the daily assignments for this case should be considered as preliminary efforts and rough drafts. The objective of these assignments is simply to help you remember the overall approach to software development. Several assignments have been listed for each day to allow your instructor to select those that best meet the objectives of the course.

Day 0: Define the Vision

The primary purpose of pre-project activities is to define a vision for the new system. Either by yourself or with another class member, brainstorm all the neat functions this geocaching system might do. Keep it at a very high level. You just want to think of the major functions that Wayne might want the system to do for him. These activities closely relate to Core Process 1: Identify the problem or need and obtain approval to proceed.

Assignment D0-1: Write a rough draft of the System Vision Document based on your brainstorming ideas. [Hint: Think of what Wayne wants the system to do and why this is a benefit to him.]

Day 1: Plan the Project

Obviously, this is a small project and does not require an elaborate project plan. Based on the scope and vision you described in the System Vision Document, divide the project into at least two separate subsystems or

(continued on page 30)

(continued from page 29)

sets of functions that can be done in two different iterations. For example, perhaps a first version can run on a laptop, with a second version that includes mobile components for a smartphone. Or perhaps the first version can maintain a history of past geocache hunts, and the second version can allow Wayne to record geocaches that he creates. Or the first version can create a simple database system, and the second version can enhance it by linking to photo albums and/or blog posts. As you can see, there are often many different ways that a project can be partitioned. These activities are related to Core Process 2: Plan and monitor the project—what to do, how to do it, and who does it.

Assignment D1-1: Divide the system into at least two separate components or subsystems, which can be supported with two iterations. Briefly describe each.

Assignment D1-2: Create a Work Breakdown Structure that lists all the steps to complete the first iteration. Put a time estimate on each step. [Hint: Use the one in this chapter as a model.]

Day 2: Define and Understand the Requirements

We often describe the activities on Day 2 and Day 3 with the word *understanding*. On Day 2, we want to get an overall view of what the system needs to do for Wayne. As you learned in this chapter, there are two primary areas we focus on to obtain this high-level understanding of the system: a list of use cases and a list of object classes. We could document this information in lists, but diagrams provide a visual representation that is often easier to remember and understand. These activities support Core Process 3: Discover and understand the details of the problem or the need.

Assignment D2-1: Identify a few use cases that apply to one subsystem. [Hint: Think of what Wayne plans to do with the system. He will use the system to "do what"?]

Assignment D2-2: Try to identify the classes that apply to the first project iteration. [Hint: Think of "information things" that Wayne wants the system to "remember."]

Assignment D2-3: Create a simple use case diagram from the list of use cases. [Hint: Drawing by hand is fine. Use the one in this chapter as a model.]

Assignment D2-4: Create a simple class diagram from the list of classes. [Hint: Drawing by hand is fine. Use the one in this chapter as a model. Think of some other pieces of information that apply to each class.]

Day 3: Define the User Experience

These activities are a continuation of what we began in Day 2. The objective here is to further understand what Wayne will need and how he will actually use the system. We will determine exactly how each use case works—what steps and options are available with the use case and even what the display and data entry screens will look like. Unfortunately, this often requires a lot of work. For this case, though, let us keep it simple. These activities primarily support Core Process 3: Discover and understand the details of the problem or the need.

Assignment D3-1: Select a single use case and then identify the individual steps required to perform the use case. [Hint: Think of what Wayne does and how the system responds.]

Assignment D3-2: Make a workflow diagram of the selected use case. [Hint: Drawing by hand is fine. Each step from D3-1 goes in an oval. Connect the ovals with arrows.]

Assignment D3-3: Sketch out one of the screens that will be required to support a use case. The screen should allow for data entry and display of information. [Hint: Don't make it elaborate. Focus only on the input and output data fields that apply to only one use case.]

Day 4: Develop the Architectural Design

The high-level architectural design of the system generally includes decisions about how the system will be built and what the database will look like. Design is a technical activity that requires experience in programming, database development, and system architecture. These activities support Core Process 4: Design the system components that solve the problem or satisfy the need.

Assignment D4-1: Design a preliminary database schema for the classes in this iteration. [Hint: Each class becomes a table. The attributes become table columns.]

Assignment D4-2: Decide whether you will build a desktop system or a browser-based system. Write a couple of paragraphs listing the pros and cons of each alternative in order to defend your decision. [Hint: Either option is valid. Think of reasons to support your decision.]

Day 5: Develop the Detailed Design and Program the System

You have learned how to do these activities in your programming classes. You probably have had many class projects where you designed a system and then

(continued on page 31)

(continued from page 30)

programmed it. These kind of activities support Core Process 5: Build, test, and integrate system components.

Assignment D5-1: Write a paragraph describing what programming language(s) you would recommend and what development environment you prefer. For this answer, draw on your previous programming and development experiences. [Hint: There are many valid solutions. Give reasons for your preference.]

Day 6: Test and Deploy the System

You may have had opportunities to perform comprehensive testing of your programming class projects, especially if you have developed systems that integrated with other systems. These activities support Core Process 6: Complete system tests and then deploy the solution. Obviously, you can only do this if you have programmed the system.

Assignment D6-1: Write a paragraph describing the difference between programmer testing and user testing. [Hint: Why is it hard to test your own work? What do the users know that you don't know?]

Assignment D6-2: Write a paragraph describing all the issues that might need to be addressed to deploy this system. [Hint: You might want to search the Internet to learn about deployment issues.]

Assignment D6-3: Look at *www.geocaching.com*, which is a commercial Web site. What other issues need to be addressed to deploy this type of Web site? [Hint: Think about all the issues related to security, robustness, financial protection, high volumes, up-time, different browsers, and so forth.]

PART 2

Systems Analysis Activities

2

Investigating System Requirements

Chapter Outline

- The RMO Consolidated Sales and Marketing System Project
- Systems Analysis Activities
- What Are Requirements?
- Models and Modeling
- Stakeholders
- Information-Gathering Techniques
- Documenting Workflows with Activity Diagrams

Learning Objectives

After reading this chapter, you should be able to:

- Describe the activities of systems analysis
- Explain the difference between functional and nonfunctional requirements
- Describe the role of models in systems analysis
- Identify and understand different kinds of stakeholders and their contributions to requirements definition
- Describe information-gathering techniques and determine when each is best applied
- Develop activity diagrams to model workflows

Mountain Vista Motorcycles

Amanda Lamy, president and majority stockholder of Mountain Vista Motorcycles (MVM), is an avid motorcycle enthusiast and businesswoman. MVM is headquartered in Denver and has locations throughout the western United States and Canada. Since the late-1990s, the market for motorcycles has grown tremendously. Amanda expects that the market will continue to be strong throughout the 2010s, although she is concerned about the "graying" of a significant portion of MVM's customer base.

The demographics of the motorcycle market are an interesting study in contrasts. At present, the majority of customers are over 50 years of age, male professionals or businesspeople, and partly or fully retired. They have substantial disposable income, lots of free time, and tend to own multiple expensive motorcycles from such manufacturers as Harley-Davidson, Honda, Ducati, and Moto Guzi. Older customers are generally comfortable with Internet and Web technology but are not significant users of social media technology. Although many own smartphones, they tend to use them primarily for voice, e-mail, and texting.

Male customers under 30 years of age tend to buy sport and dirt bikes, typically from such manufacturers as Suzuki and Kawasaki. They buy less expensive bikes than older customers and are more likely to buy parts and supplies from MVM to service their own bikes. Female customers under 30 years of age tend to buy motor scooters and smaller "commuter" motorcycles. Customers in the 30–50 age range include men and women who buy bikes of many types from many manufacturers. Comfort with and use of Internet technology, social media, and portable computing devices such as smartphones and iPads is very high with customers under 50 years of age, especially with customers under age 30.

Amanda is convinced that the key to long-term success in the motorcycle market is to build an active community of motorcycle enthusiasts at each MVM location that includes a broad spectrum of customers. In essence, each location needs to be seen as a hub of local motorcycle-related activity and information in physical and virtual terms. On the physical side, MVM has added activity and event-oriented pages to its Web sites, sponsored rallies and clubs, added meeting rooms and small coffee shops in some locations, and colocated with bars and restaurants that feature motorcycle-related themes and entertainment. These efforts have yielded good results with older customers but less so with younger customers.

Amanda is concerned about the lack of participation by younger customers and is sure that MVM's lack of presence in social media and virtual relationships is a significant factor. She and her senior staff, most of whom are older, are unsure how to attract younger customers. They have little knowledge of and no experience creating modern technology-based virtual communities.

MVM's chief information officer is starting to develop a project plan for a virtual community oriented toward younger customers. If the plan were for developing a traditional information system, she would use such standard approaches as interviewing internal users and managers and having her development staff write specifications, generate storyboards and screen layouts, and develop prototypes. But few of the intended virtual community users are MVM employees, and none of her staff members fully comprehends how to successfully use social media and other techniques for building virtual societies. Traditional methods of defining and refining requirements seem inadequate to the task.

Overview

In Chapter 1, you saw the System Development Life Cycle (SDLC) being employed by Ridgeline Mountain Outfitters (RMO) for a small information system application called the Tradeshow System. Development of that system followed the six core processes of the SDLC:

1. Identify the problem or need and obtain approval to proceed.
2. Plan and monitor the project—what to do, how to do it, and who does it.
3. Discover and understand the details of the problem or the need.
4. Design the system components that solve the problem or satisfy the need.
5. Build, test, and integrate system components.
6. Complete system tests and then deploy the solution.

In this chapter, we start expanding the scope and detail of the SDLC processes to cover a wider range of concepts, tools, and techniques. The extra depth and detail are needed to tackle larger and more complex projects. This chapter concentrates on systems analysis activities (the third core process listed), and the next few chapters follow up on that with detailed discussions of models developed during those systems analysis activities. Subsequent chapters expand the discussion of other core SDLC processes.

The RMO Consolidated Sales and Marketing System Project

Ridgeline Mountain Outfitters has an elaborate set of information systems applications developed over the years to support operations and management. However, there is a growing gap between customer expectations, modern technological capabilities, and existing RMO systems that support sales and customer interaction. This section reviews the existing system inventory and introduces the proposed Consolidated Sales and Marketing System that will update and enhance sales and marketing.

Existing RMO Information Systems and Architecture

RMO's Information Systems Department has always been forward looking. In past years, the department, in conjunction with corporate strategic plans, has developed five-year plans for development and deployment of new technology and information systems. The planning process has been an excellent tool to help the organization stay current with new trends and technology capabilities. However, the plans themselves have had mixed success. One of the complexities with long-range IT and software development plans is that technology changes rapidly and moves in unexpected directions. For example, the Tradeshow System described in Chapter 1 was made possible by the availability of powerful and flexible handheld devices and the widespread availability of Wi-Fi and Internet connections. Both technologies reached a fairly mature level in just a couple of years, which created an opportunity for RMO to optimize this important business process.

Historically, RMO has adopted new technology as soon as it became cost-effective. Past examples include adoption of smaller servers and desktop computing, interconnection of locations with dedicated telecommunications links, and such Web-based technologies as customer-oriented Web sites and browser-based user interfaces for internal systems. At present, RMO has a disparate collection of computers dispersed across home offices, retail stores, telephone centers, order fulfillment/shipping centers, and warehouses—everything connected by a complex set of local area networks (LANs), wide area networks (WANs), and virtual private networks (VPNs). The term **technology architecture** describes the set of computing hardware, network hardware and topology, and system software—such as operating and database management systems—employed by an organization. RMO's technology architecture is modern but not state of the art, which is consistent with its goal of adopting only cost-effective technology.

The term **application architecture** describes how software resources are organized and constructed to implement an organization's information systems. It describes the organization of software into modules and subsystems and includes supporting technologies (such as programming languages and development environments), architectural approaches (such as service-oriented architecture), and user-interface technologies (such as mobile computing displays, touch screen technology, and voice recognition).

Currently, the major RMO systems consist of:

■ Supply Chain Management (SCM)—This application was deployed five years ago as a client/server application using Java and Oracle. Currently, it supports inventory control, purchasing, and distribution, although integration of functions needs improvement. The new Tradeshow System will interface with this system.

■ Phone/Mail Order System—A modest client/server application developed 12 years ago by using Visual Studio and Microsoft SQL Server as a quick

technology architecture a set of computing hardware, network hardware and topology, and system software employed by an organization

application architecture the organization and construction of software resources to implement an organization's information systems

solution to customer demand for catalog phone and mail orders. It is integrated with the SCM and has reached capacity.

■ Retail Store System (RSS)—A retail store package with point-of-sale processing. It was upgraded eight years ago from overnight batch to real-time inventory updates to/from the SCM.

■ Customer Support System (CSS)—This system was first deployed 15 years ago as a Web-based catalog to support customer mail and phone orders. Four years later, it was upgraded to an Internet storefront, supporting customer inquiries, shopping cart, order tracking, shipping, back orders, and returns.

All organizations—including RMO—face a difficult challenge keeping all their information systems current and effective. Because development resources are limited, an organization's technology and application architecture and its information system inventory will include a mix of old and new. Older systems were often designed for outdated operational methods and typically lack modern technologies and features that some competitors have adopted to improve efficiency or competitiveness. Such is the case with RMO's existing customer-facing systems, which have several shortcomings, including:

■ Treating phone, Web, and retail sales as separate systems rather than as an integrated whole
■ Employing outdated Web-based storefront technology
■ Not supporting modern technologies and customer interaction modes, including mobile computing devices and social networking

Rather than incrementally update the existing sales systems, RMO plans to replace them, as shown in **Figure 2-1**.

FIGURE **2-1** *Proposed application architecture for RMO*

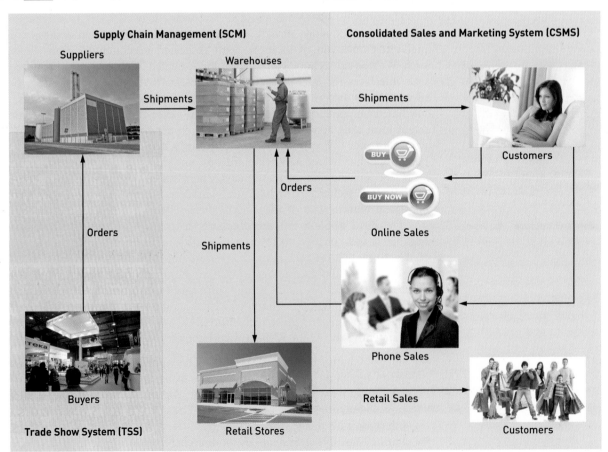

The New Consolidated Sales and Marketing System

The goals of the Consolidated Sales and Marketing System (CSMS) are to modernize the technology and functionality of the CSS and to add more customer-oriented functionality. On the technology side, the CSMS will incorporate current Web standards and be built under the assumption of high-bandwidth customer Internet connections and high-resolution displays. Updating the technology will enable a higher degree of interactivity, richer graphics, and a streamlined interface.

Key additions to system functionality will be support for mobile computing devices, incorporation of customer feedback and comments into product information, and integration of social networking functions. Unlike the CSS, the CSMS will support smartphones and tablet computers with interfaces specifically designed for each platform and with downloadable apps. Customer feedback will be captured directly through the Internet storefront, from RMO-supported comment forums and blogs, and mined from Facebook and Twitter. RMO will develop a complete presence in each social networking venue and enable system users to share purchases, recommendations, coupons, and store credits using those venues.

The new CSMS will also have four subsystems:

■ The Sales subsystem provides such basic functions as searching the online catalog and purchasing items and paying for them online. However, it has many new capabilities to assist the shopper making purchases. The system will provide specific suggestions about accessories that go with the purchased item. Images and videos of animated models will be available to help the customer see how various items and accessory packages will look together. The system will also provide information to shoppers about related purchases made by other shoppers. Customer ratings and comments are available for viewing. Finally, key social networking components will permit shoppers to network with their friends by sending messages to ask their opinions about particular merchandise items.

■ The Order Fulfillment subsystem will perform all the normal tasks of shipping items and allowing customers to track the status of their orders as well as the shipments. In addition, as part of order fulfillment, customers can rate and make comments about particular merchandise and their overall shopping experience. They may also make suggestions directly to RMO about the attractiveness of the Web site and the quality of the service they received.

■ The Customer Account subsystem provides all those services that enhance the customer experience. Customers can view and maintain their account information. They also can "link up" with friends who are also customers to share experiences and opinions on merchandise. The system will keep track of detailed shipping addresses as well as payment preferences and information. RMO also instituted a Mountain Bucks program wherein customers accumulate credits that can be used to redeem prizes as well as purchase merchandise. Customers may use these Mountain Bucks for themselves or they may transfer them to other people in their family/friends group. This is a great opportunity to combine accumulated bucks to obtain expensive merchandise.

■ The Marketing subsystem is primarily for employees to set up the information and services for customers. In this subsystem, employees can enter information about all the merchandise offered by RMO. This subsystem is also fed by the SCM system to maintain accurate data on the inventory in stock and anticipated arrival dates of items on order. Employees also set up the various promotional packages and seasonal catalogs by using the functions of this subsystem. RMO is experimenting with a new idea to enhance customer satisfaction: It is building partner relationships with other retailers so that customers can earn "combined" points with RMO purchases or a partner retailer purchase. These points can be used at RMO or transferred and used at the partner. For example, because RMO sells outdoor and sporting clothing, it has partnered with various sporting goods stores.

That way, a person can buy sporting equipment at the sporting equipment store and get promotional discounts for clothing at RMO. The success of this new venture has yet to be proven, but RMO anticipates that it will enhance the shopping experience of all its customers.

In later chapters, more details will be given about the capabilities of the new CSMS system. Detailed explanations will also describe the information that must be maintained in the database to support these functions.

Systems Analysis Activities

The callout on the left side of **Figure 2-2** lists the activities of the third core process, which is to discover and understand the details. This core process also goes by the name *systems analysis*. By completing these activities, the analyst defines in great detail what the information system needs to accomplish to provide the organization with the desired benefits. In essence, analysis activities are a second and more thorough pass at defining the problem and need. The first pass generates only enough detail to decide whether a new or upgraded system is warranted and feasible. The second (analysis) pass assumes that the organization is committed to the project. Thus, considerably more time and resources are invested to produce a much more detailed description of what the system will do.

Although we concentrate only on analysis activities in this chapter, keep in mind that they are usually intermixed with design, implementation, and other activities during the system development life cycle. For example, as shown in Figure 2-2, analysis activities are most intensive in the second iteration but occur in varying degrees during all project iterations except the last. This pattern is typical because an analyst will often concentrate on one part of a system, defining requirements only for that part and simultaneously designing and implementing software to satisfy those requirements. Organizing development activities in this iterative manner enables development to be broken into smaller steps and enables users to validate requirements by testing and observing a functional system. The following sections briefly describe analysis activities, and the remainder of this chapter expands the discussion of information gathering and defining system requirements.

Gather Detailed Information

Systems analysts obtain information from people who will be using the system, either by interviewing them or by watching them work. They obtain additional information by reviewing planning documents and policy statements. Analysts also study existing systems, including their documentation. They also frequently obtain additional information by looking at what other companies (particularly vendors) have done when faced with a similar business need. They try to

FIGURE **2-2** *Analysis activities*

Core Processes	Iterations					
	1	2	3	4	5	6
Identify problem and obtain approval						
Plan and monitor the project						
Discover and understand details						
Design system components						
Build, test, and integrate system components						
Complete system tests and deploy solution						

Analysis activities
- Gather detailed information
- Define requirements
- Prioritize requirements
- Develop user-interface dialogs
- Evaluate requirements with users

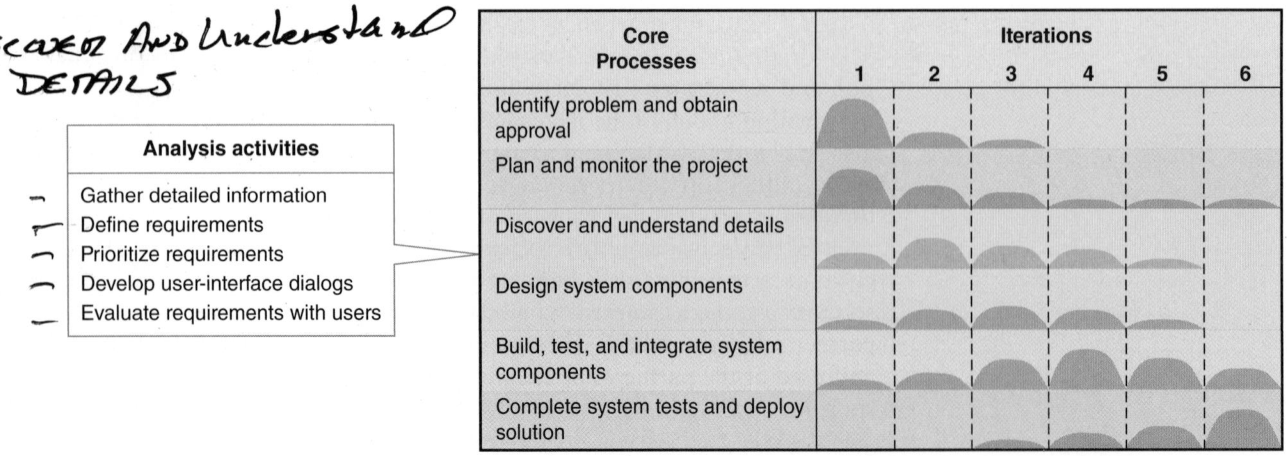

understand an existing system by identifying and understanding the activities of all the current and future users and by identifying all the present and future locations where work occurs and all the system interfaces with other systems, both inside and outside the organization. In short, analysts need to talk to nearly everyone who will use the new system or has used similar systems, and they must read nearly everything available about the existing system. Later in this chapter, we discuss how to identify and extract information from all these people.

Beginning analysts often underestimate how much there is to learn about the work the user performs. The analyst must become an expert in the business area the system will support. For example, if you are implementing an order-entry system, you need to become an expert on the way orders are processed (including related accounting procedures). If you are implementing a loan-processing system, you need to become an expert on the rules used for approving credit. If you work for a bank, you need to think of yourself as a banker. The most successful analysts become experts in their organization's main business.

Define Requirements

DEFINE

The analyst uses information gathered from users and documents to define requirements for the new system. System requirements include the functions the system must perform (functional requirements) and such related issues as user-interface formats and requirements for reliability, performance, and security (nonfunctional requirements). We further discuss the distinction between functional and nonfunctional requirements a bit later in this chapter.

Defining requirements is not just a matter of writing down facts and figures. Instead, the analyst creates models to record requirements, reviews the models with users and others, and refines and expands the models to reflect new or updated information. Building and refining requirements models occupies much of the analyst's time. This chapter and the next two chapters explain in considerable depth how to create requirements models.

Prioritize Requirements

PRIORITIZE

Once the system requirements are well understood, it is important to establish which requirements are most crucial for the system. Sometimes, users suggest additional system functions that are desirable but not essential. However, users and analysts need to ask themselves which functions are truly important and which are fairly important but not absolutely required. Again, an analyst who understands the organization and the work done by the users will have more insight toward answering these questions.

Why prioritize the functions requested by the users? Resources are always limited, and the analyst must always be prepared to justify the scope of the system. Therefore, it is important to know what is absolutely required. Unless the analyst carefully evaluates priorities, system requirements tend to expand as users make more suggestions (a phenomenon called *scope creep*). Requirements priorities also help to determine the number, composition, and ordering of project iterations. High-priority requirements are often incorporated into early project iterations so analysts and users have ample opportunity to refine those parts of the system. Also, a project with many high-priority requirements will typically have many iterations.

Develop User-Interface Dialogs

When a new system is replacing an old system that does similar work, users are usually quite sure about their requirements and the desired form of the user interface. But many system development projects break new ground by automating functions that were previously performed manually or by implementing functions that were not performed in the past. In either case, users tend to be uncertain about many aspects of system requirements. Such requirements models as use cases, activity diagrams, and interaction diagrams can be developed based on user input, but it is often difficult for users to interpret and validate such abstract models.

In comparison, user validation of an interface is much simpler and more reliable because the user can see and feel the system. To most users, the user interface is all that matters. Thus, developing user-interface dialogs is a powerful method of eliciting and documenting requirements. Analysts can develop user interfaces via abstract models, such as storyboards (covered in more detail in Chapter 7), or they can develop user-interface prototypes on the actual input/output devices that users will use (e.g., a computer monitor, iPad, or cell phone). A prototype interface can serve as a requirement and a starting point for developing a portion of the system. In other words, a user-interface prototype developed in an early iteration can be expanded in later iterations to become a fully functioning part of the system.

Evaluate Requirements with Users

Ideally, the activities of evaluating requirements with users and documenting the requirements are fully integrated. But in practice, users generally have other responsibilities besides developing a new system. Thus, analysts usually use an iterative process in which they elicit user input, work alone to model requirements, return to the user for additional input or validation, and then work alone to incorporate the new input and refine the models. Prototypes of user interfaces and other parts of the system may also be developed when "paper" models are inadequate or when users and analysts need to prove that chosen technologies will do what they are supposed to do. Also, if the system will include new or innovative technology, the users may need help visualizing the possibilities available from the new technology when defining what they require. Prototypes can fill that need. The processes of eliciting requirements, building models and prototypes, and evaluating them with users may repeat many times until requirements models and prototypes are complete and accurate.

What Are Requirements?

As you can see from the previous section, requirements and models that represent them are a key focus of analysis phase activities. Most of the analyst's time is devoted to requirements: gathering information about them, formalizing them by using models and prototypes, refining and expanding them, prioritizing them, and generating and evaluating alternatives. But to fully understand those activities, you need to answer a fundamental question: What are requirements?

system requirements the activities a system must perform or support and the constraints that the system must meet

functional requirements the activities that the system must perform

System requirements are all the activities the new system must perform or support and the constraints that the new system must meet. Generally, analysts divide system requirements into two categories: functional and nonfunctional requirements. **Functional requirements** are the activities that the system must perform (i.e., the business uses to which the system will be applied). For example, if you are developing a payroll system, the required business uses might include such functions as "generate electronic fund transfers," "calculate commission amounts," "calculate payroll taxes," "maintain employee-dependent information," and "report tax deductions to the IRS." The new system must handle all these functions. Identifying and describing all these business uses require a substantial amount of time and effort because the list of functions and their relationships can be very complex.

Functional requirements are based on the procedures and rules that the organization uses to run its business. Sometimes, they are well documented and easy to identify and describe. An example might be the following: "All new employees must fill out a W-4 form to enter information about their tax withholding in the payroll system." Other business rules might be more obtuse or difficult to find. An example from RMO might be the following: "Air shipping charges are reduced by 50 percent for orders over $200 that weigh less than two pounds." Discovering such rules is critical to the final design of the system. If this rule were not discovered, customers that had relied on it in the past might become angry. Modifying the system after customers start complaining would be much more difficult and expensive than building in the rule from the start.

FIGURE
FURPS+

Requirement categories	FURPS + categories	Example requirements
Functional	Functions	Business rules and processes
Nonfunctional	Usability Reliability Performance Security + Design constraints 　Implementation 　Interface 　Physical 　Support	User interface, ease of use Failure rate, recovery methods Response time, throughput Access controls, encryption Hardware and support software Development tools, protocols Data interchange formats Size, weight, power consumption Installation and updates

nonfunctional requirements system characteristics other than the activities it must perform or support

FURPS a requirements classification framework (acronym stands for functionality, usability, reliability, performance, and security)

usability requirements operational characteristics related to users, such as the user interface, related work procedures, online help, and documentation

reliability requirements requirements that describe system dependability

performance requirements operational characteristics related to measures of workload, such as throughput and response time

security requirements requirements that describe how access to the application will be controlled and how data will be protected during storage and transmission

FURPS+ an extension of FURPS that includes design constraints as well as implementation, interface, physical, and supportability requirements

design constraints restrictions to which the hardware and software must adhere

Nonfunctional requirements are characteristics of the system other than those activities it must perform or support. It is not always easy to distinguish functional from nonfunctional requirements. One way to do so is to use a framework for identifying and classifying requirements. There have been many such frameworks developed over time; the most widely used today is called FURPS+ (see **Figure 2-3**). FURPS is an acronym that stands for functionality, usability, reliability, performance, and security. The "F" in FURPS is equivalent to the functional requirements defined previously. The remaining FURPS categories describe nonfunctional requirements:

■ **Usability requirements** describe operational characteristics related to users, such as the user interface, related work procedures, online help, and documentation. For example, the user interface for a smartphone app should behave similarly to other apps when responding to such gestures as two-finger slides, pinching, and expanding. Additional requirements might include menu format, color schemes, use of the organization's logo, and multingual support.

■ **Reliability requirements** describe the dependability of a system—how often a system exhibits such behaviors as service outages and incorrect processing and how it detects and recovers from those problems.

■ **Performance requirements** describe operational characteristics related to measures of workload, such as throughput and response time. For example, the client portion of a system might be required to have a one-half-second response time to all button presses, and the server might need to support 100 simultaneous client sessions (with the same response time).

■ **Security requirements** describe how access to the application will be controlled and how data will be protected during storage and transmission. For example, the application might be password protected, encrypt locally stored data with 1024-bit keys, and use secure HTTP for communication among client and server nodes.

FURPS+ is an extension of FURPS that adds additional categories, including design constraints as well as implementation, interface, physical, and supportability requirements—all these additional categories summarized by the plus sign. Here are short descriptions of each category:

■ **Design constraints** describe restrictions to which the hardware and software must adhere. For example, a cell phone application might be required to use the Android operating system, consume no more than 30MB of flash memory storage, consume no more than 10MB of system memory while running, and operate on CPUs rated at 1 GHz or higher.

implementation requirements constraints such as required programming languages and tools, documentation method and level of detail, and a specific communication protocol for distributed components

interface requirements required interactions among systems

physical requirements characteristics of hardware such as size, weight, power consumption, and operating conditions

supportability requirements how a system is installed, configured, monitored, and updated

■ **Implementation requirements** describe constraints such as required programming languages and tools, documentation method and level of detail, and a specific communication protocol for distributed components.

■ **Interface requirements** describe interactions among systems. For example, a financial reporting system for a publicly traded company in the United States must generate data for the Securities and Exchange Commission (SEC) in a specific XML format. The system might also supply data directly to stock exchanges and bond rating agencies and automatically generate Twitter messages, RSS feeds, and Facebook updates.

■ **Physical requirements** describe such characteristics of hardware as size, weight, power consumption, and operating conditions. For example, a system that supports battlefield communications might have such requirements as weighing less than 200 grams, being no larger than 5 centimeters cubed, and operating for 48 hours on a fully charged 1200 milliwatt lithium ion battery.

■ **Supportability requirements** describe how a system is installed, configured, monitored, and updated. For example, requirements for a game installed on a home PC might include automatic configuration to maximize performance on existing hardware, error reporting, and download of updates from a support server.

As with any set of requirements categories, FURPS+ has some gray areas and some overlaps among its categories. For example, is a requirement that a battlefield communications device survive immersion in water and operate across a temperature range of –20°C to 50°C a performance or physical requirement? Is a restriction to use no more than 100 MB of memory a performance or design requirement? Is a requirement to secure communication between workstations and servers with 1024-bit encryption a performance, design, or implementation requirement? The answers to such questions are not important. What is important is that all requirements be identified and precisely stated early in the development process and that inconsistencies or trade-offs among them be resolved.

Models and Modeling

Modeling is an important part of analysis and design. Analysts build models to describe system requirements and use those models to communicate with users and designers. By developing a model and reviewing it with a user, an analyst demonstrates an understanding of the user's requirements. If the user spots errors or omissions, they are incorporated into the model before it becomes the basis for subsequent design and implementation activities. **Figure 2-4** summarizes the key reasons for building and using models.

Designers construct high-level and detailed models to describe system components and their interactions. Design models serve as a scratch pad for evaluating design alternatives and as a way to communicate the final design to programmers, vendors, and others who will build, acquire, and assemble components to create the final system. In general, models built during one SDLC activity are "consumed" during other activities.

model representation of some aspect of a system

A **model** is a representation of some aspect of the system being built. There are dozens of different models that an analyst or designer might develop and use (see **Figure 2-5**). Although this book emphasizes models and techniques for creating models, it is important to remember that system projects vary in the number of models required and in their formality. Smaller, simpler system projects will not need models showing every system detail, particularly when

FIGURE

FIGURE **2-4**
Reasons for modeling

❏ Learning from the modeling process

❏ Reducing complexity by abstraction

❏ Remembering all the details

❏ Communicating with other development team members

❏ Communicating with a variety of users and stakeholders

❏ Documenting what was done for future maintenance/enhancement

the project team has experience with the type of system being built. Sometimes, the key models are created informally in a few hours. Although models are often created by using specialized software tools, useful and important models are sometimes drawn quickly over lunch on a paper napkin or in an airport waiting room on the back of an envelope! As with any development activity, an iterative approach is used for creating models. The first draft of a model has some but not all details worked out. The next iteration might fill in more details or correct previous misconceptions.

Analysis and design models can be grouped into three generic types:

textual models text-based system models such as memos, reports, narratives, and lists

■ **Textual models**—Analysts use such textual models as memos, reports, narratives, and lists to describe requirements that are detailed and are difficult to represent in other ways. The event list in Figure 2-5 is one example of a textual model. Narrative description is often the best way to initially record information gathered verbally from stakeholders, such as during an interview. In many cases, narratives and other textual models are later converted into a graphical format.

FIGURE **2-5**
Some analysis and design models

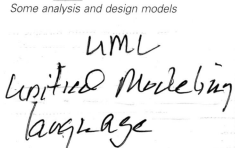

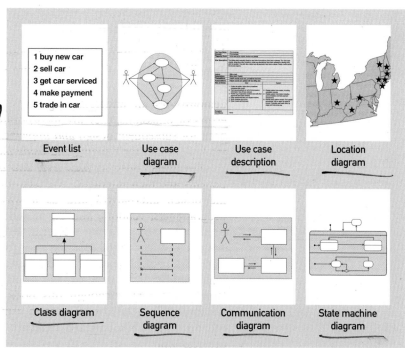

Event list	Use case diagram	Use case description	Location diagram
Class diagram	Sequence diagram	Communication diagram	State machine diagram

graphical models *system models that use pictures and other graphical elements*

mathematical models *system models that describes requirements numerically or as mathematical expressions*

Unified Modeling Language (UML) *standard set of model constructs and notations defined by the Object Management Group*

OBJECT Mngt Group (OMG)

stakeholders *persons who have an interest in the successful implementation of the system*

internal stakeholders *persons within the organization who interact with the system or have a significant interest in its operation or success*

external stakeholders *persons outside the organization's control and influence who interact with the system or have a significant interest in its operation or success*

- **Graphical models**—Graphical models make it easier to understand complex relationships that are difficult to follow when described as a list or narrative. Recall the old saying that a picture is worth a thousand words. In system development, a carefully constructed graphical model might be worth a million words! Some graphical models actually look similar to a real-world part of the system, such as a screen design or a report layout design. However, the graphical models developed during analysis activities typically represent more abstract things, such as external agents, processes, data, objects, messages, and connections.

- **Mathematical models**—Mathematical models are one or more formulas that describe technical aspects of a system. Analysts often use mathematical models to represent functional requirements for scientific and engineering applications and occasionally use them to describe business system requirements in areas such as accounting and inventory control. Analysts and designers use mathematical models to describe requirements and operational parameters such as network throughput or database query response time.

Many graphical models used in system development are drawn according to the notation specified by the **Unified Modeling Language (UML)**. In Figure 2-5, the use case diagram, class diagram, sequence diagram, communication diagram, and state machine diagram are UML graphical models. UML is the standard set of model constructs and notations defined by the Object Management Group (OMG), a standards organization for system development. By using UML, analysts and end users are able to depict and understand a variety of specific diagrams used in a system development project. Prior to UML, there was no standard, so diagrams could be confusing, and they varied from company to company (and from book to book).

We expand the discussion of models later in this chapter with a detailed look at one type of graphical model: the workflow diagram. In later chapters, you learn how to develop many of other types of analysis and design models.

Stakeholders

Stakeholders are your primary source of information for system requirements. **Stakeholders** are all the people who have an interest in the successful implementation of the system. Depending on the nature and scope of the system, this can be a small or a large, diverse group. For example, when implementing a comprehensive accounting system for a publicly traded corporation in the United States, the stakeholders include bookkeepers, accountants, managers and executives, customers, suppliers, auditors, investors, the SEC, and the Internal Revenue Service (IRS). Each stakeholder group interacts with the system in different ways, and each has a unique perspective on system requirements. Before gathering detailed information, the analyst identifies every type of stakeholder who has an interest in the new system and ensures that critical people from each stakeholder category are available to serve as the business experts.

One useful way to help identify all the interested stakeholders is to consider two characteristics by which they vary: internal stakeholders versus external stakeholders, and operational stakeholders versus executive stakeholders (see **Figure 2-6**). **Internal stakeholders** are those within the organization who interact with the system or have a significant interest in its operation or success. You may be tempted to define internal stakeholders as employees of an organization, but some organizations—such as nonprofits and educational institutions—have internal users (e.g., volunteers and students) who are not employees. **External stakeholders** are those outside the organization's control and influence, although this distinction can also be fuzzy, such as when an organization's strategic partners (e.g., suppliers and shipping companies) interact directly with internal systems.

FIGURE **2-6**

Stakeholders of a comprehensive accounting system for a publicly traded company

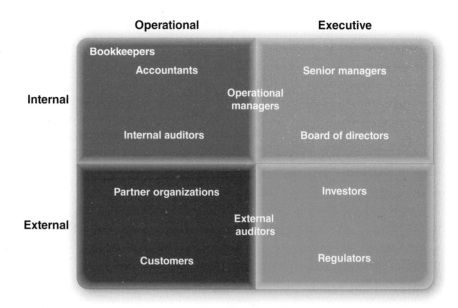

operational stakeholders _persons who regularly interact with a system in the course of their jobs or lives_

executive stakeholders _persons who don't interact directly with the system but who either use information produced by the system, or have a significant financial or other interest in its operation and success_

client _person or group that provides the funding for a system development project_

Operational stakeholders are those who regularly interact with a system in the course of their jobs or lives. Examples include bookkeepers interacting with an accounting or billing system, factory workers interacting with a production scheduling system, customers interacting with an Internet storefront, and patients who interact with a health care Web site, Facebook page, or Twitter newsfeed. Operational users are a key source of requirements information, especially as it pertains to user interfaces and related functionality. **Executive stakeholders** are those who do not interact directly with the system but who either use information produced by the system or have a significant financial or other interest in its operation and success. Examples include an organization's senior managers and board of directors, regulatory agencies, and taxing authorities.

Including such stakeholders in analysis activities is critical because the requirements information they possess may not be obvious or widely known in the organization. In addition, system requirements imposed by executive stakeholders, especially external ones, often have significant legal and financial implications. For example, consider the potential effects of IRS regulations on an accounting system or the effects of federal and state privacy laws on a social networking system.

Two other stakeholder groups that do not neatly fall into the categories just described deserve special attention. The **client** is the person or group that provides the funding for the project. In many cases, the client is senior management. However, clients may also be a separate group, such as a corporation's board of directors, executives in a parent company, or the board of regents of a university. The project team includes the client in its list of important stakeholders because the team must provide periodic status reviews to the client throughout development. The client or a direct representative on a steering or oversight committee also usually approves stages of the project and releases funds.

An organization's technical and support staff are also stakeholders in any system. The technical staff includes people who establish and maintain the computing environment of the organization. Support staff provide user training, troubleshooting, and related services. Both groups should provide guidance in such areas as programming language, computer platforms, network interfaces, and existing systems and their support issues. Any new system must fit within an organization's existing technology architecture, application architecture, and support environment. Thus, technical and support representatives are important stakeholders.

The Stakeholders for RMO

As a starting point for identifying CSMS stakeholders, it is helpful to develop a list of CSS stakeholders, which include:

■ Phone/mail sales order clerks
■ Warehouse and shipping personnel
■ Marketing personnel who maintain online catalog information
■ Marketing, sales, accounting, and financial managers
■ Senior executives
■ Customers
■ External shippers (e.g., UPS and FedEx)

Because the CSMS will take over existing functions of the CSS, the list of CSMS stakeholders includes all the stakeholders in the CSS list. However, there are some subtle differences. For example, the inclusion of social networking functions in the CSMS and the planned ability to share Mountain Bucks expands the concept of who is a customer. While the old CSS was intended for use by potential customers visiting the Web site, the new system potentially interacts with a much larger group of external stakeholders, including friends and family of existing customers and potentially all users of popular social networking sites. In essence, the stakeholder group "Customers" is much larger, more diverse, and includes people who have not purchased from RMO. Ensuring that the viewpoints and requirements of this diverse group are represented during analysis activities will be a considerable challenge.

Expanded functionality for sales promotions with partner organizations creates an entirely new group of external stakeholders within those partner organizations. At this point, it is unclear whether that group will include operational stakeholders, executive stakeholders, or both and exactly how those stakeholders will interact with the system. Once again, ensuring adequate input from those stakeholders will be a significant challenge, especially because the portions of the system used by those stakeholders will not be based on an existing system.

RMO is a privately held company; John and Liz Blankens are the owners, and they hold two senior management positions. This is significant because the key operational systems of any publicly traded company "inherit" many external stakeholders due to the flow of information from those systems to the organization's financial reports. RMO is audited by an external accounting firm, primarily to ensure access to bank loans and other private financing.

As owners and senior managers, John and Liz are the primary clients, but so are other senior executives who form a collaborative decision-making body. In addition, existing technical and support staff are key stakeholders. **Figure 2-7** summarizes the internal managerial stakeholders in the form of an organization chart.

Information-Gathering Techniques

Techniques for gathering detailed requirements information include:

■ Interviewing users and other stakeholders
■ Distributing and collecting questionnaires
■ Reviewing inputs, outputs, and documentation
■ Observing and documenting business procedures
■ Researching vendor solutions
■ Collecting active user comments and suggestions

All these methods have proven to be effective, although some are more efficient than others. In most cases, analysts combine methods to increase their effectiveness and efficiency and to provide a comprehensive fact-finding approach.

FIGURE **2-7**

RMO management stakeholders involved in the CSS requirements definition

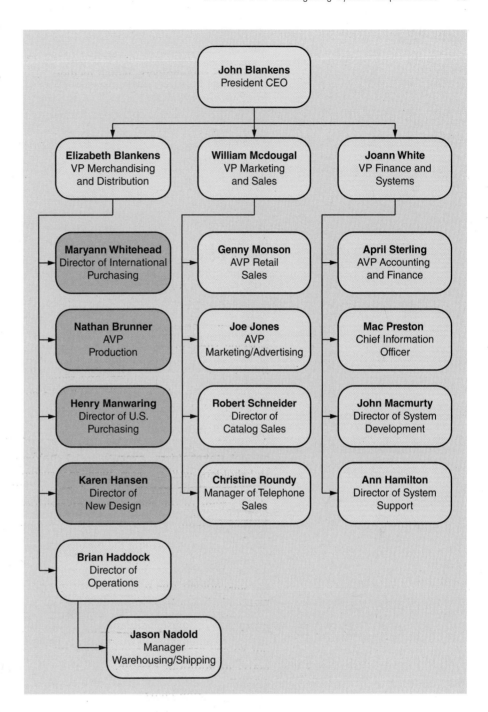

Interview Users and Other Stakeholders

Interviewing users and other stakeholders is an effective way to understand business functions and business rules. Unfortunately, it is also the most time-consuming and resource-expensive option. In this method, systems analysts:

- Prepare detailed questions.
- Meet with individuals or groups of users.
- Obtain and discuss answers to the questions.
- Document the answers.
- Follow up as needed in future meetings or interviews.

Obviously, this process may take some time, so it usually requires multiple sessions with each of the users or user groups.

FIGURE **2-8**
Themes for information-gathering questions

Theme	Questions to users
What are the business operations and processes?	What do you do?
How should those operations be performed?	How do you do it? What steps do you follow? How could they be done differently?
What information is needed to perform those operations?	What information do you use? What inputs do you use? What outputs do you produce?

Question Themes

Whether in informal meetings, formal interviews, or as part of a questionnaire or survey, analysts ask questions. But which questions should analysts ask? **Figure 2-8** shows three major themes that guide the analysts when they are asking questions to define system requirements; it also shows sample questions that arise from those themes.

What Are the Business Processes?   The analyst must obtain a comprehensive list of all the business processes. In most cases, the users provide answers in terms of the current system, so the analyst must carefully discern which of those functions are fundamental (i.e., which will remain and which may possibly be eliminated with an improved system). For example, sales clerks might indicate that the first task they perform when a customer places an order is to check the customer's credit history. In the new system, sales clerks might never need to perform that function; the system might perform the check automatically. The function remains a system requirement, but the method of carrying out the function and its timing are changed.

How Are the Business Processes Performed?   Again, the focus starts with the current system but gradually moves to the new system. The goal is to determine how the new system should support the function rather than how it supports it now. The analyst must be able to help the user visualize new and more efficient approaches to performing the business processes made possible by new technology or processes.

What Information Is Required?   Some information inputs are formal; others are informal. When questioning the user, the analyst should specifically ask about exceptions or unusual situations in order to identify additional (nonroutine) information requirements. In this theme and the previous one, detail is the watchword. An analyst must understand the nitty-gritty detail to develop a correct solution.

Question Types

Questions can be roughly divided into two types:

open-ended questions questions that encourage discussion or explanation.

closed-ended questions questions that elicit specific facts

- **Open-ended questions**—questions such as "How do you do this function?" that encourage discussion and explanation
- **Closed-ended questions**—questions such as "How many forms a day do you process?" that are used to get specific facts

Generally, open-ended questions help to start a discussion and enable a large number of requirements to be uncovered fairly quickly. Note that all the questions in the previous section are open ended. A discussion that starts with open-ended questions usually shifts gradually to closed-ended questions that elicit or confirm specific details of a business process.

Focus of Questions—Current System or New?

A significant question that faces all analysts is how much effort to expend studying and documenting the existing system (if one exists). Excess attention to an existing system can consume considerable time and can result in simply updating that system with newer technology. As a result, no matter how inefficient the current system is, system developers simply reimplement the procedures that are already in place. On the other hand, if a new system inherits many or all of the requirements of an existing system, then an analyst risks missing important requirements through insufficient study of the existing system.

To minimize both risks, analysts must balance the review of current business functions with discovery of new system requirements. It is still critical to have a complete correct set of system requirements, but in today's fast-paced world, there is no time or money to review all the old systems and document all the inefficient procedures. In fact, in today's development environment, one of the most valuable capabilities that a good system developer can bring is a new perspective to the problem.

Interview Preparation, Conduct, and Follow-Up

Figure 2-9 is a sample checklist that summarizes the major points to be covered; it is useful in preparing for, conducting, and following up an interview.

Preparing for the Interview Every successful interview requires preparation. The first and most important step in preparing for an interview is to establish its objective. In other words, what do you want to accomplish with this interview? Write down the objective so it is firmly established in your mind. The second step is to determine which stakeholders should be involved in the interview. A small number of interviewees is generally best when the objective is narrow or of a fact-finding nature. Larger groups are better if the objective is more open ended, such as when generating and evaluating new ideas. However, it can be difficult to manage a large group meeting to ensure high-quality input from all participants. If possible, have at least two analysts involved in every interview, and have them compare notes afterward to ensure accuracy.

The next step is to prepare detailed questions to be used in the interview. Write down a list of specific questions, and prepare notes based on the forms or reports received earlier. Usually, you should prepare a list of questions that are consistent with the objective of the interview. Open-ended questions and closed-ended questions are appropriate. Generally, open-ended questions help

FIGURE **2-9**

Sample checklist to prepare for user interviews

Checklist for Conducting an Interview

Before
- ❑ Establish the objective for the interview.
- ❑ Determine correct user(s) to be involved.
- ❑ Determine project team members to participate.
- ❑ Build a list of questions and issues to be discussed.
- ❑ Review related documents and materials.
- ❑ Set the time and location.
- ❑ Inform all participants of objective, time, and locations.

During
- ❑ Arrive on time.
- ❑ Look for exception and error conditions.
- ❑ Probe for details.
- ❑ Take thorough notes.
- ❑ Identify and document unanswered items or open questions.

After
- ❑ Review notes for accuracy, completeness, and understanding.
- ❑ Transfer information to appropriate models and documents.
- ❑ Identify areas needing further clarification.
- ❑ Thank the participants.
- ❑ Follow up on open and unanswered questions.

get the discussion started and encourage the user to explain all the details of the business process and the rules.

The last step is to make the final interview arrangements and then communicate those arrangements to all participants. A specific time and location should be established. If possible, a quiet location should be chosen, to avoid interruptions. Each participant should know the objective of the meeting and, when appropriate, should have a chance to preview the questions or materials to be used. Interviews consume a substantial amount of time, and they can be made more efficient if each participant knows beforehand what is to be accomplished.

Conducting the Interview   The usual rules of workplace meetings apply during stakeholder interviews: plan ahead, arrive early, and ensure that the room is prepared and that needed resources are available. Limit the time of the interview for the benefit of the analyst(s) and stakeholder(s); stakeholders have other responsibilities, and the analysts can absorb only so much information at one time. It is better to have several shorter interviews than one long interview. A series of interviews provides an opportunity to absorb the material and then go back to get clarification later.

Look for exception and error conditions. Look for opportunities to ask "what if" questions. "What if it doesn't arrive? What if the signature is missing? What if the balance is incorrect? What if two orders are exactly the same?" The essence of good systems analysis is understanding all the "what ifs." Make a conscious effort to identify all the exception conditions and then ask about them. More than any other skill, the ability to think of the exceptions will help you discover the detailed business rules. It is a hard skill to teach from a textbook; experience will hone this skill. You will teach yourself this skill by conscientiously practicing it.

Probe for details. In addition to looking for exception conditions, the analyst must probe to ensure a complete understanding of all procedures and rules. One of the most difficult skills to learn as a new systems analyst is how to get enough details. Frequently, it is easy to get a general overview of how a process works. But do not be afraid to ask detailed questions until you thoroughly understand how the process works and what information is used. You cannot do effective systems analysis by glossing over the details.

Take careful notes. It is a good idea to take handwritten notes. Usually, tape recorders make users nervous. However, taking notes signals that you think the information you are obtaining is important, and the user is complimented. If two analysts conduct each interview, they can compare notes later. Identify and document in your notes any unanswered questions or outstanding issues that were not resolved. A good set of notes provides the basis for building the analysis models as well as establishing a basis for the next interview session.

Figure 2-10 is a sample agenda for an interview session. Obviously, you do not need to conform exactly to a particular agenda. However, as with the interview checklist shown in Figure 2-9, this figure will help prod your memory on issues and items that should be discussed in an interview. Make a copy and use it. As you develop your own style, you can modify the checklist to the way you like to work.

Following Up the Interview   Follow-up is an important part of each interview. The first task is to absorb, understand, and document the information that was obtained. Generally, analysts document the details of the interview by constructing models of the business processes and writing textual descriptions of nonfunctional requirements. These tasks should be completed as soon after the interview as possible and the results distributed to the interview participants for validation. If the modeling methods are complex or unfamiliar to the users, the analyst should schedule follow-up meetings to explain and verify the models, as described in the last section of this chapter.

FIGURE **2-10**
Sample interview session agenda with follow-up information

Discussion and Interview Agenda

Setting

Objective of Interview
Determine processing rules for sales commission rates

Date, Time, and Location
April 21, 2012, at 9:00 a.m. in William McDougal's office

User Participants (names and titles/positions)
William McDougal, vice president of marketing and sales, and several of his staff

Project Team Participants
Mary Ellen Green and Jim Williams

Interview/Discussion

1. Who is eligible for sales commissions?
2. What is the basis for commissions? What rates are paid?
3. How is commission for returns handled?
4. Are there special incentives? Contests? Programs based on time?
5. Is there a variable scale for commissions? Are there quotas?
6. What are the exceptions?

Follow-Up

Important decisions or answers to questions
See attached write-up on commission policies

Open items not resolved with assignments for solution
See Item numbers 2 and 3 on open items list

Date and time of next meeting or follow-up session
April 28, 2012, at 9:00 a.m.

During the interview, you probably asked some "what if" questions that the users could not answer. They are usually policy questions raised by the new system that management has not considered before. It is extremely important that these questions not get lost or forgotten. **Figure 2-11** shows a sample table for tracking outstanding or unresolved issues for RMO. The table includes questions posed by users or analysts and responsibilities assigned for resolving the issues. If several teams are working, a combined list can be maintained. Other columns that might be added to the list are an explanation of the problem's resolution and the date resolved.

FIGURE **2-11** *Sample open-items list*

ID	Issue title	Date identified	Target end date	Responsible project person	User contact	Comments
1	Partial shipments	6-12-2012	7-15-2012	Jim Williams	Jason Nadold	Ship partials or wait for full shipment?
2	Returns and commissions	7-01-2012	9-01-2012	Jim Williams	William McDougal	Are commissions recouped on returns?
3	Extra commissions	7-01-2012	8-01-2012	Mary Ellen Green	William McDougal	How to handle commissions on special promotions?

Finally, make a list of new questions based on areas that need further elaboration or that are missing information. This list will prepare you for the next interview.

Distribute and Collect Questionnaires

Questionnaires enable analysts to collect information from a large number of stakeholders. Even if the stakeholders are widely distributed geographically, they can still help define requirements through questionnaires. Questionnaires are often used to obtain preliminary insight into stakeholder information needs, which helps to determine areas that need further research by using other methods.

Figure 2-12 is a sample questionnaire showing three types of questions. The first part has closed-ended questions to determine quantitative information.

FIGURE **2-12** *Sample questionnaire*

RMO Questionnaire

This questionnaire is being sent to all telephone-order sales personnel. As you know, RMO is developing a new customer support system for order taking and customer service.

The purpose of this questionnaire is to obtain preliminary information to assist in defining the requirements for the new system. Follow-up discussions will be held to permit everybody to elaborate on the system requirements.

Part I. Answer these questions based on a typical four-hour shift.
1. How many phone calls do you receive?_____
2. How many phone calls are necessary to place an order for a product?_____
3. How many phone calls are for information about RMO products, that is, questions only?_____
4. Estimate how many times during a shift customers request items that are out of stock._____
5. Of those out-of-stock requests, what percentage of the time does the customer desire to put the item on back order?_____%
6. How many times does a customer try to order from an expired catalog?_____
7. How many times does a customer cancel an order in the middle of the conversation?_____
8. How many times does an order get denied due to bad credit?_____

Part II. Circle the appropriate number on the scale from 1 to 7 based on how strongly you agree or disagree with the statement.

Question	Strongly Agree					Strongly Disagree	
It would help me do my job better to have longer descriptions of products available while talking to a customer.	1	2	3	4	5	6	7
It would help me do my job better if I had the past purchase history of the customer available.	1	2	3	4	5	6	7
I could provide better service to the customer if I had information about accessories that were appropriate for the items ordered.	1	2	3	4	5	6	7
The computer response time is slow and causes difficulties in responding to customer requests.	1	2	3	4	5	6	7

Part III. Please enter your opinions and comments.

Please briefly identify the problems with the current system that you would like to see resolved in a new system.
_____
_____
_____

The second part consists of opinion questions in which respondents are asked whether they agree or disagree with the statement. Both types of questions are useful for tabulating and determining quantitative averages. The third part requests an explanation of a procedure or problem. Questions such as these are good as a preliminary investigation to help direct further fact-finding activities.

Questionnaires are not well suited to helping you learn about processes, workflows, or techniques. Open-ended questions such as "How do you do this process?" are best answered by using interviews or observation. Although a questionnaire can contain a very limited number of open-ended questions, stakeholders frequently do not return questionnaires that contain many open-ended questions.

Review Inputs, Outputs, and Procedures

There are two sources of information about inputs, outputs, and procedures. One source is external to the organization—industry-wide professional organizations and other companies. It may not be easy to obtain information from other companies, but they are a potential source of important information. Sometimes, industry journals and magazines report the findings of "best practices" studies. The project team would be negligent in its duties if its members were not familiar with best practice information.

The second source of inputs, outputs, and procedures is existing business documents and procedure descriptions within the organization. Reviewing internal documents and procedures serves two purposes. First, it is a good way to get a preliminary understanding of the processes. Second, existing inputs, outputs, and documents can serve as visual aids for the interview and as the working documents for discussion (see **Figure 2-13**). Discussion can focus on a specific input or output, its objective, its distribution, and its information content. The discussion should also include specific business events that initiate the use of an input or generation of an output. Several different business events

FIGURE 2-13 *RMO mail-order form used as a visual aid during an interview*

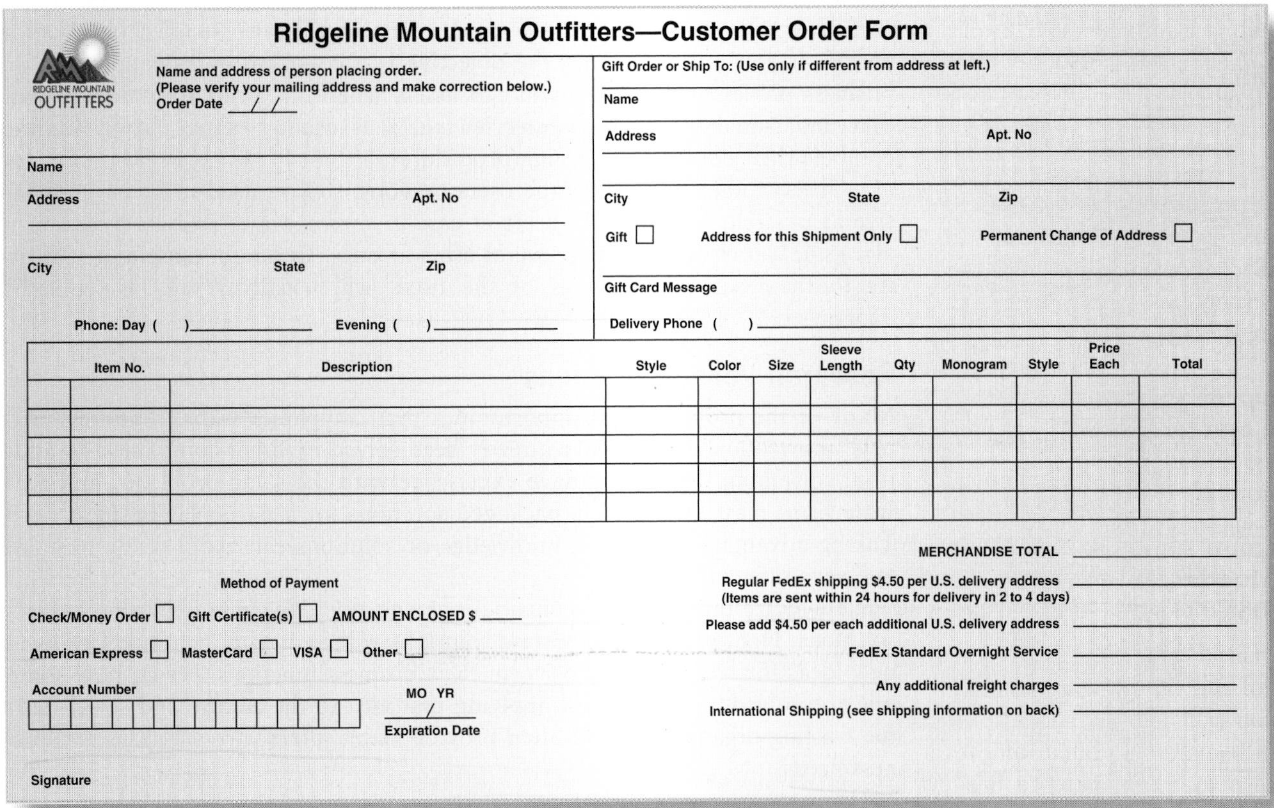

Ridgeline Mountain Outfitters—Customer Order Form

RIDGELINE MOUNTAIN OUTFITTERS

Name and address of person placing order.
(Please verify your mailing address and make correction below.)
Order Date / /

Name
Address Apt. No
City State Zip
Phone: Day ( ) Evening ( )

Gift Order or Ship To: (Use only if different from address at left.)
Name
Address Apt. No
City State Zip
Gift ☐ Address for this Shipment Only ☐ Permanent Change of Address ☐
Gift Card Message
Delivery Phone ( )

Item No.	Description	Style	Color	Size	Sleeve Length	Qty	Monogram	Style	Price Each	Total

MERCHANDISE TOTAL
Regular FedEx shipping $4.50 per U.S. delivery address (Items are sent within 24 hours for delivery in 2 to 4 days)
Please add $4.50 per each additional U.S. delivery address
FedEx Standard Overnight Service
Any additional freight charges
International Shipping (see shipping information on back)

Method of Payment
Check/Money Order ☐ Gift Certificate(s) ☐ AMOUNT ENCLOSED $
American Express ☐ MasterCard ☐ VISA ☐ Other ☐
Account Number
MO YR /
Expiration Date
Signature

might require the same form, and specific information about the event and the business process is critical. It is always helpful to have screens and forms that have been filled out with real information to ensure that the analyst obtains a correct understanding of the data content.

Reviewing the documentation of existing procedures helps identify business rules that may not come up in the interviews. Analyzing formal procedure documentation also helps reveal discrepancies and redundancies in the business processes. However, procedure documents frequently are not kept up to date, and they commonly include errors. To ensure that the assumptions and business rules that derive from the existing documentation are correct, analysts should review them with the users.

Observe and Document Business Processes

Firsthand experience is invaluable to understand exactly what occurs within business processes. More than any other activity, observing a business process in action will help you understand the business functions. However, while observing existing processes, you must also be able to visualize the new system's associated business processes. That is, as you observe the current business processes to understand the fundamental business needs, you should never forget that the processes could and often should change to be made more efficient. Do not get locked into believing there is only one way of performing the process.

You can observe a business process in many ways, ranging from a quick walk-through of an office or plant to doing the work yourself. A quick walk-through gives a general understanding of the layout of the office, the need for and use of computer equipment, and the general workflow. Spending several hours observing users at their jobs helps you understand the details of using the computer system and carrying out business functions. Being trained as a user and actually doing the job enables you to discover the difficulties of learning new procedures, the importance of a system that is easy to use, and the stumbling blocks and bottlenecks of existing procedures and information sources.

It is not necessary to observe all processes at the same level of detail. A quick walk-through may be sufficient for one process, whereas a process that is critical or more difficult to understand might require an extended observation period. If you remember that the objective is a complete understanding of the business processes and rules, you can assess where to spend your time to gain that understanding. As with interviewing, it is usually better if two analysts combine their efforts in observing procedures.

Observation often makes the users nervous, so you need to be as unobtrusive as possible. You can put users at ease in several ways, such as by working alongside a user or observing several users at once. Common sense and sensitivity to the needs and feelings of the users will usually result in a positive experience.

Research Vendor Solutions

Many of the problems and opportunities that companies want to address with new information systems have already been solved by other companies. In addition, consulting firms often have experience with the same problems, and software firms may have already packaged solutions for a particular business need. Taking advantage of existing knowledge or solutions can avoid costly mistakes and save time and money.

There are three positive contributions and one danger in exploring existing solutions. First, researching existing solutions will frequently help users generate new ideas for how to better perform their business functions. Seeing how someone else solved a problem and applying that idea to the culture and structure of the existing organization will often provide viable alternative solutions for business needs.

Second, some of these solutions are excellent and state of the art. Without this research, the development team may create a system that is obsolete even before it is designed. Companies need solutions that not only solve basic business problems but that are up to date with competitive practices.

Third, it is often cheaper and less risky to buy a solution rather than to build it. If the solution meets the needs of the company and can be purchased, then that is usually a safer, quicker, and less expensive route.

The danger in exploring existing solutions is that the users and even the systems analysts may want to buy one of the alternatives immediately. But if a solution, such as a packaged software system, is purchased too early in the process, the company's needs may not be thoroughly investigated. Too many companies have purchased systems only to find out later that they only support half the functions that were needed. Do not rush into a purchase decision until requirements are fully defined and all viable alternatives have been thoroughly investigated.

Collect Active User Comments and Suggestions

As discussed in Chapter 1 and earlier in this chapter, system development normally proceeds with analysis, design, and other activities spread across multiple iterations. Portions of the system are constructed and tested during each iteration. Users and other stakeholders perform the initial testing of system functions during the iteration in which those functions are implemented. They also test and use those same functions during later iterations.

User feedback from initial and later testing is a valuable source of requirements information. Interviews, discussions, and model reviews are an imperfect way of eliciting complete and accurate requirements. The phrase "I'll know it when I see it" applies well to requirements definition. Users often cannot completely or accurately state their requirements until they can interact with a live system that implements those requirements. Based on those interactions, users can develop concrete suggestions for improvement and identify missing or poorly implemented requirements.

Documenting Workflows with Activity Diagrams

As you gather information about business processes, you will need to document your results. One effective way to capture this information is with diagrams. Eventually, you may want to use diagrams to describe the workflows of the new system, but for now, let us focus on how we would document the current business workflows.

A **workflow** is the sequence of processing steps that completely handles one business transaction or customer request. Workflows may be simple or complex. Complex workflows can be composed of dozens or hundreds of processing steps and may include participants from different parts of an organization.

An **activity diagram** describes the various user (or system) activities, the person who does each activity, and the sequential flow of these activities.

Figure 2-14 shows the basic symbols used in an activity diagram. The ovals represent the individual activities in a workflow. The connecting arrows represent the sequence between the activities. The black circles denote the beginning and the ending of the workflow. The diamond is a decision point at which the flow of the process will either follow one path or another. The heavy solid line is a **synchronization bar**, which either splits the path into multiple concurrent paths or recombines concurrent paths. The **swimlane heading** represents an agent who performs the activities. Because it is common in a workflow to have different agents (i.e., people) performing different steps of the workflow process,

workflow sequence of processing steps that completely handles one business transaction or customer request

activity diagram describes user (or system) activities, the person who does each activity, and the sequential flow of these activities

synchronization bar activity diagram component that either splits a control path into multiple concurrent paths or recombines concurrent paths

swimlane heading activity diagram column containing all activities for a single agent or organizational unit

FIGURE **2-14**

Activity diagram symbols

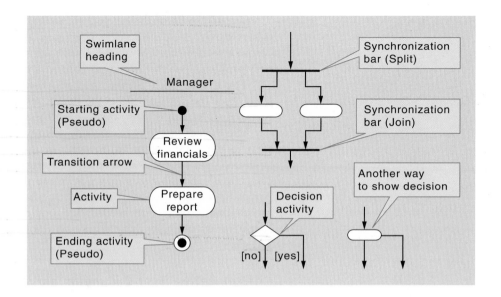

the swimlane symbol divides the workflow activities into groups showing which agent performs which activity.

Figure 2-15 is an activity diagram that describes the order fulfillment process for the current RMO CSMS. Processing begins when the customer has completed the order checkout process. The diagram describes the back-and-forth flow of information and control between the Order subsystem, Inventory subsystem, warehouse(s), and shipper. The diagram is simplified because it omits many error-handling pathways, including what happens if enough item stock is on hand to fulfill part of an order.

Figure 2-16 illustrates another workflow diagram, which demonstrates some new concepts. In this example, a customer is ordering a product that has to be manufactured specifically to match customer specifications. The salesperson sends the order to Engineering, and the diagram uses a new symbol to emphasize the transmission of the document between Sales and Engineering. After Engineering develops the specifications, two concurrent activities happen: Purchasing orders the materials, and Production writes the program for the automated milling machines. These two activities are completely independent and can occur at the same time. Notice that one synchronization bar splits the path into two concurrent paths and that another synchronization bar reconnects them. Finally, Scheduling puts the order on the production schedule.

Creating activity diagrams to document workflows is straightforward. The first step is to identify the agents to create the appropriate swimlanes. Next, follow the various steps of the workflow and then make appropriate ovals for the activities. Connect the activity ovals with arrows to show the workflow. Here are a couple guidelines:

■ Use a decision symbol to represent an either/or situation—one path or the other path but not both. As a shorthand notation, you can merge an activity (by using an oval) and a decision (by using a diamond) into a single oval with two exit arrows, as indicated on the right in Figure 2-14. This notation represents a decision (either/or) activity. Wherever you have an activity that reads "verify" or "check," you will probably require a decision—one for the "accept" path and one for the "reject" path.

■ Use synchronization bars for parallel paths—situations in which both paths are taken. Include a beginning and an ending synchronization bar. You can also use synchronization bars to represent a loop, such as a "do while" programming loop. Put the bar at the beginning of the loop and then describe it as "for every." Put another synchronization bar at the end of the loop with the description "end for every."

FIGURE **2-15** *Simple activity diagram for online checkout*

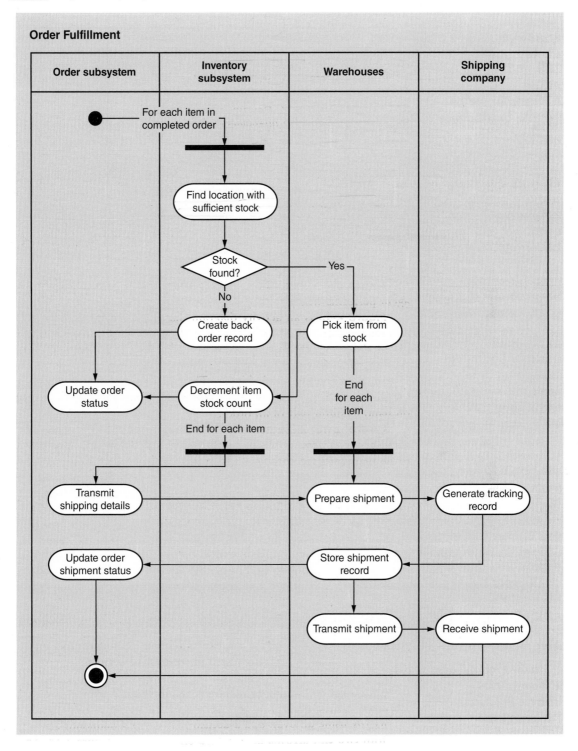

FIGURE **2-16** *Activity diagram showing concurrent paths*

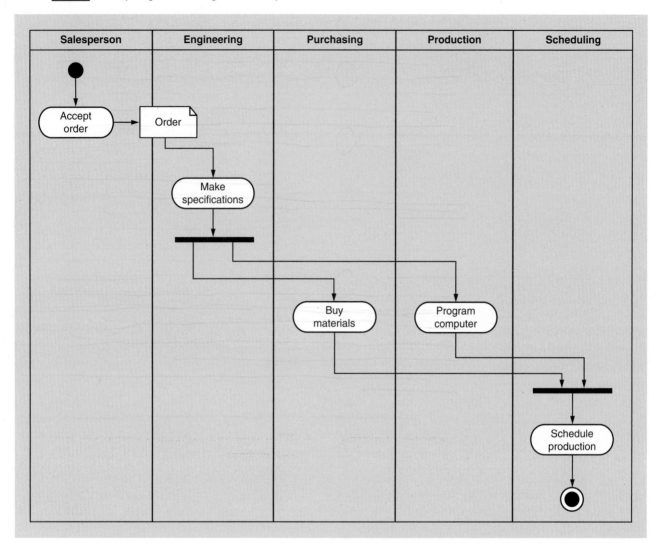

Chapter Summary

There are five primary activities of systems analysis:

- Gather detailed information.
- Define requirements.
- Prioritize requirements.
- Develop user-interface dialogs.
- Evaluate requirements with users.

Functional requirements are those that explain the basic business functions that the new system must support. Nonfunctional requirements involve the system's objectives with regard to technology, performance, usability, reliability, and security.

Mathematical, descriptive, and graphical models are developed to document requirements and as an aid in evaluating requirements with users and other stakeholders. Stakeholders include internal and external users

of the system and other persons or organizations that have a vested interest in the system.

Analysts use many techniques to gather information about requirements, including:

- Interviews
- Questionnaires
- Documentation, input, and output reviews
- Process observation and documentation
- Vendor solution research
- Active comments and suggestions from users

Workflow diagrams are a key modeling technique often used as an early requirements model. Workflow diagrams graphically model the steps of a business process and the participants who perform them. Other models and diagrams are covered in later chapters.

Key Terms

activity diagram   57

application architecture   37

client   47

closed-ended questions   50

design constraints   43

executive stakeholders   47

external stakeholders   46

functional requirements   42

FURPS   43

FURPS+   43

graphical models   46

implementation requirements   44

interface requirements   44

internal stakeholders   46

mathematical models   46

model   44

nonfunctional requirements   43

open-ended questions   50

operational stakeholders   47

performance requirements   43

physical requirements   44

reliability requirements   43

security requirements   43

stakeholders   46

supportability requirements   44

swimlane heading   57

synchronization bar   57

system requirements   42

technology architecture   37

textual models   45

unified modeling
    language (UML)   46

usability requirements   43

workflow   57

Review Questions

1. List and briefly describe the five activities of systems analysis.

2. What are three types of models?

3. What is the difference between functional requirements and nonfunctional requirements?

4. Describe the steps in preparing for, conducting, and following up an interview session.

5. What are the benefits of doing vendor research during information-gathering activities?

6. What types of stakeholders should you include in fact finding?

7. Describe the open-items list and then explain why it is important.

8. List and briefly describe the six information-gathering techniques.

9. What is the purpose of an activity diagram?

10. Draw and explain the symbols used on an activity diagram.

Problems and Exercises

1. Provide an example of each of the three types of models that might apply to designing a car, a house, and an office building.

2. One of the toughest problems in investigating system requirements is to make sure they are complete and comprehensive. How would you ensure that you get all the right information during an interview session?

3. One of the problems you will encounter during your investigation is "scope creep" (i.e., user requests for additional features and functions). Scope creep happens because users sometimes have many unsolved problems and the system investigation may be the first time anybody has listened to

their needs. How do you keep the system from growing and including new functions that should not be part of the system?

4. What would you do if you got conflicting answers for the same procedure from two different people you interviewed? What would you do if one was a clerical person and the other was the department manager?

5. You have been assigned to resolve several issues on the open-items list, and you are having a hard time getting policy decisions from the user contact. How can you encourage the user to finalize these policies?

6. In the running case of RMO, assume that you have set up an interview with the manager of the

shipping department. Your objective is to determine how shipping works and what the information requirements for the new system will be. Make a list of questions—open ended and closed ended—that you would use. Include any questions or techniques you would use to ensure you find out about the exceptions.

7. Develop an activity diagram based on the following narrative. Note any ambiguities or questions that you have as you develop the model. If you need to make assumptions, also note them.

The purchasing department handles purchase requests from other departments in the company. People in the company who initiate the original purchase request are the "customers" of the purchasing department. A case worker within the purchasing department receives the request and monitors it until it is ordered and received.

Case workers process requests for the purchase of products under $1,500, write a purchase order, and then send it to the approved vendor. Purchase requests over $1,500 must first be sent out for bid from the vendor that supplies the product. When the bids return, the case worker selects one bid and then writes a purchase order and sends it to the vendor.

8. Develop an activity diagram based on the following narrative. Note any ambiguities or questions that you have as you develop the model. If you need to make assumptions, also note them.

The shipping department receives all shipments on outstanding purchase orders. When the clerk in the shipping department receives a shipment, he or she finds the outstanding purchase order for those items. The clerk then sends multiple copies of the shipment packing slip. One copy goes to

Purchasing, and the department updates its records to indicate that the purchase order has been fulfilled. Another copy goes to Accounting so a payment can be made. A third copy goes to the requesting in-house customer so he or she can receive the shipment.

After payment is made, the accounting department sends a notification to Purchasing. After the customer receives and accepts the goods, he or she sends notification to Purchasing. When Purchasing receives these other verifications, it closes the purchase order as fulfilled and paid.

9. Conduct a fact-finding interview with someone involved in a procedure that is used in a business or organization. This person could be someone at the university, in a small business in your neighborhood, in the student volunteer office at the university, in a doctor's or dentist's office, or in a volunteer organization. Identify a process that is done, such as keeping student records, customer records, or member records. Make a list of questions and then conduct the interview. Remember, your objective is to understand that procedure thoroughly (i.e., to become an expert on that single procedure).

10. Using RMO and the CSMS as your guide, develop a list of all the procedures that may need to be researched. You may want to think about the exercise in the context of your experience with such retailers as L.L. Bean, Lands' End, or Amazon.com. Check out the Internet marketing done on the retailers' Web sites and then think about the underlying business procedures that are required to support those sales activities. List the procedures and then describe your understanding of each.

Case Study

John and Jacob, Inc.: Online Trading System

John and Jacob, Inc. is a regional brokerage firm that has been successful over the last several years. Competition for customers is intense in this industry. The large national firms have very deep pockets, with many services to offer clients. Severe competition also comes from discount and Internet trading companies. However, John and Jacob has been able to cultivate a substantial customer base from upper-middle-income clients in the northeastern United States. To maintain a competitive edge with its customers, John and Jacob is in the process of modernizing its online trading system. The modernization will add new features to the existing system and expand the range of interfaces beyond desktop and laptop computers to include tablet computers and

smartphones. The system will add Twitter messaging in addition to continued support for traditional e-mail.

Edward Finnigan, the project manager, is in the process of identifying all the groups of people who should be included in the development of the system requirements. He is not quite sure exactly who should be included. Here are the issues he is considering:

■ **Users:** The trading system will be used by customers and by staff in each of the company's 30 trading offices. Obviously, the brokers who are going to use the system need to have input, but how should this be done? Edward also is not sure what approach would be best to ensure that the requirements are complete yet not require tremendous amounts of time. Including all the offices would increase enthusiasm and support for the system, but it would take a lot of time. Involving more brokers would bring divergent opinions that would have to be reconciled.

■ **Customers:** The trading system will also include trade order entry, investment analysis reports, trade confirmations, standard and customized reporting, and customer statements. Edward wonders how to involve John and Jacob customers in the development of system requirements. Edward is sensitive to this issue because many brokers have told him that many customers are unhappy with the current system, and customer complaints are sometimes posted to the public comments area of the current Web site. He would like to involve customers, but he does not know how.

■ **Other stakeholders:** Edward knows he should involve other stakeholders to help define system requirements. He is not quite sure whom he should contact. Should he go to senior executives? Should he contact middle management? Should he include such back-office functions as accounting and investing? He is

not quite sure how to get organized or how to decide who should be involved.

Answer the following questions:

1. What is the best method for Edward to involve the brokers (users) in the development of the updated online trading system? Should he use a questionnaire? Should he interview the brokers in each of the company's 30 offices or would one or two brokers representing the entire group be better? How can Edward ensure that the information about requirements is complete yet not lose too much time doing so?

2. Concerning customer input for the new system, how can Edward involve customers in the process? How can he interest them in participating? What methods can Edward use to ensure that the customers he involves are representative of John and Jacob's entire customer group?

3. As Edward considers what other stakeholders he should include, what are some criteria he should use? Develop some guidelines to help him build a list of people to include.

RUNNING CASES

Community Board of Realtors

The real estate business relies on an extensive amount of information used in the buying and selling of real property. Most communities of real estate agents and brokers have formed cooperative organizations to help consolidate and distribute information on the real estate profession, real estate trends, properties in the community, historical records of property sales, and current listings of properties for sale. These organizations are usually referred to as the Community Board of Realtors.

Research your local Community Board of Realtors to answer these questions:

1. Who are the stakeholders for the issues related to real estate in your community, and what are their main interests?

2. What types of information does the board collect and make available to its members and to the community?

3. Research the real estate industry in at least two countries other than the United States. For each of these countries, what are some of the cultural and legal issues that differ from those in the United States? If you were working on support for an international real estate cooperative system, in what ways would the information collection activity process be complicated?

The Spring Breaks 'R' Us Travel Service

Spring Breaks 'R' Us (SBRU) is an online travel service that books spring break trips to resorts for college students. Students have booked spring break trips for decades, but changes in technology have transformed the travel business in recent years.

SBRU moved away from having campus reps with posted fliers and moved to the Web early on. The basic idea is to get a group of students to book a room at a resort for one of the traditional spring break weeks. SBRU contracts with dozens of resorts

(continued on page 64)

(continued from page 63)

in key spring break destinations in Florida, Texas, the Caribbean, and Mexico. Its Web site shows information on each resort and includes prices, available rooms, and special features. Students can research and book a room, enter contract information, and pay deposits and final payments through the system. SBRU provides updated booking information, resort information updates, and travel information for booked students when they log in to the site.

The resorts also need access to information from SBRU. They need to know about their bookings for each week, the room types that are booked, and so forth. Before the spring break booking season starts, they need to enter information on their resorts, including prices and special features. Resorts need to be paid by SBRU for the bookings, and they need to be able to report and collect for damages caused by spring-breakers during their stay.

SBRU has recently decided to upgrade its system to provide social networking features for students. It is currently researching possibilities and collecting information from prospective customers about desirable features and functions. From the business standpoint, the idea is to increase bookings by enhancing the experience before, during, and after the trip.

1. Who are the stakeholders for SBRU? For each type of stakeholder, what aspects of the SBRU booking system are of particular interest?
2. What are the main functional requirements for the major subsystem areas (i.e., resort relations, student booking, accounting and finance, and social networking)?
3. Describe some usability requirements for students, booking interactions, and social networking interactions.
4. Assuming that social networking at the resorts will require wireless communication and connection to the Internet, what are some reliability requirements that resorts might be asked to maintain? What are some performance requirements? Is this a bigger issue because resorts are in international locations?
5. What are some security requirements? Is there any reason why students in Europe, Asia, or other locations could not book rooms through SBRU? What issues might be anticipated?
6. To collect information on functional requirements for the social networking subsystem, what are some techniques that might be used? Be specific and include some sample questions you might ask by using various techniques.

On the Spot Courier Services

As an employee of a large international courier and shipping service, Bill Wiley met almost every day with many companies that shipped and received packages. He was frequently asked if his company could deliver local packages on the same day. Over several months, he observed that there appeared to be a substantial need for courier services in the city in which he lived. He decided that he would form his own courier delivery company called On the Spot to fill this need.

Bill began by listing his mobile telephone number in the Yellow Pages. He also sent letters to all those companies that had requested same-day courier service that his prior company had not been able to serve. He hoped that, through good service and word-of-mouth advertising, his business would grow. He also began other advertising and marketing activities to promote his services.

At first, Bill received delivery requests on his business mobile phone. However, it was not long before his customers were asking if he had a Web site where they could place orders for shipments. He knew that if he could get a Web presence he could increase his exposure and help his business grow.

After he had been in business only a few short months, Bill discovered he needed to have additional help. He hired another person to help with the delivery and pickup of packages. It was good to see the business grow, but another person added to the complexity of coordinating pickups and deliveries. With the addition of a new person, he could no longer "warehouse" the packages out of his delivery van. He now needed a central warehouse where he could organize and distribute packages for delivery. He thought that if his business grew enough to add one more delivery person he would also need someone at the warehouse to coordinate the arrival and distribution of all the packages.

1. Who are the stakeholders for On the Spot? How involved should On the Spot's customers be in system definition? As the business grows, who else might be potential stakeholders and interested in system functions?
2. If you were commissioned to build a system for Bill, how would you determine the requirements? Be specific in your answer. Make a list of the questions you need answered.

(continued on page 65)

(continued from page 64)

3. What technology and communication requirements do you see? What are the hardware requirements, and what kind of equipment will provide viable options to the system? What would you recommend to Bill?

4. What are the primary functional requirements for the system as described so far in the case?

Sandia Medical Devices

Medical monitoring technology has advanced significantly in the last decade. Monitoring that once required a visit to a health-care facility can now be performed by devices located in a patient's home or carried or worn at all times. Examples include glucose level (blood sugar), pulse, blood pressure, and electrocardiogram (EKG). Measurements can be transmitted via telephone, Internet connection, and wireless data transmission standards, such as Bluetooth. A particularly powerful technology combination is a wearable device that records data periodically or continuously and transmits it via Bluetooth to a cell phone app. The cell phone app can inform the patient of problems and can automatically transmit data and alerts to a central monitoring application (see **Figure 2-17**).

Health-care providers and patients incur significant costs when glucose levels are not maintained within acceptable tolerances. Short-term episodes of very high or very low glucose often result in expensive visits to urgent care clinics or hospitals. In addition, patients with frequent but less severe episodes of high or low glucose are more susceptible to such expensive

long-term complications as vision, circulatory, and kidney problems.

Sandia Medical Devices (SMD), an Albuquerque manufacturer of portable and wearable medical monitoring devices, has developed a glucose monitor embedded in a wristband. The device is powered by body heat and senses glucose levels from minute quantities of perspiration. SMD is developing the Real-Time Glucose Monitoring (RTGM) device in partnership with New Mexico Health Systems (NMHS), a comprehensive health delivery service with patients throughout New Mexico, The system's vision statement reads as follows:

RTGM will enable patients and their healthcare providers to continuously monitor glucose levels, immediately identify short- and long-term medical dangers, and rapidly respond to those dangers in medically appropriate ways.

SMD will develop the initial prototype software for smartphones with Bluetooth capability running the Google Android operating system. If successful,

FIGURE **2-17** *Data movement among devices and users*

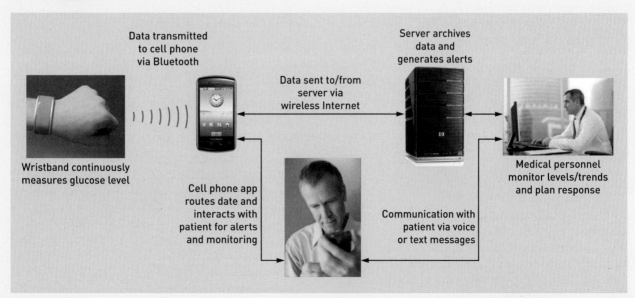

Data transmitted to cell phone via Bluetooth

Server archives data and generates alerts

Data sent to/from server via wireless Internet

Wristband continuously measures glucose level

Cell phone app routes date and interacts with patient for alerts and monitoring

Communication with patient via voice or text messages

Medical personnel monitor levels/trends and plan response

(continued on page 66)

(continued from page 65)

NMHS and its patients will have free use of the software and SMD will resell the software to other health systems worldwide.

1. Who are RTGM's stakeholders? Should NMHS's patients be included in defining the system requirements? Why or why not? Should RTGM interact with medical professionals other than physicians? Why or why not?
2. If you were the lead analyst for RTGM, how would you determine the requirements? Be specific in your answer. List several questions you need answered.
3. What are the primary functional requirements for the system as described so far in the case?
4. Are the parameters for alerting patients and medical personnel the same for every patient? Can they vary over time for the same patient? What are the implications for the system's functional requirements?
5. Briefly describe some possible nonfunctional requirements for RTGM.

Further Resources

Soren Lauesen, *Software Requirements: Styles and Techniques*. Addison-Wesley, 2002.

Stan Magee, *Guide to Software Engineering Standards and Specifications*. Artech House, 1997.

Suzanne Robertson and James Robertson, *Mastering the Requirements Process*, Second Edition. Addison-Wesley, 2006.

Karl Wiegers, *Software Requirements*. Microsoft Press, 2003.

Karl Wiegers, *More About Software Requirements: Thorny Issues and Practical Advice*. Microsoft Press, 2006.

Ralph Young, *The Requirements Engineering Handbook*. Artech House, 2003.

3

Use Cases

Chapter Outline

- Use Cases and User Goals
- Use Cases and Event Decomposition
- Use Cases and CRUD
- Use Cases in the Ridgeline Mountain Outfitters Case
- User Case Diagrams

Learning Objectives

After reading this chapter, you should be able to:

- Explain why identifying use cases is the key to defining functional requirements
- Describe the two techniques for identifying use cases
- Apply the user goal technique to identify use cases
- Apply the event decomposition technique to identify use cases
- Apply the CRUD technique to validate and refine the list of use cases
- Describe the notation and purpose for the use case diagram
- Draw use case diagrams by actor and by subsystem

OPENING CASE

Waiters on Call Meal-Delivery System

Waiters on Call is a restaurant meal-delivery service started in 2008 by Sue and Tom Bickford. The Bickfords worked for restaurants while in college and always dreamed of opening their own restaurant. Unfortunately, the initial investment was always out of reach. The Bickfords noticed that many restaurants offer takeout food and that some restaurants—primarily pizzerias—offer home-delivery service. However, many people they met seemed to want home delivery with a wider food selection.

Sue and Tom conceived Waiters on Call as the best of both worlds: a restaurant service without the high initial investment. They contracted with a variety of well-known restaurants in town to accept orders from customers and to deliver the complete meals. After preparing the meal to order, the restaurant charges Waiters on Call a wholesale price, and the customer pays retail plus a service charge and tip. Waiters on Call started modestly, with only two restaurants and one delivery driver working the dinner shift. Business rapidly expanded, and the Bickfords realized they needed a custom computer system to support their operations. They hired a consultant, Sam Wells, to help them define what sort of system they needed.

"What sort of events happen when you are running your business that make you want to reach for a computer?" asked Sam. "Tell me about what usually goes on."

"Well," answered Sue, "when a customer calls in wanting to order, I need to record it and get the information to the right restaurant. I need to know which driver to ask to pick up the order, so I need drivers to call in and tell me when they are free. Perhaps this could be included as a smartphone or iPad app. Sometimes, customers call back wanting to change their orders, so I need to get my hands on the original order and notify the restaurant to make the change."

"Okay, how do you handle the money?" queried Sam.

Tom jumped in. "The drivers get a copy of the bill directly from the restaurant when they pick up the meal. The bill should agree with our calculations. The drivers collect that amount plus a service charge. When drivers report in at

closing, we add up the money they have and compare it with the records we have. After all drivers report in, we need to create a deposit slip for the bank for the day's total receipts. At the end of each week, we calculate what we owe each restaurant at the agreed-to wholesale price and send each a statement and check."

"What other information do you need to get from the system?" continued Sam.

"It would be great to have some information at the end of each week about orders by restaurant and orders by area of town—things like that," Sue said. "That would help us decide about advertising and contracts with restaurants. Then, we need monthly statements for our accountant."

Sam made some notes and sketched some diagrams as Sue and Tom talked. Then, after spending some time thinking about it, he summarized the situation for Waiters on Call. "It sounds to me like you need a system to use whenever these events occur:

- A customer calls in to place an order, so you need to *Record an order*.
- A driver is finished with a delivery, so you need to *Record delivery completion*.
- A customer calls back to change an order, so you need to *Update an order*.
- A driver reports for work, so you need to *Sign in the driver*.
- A driver submits the day's receipts, so you need to *Reconcile driver receipts*.

"Then, you need the system to produce information at specific points in time—for example, when it is time to:

- *Produce an end-of-day deposit slip.*
- *Produce end-of-week restaurant payments.*
- *Produce weekly sales reports.*
- *Produce monthly financial reports.*

"Am I on the right track?"

Sue and Tom quickly agreed that Sam was talking about the system in a way they could understand. They were confident that they had found the right consultant for the job.

Overview

Chapter 2 described the systems analysis activities used in system development and then introduced the many tasks and techniques involved when completing the first analysis activity—gathering information about the system, its stakeholders, and its requirements. An extensive amount of information is required to properly define the system's functional and nonfunctional requirements. This chapter, like Chapter 4 and Chapter 5, presents techniques for documenting the functional requirements by creating a variety of models. These models are created as part of the analysis activity *Define functional requirements*, although remember that the analysis activities are actually done in parallel and in each iteration of the project.

use case an activity that the system performs, usually in response to a request by a user

Virtually all newer approaches to system development begin the requirements modeling process with the concept of a use case. A **use case** is an activity the system performs, usually in response to a request by a user. In Chapter 1, the RMO Tradeshow System example had a list of uses that included *Look up supplier*, *Enter/update product information*, and *Upload product information*. Two techniques are recommended for identifying use cases: the user goal technique and the event decomposition technique. An additional technique, known as the CRUD technique, is often used to validate and enhance the list of use cases. These techniques are described in the following sections.

Use Cases and User Goals

user goal technique a technique to identify use cases by determining what specific goals or objectives must be completed by a user

One approach to identifying use cases, called the **user goal technique**, is to ask users to describe their goals for using the new or updated system. The analyst first identifies all the users and then conducts a structured interview with each user. By focusing on one type of user at a time, the analyst can systematically address the problem of identifying use cases.

During the interview, the analyst guides the user to identify specific ways that a computer system can help the user perform his or her assigned tasks. The overarching objective is to identify how a system can improve the user's performance and productivity. Subsidiary goals might include streamlining tasks the user currently performs or enabling the user to perform new tasks that are not possible or practical with the current system. As these goals are uncovered and described, the analyst probes for specific requests from the user and desired responses from the proposed system, which the analyst documents as use cases. Although the user is the ultimate source of this information, he or she often requires guidance from the analyst to think beyond the boundaries of the ways they currently approach their jobs.

Consider various user goals for the RMO Consolidated Sales and Marketing System (CSMS) introduced in Chapter 2. In an example like this, the analyst might talk to the people in the shipping department to identify their specific goals. These might include: *Ship items*, *Track shipment*, and *Create item return*. When talking with people in the marketing department, goals identified might include *Add/update product information*, *Add/update promotion*, and *Produce sales history report*. When considering the goals of the prospective customer, the analyst might ask a number of people to think about the system from the customer's viewpoint and to imagine the value-added features and functions that would make RMO appealing and useful to customers. Focus groups might be formed to uncover the wants and needs of potential customers. Potential customer goals identified might include *Search for item*, *Fill shopping cart*, and *View product comments and ratings*. **Figure 3-1** lists a few of the user goals for potential users of the CSMS.

The user goal technique for identifying use cases includes these steps:

1. Identify all the potential users for the new system.
2. Classify the potential users in terms of their functional role (e.g., shipping, marketing, sales).
3. Further classify potential users by organizational level (e.g., operational, management, executive).

FIGURE **3-1**
Identifying use cases with the user goal technique

User	User goal and resulting use case
Potential customer	Search for item Fill shopping cart View product rating and comments
Marketing manager	Add/update product information Add/update promotion Produce sales history report
Shipping personnel	Ship items Track shipment Create item return

4. For each type of user, interview them to find a list of specific goals they will have when using the new system. Start with goals they currently have and then get them to imagine innovative functions they think would add value. Encourage them to state each goal in the imperative verb-noun form, such as *Add customer*, *Update order*, and *Produce month end report*.
5. Create a list of preliminary use cases organized by type of user.
6. Look for duplicates with similar use case names and resolve inconsistencies.
7. Identify where different types of users need the same use cases.
8. Review the completed list with each type of user and then with interested stakeholders.

Use Cases and Event Decomposition

The most comprehensive technique for identifying use cases is the event decomposition technique. The **event decomposition technique** begins by identifying all the business events that will cause the information system to respond, and each event leads to a use case. Starting with business events helps the analyst define each use case at the right level of detail. For example, one analyst might identify a use case as typing in a customer name on a form. A second analyst might identify a use case as the entire process of adding a new customer. A third analyst might even define a use case as working with customers all day, which could include adding new customers, updating customer records, deleting customers, following up on late-paying customers, or contacting former customers. The first example is too narrow to be useful. The second example defines a complete user goal, which is the right level of analysis for a use case. Working with customers all day—the third example— is too broad to be useful.

The appropriate level of detail for identifying use cases is one that focuses on **elementary business processes (EBPs)**. An EBP is a task that is performed by one person in one place in response to a business event, adds measurable business value, and leaves the system and its data in a stable and consistent state. In Figure 3-1, the RMO CSMS customer goals that will become use cases are *Search for item*, *Fill shopping cart*, *View product ratings and comments*, and so forth. These use cases are good examples of elementary business processes. To fill a shopping cart is in response to the business event "Customer wants to shop." There is one person filling the cart, and there is measurable value for the customer as items are added to the cart. When the customer stops adding items and moves to another task, the system remembers the current cart and is ready to switch to the new task.

event decomposition technique   a technique to identify use cases by determining the external business events to which the system must respond

elementary business processes (EBPs)   the most fundamental tasks in a business process, which leaves the system and data in a quiescent state; usually performed by one person in response to a business event

[handwritten margin notes:]
@ specific time & place
Can be precisely defined identified
must be remembered

Note that each EBP (and thus each use case) occurs in response to a business event. An **event** occurs at a specific time and place, can be described, and should be remembered by the system. Events drive or trigger all processing that a system does, so listing events and analyzing them makes sense when you need to define system requirements by identifying use cases.

Event Decomposition Technique

As stated previously, the event decomposition technique focuses on identifying the events to which a system must respond and then determining how a system must respond (i.e., the system's use cases). When defining the requirements for a system, it is useful to begin by asking, "What business events occur that will require the system to respond?" By asking about the events that affect the system, you direct your attention to the external environment and look at the system as a black box. This initial perspective helps keep your focus on a high-level view of the system (looking at the scope) rather than on the inner workings of the system. It also focuses your attention on the system's interfaces with outside people and other systems.

Some events that are important to a retail store's charge account processing system are shown in **Figure 3-2**. The functional requirements are defined by use cases based on six events. A customer triggers three events: "customer pays a bill," "customer makes a charge," and "customer changes address." The system responds with three use cases: *Record a payment*, *Process a charge*, or *Maintain customer data*. Three other events are triggered inside the system by reaching a point in time: "time to send out monthly statements," "time to send late notices," and "time to

FIGURE **3-2** *Events in a charge account processing system that lead to use cases*

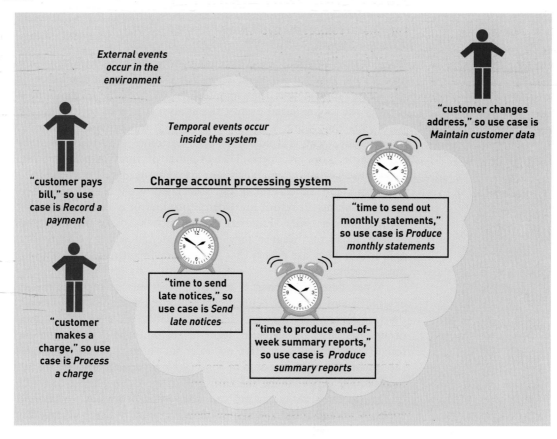

produce end-of-week summary reports." The system responds with use cases that carry out what it is time to do: *Produce monthly statements*, *Send late notices*, and *Produce summary reports*. Describing this system in terms of events keeps the focus of the charge account system on the business requirements and the elementary business processes. The next step is to divide the work among developers: One analyst might focus on the events triggered by people, and another analyst might focus on events triggered by reaching a point in time. The system is decomposed in a way that allows it to be understood in detail. The result is a list of use cases triggered by business events at the right level of analysis.

The importance of the concept of events for defining functional requirements was first emphasized for real-time systems in the early 1980s. Real-time systems must react immediately to events in the environment. Early real-time systems include manufacturing process control systems and avionics guidance systems. For example, in process control, if a vat of chemicals is full, then the system needs to *Turn off the fill valve*. The relevant event is "vat is full," and the system needs to respond to that event immediately. In an airplane guidance system, if the plane's altitude drops below 5,000 feet, then the system needs to *Turn on the low-altitude alarm*.

Most information systems now being developed are so interactive that they can be thought of as real-time systems. In fact, people expect a real-time response to almost everything. Thus, use cases for business systems are often identified by using the event decomposition technique.

Types of Events

There are three types of events to consider when using the event decomposition technique to identify use cases: external events, temporal events, and state events (also called *internal events*). The analyst begins by trying to identify and list as many of these events as possible, refining the list while talking with system users.

External Events

external event an event that occurs outside the system, usually initiated by an external agent

actor an external agent; a person or group that interacts with the system by supplying or receiving data

An **external event** is an event that occurs outside the system—usually initiated by an external agent or actor. An external agent (or **actor**) is a person or organizational unit that supplies or receives data from the system. To identify the key external events, the analyst first tries to identify all the external agents that might want something from the system. A classic example of an external agent is a customer. The customer may want to place an order for one or more products. This event is of fundamental importance to an order-processing system, such as the one needed by Ridgeline Mountain Outfitters. But other events are associated with a customer. Sometimes, a customer wants to return an ordered product or a customer needs to pay the invoice for an order. External events such as these are the types that the analyst looks for because they begin to define what the system needs to be able to do. They are events that lead to important transactions that the system must process.

When describing external events, it is important to name the event so the external agent is clearly defined. The description should also include the action that the external agent wants to pursue. Thus, the event "Customer places an order" describes the external agent (a customer) and the action that the customer wants to take (to place an order for some products) that directly affects the system. Again, if the system is an order-processing system, the system needs to create the order for the customer.

Important external events can also result from the wants and needs of people or organizational units inside the company (e.g., management requests for information). A typical event in an order-processing system might be

FIGURE **3-3**
External event checklist

> External events to look for include:
> √ External agent wants something resulting in a transaction
> √ External agent wants some information
> √ Data changed and needs to be updated
> √ Management wants some information

"Management wants to check order status." Perhaps managers want to follow up on an order for a key customer; the system must routinely provide that information.

Another type of external event occurs when external entities provide new information that the system simply needs to store for later use. For example, a regular customer reports a change in address, phone, or employer. Usually, one event for each type of external agent can be described to handle updates to data, such as "Customer needs to update account information." **Figure 3-3** provides a checklist to help in identifying external events.

Temporal Events

temporal event *an event that occurs as a result of reaching a point in time*

A second type of event is a **temporal event**—an event that occurs as a result of reaching a point in time. Many information systems produce outputs at defined intervals, such as payroll systems that produce a paycheck every two weeks (or each month). Sometimes, the outputs are reports that management wants to receive regularly, such as performance reports or exception reports. These events are different from external events in that the system should automatically produce the required output without being told to do so. In other words, no external agent or actor is making demands, but the system is supposed to generate information or other outputs when they are needed.

The analyst begins identifying temporal events by asking about the specific deadlines that the system must accommodate. What outputs are produced at that deadline? What other processing might be required at that deadline? The analyst usually identifies these events by defining what the system needs to produce at that time. In a payroll system, a temporal event might be named "Time to produce biweekly payroll." The event defining the need for a monthly summary report might be named "Time to produce monthly sales summary report." **Figure 3-4** provides a checklist to use in identifying temporal events.

Temporal events do not have to occur on a fixed date. They can occur after a defined period of time has elapsed. For example, a bill might be given to a customer when a sale has occurred. If the bill has not been paid within 15 days, the system might send a late notice. The temporal event "Time to send late notice" might be defined as a point 15 days after the billing date.

State Events

state event *an event that occurs when something happens inside the system that triggers some process*

A third type of event is a **state event**—an event that occurs when something happens inside the system that triggers the need for processing. State events are also called *internal events*. For example, if the sale of a product results in an adjustment to an inventory record and the inventory in stock drops below a reorder point, it is

FIGURE **3-4**
Temporal event checklist

> Temporal events to look for include:
> √ Internal outputs needed
>   √ Management reports (summary or exception)
>   √ Operational reports (detailed transactions)
>   √ Internal statements and documents (including payroll)
> √ External outputs needed
>   √ Statements, status reports, bills, reminders

necessary to reorder. The state event might be named "Reorder point reached." Often, state events occur as a consequence of external events. Sometimes, they are similar to temporal events, except the point in time cannot be defined.

Identifying Events

It is not always easy to define the events that affect a system, but some guidelines can help an analyst think through the process.

Events versus Prior Conditions and Responses

It is sometimes difficult to distinguish between an event and part of a sequence of prior conditions that leads up to the event. Consider a customer buying a shirt from a retail store (see **Figure 3-5**). From the customer's perspective, this purchase involves a long sequence of events. The first event might be that the customer wants to get dressed. Then, the customer wants to wear a striped shirt. Next, the striped shirt appears to be worn out. Then, the customer decides to drive to the mall. Then, he decides to go into Sears. Then, he tries on a striped shirt. Then, the customer decides to leave Sears and go to Walmart to try on a shirt. Finally, the customer wants to purchase the shirt. The analyst has to think through such a sequence to arrive at the point at which an event directly affects the system. In this case, the system is not affected until the customer is in the store, has a shirt in hand ready to purchase, and says "I want to buy this shirt."

In other situations, it is not easy to distinguish between an external event and the system's response. For example, when the customer buys the shirt, the system requests a credit card number and then the customer supplies the credit card. Is the act of supplying the credit card an event? In this case, no. It is part of the interaction that occurs while completing the original transaction.

The way to determine whether an occurrence is an event or part of the interaction following the event is by asking whether any long pauses or intervals occur (i.e., can the system transaction be completed without interruption?). Or is the system at rest again, waiting for the next transaction? After the customer wants to buy the shirt, the process continues until the transaction is complete. There are no significant stops after the transaction begins. After the transaction is complete, the system is at rest, waiting for the next transaction to begin. The EBP concept defined earlier describes this as leaving the system and its data in a consistent state.

On the other hand, separate events occur when the customer buys the shirt by using his store credit card account. When the customer pays the bill at the

FIGURE **3-5**

Sequence of actions that lead up to only one event affecting the system

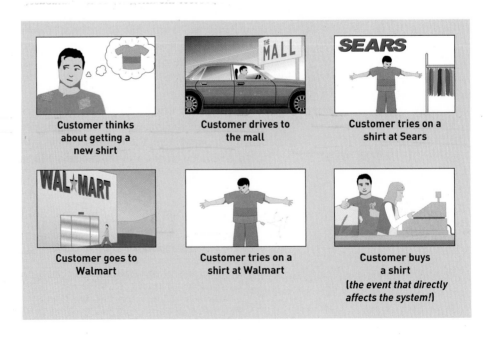

| Customer thinks about getting a new shirt | Customer drives to the mall | Customer tries on a shirt at Sears |
| Customer goes to Walmart | Customer tries on a shirt at Walmart | Customer buys a shirt (the event that directly affects the system!) |

end of the month, is the processing part of the interaction involving the purchase? In this case, no. The system records the transaction and then does other things. It does not halt all processes to wait for the payment. A separate event occurs later that results in sending the customer a bill. (This is a temporal event: "Time to send monthly bills.") Eventually, another external event occurs ("Customer pays the bill").

The Sequence of Events: Tracing a Transaction's Life Cycle

It is often useful in identifying events to trace the sequence of events that might occur for a specific external agent or actor. In the case of Ridgeline Mountain Outfitters' new CSMS, the analyst might think through all the possible transactions that might result from one new customer (see **Figure 3-6**). First, the customer wants a catalog or asks for some information about item availability, resulting in a name and address being added to the database. Next, the customer might want to place an order. Perhaps he or she will want to change the order—for example, correcting the size of the shirt or buying another shirt. Next, the customer might want to check the status of an order to find out the shipping date. Perhaps the customer has moved and wants an address change recorded for future catalog mailings. Finally, the customer might want to return an item. Thinking through this type of sequence can help identify events.

Technology-Dependent Events and System Controls

Sometimes, the analyst is concerned about events that are important to the system but do not directly concern users or transactions. Such events typically involve design choices or system controls. During analysis, the analyst should temporarily ignore these events. However, they are important later for design.

Some examples of events that affect design issues include external events that involve actually using the physical system, such as logging on. Although important to the final operation of the system, such implementation details should be deferred. At this stage, the analyst should focus only on the functional requirements (i.e., the work that the system needs to complete). A functional requirements model does not need to indicate how the system is actually implemented, so the model should omit the implementation details.

system controls checks or safety procedures to protect the integrity of the system and the data

Most of these events involve **system controls**, which are checks or safety procedures put in place to protect the integrity of the system. For example, logging on to a system is required because of system security controls. Other controls protect the integrity of the database, such as backing up the data every day.

FIGURE **3-6** *The sequence of "transactions" for one specific customer resulting in many events*

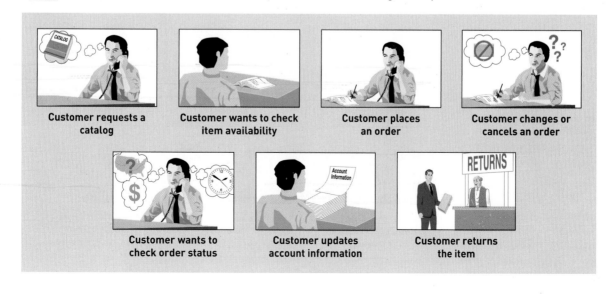

Customer requests a catalog

Customer wants to check item availability

Customer places an order

Customer changes or cancels an order

Customer wants to check order status

Customer updates account information

Customer returns the item

These controls are important to the system, and they will certainly be added to the system during design. But spending time on these controls during analysis only adds details to the requirements model that users are not typically very concerned about; they trust the system developers to take care of such details.

One technique used to help decide which events apply to controls is to assume that technology is perfect. The **perfect technology assumption** states that events should be included during analysis only if the system would be required to respond under perfect conditions (i.e., with equipment never breaking down, capacity for processing and storage being unlimited, and people operating the system being completely honest and never making mistakes). By pretending that technology is perfect, analysts can eliminate events like "Time to back up the database" because they can assume that the disk will never crash. Again, during design, the project team adds these controls because technology is obviously not perfect. **Figure 3-7** lists some examples of events that can be deferred until the design phase.

perfect technology assumption the assumption that a system runs under perfect operating and technological conditions

Using the Event Decomposition Technique

To summarize, the event decomposition technique for identifying use cases includes these steps:

1. Consider the external events in the system environment that require a response from the system by using the checklist shown in Figure 3-3.
2. For each external event, identify and name the use case that the system requires.
3. Consider the temporal events that require a response from the system by using the checklist shown in Figure 3-4.
4. For each temporal event, identify and name the use case that the system requires and then establish the point of time that will trigger the use case.
5. Consider the state events that the system might respond to, particularly if it is a real-time system in which devices or internal state changes trigger use cases.
6. For each state event, identify and name the use case that the system requires and then define the state change.
7. When events and use cases are defined, check to see if they are required by using the perfect technology assumption. Do not include events that involve such system controls as login, logout, change password, and backup or restore the database, as these are put in as system controls.

FIGURE **3-7** *Events and functions deferred until design*

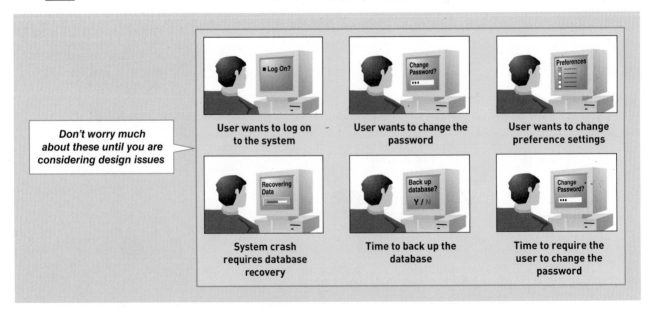

Use Cases and CRUD

[handwritten margin notes: CREATE, READ/REPORT, Update, DELETE]

Another important technique used to validate and refine use cases is the **CRUD technique.** "CRUD" is an acronym for Create, Read or Report, Update, and Delete, and it is often introduced with respect to database management. The analyst starts by looking at the types of data stored by the system, which are modeled as data entities or domain classes, as described in Chapter 4. In the RMO Tradeshow System discussed in Chapter 1, the types of data included Supplier, Contact, Product, and ProductPicture. In the RMO CSMS, the types of data include Customer, Sale, Inventory Item, Promotion, and Shipment. To validate and refine use cases, the analyst looks at each type of data and verifies that use cases have been identified that create the data, read or report on the data, update the data, and delete (or archive) the data.

The CRUD technique is most useful when used as a cross-check along with the user goal technique. Users will focus on their primary goals, and use cases that update or archive data will often be overlooked. The CRUD technique makes sure all possibilities are identified. Sometimes, data entities or domain classes are shared by a set of integrated applications. For example, RMO has a supply chain management application that is responsible for managing inventory levels and adding products. The RMO CSMS will not need to create or delete products, but it will need to look up and update product information. It is important to identify the other application that is responsible for creating, updating, or deleting the data to be clear about the scope of each system. **Figure 3-8** shows an example of potential use cases based on the CRUD technique for RMO Customer data.

Note in Figure 3-8 that the analyst has not blindly added a use case to create, read/report, update, and delete instances of a customer. The CRUD technique is best used to take already identified use cases and verify that there are use cases for create, read, update, and delete as a cross-check.

Another use of the CRUD technique is to summarize all use cases and all data entities/domain classes to show the connection between use cases and data. In **Figure 3-9**, some of the use cases are matched with data entities/domain classes by including "C," "R," "U," or "D" in the cell corresponding to the role of the use

FIGURE **3-8**

Verifying use cases with the CRUD technique

Data entity/domain class	CRUD	Verified use case
Customer	Create	Create customer account
	Read/report	Look up customer Produce customer usage report
	Update	Process account adjustment Update customer account
	Delete	Update customer account (to archive)

FIGURE **3-9**   *CRUD table showing use cases and corresponding data entities/domain classes*

[handwritten: ENTITY]
[handwritten margin note: USE CASE]

Use case vs. entity/domain class	Customer	Account	Sale	Adjustment
Create customer account	C	C		
Look up customer	R	R		
Produce customer usage report	R	R	R	
Process account adjustment	R	U	R	C
Update customer account	UD (archive)	UD (archive)		

case in terms of data. For example, the use case *Create customer account* actually creates customer data and account data, so the "C" is included in those two cells. The use case *Process account adjustment* reads information about the sale, reads information about the customer, updates the account, and creates an adjustment.

The CRUD technique for validating and refining use cases includes these steps:

1. Identify all the data entities or domain classes involved in the new system. Chapter 4 discusses these in more detail.
2. For each type of data (data entity or domain class), verify that a use case has been identified that creates a new instance, updates existing instances, reads or reports values of instances, and deletes (archives) an instance.
3. If a needed use case has been overlooked, add a new use case and then identify the stakeholders.
4. With integrated applications, make sure it is clear which application is responsible for adding and maintaining the data and which system merely uses the data.

Use Cases in the Ridgeline Mountain Outfitters Case

The RMO CSMS involves a variety of use cases, many of them just discussed. The analysts working on the new system have used all three techniques for identifying, validating, and refining use cases. The initial system vision (discussed in Chapter 2) identified four subsystems: the Sales subsystem, the Order Fulfillment subsystem, the Customer Account subsystem, and the Marketing subsystem. As work progressed, the analysts combined reports required by each subsystem into a fifth subsystem called the Reporting subsystem. In a system this size, the analyst should organize the use cases by subsystem to help track which subsystem is responsible for each use case. The analyst should also identify which use cases involve more than one type of user.

It is important to recognize that this list of use cases will continue to evolve as the project progresses. Additional use cases will be added, some might be eliminated, and some might be combined. It is helpful to immediately describe some of the details of each use case, preferably in one sentence. This brief description is usually expanded to record more of the details when the developers are designing and implementing the use case (see Chapter 5). Some examples of **brief use case descriptions** are shown in **Figure 3-10**. **Figures 3-11a** through **Figure 3-11e** show the initial list of use cases for the RMO CSMS along with the users. Note that many use cases have more than one user.

Use Case Diagrams

Sometimes, it is useful to create diagrams that more graphically show use cases and how they are organized. The **use case diagram** is the UML model used to graphically show the use cases and their relationship to users. Recall from Chapter 2 that UML is the standard set of diagrams and model constructs used

brief use case description an often one-sentence description that provides a quick overview of a use case

use case diagram the UML model used to graphically show use cases and their relationships to actors

FIGURE **3-10**
Use cases and brief descriptions

Use case	Brief use case description
Create customer account	User/actor enters new customer account data, and the system assigns account number, creates a customer record, and creates an account record.
Look up customer	User/actor enters customer account number, and the system retrieves and displays customer and account data.
Process account adjustment	User/actor enters order number, and the system retrieves customer and order data; actor enters adjustment amount, and the system creates a transaction record for the adjustment.

FIGURE **3-11A,B**
Use cases and users/actors by CSMS subsystem

CSMS sales subsystem	
Use cases	**Users/actors**
Search for item	Customer, customer service representative, store sales representative
View product comments and ratings	Customer, customer service representative, store sales representative
View accessory combinations	Customer, customer service representative, store sales representative
Fill shopping cart	Customer
Empty shopping cart	Customer
Check out shopping cart	Customer
Fill reserve cart	Customer
Empty reserve cart	Customer
Convert reserve cart	Customer
Create phone sale	Customer service representative
Create store sale	Store sales representative

CSMS order fulfillment subsystem	
Use cases	**Users/actors**
Ship items	Shipping
Manage shippers	Shipping
Create backorder	Shipping
Create item return	Shipping, customer
Look up order status	Shipping, customer, management
Track shipment	Shipping, customer, marketing
Rate and comment on product	Customer
Provide suggestion	Customer
Review suggestions	Management
Ship items	Shipping
Manage shippers	Shipping

FIGURE **3-11C,D,E**
*Use cases and users/actors by
CSMS subsystem* (continued)

CSMS Customer account subsystem	
Use cases	**Users/actors**
Create/update customer account	Customer, customer service representative, store sales representative
Process account adjustment	Management
Send message	Customer
Browse messages	Customer
Request friend linkup	Customer
Reply to linkup request	Customer
Send/receive points	Customer
View "mountain bucks"	Customer
Transfer "mountain bucks"	Customer

CSMS marketing subsystem	
Use cases	**Users/actors**
Add/update product information	Merchandising, marketing
Add/update promotion	Marketing
Add/update accessory package	Merchandising
Add/update business partner link	Marketing

CSMS reporting subsystem	
Use cases	**Users/actors**
Produce daily transaction summary report	Management
Produce sales history report	Management, marketing
Produce sales trends report	Marketing
Produce customer usage report	Marketing
Produce shipment history report	Management, shipping
Produce promotion impact report	Marketing
Produce business partner activity report	Management, marketing

in system development. You saw an example of a use case diagram in Chapter 1. The notation is fairly simple.

Use Cases, Actors, and Notation

Implied in most use cases is a person who uses the system, which we have referred to up to this point as the user. In UML, that person is called an actor. An actor is always outside the automation boundary of the system but may be part of the manual portion of the system. Sometimes, the actor for a use case is not a person; instead, it can be another system or device that receives services from the system.

Figure 3-12 shows the basic parts of a use case diagram. A simple stick figure is used to represent an actor. The stick figure is given a name that characterizes the role the actor is playing. The use case itself is represented by an oval with the name of the use case inside. The connecting line between the actor and the use case indicates that the actor is involved with that use case. Finally, the **automation boundary,** which defines the border between the computerized portion of the application and the people operating the application, is shown as a rectangle containing the use case. The actor's communication with the use case crosses the automation boundary. The example in Figure 3-12 shows the actor as a shipping clerk and the use case *Ship items*.

Use Case Diagram Examples

Figure 3-13 shows a more complete use case diagram for a subsystem of the RMO CSMS: the Customer Account subsystem. The information in Figure 3-11c is recast as a single use case diagram to visually highlight the use cases and actors for an individual subsystem. This diagram would be useful to take to a meeting to review the use cases and actors for the subsystem. In this example, the customer, customer service representative, and store sales representative are all allowed to access the system directly. As indicated by the relationship lines, each actor can use the use case *Create/update customer account*. The customer might do this when checking out online. The customer service representative might do this when talking to a customer on the phone. The store sales representative might do this when dealing with the customer in a store. Only a member of management can process an account adjustment. The other use cases are included only for the customer.

There are many ways to organize use case diagrams for communicating with users, stakeholders, and project team members. One way is to show all use cases that are invoked by a particular actor (i.e., from the user's viewpoint). This approach is often used during requirements definition because the systems analyst may be working with a particular user and identifying all

automation boundary the boundary between the computerized portion of the application and the users who operate the application but are part of the total system

FIGURE **3-12**

A simple use case with an actor

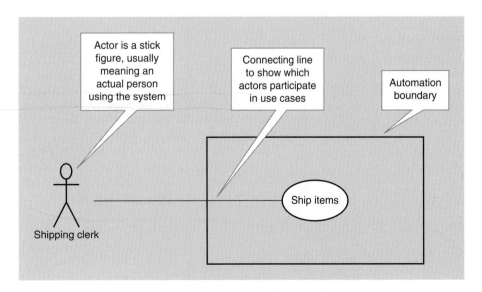

FIGURE **3-13**

A use case diagram of the Customer Account subsystem for RMO, showing all actors

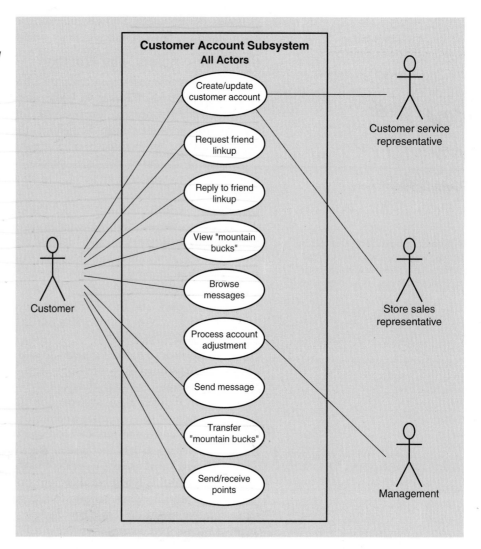

the functions that user performs with the system. **Figure 3-14** illustrates this viewpoint, showing all the use cases involving the customer for the Sales subsystem. **Figure 3-15** shows use cases involving the customer service representative and the store sales representative for the Sales subsystem. Analysts can expand this approach to include all the use cases belonging to a particular department regardless of the subsystem or all use cases important to a specific stakeholder.

≪*includes*≫ *Relationships*

Frequently during the development of a use case diagram, it becomes apparent that one use case might use the services of another use case. For example, in the Sales subsystem use case diagram shown in Figure 3-14, the customer might search for an item, view product comments and ratings, and view accessory combinations before beginning to fill the shopping cart. However, while filling the shopping cart, the customer might also search for an item, view product comments, and view accessories. Therefore, one use case uses, or "includes," another use case. **Figure 3-16** shows a use case diagram emphasizing this aspect of these use cases. *Fill shopping cart* also includes *Search for item*, *View product comments and ratings*, and *View accessory combinations*. Thus, the Customer can view comments initially and also while carrying out the *Fill shopping cart* use case. The relationship between these use cases is denoted by the dashed connecting line with the arrow that

CHAPTER 3 ■ Use Cases **83**

FIGURE **3-14**

All use cases involving the customer actor for the Sales subsystem

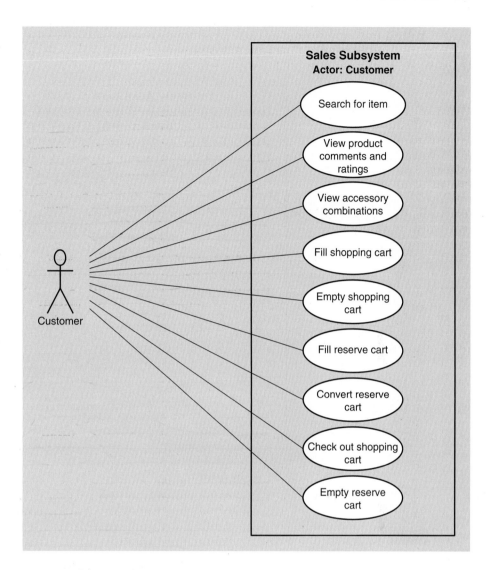

≪includes≫ relationship a relationship between use cases in which one use case is stereotypically included within the other use case

points to the use case that is included. The relationship is read *Fill shopping cart includes Search for item*. Sometimes, this relationship is referred to as the **≪includes≫ relationship** or the ≪uses≫ relationship. Note that the word "includes" is enclosed within guillemets in the diagram; this is the way to refer to a stereotype in UML. It means that the relationship between one use case and another use case is a stereotypical ≪includes≫ relationship.

Developing a Use Case Diagram

Analysts create a variety of use case diagrams to communicate with users, stakeholders, management, and team members. The steps to develop use case diagrams are:

1. Identify all the stakeholders and users who would benefit by having a use case diagram.
2. Determine what each stakeholder or user needs to review in a use case diagram. Typically, a use case diagram might be produced for each subsystem, for each type of user, for use cases with the ≪includes≫ relationship, and for use cases that are of interest to specific stakeholders.
3. For each potential communication need, select the use cases and actors to show and draw the use case diagram. There are many software packages that can be used to draw use case diagrams.
4. Carefully name each use case diagram and then note how and when the diagram should be used to review use cases with stakeholders and users.

FIGURE **3-15** *Use cases involving the customer service representative and store sales representative for the Sales subsystem*

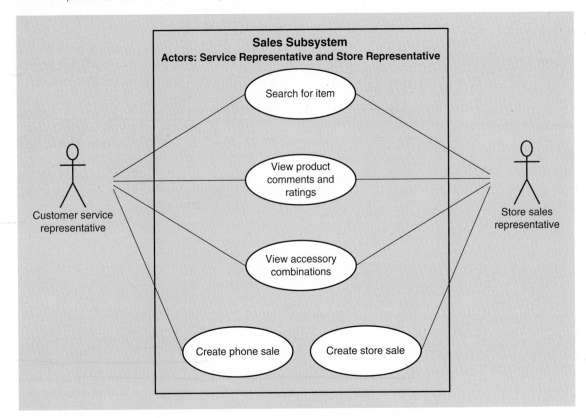

FIGURE **3-16**

A use case diagram of the Fill shopping cart ≪*includes*≫ *relationships*

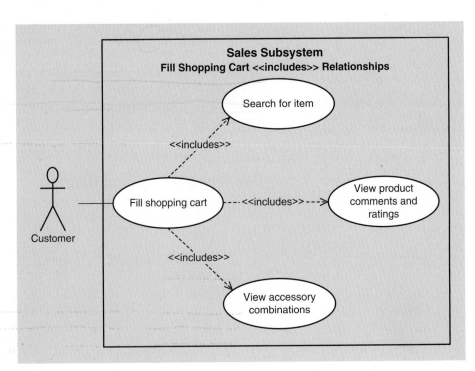

Chapter Summary

This chapter is the first of three chapters that present techniques for modeling a system's functional requirements. A key early step in the modeling process is to identify and list the use cases that define the functional requirements for the system. Use cases can be identified by using the user goal technique and the event decomposition technique. The user goal technique begins by identifying types of system end users, called *actors*. Then, users are asked to list specific user goals they have when using the system to support their work. The event decomposition technique begins by identifying the events that require a response from the system. An event is something that can be described, something that occurs at a specific time and place, and something worth remembering. External events occur outside the system—usually triggered by someone who interacts with the system. Temporal events occur at a defined point in time, such as the end of a workday or the end of every month. State or internal events occur based on an internal system change. For each event, a use case is identified and named. The event decomposition technique helps ensure that each use case is identified at the elementary business process (EBP) level of detail. Use cases are validated and refined by using the CRUD technique—"CRUD" being an acronym for Create, Read or Report, Update, and Delete.

Each use case identified by the analyst is further documented by a brief use case description and by identifying the actors. UML use case diagrams are drawn to document use cases and their actors. Many different use case diagrams are drawn based on the need to review use cases with various stakeholders, users, and team members.

Key Terms

actor 72

automation boundary 81

brief use case descriptions 78

CRUD technique 77

elementary business processes (EBPs) 70

event 71

event decomposition technique 70

external event 72

≪includes≫ relationship 83

perfect technology assumption 76

state event 73

system controls 75

temporal event 73

use case 69

use case diagram 78

user goal technique 69

Review Questions

1. What are the six activities of systems analysis, and which activity is discussed beginning with this chapter?

2. What is a use case?

3. What are the two techniques used to identify use cases?

4. Describe the user goal technique for identifying use cases.

5. What are some examples of users with different functional roles and at different operational levels?

6. What are some examples of use case names that correspond to your goals as a student going through the college registration process? Be sure to use the verb-noun naming convention.

7. What is the overarching objective of asking users about their specific goals?

8. How many types of users can have the same user goals for using the system?

9. Describe the event decomposition technique for identifying use cases.

10. Why is the event decomposition technique considered more comprehensive than the user goal technique?

11. What is an elementary business process (EBP)?

12. Explain how the event decomposition technique helps identify use cases at the right level of analysis.

13. What is an event?

14. What are the three types of events?

15. Define an external event and then give an example that applies to a checking account system.

16. Define a temporal event and then give an example that applies to a checking account system.

17. What are system controls, and why are they not considered part of the users' functional requirements?

18. What is the perfect technology assumption?

19. What are three examples of events that are system controls in a typical information system that should not be included as a use case because of the perfect technology assumption?

20. What are the four operations that make up the CRUD acronym?

21. What is the main purpose of using the CRUD technique?

22. What is a brief use case description?

23. What is UML?

24. What is the purpose of UML use case diagrams?

25. What is another name for "actor" in UML, and how is it represented on a use case diagram?

26. What is the automation boundary on a use case diagram, and how is it represented?

27. How many actors can be related to a use case on a use case diagram?

28. Why might a systems analyst draw many different use case diagrams when reviewing use cases with end users?

29. What is the ≪includes≫ relationship between two use cases?

Problems and Exercises

1. Review the external event checklist in Figure 3-3 and then think about a university course registration system. What is an example of an event of each type in the checklist? Name each event by using the guidelines for naming an external event.

2. Review the temporal event checklist in Figure 3-4. Would a student grade report be an internal or external output? Would a class list for the instructor be an internal or external output? What are some other internal and external outputs for a course registration system? Using the guidelines for naming temporal events, what would you name the events that trigger these outputs?

3. Consider the following sequence of actions taken by a customer at a bank. Which action is the event the analyst should define for a bank account transaction-processing system? (1) Kevin gets a check from Grandma for his birthday. (2) Kevin wants a car. (3) Kevin decides to save his money. (4) Kevin goes to the bank. (5) Kevin waits in line. (6) Kevin makes a deposit in his savings account. (7) Kevin grabs the deposit receipt. (8) Kevin asks for a brochure on auto loans.

4. Consider the perfect technology assumption, which states that use cases should be included during analysis only if the system would be required to respond under perfect conditions. Could any of the use cases listed for the RMO CSMS be eliminated based on this assumption? Explain. Why are such use cases as *Log on to the system* and *Back up the database* required only under imperfect conditions?

5. Visit some Web sites of car manufacturers, such as Honda, BMW, Toyota, and Acura. Many of these sites have a use case that is typically named *Build and price a car*. As a potential customer, you can select a car model, select features and options, and get the car's suggested price and list of specifications. Write a brief use case description for this use case (see Figure 3-10).

6. Again looking at a Web site for one of the car manufacturers, consider yourself a potential buyer and then identify all the use cases included on the site that correspond to your goals.

7. Set up a meeting with a librarian. During your meeting, ask the librarian to describe the situations that come up in the library to which the book checkout system needs to respond. List these external events. Now ask about points in time, or deadlines, that require the system to produce a statement, notice, report, or other output. List these temporal events. Does it seem natural for the librarian to describe the system in this way? List each event and then name the resulting use case.

8. Again considering the library, ask some students what their goals are in using the library system. Also ask some library employees about their goals in using the system. Name these goals as use cases (verb-noun) and discuss whether student users have different goals than employee users.

9. Visit a restaurant or the college food service to talk to a server (or talk with a friend who is a food server). Ask about the external events and temporal events, as you did in exercise 7. What are the events and resulting use cases for order processing at a restaurant?

10. Review the procedures for course registration at your university and then talk with the staff in advising, in registration, and in your major department. Think about the sequence that goes on over an entire semester. What are the events that students trigger? What are the events that your own

department triggers? What are the temporal events that result in information going to students? What are the temporal events that result in information going to instructors or departments? List all the events and the resulting use cases that should be included in the system.

11. Refer to the RMO CSMS Order Fulfillment subsystem shown in Figure 3-11. Draw a use case diagram that shows all actors and all use cases. Use a drawing tool such as Microsoft Visio if it is available.

12. Again for the Order Fulfillment subsystem, draw a use case diagram showing just the use cases for the shipping department in preparation for a meeting with them about the system requirements. Use a drawing tool such as Microsoft Visio if it is available.

13. Refer to the RMO CSMS Marketing subsystem shown in Figure 3-11. Draw a use case diagram that

shows all actors and all use cases. Use a drawing tool such as Microsoft Visio if it is available.

14. Refer to the RMO CSMS Reporting subsystem shown in Figure 3-11. These reports were identified by asking users about temporal events, meaning points in time that require the system to produce information of value. In most actual systems today, an actor is assigned responsibility for producing the reports or other outputs when they are due. Recall that the actor is part of the system—the manual, nonautomated part. Thus, this is one way the "system" can be responsible for producing an output at a point in time. In the future, more outputs will be produced automatically. Draw a use case diagram that shows the use cases and actors, as shown in Figure 3-11. Use a drawing tool such as Microsoft Visio if it is available.

Case Study

The State Patrol Ticket-Processing System

The purpose of the State Patrol ticket-processing system is to record moving violations, keep records of the fines paid by drivers when they plead guilty or are found guilty of moving violations, and notify the court that a warrant for arrest should be issued when such fines are not paid in a timely manner. A separate State Patrol system records accidents and the verification of financial responsibility (insurance). A third system uses ticket and accident records to produce driving record reports for insurance companies. Finally, a fourth system issues, renews, or suspends driver's licenses. These four systems are obviously integrated in that they share access to the same database; otherwise, they are operated separately by different departments of the State Patrol.

When an officer gives a ticket to a driver, a copy of the ticket is turned in and entered into the system. A new ticket record is created, and relationships to the correct driver, officer, and court are established in the database. If the driver pleads guilty, he or she mails in the fine in a preprinted envelope with the ticket number on it. In some cases, the driver claims innocence and wants a court date. When the envelope is returned without a check and the trial request box has an "X" in it, the system does the following: notes the plea on the ticket record; looks up driver, ticket, and officer information; and sends a ticket details report to the appropriate court. A trial date questionnaire form is also produced at the same time and is mailed to the driver. The instructions on the questionnaire tell the driver to fill in convenient dates and mail the questionnaire directly to the court. Upon receiving this information, the court schedules a trial date and notifies the driver of the date and time.

When the trial is completed, the court sends the verdict to the ticketing system. The verdict and trial date are recorded for the ticket. If the verdict is innocent, the system that produces driving record reports for insurance companies will ignore the ticket. If the verdict is guilty, the court gives the driver another envelope with the ticket number on it for mailing in the fine.

If the driver fails to pay the fine within the required period, the ticket-processing system produces a warrant request notice and sends it to the court. This happens if the driver does not return the original envelope within two weeks or does not return the court-supplied envelope within two weeks of the trial date. What happens then is in the hands of the court. Sometimes, the court requests that the driver's license be suspended, and the system that processes drivers' licenses handles the suspension.

1. To what events must the ticket-processing system respond? List each event, the type of event, and the resulting use case.
2. Write a brief use case description for each use case.
3. The portion of the database used with the ticket-processing system involves driver data, ticket data, officer data, and court data. Driver data, officer data, and court data are read by the system, and the ticket-processing system creates and updates ticket data. In an integrated system like the ticket-processing system, some domain classes are created by and updated by other systems, as described in this case. Create a table with systems down the rows and the four types of data (domain classes) across the columns. Indicate C, R, U, or D for each domain class and each system.

Community Board of Realtors

One of the functions of the Board of Realtors introduced in Chapter 2 is to provide a Multiple Listing Service (MLS) system that supplies information that local real estate agents use to help them sell houses to their customers. During the month, agents list houses for sale (listings) by contracting with homeowners. The agent works for a real estate office, which sends information on the listing to the MLS. Therefore, any agent in the community can get information on the listing.

Information on a listing includes the address, year built, square feet, number of bedrooms, number of bathrooms, owner name, owner phone number, asking price, and status code. At any time during the month, an agent might directly request information on listings that match customer requirements, so the agent contacts the MLS with the request. Information is provided on the house, on the agent who listed the house, and on the real estate office for which the agent works. For example, an agent might want to call the listing agent to ask additional questions or call the homeowner directly to make an appointment to show the house. Once each week, the MLS produces a listing book that contains information on all listings. These books are sent to some real estate agents. Some real estate agents want the books (which are easier to flip through), so they are provided even though the information is often out of date. Sometimes, agents and owners decide to change information about a listing, such as reducing the price, correcting previous information on the house, or indicating that the house is sold. The real estate office sends in these change requests to the MLS when the agent asks the office to do so.

1. To what events must the MLS system respond? List each event, the type of event, and the resulting use case. Be sure to consider all the use cases that would be needed to maintain the data in the MLS system, thinking in terms of the CRUD technique.
2. Draw a use case diagram based on the actors and use cases you identified in question 1.
3. Given the information available in the system, consider yourself a potential customer looking for real estate. List as many specific use cases you would like to see based on your specific goals.
4. Draw a use case diagram for all the use cases for the potential customer you identified in question 3.

The Spring Breaks 'R' Us Travel Service

Spring Breaks 'R' Us (SBRU), introduced in Chapter 2, includes many use cases that make up the functional requirements. Consider the following description of the Booking subsystem. A few weeks before Thanksgiving break, it is time to open the system to new bookings. Students usually want to browse through the resorts and do some planning. After that, when a student or group of students wants to book a trip, the system allows it. Sometimes, a student needs to be added or dropped from the group or a group changes size and needs a different type of room. One month before the actual trip, it is time for the system to send out final payment requirement notices. Students cancel the booking or they pay their final bills. Students often want to look up their booking status and check on resort details. When they arrive at the resort, they need to check in; and when they leave, they need to check out.

1. Using the event decomposition technique for each event you identify in the description here, name the event, state the type of event, and name the resulting use case. Draw a use case diagram for these use cases.
2. Consider the new Social Networking subsystem that SBRU is researching. Think in terms of the user goal technique to identify as many use cases as you can think of that you would like to have in the system. SBRU is guessing you might want to join, send messages, and so forth, but there must be many interesting and useful things the system could do before, during, and after the trip. Draw a use case diagram for these use cases.

VISIO - USE CASE DIAGRAM
EXCEL - EVENT TABLE

(continued on page 89)

(continued from page 88)

On the Spot Courier Services

Recall the On the Spot courier service introduced in Chapter 2. The details of the package pickup and delivery process are described here.

When Bill got an order, at first, only on his phone, he recorded when he received the call and when the shipment would be ready for pickup. Sometimes, customers wanted immediate pickup; sometimes, they were calling to schedule a later time in the day for pickup.

Once he arrived at the pickup location, Bill collected the packages. It was not uncommon for the customer to have several packages for delivery. In addition to the name and address of the delivery location, he also recorded the time of pickup. He noted the desired delivery time, the location of the delivery, and the weight of the package to determine the courier cost. When he picked up the package, he printed out a label with his portable printer that he kept in the delivery van.

At first, Bill required customers to pay at the time of pickup, but he soon discovered that there were some regular customers who preferred to receive a monthly bill for all their shipments. He wanted to be able to accommodate those customers. Bills were due and payable upon receipt.

To help keep track of all the packages, Bill decided that he needed to scan each package as it was sorted in the warehouse. This would enable him to keep good control of his packages and avoid loss or delays.

The delivery of a package was fairly simple. Upon delivery, he would record information about when the delivery was made and who received it. Because some of the packages were valuable, it was necessary in those instances to have someone sign for the package.

1. From this description as well as the information from Chapter 2, identify all the actors that will be using the system.
2. Using the actors that you identified in question 1, develop a list of use cases based on the user goal technique. Draw a use case diagram for these use cases.
3. Using the event decomposition technique for each event you identify in the description here, name the event, state the type of event, and name the resulting use case. Draw a use case diagram for these use cases.

Sandia Medical Devices

Recall the Sandia Medical Devices Real-Time Glucose Monitoring (RTGM) system introduced in Chapter 2. As the project began, interviews with patients and physicians about potential RTGM capabilities and interaction modes identified several areas of concern that will need to be incorporated into the system requirements and design. The relevant patient concerns include:

■ **Viewing and interpreting data and trends:** Patients want to be able to view more than their current glucose level. They would like the ability to see past glucose levels over various time periods, with a specific focus on time periods during which their glucose was within and outside of acceptable ranges. A graphical view of the data is preferred, although some patients also want to be able to see actual numbers.

■ **Additional data:** Some patients want to be able to enter text notes or voice messages to supplement glucose level data. For example, patients who see a

high glucose alert might record voice messages describing how they feel or what they had recently eaten. Some patients thought that sharing such information with their health-care providers might be valuable, but others only wanted such information for themselves.

Physicians expressed these concerns:

■ They do not want to be the "first line of response" to all alerts. They prefer that nurses or physician assistants be charged with that role and that they be notified only when frontline personnel determine that an emergency situation exists.
■ They want to be able to monitor and view past patient data and trends in much the same way as described for patients.
■ They want all their actions to be logged and for patient-specific responses to be stored as part of the patient's electronic medical record.

(continued on page 90)

(continued from page 89)

Perform the following tasks by using the information here as well as the system description in Chapter 2:

1. Identify all the actors that will use RTGM.
2. Using the actors that you identified in question 1, develop a list of use cases based on the user goal technique. Draw a use case diagram for these use cases.

3. Using the event decomposition technique for each event you identified in the description, name the event, state the type of event, and name the resulting use case. Draw a use case diagram for these use cases.

Further Resources

Classic and more recent texts include:

Craig Larman, *Applying UML and Patterns* (3rd ed.). Prentice-Hall, 2005.

Grady Booch, Ivar Jacobson, and James Rumbaugh, *The Unified Modeling Language User Guide*. Addison-Wesley, 1999.

Ed Yourdon, *Modern Structured Analysis*. Prentice Hall, 1989.

Stephen McMenamin and John Palmer, *Essential Systems Analysis*. Prentice Hall, 1984.

4

Domain Modeling

Chapter Outline

- "Things" in the Problem Domain
- The Entity-Relationship Diagram
- The Domain Model Class Diagram

Learning Objectives

After reading this chapter, you should be able to:

- Explain how the concept of "things" in the problem domain also defines requirements
- Identify and analyze data entities and domain classes needed in the system
- Read, interpret, and create an entity-relationship diagram
- Read, interpret, and create a domain model class diagram
- Understand the domain model class diagram for the RMO Consolidated Sales and Marketing System

Waiters on Call Meal-Delivery System (Part 2)

Recall that Waiters on Call has been working with Sam Wells on the requirements for its meal-delivery system. Sue and Tom Bickford want a new system that will automate and improve their specialty business of providing customer-ordered, home-delivered meals prepared by a variety of local restaurants. Sam did a great job of identifying the use cases required for the delivery service, which impressed the Bickfords. And while working on the use cases, he continued to note all the business terms and concepts that the Bickfords used as they described their operations. He followed up with questions about the types of things they work with each day, which they answered.

"Based on what you've told me," Sam said, "I assume you will need the system to store information about the following types of things, which we call data entities or domain classes: restaurants, menu items, customers, and orders. I also think you're going to need to store information about the following types of things: drivers, addresses, routes, and order payments."

The Bickfords readily agreed and added that it was important to know what route a restaurant was on and how far it might be to the customer's address. They wanted drivers to be assigned to a route based on the distances from place to place.

"Yes, we need to decide how things need to be associated in the system," Sam agreed. "Can you tell me if drivers pick up orders from several restaurants when they go out? Can you tell me how many items are usually included in one order? Do you note pickup times and delivery times? Do you need to plan the route so that hot dishes are delivered first?"

The Bickfords were further reassured that they had picked an analyst who was aware of the needs of their business.

Overview

Chapter 3 focused on identifying use cases to define the functional requirements for an information system. In this chapter, we focus on another key concept that defines requirements: things in the problem domain of system users. You first learned about these as data entities or domain classes when we discussed use cases. You might have learned about these when studying database management, as they define the sources for the tables used in a relational database management system. Nearly all approaches to system development include identifying and modeling data entities or domain classes as an important task in the analysis activity *Define functional requirements*.

"Things" in the Problem Domain

problem domain   the specific area (or domain) of the user's business need (or problem) that is within the scope of the new system

Domain classes or data entities are what end users deal with when they do their work—for example, products, sales, shippers, shipments, and customers. These are often referred to as "things" in the context of a system's problem domain. The **problem domain** is the specific area of the user's business that is included within the scope of the new system. The new system involves working with and remembering these "things." For example, some information systems need to store information about customers and products, so it is important for the analyst to identify lots of information about those two things. Often, things are related to the people who interact with the system or to other stakeholders. For example, a customer is a person who places an order, but the system needs to store information about that customer, so a customer is also a thing in the problem domain. However, things are sometimes distinct from people. For example, the system may need to store information about products, shipments, and warehouses, but these are not persons.

There are many techniques for identifying the important things in the problem domain. Two of them are introduced in this chapter: the brainstorming technique and the noun technique.

The Brainstorming Technique

As with use cases, an analyst should ask the users to discuss the types of things they work with routinely. The analyst can ask about several types of things to help identify them. Many things are tangible and therefore more easily identified, but others are intangible. Different types of things are important to different users, so it is important to involve all types of users to help identify problem domain things. The **brainstorming technique** is useful for working with users to identify things in the problem domain.

brainstorming technique a technique to identify problem domain objects in which developers work with users in an open group setting

Figure 4-1 shows some types of things to consider. Tangible things are often the most obvious, such as an airplane, a book, or a vehicle. In the Ridgeline Mountain Outfitters case, a product in the warehouse and a vehicle in the fleet are tangible things of importance. Another common type of thing in an information system is a role played by a person, such as an employee, a customer, a doctor, or a patient. The role of customer is obviously a very important one in the Ridgeline Mountain Outfitters case. Many things in the problem domain can fit into more than one type. For example, a vehicle is a device and a tangible thing. Either way, the important point is to identify potential things in the problem domain.

Other types of things can include organizational units, such as a division, department, or workgroup. Similarly, a site or location, such as a warehouse, a store, or a branch office, might be an important thing in a system. Finally, information about an incident or an interaction can be a thing—information about an order, a service call, a contract, or an airplane flight. A sale, a shipment, and a return are all important incidents in the RMO case. Sometimes, these incidents are thought of as associations between things. For example, a sale is an association between a customer and an item of inventory. Initially, the analyst might simply list all these as things and then make adjustments as required by different approaches to analysis and design.

The analyst identifies these types of things by thinking about each use case, talking to users, and asking questions. For example, for each use case, what types of things are affected that the system needs to know about and store information about? The types of things shown in Figure 4-1 can be used to systematically brainstorm about what types of things might be involved in each use case. When a customer wants to buy from the Web site, the system needs to store information about the customer, the items ordered, the details about the sale itself—such as the date and payment terms—and the location of the items

FIGURE **4-1** *Types of things to use for the brainstorming technique*

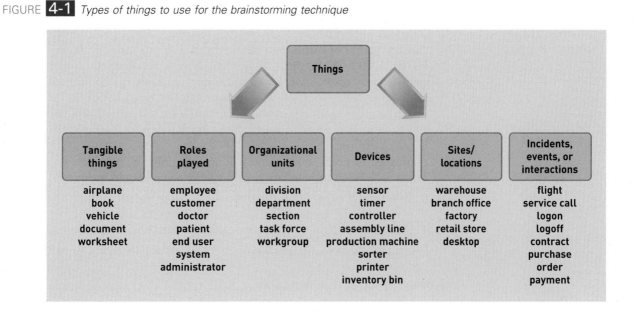

to be shipped. For that one use case, the analyst can define tangible things (items ordered), roles played (customer), incidents or events (the sale), sites/locations (warehouse), and organizational units (shipping).

Here are the steps to follow when using the brainstorming technique:

1. Identify a user and a set of use cases.
2. Brainstorm with the user to identify things involved when carrying out the use case—that is, things about which information should be captured by the system.
3. Use the types of things (categories) to systematically ask questions about potential things, such as the following: Are there any tangible things you store information about? Are there any locations involved? Are there roles played by people that you need to remember?
4. Continue to work with all types of users and stakeholders to expand the brainstorming list.
5. Merge the results, eliminate any duplicates, and compile an initial list.

The Noun Technique

noun technique a technique to identify problem domain objects by finding and classifying the nouns in a dialog or description

Another useful procedure for identifying things in the problem domain is called the **noun technique**. Recall that a noun is a person, place, or thing. Therefore, identifying nouns might help you identify what needs to be stored by the system. Begin by listing all the nouns that users mention when talking about the system. Nouns used to describe events, use cases, and the actors are potential things. Next, add to the list any additional nouns that appear in information about the existing system or that come up in discussions with stakeholders about the problem domain of the system. The list of nouns will become quite long, so the list will need to be refined. How the noun technique differs from the brainstorming technique is that the analyst lists all nouns without thinking too much about them and without talking much to users. Only later will the list be refined based on consultation with stakeholders and users.

Here are the steps to follow when using the noun technique:

1. Using the use cases, actors, and other information about the system—including inputs and outputs—identify all nouns. For the RMO CSMS, the nouns might include the following: customer, product item, sale, confirmation, transaction, shipping, bank, change request, summary report, management, transaction report, accounting, back order, back order notification, return, return confirmation, fulfillment reports, prospective customer, marketing, customer account, promotional materials, charge adjustment, sale details, merchandising, and customer activity reports.
2. Using other information from existing systems, current procedures, and current reports or forms, add items or categories of information needed. For the RMO CSMS, these might include more detailed information, such as price, size, color, style, season, inventory quantity, payment method, and shipping address. Some of these items might be additional things, and some might be more specific pieces of information (called attributes) about things you have already identified. Refine the list and then record assumptions or issues to explore.
3. As this list of nouns builds, you will need to refine it. Ask these questions about each noun to help you decide whether you should include it:

   ■ Is it a unique thing the system needs to know about?
   ■ Is it inside the scope of the system I am working on?
   ■ Does the system need to remember more than one of these items?

   Ask these questions about each noun to decide whether you should exclude it:

   ■ Is it really a synonym for some other thing I have identified?

- Is it really just an output of the system produced from other information I have identified?
- Is it really just an input that results in recording some other information I have identified?

Ask these questions about each noun to decide whether you should research it:

- Is it likely to be a specific piece of information (attribute) about some other thing I have identified?
- Is it something I might need if assumptions change?

4. Create a master list of all nouns identified and then note whether each one should be included, excluded, or researched further.
5. Review the list with users, stakeholders, and team members and then refine the list of things in the problem domain.

Figure 4-2 lists some of the nouns from the RMO CSMS, with notes about each one. As with the brainstorming technique, the initial list developed from this table is just a start. Much more work is needed to refine the list and define more information about each item in the list.

Attributes of Things

The noun technique involves listing all the nouns that come up in discussions or documents about the requirements. As discussed previously, many of these nouns are actually attributes. Most information systems store and use specific pieces of information about each thing, as shown for some nouns in Figure 4-2. The specific pieces of information are called **attributes**. For example, a customer has a name, a phone number, a credit limit, and so on. Each of these details is an attribute. The analyst needs to identify the attributes of each thing that the system needs to store. One attribute may be used to identify a

attributes descriptive pieces of information about things or objects

FIGURE 4-2

Partial list of "things" based on nouns for RMO

Identified noun	Notes on including noun as a thing to store
Accounting	We know who they are. No need to store it.
Back order	A special type of order? Or a value of order status? Research.
Back-order information	An output that can be produced from other information.
Bank	Only one of them. No need to store.
Catalog	Yes, need to recall them, for different seasons and years. Include.
Catalog activity reports	An output that can be produced from other information. Not stored.
Catalog details	Same as catalog? Or the same as product items in the catalog? Research.
Change request	An input resulting in remembering changes to an order.
Charge adjustment	An input resulting in a transaction.
Color	One piece of information about a product item.
Confirmation	An output produced from other information. Not stored.
Credit card information	Part of an order? Or part of customer information? Research.
Customer	Yes, a key thing with lots of details required. Include.
Customer account	Possibly required if an RMO payment plan is included. Research.
Fulfillment reports	An output produced from information about shipments. Not stored.
Inventory quantity	One piece of information about a product item. Research.
Management	We know who they are. No need to store.
Marketing	We know who they are. No need to store.
Merchandising	We know who they are. No need to store.

identifier or **key**   an attribute the value of which uniquely identifies an individual thing or object

compound attribute   an attribute that consists of multiple pieces of information but is best treated in the aggregate

association   a term, in UML, that describes a naturally occurring relationship between specific things, sometimes called a relationship

specific thing, such as a Social Security number for an employee or an order number for a purchase. The attribute that uniquely identifies the thing is called an **identifier** or **key**. Sometimes, the identifier is already established (a Social Security number, vehicle ID number, or product ID number). Sometimes, the system needs to assign a specific identifier (an invoice number or transaction number).

A system may need to remember many similar attributes. For example, a customer has several names: a first name, a middle name, a last name, and possibly a nickname. A **compound attribute** is an attribute that contains a collection of related attributes, so an analyst may choose one compound attribute to represent all these names, perhaps naming it *Customer full name*. A customer might also have several phone numbers: home phone number, office phone number, fax phone number, and cell phone number. The analyst might start out by describing the most important attributes but later add to the list. Attribute lists can get quite long. Some examples of attributes of a customer and the values of attributes for specific customers are shown in **Figure 4-3**.

Associations Among Things

After recording and refining the list of things and determining potential attributes, the analyst needs to research and record additional information. Many important relationships among things are important to the system. An **association** is a naturally occurring relationship between specific things, such as *an order is placed by a customer* and *an employee works in a department* (see **Figure 4-4**). *Is placed by* and *works in* are two associations that naturally occur between specific things. Information systems need to store information about employees and about departments, but equally important is storing information about the specific associations; for example, John works in the

FIGURE **4-3** *Attributes and values*

All customers have these attributes:	Each customer has a value for each attribute:		
Customer ID	101	102	103
First name	John	Mary	Bill
Last name	Smith	Jones	Casper
Home phone	555-9182	423-1298	874-1297
Work phone	555-3425	423-3419	874-8546

FIGURE **4-4**

Associations naturally occur among things

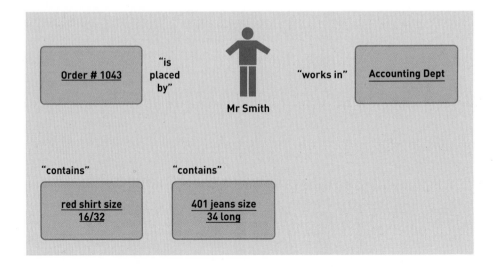

relationship   a term that describes a naturally occurring association between specific things, sometimes called an association

accounting department and Mary works in the marketing department. Similarly, it is quite important to store the fact that Order 1043 for a shirt was placed by John Smith. In database management, the term **relationship** is often used in place of association, which is the term used when modeling in UML. We will use *association* in this book because we emphasize UML diagrams and terms.

Associations between things apply in two directions. For example, *a customer places an order* describes the association in one direction. Similarly, *an order is placed by a customer* describes the association in the other direction. It is important to understand the association in both directions because sometimes it might seem more important for the system to record the association in one direction than in the other. For example, Ridgeline Mountain Outfitters definitely needs to know what items a customer ordered so the shipment can be prepared. However, it might not be initially apparent that the company needs to know about the customers who have ordered a particular item. What if the company needs to notify all customers who ordered a defective or recalled product? Knowing this information would be very important, but the operational users might not immediately recognize that fact.

cardinality   a measure of the number of links between one object and another object in a relationship

multiplicity   a measure, in UML, of the number of links between one object and another object in an association

It is also important to understand the nature of each association in terms of the number of links for each thing. For example, a customer might place many different orders, but an order is placed by only one customer. In database management, the number of links that occur is referred to as the **cardinality** of the association. Cardinality can be one-to-one or one-to-many. The term **multiplicity** is used to refer to the number of links in UML and should be used when discussing UML models. Multiplicity is established for each direction of the association. **Figure 4-5** lists examples of cardinality/multiplicity associated with an order.

It is important to describe not just the multiplicity but also the range of possible values of the multiplicity (the minimum and maximum multiplicity). For example, a particular customer might not ever place an order. In this case, there are zero associations. Alternatively, the customer might place one order, meaning one association exists. Finally, the customer might place two, three, or even more orders. The relationship for a customer placing an order can have a range of zero, one, or more, usually indicated as zero or more. The zero is the minimum multiplicity, and more is the maximum multiplicity. These terms are referred to as **multiplicity constraints**.

multiplicity constraints   the actual numeric count of the constraints on objects allowed in an association

In some cases, at least one association is required (a mandatory as opposed to optional association). For example, the system might not record any information about a customer until the customer places an order. Therefore, the multiplicity would read *customer places one or more orders*.

A one-to-one association can also be refined to include minimum and maximum multiplicity. For example, an order is placed by one customer; it is impossible to have an order if there is no customer. Therefore, one is the minimum multiplicity, making the association mandatory. Because there cannot be more than one customer for each order, one is also the maximum multiplicity.

FIGURE **4-5**
Multiplicity/cardinality of associations

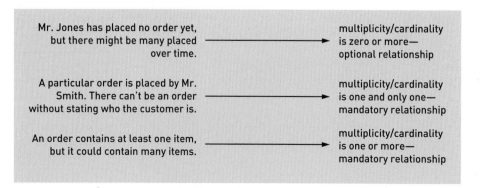

Mr. Jones has placed no order yet, but there might be many placed over time.	→ multiplicity/cardinality is zero or more— optional relationship
A particular order is placed by Mr. Smith. There can't be an order without stating who the customer is.	→ multiplicity/cardinality is one and only one— mandatory relationship
An order contains at least one item, but it could contain many items.	→ multiplicity/cardinality is one or more— mandatory relationship

binary associations   associations between exactly two distinct types of things

unary association   an association between two instances of the same type of thing

ternary association   an association between exactly three distinct types of things

***n*-ary association**   an association between *n* distinct types of things

data entities   the term used in an ER diagram to describe sets of things or individual things

entity-relationship diagram (ERD)   a diagram consisting of data entities (i.e., sets of things) and their relationships

Sometimes, such an association is read as *an order must be placed by one and only one customer*.

The associations described here are between two different types of things—for example, a customer and an order. These are called **binary associations**. Sometimes, an association is between two things of the same type—for example, the association *is married to*, which is between two people. This type of association is called a **unary association** (and sometimes called a recursive association). Another example of a unary association is an organizational hierarchy in which one organizational unit reports to another organizational unit—the packing department reports to shipping, which reports to distribution, which reports to marketing.

An association can also be among three different types of things, when it is called a **ternary association**, or among any number of different types of things, when it is called an ***n*-ary association**. For example, one particular order might be associated with a specific customer plus a specific sales representative, requiring a ternary association.

Storing information about the associations is just as important as storing information about the specific things. It is important to have information like the name and address of each customer, but it is equally important (perhaps more so) to know what items each customer has ordered.

The Entity-Relationship Diagram

More traditional approaches to system development place a great deal of emphasis on data storage requirements for a new system and use the term **data entities** for the things about which the system needs to store information. Data storage requirements include the data entities, their attributes, and the relationships (called "associations" in UML) among the data entities. A model commonly used by traditional analysts and database analysts is called the **entity-relationship diagram (ERD)**. The ERD is not a UML diagram, but it is very commonly used and is quite similar to the UML domain model class diagram that is discussed later in this chapter.

Examples of ERD Notation

On the entity-relationship diagram, rectangles represent data entities, and the lines connecting the rectangles show the relationships among data entities. **Figure 4-6** shows an example of a simplified entity-relationship diagram with two data entities: Customer and Order. Each Customer can place many Orders, and each Order is placed by one Customer. The cardinality is one-to-many in

FIGURE **4-6**

A simple entity-relationship diagram

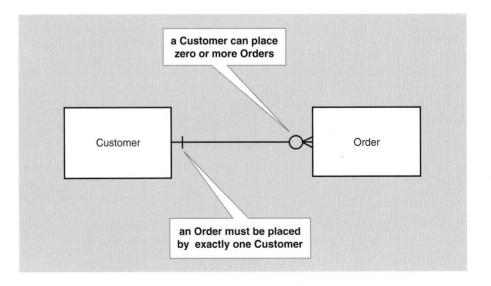

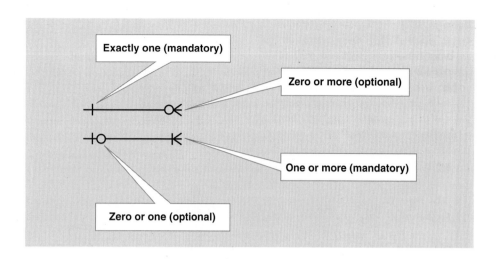

FIGURE 4-7

Cardinality symbols of ERD relationships

one direction and one-to-one in the other direction. The crow's-feet symbol on the line next to the Order data entity indicates many orders. But other symbols on the relationship line also represent the minimum and maximum cardinality constraints. See **Figure 4-7** for an explanation of ERD relationship symbols. The model in Figure 4-6 actually says that a Customer places a minimum of zero Orders and a maximum of many Orders. Reading in the other direction, the model says an Order is placed by at least one and only one Customer. This notation can express precise details about the system. The constraints reflect the business policies that management has defined, and the analyst must discover what these policies are. The analyst does not determine that two customers cannot share one order; management does.

Figure 4-8 shows the model expanded to include the order items (one or more specific items included on the order). Each order contains a minimum of one and a maximum of many items (there could not be an order if it did not contain at least one item). For example, an order might include a shirt, a pair of shoes, and a belt, and each of these items is associated with the order. This example also shows some of the attributes of each data entity: A customer has a customer number, a name, a billing address, and several phone numbers. Each order has an order ID, order date, and so on. Each order item has an item ID, quantity, and price. The attributes of the data entity are listed below the name, with the key identifier listed first, usually followed by "PK" to indicate primary key.

Figure 4-9 shows how the actual data in some transactions might look. John is a customer who has placed two orders. The first order, placed on February 4, was for two shirts and one belt. The second order, placed on March 29, was for one pair of boots and two pairs of sandals. Mary is a customer who has not yet placed an order. Recall that a customer might place zero or more orders. Therefore, Mary is not associated with any orders. Finally, Sara placed an order on March 30 for three pairs of sandals. The diagram shown in Figure 4-9 is sometimes referred to as a semantic net. A **semantic net** shows specific objects that belong to a class or data entity and the links among them.

semantic net a graphical representation of an individual data entity and its relationship with other individual data entities

FIGURE 4-8

An expanded ERD with attributes shown

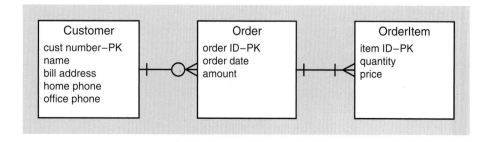

100 PART 2 ■ Systems Analysis Activities

FIGURE **4-9**

Semantic net of customers, orders, and order items consistent with the expanded ERD

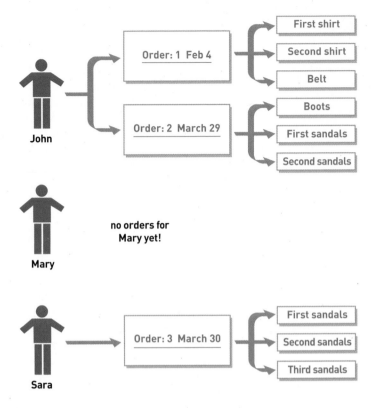

A semantic net is useful for thinking through and verifying the entities and relationships in an ERD and the classes and associations in a class diagram (discussed next).

Another example is shown in **Figure 4-10**. This ERD is for a bank that has many branches. Each branch has one or more accounts. Each account is owned by one customer and results in one or more transactions. There are a few other issues to consider in the bank example. First, there is no data entity named Bank. That is because the ERD shows data storage requirements for the bank. There is only one bank. Therefore, there is no need to include Bank in the model. This is a general rule that applies to ERDs. If the system were for state bank regulators, then Bank would be an important data entity because there are lots of banks under the state regulators' jurisdiction.

FIGURE **4-10**

An ERD for a bank with many branches

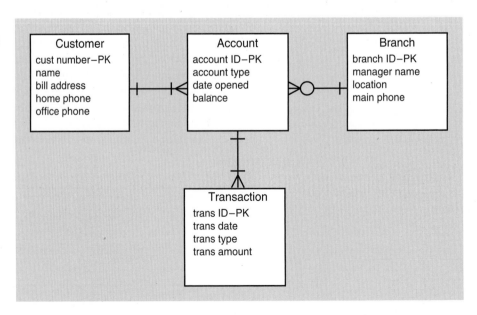

Look again at the cardinality. Note that a customer must have at least one account. The rationale for this is that the bank would not add a customer unless he or she were adding an account. Note also that the branch can have zero accounts. A branch might be added long before it opens its doors, so it is possible that it is does not have any accounts. Additionally, there might be some branches that do not have accounts, such as a kiosk at a university or airport. Note that an account must have at least one transaction. The rationale is that opening a new account requires an initial deposit, which is a transaction. It is important to recognize that questions about the cardinality and minimum and maximum cardinality constraints need to be discussed and reviewed with stakeholders.

The Domain Model Class Diagram

Many current approaches to system development use the term *class* rather than *data entity* and use concepts and notations based on UML to model the things in the problem domain. These concepts come from the object-oriented approach to systems. A **class** is a category or classification used to describe a collection of objects. Each object belongs to a class. Therefore, students Mary, Joe, and Maria belong to the class Student. Classes that describe things in the problem domain are called **domain classes**. Domain classes have attributes and associations. Multiplicity (called cardinality in an ERD) applies among classes. Initially, when defining requirements, the approach to modeling using an ERD or UML is very similar.

The UML **class diagram** is used to show classes of objects for a system. One type of UML class diagram that shows the things in the users' problem domain is called the **domain model class diagram**. Another type of UML class diagram is called the design class diagram, and it is used when designing software classes. You will learn about the design class diagram in Chapter 10.

On a class diagram, rectangles represent classes, and the lines connecting the rectangles show the associations among classes. **Figure 4-11** shows such a symbol for a single domain class: Customer. The domain class symbol is a rectangle with two sections. The top section contains the name of the class, and the bottom section lists the attributes of the class. Later, you will learn that the design class symbol includes a third section at the bottom for listing methods of the class; methods do not apply to problem domain classes.

Class names and attribute names use **camelback notation**, in which the words run together without a space or underscore. Class names begin with a capital letter; attribute names begin with a lowercase letter (see Figure 4-11). Class diagrams are drawn by showing classes and associations among classes. The examples used previously for the entity-relationship diagram are redrawn by using UML domain class diagram notation in the following section so you can compare them. Additionally, more complex issues about classes and associations are illustrated in domain model class diagrams.

---

class a category or classification of a set of objects or things

domain classes classes that describe objects from the problem domain

class diagram a diagram consisting of classes (i.e., sets of objects) and associations among the classes

domain model class diagram a class diagram that only includes classes from the problem domain

camelback notation or **camelcase notation** when words are concatenated to form a single word and the first letter of each embedded word is capitalized

---

FIGURE **4-11**

The UML domain class symbol with name and attributes

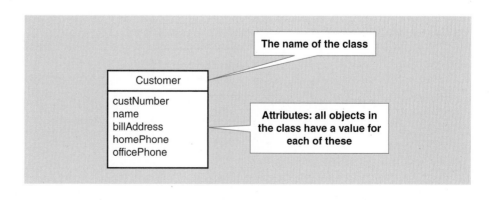

FIGURE **4-12**
A simple domain model class diagram

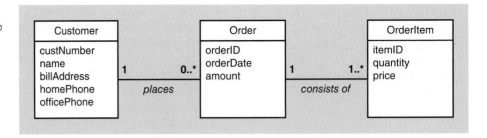

Domain Model Class Diagram Notation

Figure 4-12 shows a simplified domain model class diagram with three classes: Customer, Order, and OrderItem (just like the example of an ERD shown in Figure 4-9). Here, each class symbol includes two sections. In diagram notation, we see that each Customer can place many Orders (a minimum of zero and a maximum of many) and that each Order is placed by one Customer. The associations *places* and *consists of* can be included on the diagram for clarity, as shown in Figure 4-12, but this detail is optional. The multiplicity is one-to-many in one direction and one-to-one in the other direction. The multiplicity notation, shown as an asterisk on the line next to the Order class, indicates many orders. The other association shows that an Order consists of one or more OrderItems, and each OrderItem is associated with one Order.

See **Figure 4-13** for a summary of multiplicity notation.

Figure 4-14 shows another example of a domain model class diagram, this one for the bank with multiple branches that was discussed earlier and shown as an ERD. In this example, the UML notation for indicating an attribute that is an identifier or key is {key}.

Figure 4-15 shows an example of a domain model class diagram with a many-to-many association. At a university, courses are offered as course sections, and a student enrolls in many course sections. Each course section contains many students. Therefore, the association between CourseSection and Student is many-to-many. There are situations in which many-to-many associations are appropriate, and they can be modeled as shown.

However, on closer analysis, analysts often discover that many-to-many associations involve additional data that are important and must be stored. For example, in Figure 4-15, where is the grade that each student receives for the course stored? This is important data, and although the model indicates which course section a student took, it does not have a place for the grade. The solution is to add a domain class to represent the association between student and course section; this is called an **association class**. The association class is given the missing attribute. **Figure 4-16** shows the expanded class diagram,

association class an association that is also treated as a class; often required in order to capture attributes for the association

FIGURE **4-13**
UML notation for multiplicity of associations

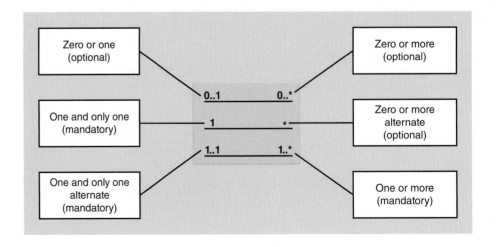

FIGURE **4-14**

A domain model class diagram for a bank

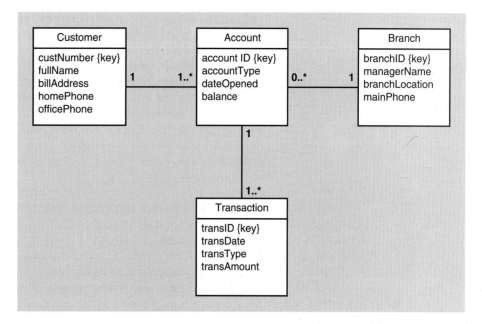

FIGURE **4-15**

A university course enrollment domain model class diagram with a many-to-many association

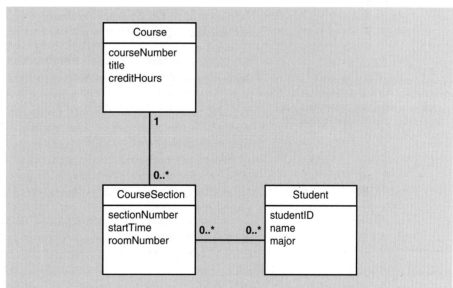

FIGURE **4-16**

A refined university course enrollment domain model class diagram with an association class

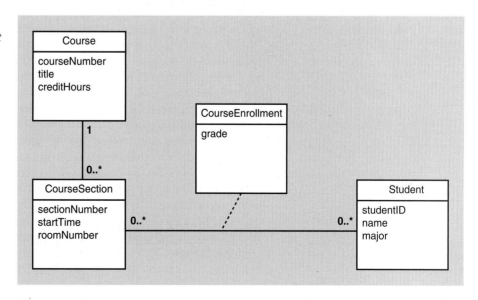

with an association class named CourseEnrollment, which has an attribute for the student's grade. A dashed line connects the association class with the association line between the CourseSection and Student classes.

Reading the association in Figure 4-16 from left to right, one course section has many course enrollments—each with its own grade—and each course enrollment applies to one specific student. Reading from right to left, one student has many course enrollments—each with its own grade—and each course enrollment applies to one specific course section. A database implemented by using this model will be able to produce grade lists showing all students and their grades in each course section as well as grade transcripts showing all grades earned by each student.

More Complex Issues about Classes of Objects

Previously, we discussed associations between domain classes. In UML, an association is one of many types of relationships, so we need to be more precise when discussing UML diagrams than when discussing ERDs. For example, the use case diagram introduced in Chapter 3 shows the ≪includes≫ relationship between use cases. With class diagrams, there are three types of relationships among classes of objects: association relationships (which we have already discussed), generalization/specialization relationships, and whole/part relationships. This section discusses generalization/specialization relationships and whole-part relationships and shows how they are represented in UML class diagrams.

Generalization/Specialization Relationships

generalization/specialization relationship a type of hierarchical relationship in which subordinate classes are subsets of objects of the superior classes; an inheritance hierarchy

Generalization/specialization relationships are based on the idea that people classify things in terms of similarities and differences. Generalizations are judgments that group similar types of things. For example, there are several types of motor vehicles: cars, trucks, and tractors. All motor vehicles share certain general characteristics, so a motor vehicle is a more general class. Specializations are judgments that group different types of things. For example, special types of cars include sports cars, sedans, and sport utility vehicles. These types of cars are similar in some ways yet different in other ways. Therefore, a sports car is a special type of car.

superclass the superior or more general class in a generalization/specialization relationship

subclass the subordinate or more specialized class in a generalization/specialization relationship

A generalization/specialization relationship is used to structure or rank these things from the more general to the more special. As discussed previously, classification refers to defining classes of things. Each class of things in the hierarchy might have a more general class above it, called a **superclass**. At the same time, a class might have a more specialized class below it, called a **subclass**. In **Figure 4-17**, the class Car has three subclasses and one superclass (MotorVehicle). UML class diagram notation uses a triangle that points to the superclass to show a generalization/specialization hierarchy.

We mentioned that people structure their understanding by using generalization/specialization relationships. In other words, people learn by refining the classifications they make about some field of knowledge. A knowledgeable banker can talk at length about special types of loans and deposit accounts. A knowledgeable merchandiser like John Blankens at Ridgeline Mountain Outfitters can talk at length about special types of outdoor activities and clothes. Therefore, when asking users about their work, the analyst is trying to understand the knowledge the user has about the work, which the analyst can represent by constructing generalization/specialization relationships. At some level, the motivation for the new CSMS project at RMO started with John's recognition that Ridgeline Mountain Outfitters might handle many special types of sales with a new system (online sales, telephone sales, and in-store sales). These special types of sales are shown in **Figure 4-18**.

inheritance the concept that specialization classes inherit the attributes of the generalization class

Inheritance allows subclasses to share characteristics of their superclass. Returning to Figure 4-17, a car is everything any other motor vehicle is

FIGURE **4-17** *Generalization/specialization relationships for motor vehicles*

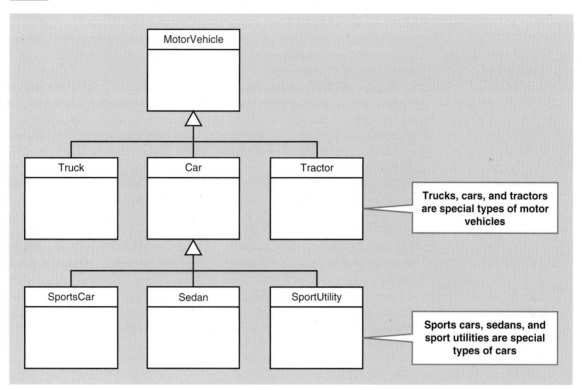

FIGURE **4-18**

Generalization/specialization relationships (inheritance) for sales

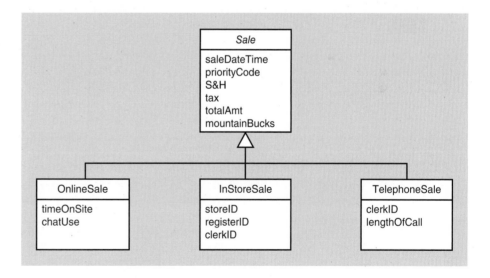

but also something special. A sports car is everything any other car is but also something special. In this way, the subclass "inherits" characteristics. In the object-oriented approach, inheritance is a key concept that is possible because of generalization/specialization hierarchies. Sometimes, these are referred to as inheritance relationships.

In Figure 4-18, attributes are included for each class. Each member of the Sale class has a saleDateTime attribute and a priorityCode attribute. Each InStoreSale has a storeID, clerkID, and registerID, but an OnlineSale and a TelephoneSale have other attributes. OnlineSale, InStoreSale, and TelephoneSale inherit the attributes from Sale, plus they have some special attributes of their own. An OnlineSale actually has eight attributes (six from Sale and two additional). An InStoreSale has nine attributes, and a TelephoneSale has eight attributes.

abstract class   a class that describes a category or set of objects but that never includes individual objects or instances

concrete class   a class that allows individual objects or instances to exist

Note that in Figure 4-18 the class name Sale is in italics; that is because it is an abstract class. An **abstract class** is a class that exists so subclasses can inherit from it. There is never an actual object simply called a Sale. Each sale must be one of the three subclasses. A **concrete class** is a class that does have actual objects. Sometimes, a superclass is abstract; sometimes, it is concrete depending on the intention of the analyst.

Figure 4-19 shows an extension of the previous example of a bank with multiple branches to indicate that there are two types of accounts: a SavingsAccount and a CheckingAccount. The abstract class *Account* is in italics, indicating that it is an abstract class. Rather than including an attribute for account type, the subclasses represent different types of accounts. Each subclass has its own special attributes that do not apply to the other subclasses. A SavingsAccount has four attributes, and a CheckingAccount has five attributes. Note that each subclass also inherits an association with a Customer, optionally a Branch, and one or more Transactions.

Whole-Part Relationships

whole-part relationship   a relationship between classes in which one class is a part or a component portion of another class

aggregation   a type of whole-part relationship in which the component parts also exist as individual objects apart from the aggregate

Another way that people structure information about things is by defining them in terms of their parts. For example, learning about a computer system might involve recognizing that the computer is actually a collection of parts: processor, main memory, keyboard, disk storage, and monitor. A keyboard is not a special type of computer; it is part of a computer, but it is also something separate. **Whole-part relationships** are used to show an association between one class and other classes that are parts of that class.

There are two types of whole-part relationships: aggregation and composition. **Aggregation** refers to a type of whole-part relationship between the

FIGURE **4-19**

An expanded domain model class diagram for the bank, with subclasses for types of accounts

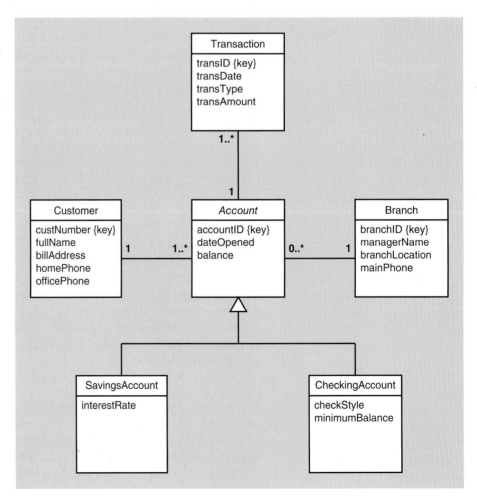

FIGURE **4-20** *Whole-part (aggregation) relationships between a computer and its parts*

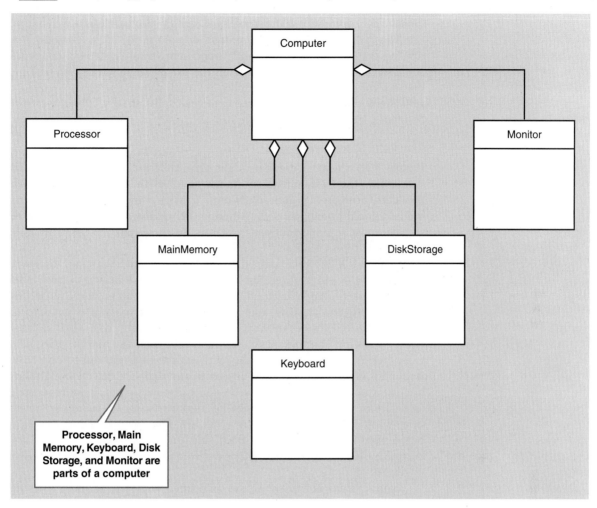

aggregate (whole) and its components (parts), where the parts can exist separately. **Figure 4-20** demonstrates the concept of aggregation in a computer system, with the UML diamond symbol representing aggregation. **Composition** refers to whole-part relationships that are even stronger, where the parts, once associated, can no longer exist separately. The UML diamond symbol is filled in to represent composition.

composition   a type of whole-part relationship in which the component parts cannot exist as individual objects apart from the total composition

Whole-part relationships—aggregation and composition—mainly allow the analyst to express subtle distinctions about associations among classes. As with any association relationship, multiplicity can apply, such as when a computer has one or more disk storage devices.

The UML class diagram examples we have seen so far are domain model class diagrams. The design class diagram is a refinement of the class diagram and is used to represent software classes in the new system. You will learn about the process of converting the domain model class diagram to a design class diagram in Chapter 10.

The Ridgeline Mountain Outfitters Domain Model Class Diagram

The RMO CSMS involves many domain classes and many complex association and generalization/specialization relationships. A domain model class diagram for an information system evolves as the project proceeds; and unlike the use case diagrams, where many diagrams are created, there is eventually only one domain model class diagram. Also, unlike the use case diagram, the domain model class diagram is not produced just for presentations. The process of developing and refining the domain model class diagram is how the analyst explores and learns about the problem domain. Therefore, the information depicted in the domain model class diagram is very detailed and rich in specific meaning.

The Ridgeline Mountain Outfitters domain model class diagram is a variation of the customer, order, and order item example shown in Figure 4-12. Most of the domain classes are from the list of nouns developed in Figure 4-2. Because the model is fairly complex, an analyst might start by focusing on one subsystem at a time to reduce the complexity. Eventually, all subsystems can be combined into one domain model.

The RMO Sales Subsystem

Figure 4-21 shows a domain model class diagram for the RMO CSMS Sales subsystem. The Sales subsystem mainly involves the customer, sale, sales items, products, promotions, and accessories. That is a good starting point, but there are additional domain classes. Additionally, recall that the association relationships are just as important as the classes, so these must be identified. There are also special types of sales and a shopping cart.

In Figure 4-21, each customer can be associated with one or more sales. Note that there are three special types of sales shown in the inheritance relationships (in-store sale, online sale, and telephone sale), as discussed in the last section. Therefore, the scope of the Sales subsystem includes in-store, online, and telephone sales processes. A customer can also be associated with an online shopping cart (OnLineCart) for any online sale. There are two special types of carts: the active cart and the on-reserve cart. The minimum multiplicity between customer and cart is zero, meaning there might not be a shopping cart involved—for example, in an in-store or telephone sale. There can be a maximum of two carts for a customer at any one time: an active cart and an on-reserve cart. The on-reserve cart can be remembered from session to session. Each sale and each cart is associated with one customer, so the subclasses inherit the association, just as they inherit the attributes of Sale.

Note that an individual sale is associated with one or more sales items. In the online cart, it is associated with one or more cart items. With an online sale, the sale is created from the cart when the customer checks out. Sale items are created from each cart item. Finally, a sales transaction is created and associated with the sale.

A sale can have one or more sale items, but what is each item? An association between each sale item and an inventory item answers the question. Each sale item is for a specific inventory item, meaning a specific size and color of the item, such as a shirt or coat. An inventory item has an attribute for the quantity on hand of that size and color. Because there are many colors and sizes (each with its own quantity), each inventory item is associated with a product item that describes the item generally (gender, description, supplier, manufacturer, and picture). Each product item is associated with many inventory items, and each inventory item is associated with many sale items.

A product item can be part of many promotions, and a promotion can include many product items, making a many-to-many association. An association class is added to store information about the price of each item in each promotion. Each product item might have many accessories, and an accessory might apply to many product items. Here, there is no defined association class for the many-to-many association. Note that this association might also be modeled as a unary (recursive) association. Finally, each product item can have many customer comments, which are reviewed during a sale.

The RMO Customer Account Subsystem

The Customer Account subsystem domain model class diagram is shown in **Figure 4-22.** Note that there are some classes repeated here that are also on the

FIGURE **4-21** *RMO Sales subsystem domain model class diagram*

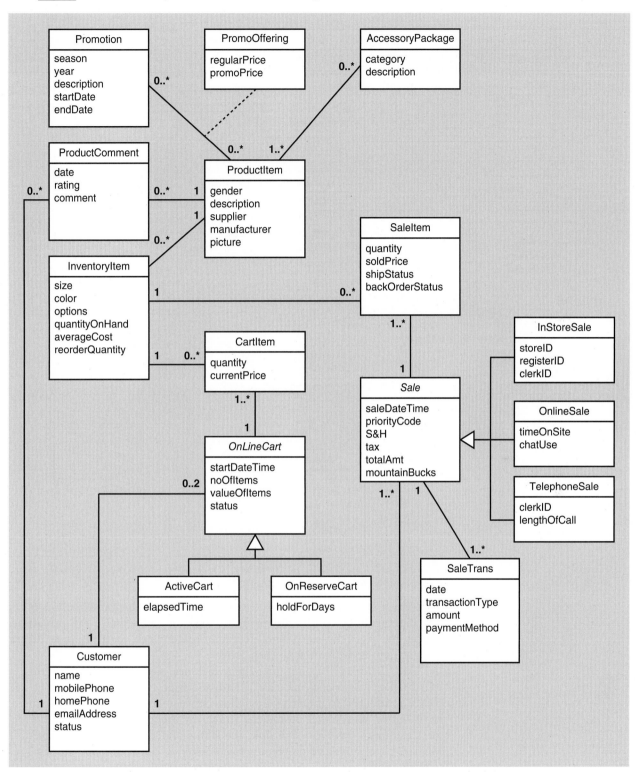

Sales subsystem domain model class diagram. For example, Customer is important to both subsystems. Note that Sale and SaleTrans are also included here. In order to make account adjustments and report on all payments and returns for a customer, all sales and sales transactions need to be referenced. Repeating domain classes in several subsystems does not mean there is redundancy.

FIGURE **4-22** *RMO Customer Account subsystem domain model class diagram*

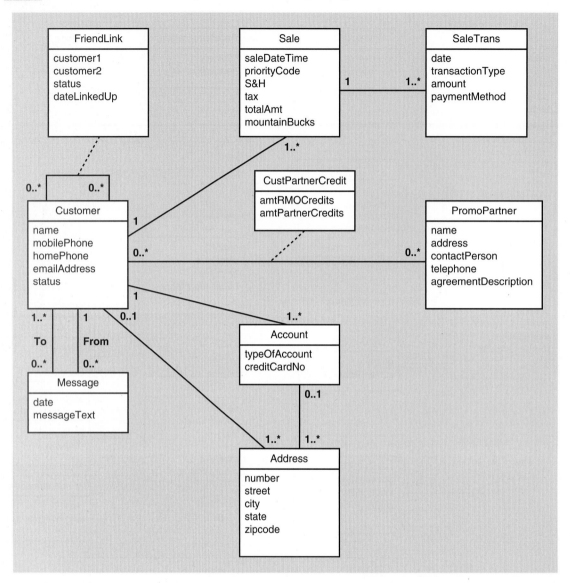

In complex domain models, it is easier to do the modeling and analysis in separate diagrams before merging them all together. Sometimes, the project team divides the work by subsystem, so each would work on a separate diagram, being sure to coordinate with each other.

The Customer Account subsystem includes messages, partner credits, and friend links. The FriendLink class is an association class, but unlike other examples, it is attached to a unary association between customers. Each customer can be linked to many other customers, shown by the association line at the top of the Customer class. For each link, the status and dateLinkedUp is stored. The Message class is handled differently. Each customer can send many messages, each to many other customers. Similarly, each customer can receive many messages.

The Complete RMO Domain Model Class Diagram

The analysts at RMO may continue to model each subsystem separately. The exercises at the end of this chapter ask you to create those other subsystem diagrams. The final domain model class diagram for the RMO CSMS is shown in **Figure 4-23**. Classes not shown before include Shipper, Shipment, ReturnItem, and Suggestion.

FIGURE **4-23** *Complete RMO CSMS domain model class diagram*

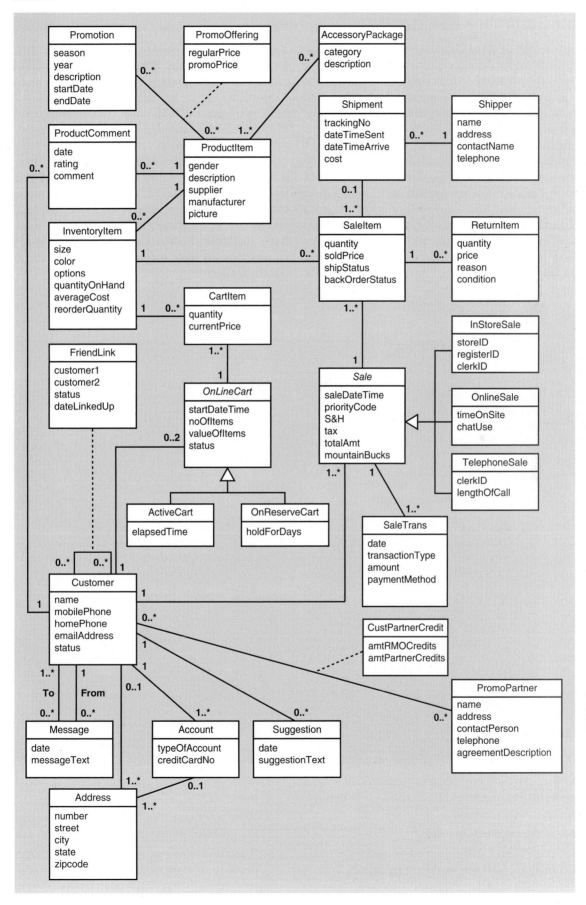

Chapter Summary

This is the second of three chapters that present techniques for modeling a system's functional requirements, highlighting the tasks that are completed during the analysis activity *Define requirements*. Use cases and things in the user's work environment are key concepts common to all approaches to system development. This chapter discusses data entities and domain classes as two terms for things in the work environment. Two techniques are demonstrated for identifying things in the problem domain: the brainstorming technique and the noun technique. The entity-relationship diagram (ERD) is used by traditional analysts and by database analysts to model things in the problem domain. An ERD shows data entities, attributes, and relationships. The UML class diagram is used for the same purpose by analysts using UML, referred to as the *domain model class diagram*. The domain model class diagram models domain classes, attributes, and associations. Multiplicity refers to the number of association links between classes. UML and the domain model class diagram can be extended to include three types of relationships: association relationships, generalization/specialization relationships (inheritance), and whole-part relationships. Additional concepts of importance in domain model class diagrams are superclasses, subclasses, abstract classes, and concrete classes. Domain classes are not software classes; therefore, they do not have methods. Design class diagrams show software classes that do have methods. Design classes are introduced in Chapter 10.

Key Terms

abstract class   106

aggregation   106

association   96

association class   102

attributes   95

binary associations   98

brainstorming technique   93

camelback notation   101

cardinality   97

class   101

class diagram   101

composition   107

compound attribute   96

concrete class   106

data entities   98

domain classes   101

domain model class diagram   101

entity-relationship diagram (ERD)   98

generalization/specialization relationships   104

identifier   96

Inheritance   104

key   96

multiplicity   97

multiplicity constraints   97

n-ary association   98

noun technique   94

problem domain   92

relationship   97

semantic net   99

subclass   104

superclass   104

ternary association   98

unary association   98

whole-part relationships   106

Review Questions

1. What are the two key concepts—one from Chapter 3 and one from this chapter—that define functional requirements?

2. What is the problem domain?

3. What is a "thing" called in models used by traditional analysts and database analysts?

4. What is a "thing" called in newer approaches that use UML?

5. What are two techniques for identifying things in the problem domain?

6. What are some examples of tangible things in the problem domain of a restaurant?

7. What are some sites or locations in the problem domain of a restaurant?

8. What are some roles played by people in the problem domain of a restaurant?

9. What are the main steps of the brainstorming technique?

10. Explain why identifying nouns helps identify things in the problem domain?

11. What are the main steps of the noun technique?

12. What is an attribute, an identifier or key, and a compound attribute?

13. What is an association, and what system development standard defines it?

14. How would you describe or name the association between a ship and a captain?

15. What is the term used for association by traditional analysts and database analysts?

16. What is multiplicity, and what is the other term used by traditional analysts and database analysts?

17. What is the minimum multiplicity for the association that reads *a customer places zero or more orders*?

18. What is the maximum multiplicity for the association that reads *an order is placed by exactly one customer*?

19. What are some examples of multiplicity constraints?

20. What are the three types of associations, and which is the most commonly used?

21. What are the three key parts of an entity-relationship diagram (ERD)?

22. Sketch a simple ERD that shows a team has zero or more players and each player is on one and only one team.

23. Sketch a semantic net that shows two teams and five players based on your ERD.

24. What is a class, a domain class, and the key parts of a class diagram?

25. What does a domain model class diagram show about system requirements, and how is it different from an ERD?

26. List appropriate UML class names by using the camelback notation for the following classes: graduate student, undergraduate major, course instructor, and final exam feedback.

27. List appropriate UML attribute names for the following attributes: student name, course grade, major name, and final exam quantity score.

28. Draw a simple domain model class diagram for the example in question 22 where a team has zero or more players and each player is on one and only one team.

29. What is an association class? Extend the domain model class diagram for teams and players about to show a record of game statistics for each player in each game.

30. In UML, what are three types of relationships found on a class diagram?

31. What is a generalization/specialization relationship, and what object-oriented terms does it illustrate?

32. Compare/contrast superclass and subclass. Compare/contrast abstract class and concrete class.

33. What is a whole-part relationship, and why does it show multiplicity?

34. Compare/contrast aggregation and composition for a whole-part relationship.

Problems and Exercises

1. Draw an entity-relationship diagram, including minimum and maximum cardinality, for the following: The system stores information about two things: cars and owners. A car has attributes for make, model, and year. The owner has attributes for name and address. Assume that a car must be owned by one owner and an owner can own many cars, but an owner might not own any cars (perhaps she just sold them all, but you still want a record of her in the system).

2. Draw a class diagram for the cars and owners described in exercise 1, but include subclasses for sports car, sedan, and minivan, with appropriate attributes.

3. Consider the domain model class diagram shown in Figure 4-16—the refined diagram showing course enrollment with an association class. Does this model allow a student to enroll in more than one course section at a time? Does the model allow a course section to contain more than one student? Does the model allow a student to enroll in several sections of the same course and get a grade for each enrollment? Does the model store information about all grades earned by all students in all sections?

4. Again consider the domain model class diagram shown in Figure 4-16. Add the following to the diagram and list any assumptions you had to make: A faculty member usually teaches many course sections, but some semesters, a faculty member may not teach any. Each course section must have at least one faculty member teaching it, but sometimes, faculty teams teach course sections. Furthermore, to make sure that all course sections are similar, one faculty member is assigned as course coordinator to oversee the course, and each faculty member can be the coordinator of many courses.

5. If the domain model class diagram you drew in exercise 4 showed a many-to-many association between faculty member and course section, a further look at the association might reveal the need to store some additional information. What might this information include? (Hint: Does the instructor have specific office hours for each course section? Do you give an instructor some sort of evaluation for each course section?) Expand the domain model class diagram to allow the system to store this additional information.

6. Consider a system that needs to store information about computers in a computer lab at a university, such as the features and location of each computer. What are the domain classes that might be included in a model? What are some of the associations among these classes? What are some of the attributes of each class? Draw a domain model class diagram for this system.

7. Consider the domain model class diagram for the RMO CSMS Sales subsystem shown in Figure 4-21. If an InStoreSale is created, how many attributes does it have? If an OnlineSale is created, how many attributes does it have? If an existing customer places a telephone order for one item, how many new objects are created overall for this transaction? Explain.

8. Again consider the domain model class diagram shown in Figure 4-21. How many attributes does an active cart object have? Can an on-reserve cart contain cart items? Explain.

9. A product item for RMO is not the same as an inventory item. A product item is something like a men's leather hunting jacket supplied by Leather 'R' Us. An inventory item is a specific size and color of the jacket—like a size medium brown leather hunting jacket. If RMO adds a new jacket to its catalog and six sizes and three colors are available in inventory, how many objects need to be added overall? Explain.

10. Consider the domain model class diagram shown in **Figure 4-24**, which includes classes for college, department, and faculty members.
    a. What kind of UML relationships are shown in the model?
    b. How many attributes does a "faculty member" have? Which (if any) have been inherited from another class?

FIGURE **4-24**

Domain model class diagram for a university

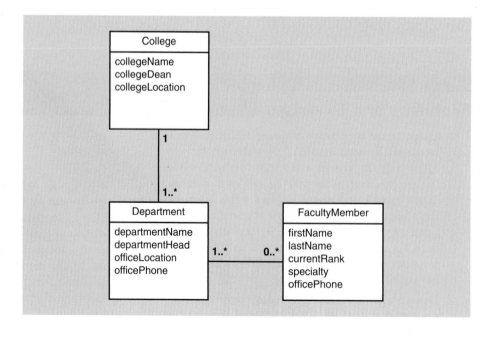

c. If you add information about one college, one department, and four faculty members, how many objects do you add to the system?

d. Can a faculty member work in more than one department at the same time? Explain.

e. Can a faculty member work in two departments at the same time, where one department is in the college of business and the other department is in the college of arts and sciences? Explain.

11. Review information about your own university. Create generalization/specialization hierarchies by using the domain model class diagram notation for (1) types of faculty, (2) types of students, (3) types of courses, (4) types of financial aid, and (5) types of housing. Include attributes for the superclass and the subclasses in each case.

12. Consider the classes involved when modeling a car and all its parts. Draw a domain model class diagram that shows the whole-part relationships involved, including multiplicity. Which type of whole-part relationships are involved?

13. Refer to the complete RMO CSMS domain model class diagram shown in Figure 4-23. Based on that model and on the discussion of subsystems in Chapter 3, draw a domain model class diagram for the CSMS Marketing subsystem.

14. Again based on the complete RMO CSMS domain model class diagram shown in Figure 4-23, draw a domain model class diagram for the CSMS Order Fulfillment subsystem.

Case Study

Metropolitan Car Service Bureau

Metropolitan Car Service Bureau needs a system that keeps car service records. The company's analyst has provided information about the problem domain in the form of notes. Your job is to use those notes to draw the domain model class diagram. The analyst's notes are as follows:

■ The Owner class has attributes name and address.

■ The Vehicle class is an abstract class that has attributes VIN, model, and model year.

■ There are two types of vehicles, cars and trucks:

  ■ Car has additional attributes for the number of doors and luxury level.

  ■ Truck has an additional attribute for cargo capacity.

■ The Manufacturer class has attributes name and location.

■ The Dealer class has attributes name and address.

A service record is an association class between each vehicle and a dealer, with attributes service date and current mileage. A warranty service record is a special type of service record with an additional attribute: eligibility verification. Each service record is associated with a predefined service type, with attributes type ID, description, and labor cost. Each service type is associated with zero or more parts, with attributes part ID, description, and unit cost. Parts are used with one or more service types.

An owner can own many vehicles, and a vehicle can be owned by many owners. An owner and a vehicle are entered into the system only when an owned vehicle is first serviced by a dealer. Vehicles are serviced many times at various dealers, which service many vehicles.

1. Draw a UML domain model class diagram for the system as described here. Be as specific and accurate as possible, given the information provided. If needed information is not given, make realistic assumptions.

2. Answer True or False to the following statements, which are based on the domain model. You may want to draw a semantic net to help you think through the questions.

  a. This domain model is for a single car dealer service department.

  b. This domain model is for a single car manufacturer.

  c. A vehicle can have service records with more than one dealer.

  d. A dealer can service vehicles from more than one manufacturer.

  e. Current mileage is recorded for service records and warranty service records.

  f. An owner can have each of his or her cars serviced by a different dealer.

  g. A warranty service for a car can include many parts.

  h. A vehicle can be made by more than one manufacturer.

Community Board of Realtors

In Chapter 3, you identified use cases for the Board of Realtors Multiple Listing Service (MLS) system, which supplies information that local real estate agents use to help them sell houses to their customers. During the month, agents list houses for sale (listings) by contracting with homeowners. Each agent works for a real estate office, which sends information on listings to the multiple listing service. Therefore, any agent in the community can get information on the listing. Much of the information is available to potential customers on the Internet.

Information on a listing includes the address, year built, square feet, number of bedrooms, number of bathrooms, owner name, owner phone number, asking price, and status code. It is also important to have information on the listing agent, such as name, office phone, cell phone, and e-mail address. Agents work through a real estate office, so it is important to know the office name, office manager name, office phone, and street address.

1. Based on the information here, draw a domain model class diagram for the MLS system. Be sure to consider what information needs to be included versus information that is not in the problem domain. For example, is detailed information about the owner, such as his employer or his credit history, required in the MLS system? Is that information required regarding a potential buyer?

2. Draw a second domain model class diagram that adds the following specifications. First, there are two types of listings: a listing for sale and a listing for lease. Additionally, a listing might include no structures, such as vacant land, or it might include more than one structure, such as a main house and a guest house, each with separate values for square footage, number of bedrooms, and number of bathrooms.

3. Draw a third domain model class diagram that assumes a listing might have multiple owners. Additionally, a listing might be shared by two or more agents, and the percentage of the commission that each agent gets from the sale can be different for each agent.

The Spring Breaks 'R' Us Travel Service

In Chapter 3, you identified use cases for the Social Networking subsystem SBRU is researching. Let us assume you were thinking about a number of potential domain classes that might be involved. For example, there would need to be information about a traveler attending a resort for a particular week. The traveler would be assigned to a room along with roommates but might also be connected to other friends. There might be different interests or hobbies a traveler can associate with in the hopes of connecting to others. The resort has many locations where a traveler might be hanging out at any given time, and a traveler can note whether the location is "liked." People might schedule a party at a location and invite specific friends.

1. For the Social Networking subsystem as described here, list the domain classes and their attributes that should be included in the Social Networking subsystem. Be creative and add those you think should be included to make the system useful and appealing.

2. Based on the domain classes you identified, draw a domain model class diagram showing domain classes with attributes and associations with multiplicity.

On the Spot Courier Services

On the Spot courier services grew and changed over the years. At first, Bill received requests for package pickups on his mobile phone, recorded that information in a log, and would then drive around to retrieve all the packages later in the day. However, he soon discovered that with another driver, it was difficult to coordinate pickups between the two of them from his van. It was not long before he reorganized his business and turned the warehouse employee into a driver. Then, he stayed in the warehouse himself, and his two employees made all the pickups and deliveries. This worked well because he could control and coordinate

(continued on page 117)

(continued from page 116)

the pickups and deliveries better. It was also easier for him to receive pickup requests working at a desk rather than trying to do it while driving a delivery van.

As he thought about how his business was growing and the services he provided to his customers, Bill began to itemize the kinds of information he would need to maintain.

Of course, he needed to maintain information about his customers. Some of his customers were businesses; some were individuals. He needed to have basic address and contact information for every customer. Also, for his corporate customers, he needed to identify a primary contact person. It was mostly his corporate customers who wanted to receive monthly statements listing all their shipments during the month and the total cost. Bill needed to distinguish which customers paid cash and which wanted monthly statements. In fact, for those that paid monthly, he needed to keep a running account of such things as when they were last billed, when they paid, and any outstanding balances. Finally, when payments were received, either for individual shipments or from monthly invoices, he needed to record information about the payment: type of payment, date, and amount. Although this was not a sophisticated billing and payment system, Bill thought it would suffice for his needs.

Next, he started thinking about his packages and shipments. At the time that a request for a pickup came in, he needed to keep track of it as some type of delivery request or delivery order. At that point in time, Bill mostly needed to know who the customer was, where the pickup location was, and what date and time the package(s) would be ready for pickup. He also recorded the date and time that he received the order. A delivery order was considered "open" until the delivery van arrived at the pickup location and the packages were all retrieved. At that point, the delivery order was satisfied.

Once the packages were retrieved, each package needed to be uniquely identified. Bill needed to know when it was picked up and which delivery person picked it up. Other important information was the "deliver to" entity name and the address. He also needed to identify the type of delivery. Some packages were high priority, requiring same-day delivery. Others were overnight. Of course, the weight and cost were recorded so the customer could either pay or have it added to the monthly invoice.

In the courier and delivery business, one of the most important information requirements is the date and time stamp. For each package, it is important to know when it was picked up, when it arrived at the warehouse, when it went back out on the delivery run, and when it was delivered. When possible, it is also important to have names associated with each of these events.

1. Using the noun technique, read through this case and identify all the nouns that may be important for this system. You may also find it helpful to read back through the case descriptions in the previous chapters.

2. Once you have identified all the nouns, identify which are classes and which are attributes of these primary classes. Begin constructing a class diagram based on the classes and attributes you have identified.

3. Now that you have identified the classes, determine what the relationships should be among the classes. Add multiplicity constraints, being especially cognizant of zero-to-many versus one-to-many differences.

4. Finalize the class diagram, including all your classes, attributes, primary keys, relationships, and multiplicity constraints.

Sandia Medical Devices

Initial discussions about the functional requirements resulted in an initial domain model class diagram for Sandia Medical Devices' Real-Time Glucose Monitoring (RTGM) system (see **Figure 4-25**). After consultations with system stakeholders, the following potential changes to the diagram are being considered:

■ Include additional medical personnel (nurses and physicians' assistants, at a minimum).

■ Include alerts sent by the system to medical personnel and messages sent by medical personnel to the patient.

1. Modify the diagram to incorporate the changes under consideration. You may need to use association classes and generalization/specialization (inheritance).

2. Is a set of abstract and concrete classes needed to represent variations among cell phones? Why or why not?

(continued on page 118)

(continued from page 117)

FIGURE **4-25**

Initial domain model class diagram for Sandia RTGM system

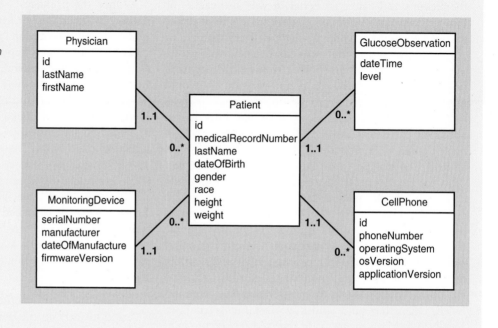

Further Resources

Classic and more recent texts include the following:

Peter Rob and Carlos Coronel, *Database Systems: Design, Implementation, and Management*, (7th ed.). Course Technology, 2007.

Craig Larman, *Applying UML and Patterns* (3rd ed.). Prentice Hall, 2005.

Grady Booch, Ivar Jacobson, and James Rumbaugh, *The Unified Modeling Language User Guide.* Addison-Wesley, 1999.

5

Extending the Requirements Models

Chapter Outline

- Use Case Descriptions
- Activity Diagrams for Use Cases
- The System Sequence Diagram—Identifying Inputs and Outputs
- The State Machine Diagram—Identifying Object Behavior
- Integrating Requirements Models

Learning Objectives

After reading this chapter, you should be able to:

- Write fully developed use case descriptions
- Develop activity diagrams to model flow of activities
- Develop system sequence diagrams
- Develop state machine diagrams to model object behavior
- Explain how use case descriptions and UML diagrams work together to define functional requirements

OPENING CASE

Electronics Unlimited: Integrating the Supply Chain

Electronics Unlimited is a warehousing distributor that buys electronic equipment from various suppliers and sells it to retailers throughout the United States and Canada. It has operations and warehouses in Los Angeles, Houston, Baltimore, Atlanta, New York, Denver, and Minneapolis. Its customers range from large nationwide retailers, such as Target, to medium-sized independent electronics stores.

Many of the larger retailers are moving toward integrated supply chains. Information systems used to be focused on processing internal data; however, today, these retail chains want suppliers to become part of a totally integrated supply chain system. In other words, the systems need to communicate between companies to make the supply chain more efficient.

To maintain its position as a leading wholesale distributor, Electronics Unlimited has to convert its system to link with its suppliers (the manufacturers of the electronic equipment) and its customers (the retailers). It is developing a completely new system that uses object-oriented techniques to provide these links. Object-oriented techniques facilitate system-to-system interfaces by using predefined components and objects to accelerate the development process. Fortunately, many of the system development staff members have experience with object-oriented development and are eager to apply the techniques and models to the system development project.

William Jones is explaining object-oriented development to the group of systems analysts who are being trained in this approach.

"We're developing most of our new systems by using object-oriented principles," he tells them. "The complexity of the new system, along with its interactivity, makes the object-oriented approach a natural way to develop requirements. It takes a little different thought process than some

of you may be used to, but the object-oriented models track very closely with the new object-oriented programming languages and frameworks."

William is just getting warmed up.

"This way of thinking about a system in terms of objects is very interesting," he adds. "It is also consistent with the object-oriented programming techniques you learned in your programming classes. You probably first learned to think about objects when you developed screens for the user interface. All the controls on the screen, such as buttons, text boxes, and drop-down boxes, are objects. Each has its own set of trigger events that activate its program functions."

"How does this apply to our situation?" one of the analysts asks.

"You just extend that thought process," William explains. "You think of such things as purchase orders and employees as objects too. We can call them the problem domain or business objects to differentiate them from screen objects, such as windows and buttons. During analysis, we have to find out all the trigger events and methods associated with each business object."

"And how do we do that?" another analyst asks.

"You continue with your fact-finding activities and build a better understanding of each use case," William says. "The way the business objects interact with each other in the use case determines how you identify the initiating activity. We refer to those activities as the messages between objects. The tricky part is that you need to think in terms of objects instead of just processes. Sometimes, it helps me to pretend I am an object. I will say, 'I am a purchase order object. What functions and services are other objects going to ask me to do?' After you get the hang of it, it works very well, and it is enlightening to see how the system requirements unfold as you develop the diagrams."

Overview

The main objective of defining requirements in system development is understanding users' needs, how the business processes are carried out, and how the system will be used to support those business processes. As we indicated in Chapter 2, system developers use a set of models to discover and understand the requirements for a new system. This activity is a key part of systems analysis in the system development process. The first step in the process for developing this understanding requires the fact-finding skills you learned in Chapter 2. Fact-finding activities are also called *discovery activities*, and obviously, discovery must precede understanding.

The models introduced in Chapters 3 and 4 focus on two primary aspects of functional requirements: the use cases and the things involved in users' work. Use cases are identified by using the user goal technique and the event decomposition technique. The UML use case diagram was introduced to show use cases and actors. An information system also needs to record and store information about things involved in the business processes. In a manual system, the information is recorded on paper and stored in a filing cabinet. In an automated system, the information is stored in electronic files or a database. The information storage requirements of a system are documented either with entity-relationship diagrams (ERDs) or with UML domain model class diagrams.

In this chapter, you learn additional techniques and models that will allow you to extend the requirements models to show additional information about the use cases and domain classes for the system. Fully developed use case descriptions, UML activity diagrams, and UML system sequence diagrams (SSDs) are introduced to show more information about each use case. Then, UML state machine diagrams are introduced; these help you show more information about domain classes. Remember, when defining requirements for a system, you will also be doing design and implementation work, as illustrated in the Trade Show application developed in Chapter 1. The next chapter begins covering system design activities.

Use Case Descriptions

A list of use cases and use case diagrams provides an overview of all the use cases for a system. Detailed information about each use case is described with a **use case description**. Brief use case descriptions were introduced in Chapter 3. A use case description lists and describes the processing details for a use case. Implied in all use cases is a person who uses the system. In UML, that person is called an actor, as shown on use case diagrams. An actor is always outside the automation boundary of the system but may be part of the manual portion of the system. By defining actors that way—as those who interact with the system—we can more precisely define the exact interactions to which the automated system must respond. This tighter focus helps define the specific requirements of the automated system itself—to refine them as we move from the event table to the use case details.

Another way to think of an actor is as a role. For example, in the RMO case, the use case *Create customer account* might involve a customer service rep talking to the customer on the phone. Or the customer might be the actor if the customer adds or updates information directly online.

To create a comprehensive, robust system that truly meets users' needs, we must understand the detailed steps of each use case. Internally, a use case includes a whole sequence of steps to complete a business process. Frequently, several variations of the business steps exist within a single use case. The use case *Create customer account* will have a separate flow of activities depending on which actor invokes the use case. The processes for a customer service representative updating information over the phone might be quite different from the processes for a customer updating the information him or herself. Each flow of activities is a valid sequence for the *Create customer account* use case. These different flows of activities are called **scenarios** or sometimes **use case instances**. Thus, a scenario is a unique set of internal activities within a use case and represents a unique path through the use case.

Brief Use Case Descriptions

Depending on an analyst's needs, use case descriptions tend to be written at two separate levels of detail: brief description and fully developed description. Some brief use case descriptions were shown in Chapter 3 (see **Figure 5-1**). A brief description can be used for very simple use cases, especially when the system to be developed is a small, well-understood application. A simple use case would normally have a single scenario and very few—if any—exception conditions. An example would be *Add product comment* or *Send message*.

use case description a textual model that lists and describes the processing details for a use case

scenarios or **use case instances** unique sets of internal activities within use cases

FIGURE **5-1**

Use cases and brief use case descriptions

Use case	Brief use case description
Create customer account	User/actor enters new customer account data, and the system assigns account number, creates a customer record, and creates an account record.
Look up customer	User/actor enters customer account number, and the system retrieves and displays customer and account data.
Process account adjustment	User/actor enters order number, and the system retrieves customer and order data; actor enters adjustment amount, and the system creates a transaction record for the adjustment.

A use case such as *Fill shopping cart* is complex enough that a fully developed description is also written.

Fully Developed Use Case Descriptions

The fully developed description is the most formal method for documenting a use case. One of the major difficulties for software developers is that they often struggle to obtain a deep understanding of the users' needs. But if you create a fully developed use case description, you increase the probability that you thoroughly understand the business processes and the ways the system must support them. **Figure 5-2** is an example of a fully developed use case description of the use case *Create customer account*.

Figure 5-2 also serves as a standard template for documenting a fully developed description for other use cases and scenarios. The first and second compartments are used to identify the use cases and the scenarios within the use cases (if needed) that are being documented. In larger or more formal projects, a unique identifier can also be added for the use case, with an extension identifying the particular scenario. Sometimes, the name of the system developer who produced the form is added.

The third compartment identifies the event that triggers the use case. The fourth compartment is a brief description of the use case or scenario. Analysts may just duplicate the brief description they constructed earlier here. The fifth compartment identifies the actor or actors. The sixth compartment identifies other use cases and the way they are related to this use case. These cross-references to other use cases help document all aspects of the users' requirements.

The seventh compartment identifies stakeholders who are interested parties other than specific actors. They might be users who don't actually invoke the use case but who have an interest in results produced from the use case. For example, in Figure 5-2, the accounting department is interested in accurately capturing billing and credit card information. Although no one in the marketing department actually creates new customer accounts, they do perform statistical analysis of the new customers and create marketing promotions. Thus, marketers have an interest in the data that are captured and stored from the *Create customer account* use case. The sales department is interested in having an easy-to-use and attractive user interface to assure sales aren't lost because of poor user experience. Considering all the stakeholders is important for system developers to ensure that they have understood all requirements.

The eighth and ninth compartments—preconditions and postconditions—provide critical information about the state of the system before and after the use case executes. **Preconditions** identify what the state of the system must be for the use case to begin, including what objects must already exist, what information must be available, and even the condition of the actor prior to beginning the use case.

Postconditions identify what must be true upon completion of the use case. Most importantly, they indicate what new objects are created or updated by the use case and how objects need to be associated. The postconditions are

precondition a condition that must be true before a use case begins

postcondition what must be true upon the successful completion of a use case

FIGURE **5-2** *Fully developed use case description for* Create customer account

Use case name:	*Create customer account.*	
Scenario:	Create online customer account.	
Triggering event:	New customer wants to set up account online.	
Brief description:	Online customer creates customer account by entering basic information and then following up with one or more addresses and a credit or debit card.	
Actors:	Customer.	
Related use cases:	Might be invoked by the *Check out shopping cart* use case.	
Stakeholders:	Accounting, Marketing, Sales.	
Preconditions:	Customer account subsystem must be available. Credit/debit authorization services must be available.	
Postconditions:	Customer must be created and saved. One or more Addresses must be created and saved. Credit/debit card information must be validated. Account must be created and saved. Address and Account must be associated with Customer.	
Flow of activities:	**Actor**	**System**
	1. Customer indicates desire to create customer account and enters basic customer information.	1.1 System creates a new customer. 1.2 System prompts for customer addresses.
	2. Customer enters one or more addresses.	2.1 System creates addresses. 2.2 System prompts for credit/debit card.
	3. Customer enters credit/debit card information.	3.1 System creates account. 3.2 System verifies authorization for credit/debit card. 3.3 System associates customer, address, and account. 3.4 System returns valid customer account details.
Exception conditions:	1.1 Basic customer data are incomplete. 2.1 The address isn't valid. 3.2 Credit/debit information isn't valid.	

important for two reasons. First, they form the basis for stating the expected results for test cases that will be used for testing the use case after it is implemented. For example, in the *Create customer account* use case, it is important to test that a customer record, address record, and account record were successfully added to the database. Second, the objects in postconditions indicate which objects involved in the use case are important for design. You will see in Chapters 10 and 11 that the design of a use case includes identifying and assigning responsibilities to objects that collaborate to complete the use case. In this situation, a customer, one or more addresses, and an account object collaborate to create a new customer account.

The tenth compartment in the template describes the detailed flow of activities of the use case. In this instance, we have shown a two-column version, identifying the steps performed by the actor and the responses required by the system. The item numbers help identify the sequence of the steps. Alternative activities and exception conditions are described in the eleventh compartment.

The numbering of exception conditions also helps tie the exceptions to specific steps in the flow of activities.

Figure 5-3 shows the use case description for the use case *Ship items*. The scenario for this description assumes they are shipping a new sale rather than backordered items from a previous sale. Notice that the use case description minimizes the description of manual work that is done in conjunction with shipping items. Some analysts put that detail in, but others don't because the emphasis is on the interaction with the computer application. In this use case, the preconditions show what existing objects must already exist before the use case can execute. They can't ship items that aren't part of an existing sale for a customer. The postconditions again indicate what to look for when stating the expected results for a test case and show the objects that will need to collaborate in the design.

FIGURE **5-3** *Fully developed use case description for* Ship items

Use case name:	*Ship items*.	
Scenario:	Ship items for a new sale.	
Triggering event:	Shipping is notified of a new sale to be shipped.	
Brief description:	Shipping retrieves sale details, finds each item and records it is shipped, records which items are not available, and sends shipment.	
Actors:	Shipping clerk.	
Related use cases	None.	
Stakeholders:	Sales, Marketing, Shipping, warehouse manager.	
Preconditions:	Customer and address must exist. Sale must exist. Sale items must exist.	
Postconditions:	Shipment is created and associated with shipper. Shipped sale items are updated as shipped and associated with the shipment. Unshipped items are marked as on back order. Shipping label is verified and produced.	
Flow of activities:	**Actor**	**System**
	1. Shipping requests sale and sale item information.	1.1 System looks up sale and returns customer, address, sale, and sales item information.
	2. Shipping assigns shipper.	2.1 System creates shipment and associates it with the shipper.
	3. For each available item, shipping records item is shipped.	3.1 System updates sale item as shipped and associates it with shipment.
	4. For each unavailable item, shipping records back order.	4.1 System updates sale item as on back order.
	5. Shipping requests shipping label supplying package size and weight.	5.1 System produces shipping label for shipment. 5.2 System records shipment cost.
Exception conditions:	2.1 Shipper is not available to that location, so select another. 3.1 If order item is damaged, get new item and updated item quantity. 3.1 If item bar code isn't scanning, shipping must enter bar code manually. 5.1 If printing label isn't printing correctly, the label must be addressed manually.	

Activity Diagrams for Use Cases

Another way to document a use case is with an activity diagram. In Chapter 2, you learned about activity diagrams as a form of workflow diagram. You learned that an activity diagram is an easily understood diagram to document the workflows of the business processes. Activity diagrams are a standard UML diagram, and they are also an effective technique to document the flow of activities for each use case.

Figure 5-4 is an activity diagram that documents the flow of activities for the *Create customer account* use case. Sometimes, an activity diagram can take the place of the flow of activities section of a use case description, and sometimes, it is created to supplement the use case description. There are two swimlanes: one for the customer and one for the system. The customer has three activities, and the system has five activities.

Activity diagrams are helpful when the flow of activities for a use case is complex. The use case *Fill shopping cart* is complex in that three other use cases might be invoked while adding items to the shopping cart. For example, the actor might search for a product and then look at product reviews before adding the item to the cart. Once an item is added, the actor might search for and view available accessories and then add one or more to the cart.

FIGURE **5-4**

Activity diagram for Create customer account *showing alternate way to model the flow of activities*

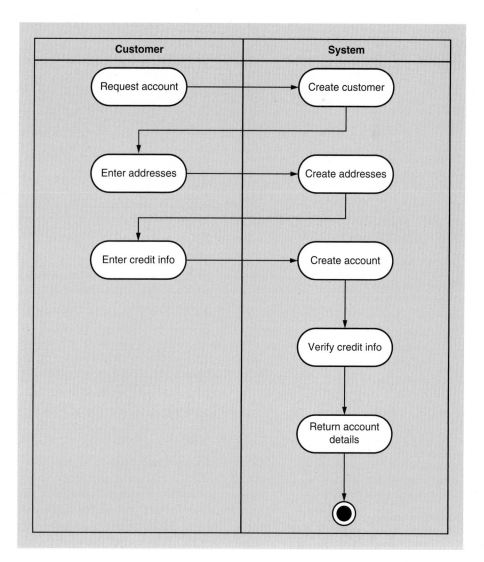

FIGURE **5-5**
Activity diagram for Fill shopping cart *showing richer user experience*

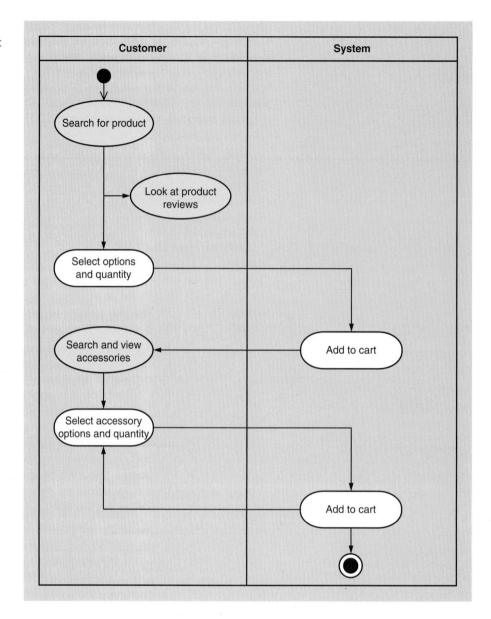

The activity diagram shown in **Figure 5-5** shows the *Fill shopping cart* use case flow of activities. The yellow ovals show the other use cases that are invoked while filling the shopping cart. The activities of the use case go in between the other use cases. The *Fill shopping cart* use case includes select options and quantities, add to cart, select accessory options and quantity, and add to cart. However, the intent of the richer user experience becomes evident when the activity diagram shows the use case in context.

The System Sequence Diagram—Identifying Inputs and Outputs

system sequence diagram (SSD) a diagram showing the sequence of messages between an external actor and the system during a use case or scenario

interaction diagram either a communication diagram or a sequence diagram that shows the interactions between objects

In the object-oriented approach, the flow of information is achieved through sending messages either to and from actors or back and forth between internal objects. A **system sequence diagram (SSD)** is used to describe this flow of information into and out of the automated system. Thus, an SSD documents the inputs and the outputs and identifies the interaction between actors and the system. An SSD is a type of **interaction diagram**.

SSD Notation

Figure 5-6 shows a generic SSD. As with a use case diagram, the stick figure represents an actor—a person (or role) that interacts with the system. In a use case diagram, the actor "uses" the system, but the emphasis in an SSD is on how the actor "interacts" with the system by entering input data and receiving output data. The box labeled :System is an object that represents the entire automated system. In SSDs and all other interaction diagrams, analysts use object notation instead of class notation. In object notation, a box refers to an individual object, not the class of all similar objects. The notation is simply a rectangle with the name of the object underlined. The colon before the underlined class name is a frequently used but optional part of the object notation. In an interaction diagram, the messages are sent and received by individual objects, not by a class. In an SSD, the only object included is one representing the entire system.

Underneath the actor and :System are vertical dashed lines called *lifelines*. A **lifeline**, or **object lifeline**, is simply the extension of that object—either actor or object—during the use case. The arrows between the lifelines represent the messages that are sent by the actor. Each arrow has an origin and a destination. The origin of the message is the actor or object that sends it, as indicated by the lifeline at the arrow's tail. Similarly, the destination actor or object of a message is indicated by the lifeline that is touched by the arrowhead. The purpose of lifelines is to indicate the sequence of the messages sent and received by the actor and object. The sequence of messages is read from top to bottom in the diagram.

A message is labeled to describe its purpose and any input data being sent. The message name should follow the verb-noun syntax to make the purpose clear. The syntax of the message label has several options; the simplest forms are shown in Figure 5-6. Remember that the arrows are used to represent a message and input data. But what is meant by the term *message* here? In a sequence diagram, a message is an action that is invoked on the destination object, much like a command. Notice in Figure 5-6 that the input message is called inquireOnItem.

lifeline or **object lifeline**   the vertical line under an object on a sequence diagram to show the passage of time for the object

FIGURE **5-6**

Sample system sequence diagram (SSD)

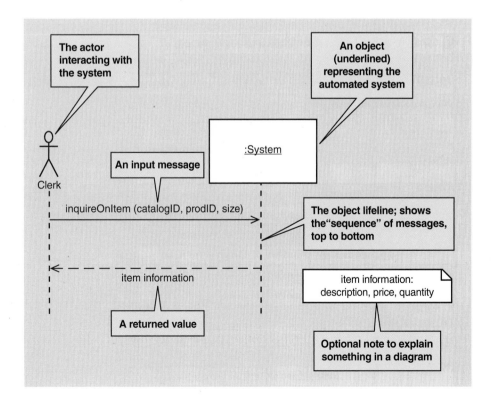

The clerk is sending a request (a message) to the system to find an item. The input data that is sent with the message is contained within the parentheses, and in this case, it is data to identify the particular item. The syntax is simply the name of the message followed by the input parameters in parentheses. This form of syntax is attached to a solid arrow.

The returned value has a slightly different format and meaning. Notice that the arrow is a dashed arrow. A dashed arrow indicates a response or an answer, and as shown in the figure, it immediately follows the initiating message. The format of the label is also different. Because it is a response, only the data that is sent on the response is noted. There is no message requesting a service—only the data being returned. In this case, a valid response might be a list of all the information returned—for example, the description, price, and quantity of an item. However, an abbreviated version is also satisfactory. In this case, the information returned is named *item information*. Additional documentation is required to show the details. In Figure 5-6, this additional information is shown as a note. A note can be added to any UML diagram to add explanations. The details of item information could also be documented in supporting narratives or even simply referenced by the attributes in the Customer class.

Frequently, the same message is sent multiple times. For example, when an actor enters items on an order, the message to add an item to an order may be sent multiple times. **Figure 5-7(a)** illustrates the notation to show this repeating operation. The message and its return are located inside a larger rectangle called

FIGURE **5-7**

Repeating message in (a) detailed loop frame notation and (b) alternate notation

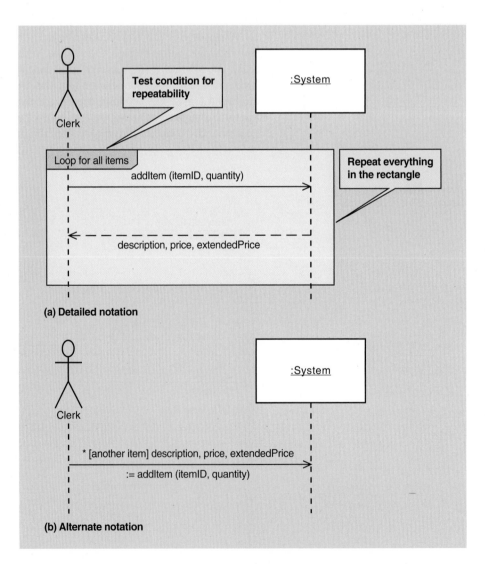

loop frame notation on a sequence diagram showing repeating messages

true/false condition part of a message between objects that is evaluated prior to transmission to determine whether the message can be sent

a **loop frame**. In a smaller rectangle at the top of the frame is the descriptive text to control the behavior of the messages within the larger rectangle. The condition loop for all items indicates that the messages in the box repeat many times or are associated with many instances.

Figure 5-7(b) shows an alternate notation. Here, the square brackets and text inside them are called a **true/false condition** for the messages. The asterisk (*) preceding the true/false condition indicates that the message repeats as long as the true/false condition evaluates to true. Analysts use this abbreviated notation for several reasons. First, a message and the returned data can be shown in one step. Note that the return data is identified as a return value on the left side of an assignment operator—the := sign. This alternative simply shows a value that is returned. Second, the true/false condition is placed on the message itself. Note that in this example, the true/false condition is used for the control of the loop. True/false conditions are also used to evaluate any type of test that determines whether a message is sent. For example, consider the true/false condition [credit card payment]. If it is true that the thing being tested is a credit card payment, the message is sent to the system to verify a credit card number. Finally, an asterisk is placed on the message itself to indicate the message repeats. Thus, for simple repeating messages, the alternate notation is shorter. However, if several messages are included within the repeat or there are multiple messages—each with its own true/false condition—the loop frame is more explicit and precise.

Here is the complete notation for a message:

[true/false condition] return-value := message-name (parameter-list)

Any part of the message can be omitted. In brief, the notation components do the following:

- An asterisk (*) indicates repeating or looping of the message.
- Brackets [ ] indicate a true/false condition. This is a test for that message only. If it evaluates to true, the message is sent. If it evaluates to false, the message isn't sent.
- Message-name is the description of the requested service. It is omitted on dashed-line return messages, which only show the return data parameters.
- Parameter-list (with parentheses on initiating messages and without parentheses on return messages) shows the data that are passed with the message.
- Return-value on the same line as the message (requires :=) is used to describe data being returned from the destination object to the source object in response to the message.

opt frame notation on a sequence diagram showing optional messages

alt frame notation on a sequence diagram showing if-then-else logic

Sequence diagrams use two additional frames to depict processing logic, as shown in **Figure 5-8**. The **opt frame** in Figure 5-8(a) is used when a message or a series of messages is optional or based on some true/false condition. The alt frame is used with if-then-else logic, as shown in Figure 5-8(b). The **alt frame** in this figure indicates that if an item is taxable, then add sales tax; otherwise, add a tax exemption code for a sales tax exemption.

Developing a System Sequence Diagram (SSD)

An SSD is usually used in conjunction with the use case descriptions to help document the details of a single use case or scenario within a use case. To develop an SSD, it is useful to have a detailed description of the use case—either in the fully developed form or as an activity diagram. These two models identify the flow of activities within a use case, but they don't explicitly identify the inputs and outputs. An SSD will provide this explicit identification of inputs and outputs. One advantage of using activity diagrams is that it is easy to identify when an input or output occurs. Inputs and outputs occur whenever an arrow in an activity diagram goes from an external actor to the computer system.

FIGURE **5-8**

Sequence diagram notation for (a) opt frame and (b) alt frame

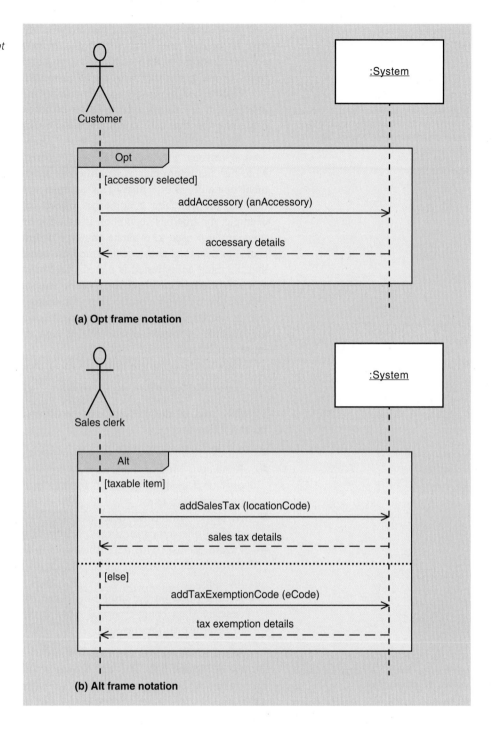

(a) Opt frame notation

(b) Alt frame notation

Recall the activity diagram for *Create customer account* shown in Figure 5-4. There are two swimlanes: the customer and the computer system. In this instance, the system boundary coincides with the vertical line between the customer swimlane and the computer system swimlane.

The development of an SSD based on an activity diagram falls into four steps:

1. **Identify the input messages**—In Figure 5-4, there are three locations with a workflow arrow crossing the boundary line between the customer and the system. At each location that the workflow crosses the automation boundary, input data is required; therefore, a message is needed.
2. **Describe the message from the external actor to the system by using the message notation described earlier**—In most cases, you will need a message

FIGURE **5-9**

SSD for the Create customer account *use case*

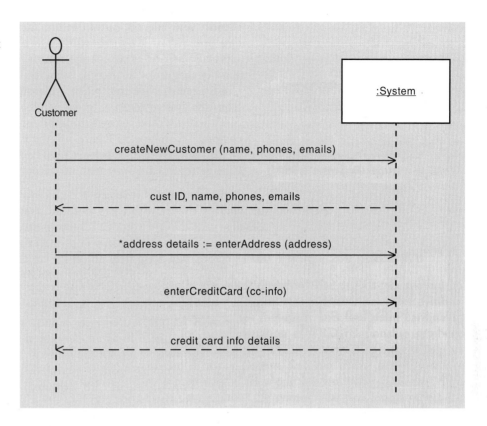

name that describes the service requested from the system and the input parameters being passed. **Figure 5-9**—the SSD for the *Create customer account* use case—illustrates the three messages based on the activity diagram. Notice that the names of the messages reflect the services that the actor is requesting of the system: createNewCustomer, enterAddress, and enterCreditCard. Other names could also have been used. For example, instead of enterAddress, the name could be createAddress. What is important is that the message name describes the service requested from the system and be in verb-noun form.

The other information required is the parameter list for each message. Determining exactly which data items must be passed in is more difficult. In fact, developers frequently find that determining the data parameters requires several iterations before a correct, complete list is obtained. The important principle for identifying data parameters is to base it on the class diagram. In other words, the appropriate attributes from the classes are listed as parameters. Looking at the attributes, along with an understanding of what the system needs to do, will help you find the right attributes. With the first message just mentioned—createNewCustomer—the parameters should include basic information about the customer, such as name, phone, and e-mail address. Note that when the system creates the customer, it assigns a new customerId and returns it with the other customer information.

In the second message—enterAddress—parameters are needed to identify the full address. Usually, that would include street address, city, state, and ZIP code. The SSD simplifies the message to show address as the parameter.

The third message—based on the activity diagram—enters the credit card information. The parameter—cc-info—represents the account number, expiration date, and security code.

3. **Identify and add any special conditions on the input messages, including iteration and true/false conditions**—In this instance, the enterAddress message is repeated for each address needed for the customer. The asterisk symbol in front of the message is shown.

4. **Identify and add the output return messages**—Remember that there are two options for showing return information: as a return value on the message itself or as a separate return message with a dashed arrow. The activity diagram can provide some clues about return messages, but there is no standard rule that when a transition arrow in the workflow goes from the system to an external actor an output always occurs. In Figure 5-4, there are three arrows from the computer system swimlane to the customer swimlane. In Figure 5-9, these are shown as return data on the dashed line. Note that they are each named with a noun that indicates what is being returned. Sometimes, no output data is returned.

Remember that the objective is discovery and understanding, so you should be working closely with users to define exactly how the workflow proceeds and exactly what information needs to be passed in and provided as output. This is an iterative process, and you will probably need to refine these diagrams several times before they accurately reflect the needs of the users.

Let us develop an SSD for the *Ship items* use case that is shown as a fully developed use case description in Figure 5-3. Note that the actor has five numbered steps in the flow of activities, so there will be five input messages in the SSD shown in **Figure 5-10**: getNextSale, setShipper, recordShippedItem, recordBackorder, and getShippingLabel. No parameter is needed for getNextSale because the system will return the information for the next sale to be shipped. The shipper is selected by the actor—probably from a list on the form or page—so the parameter is shipperID. Two messages are repeated in loops: recordShippedItem and recordBackorder. On this SSD, the loop frame notation is used. Finally, the getShippingLabel message requires two parameters: the size of the package and the weight. The system uses that information, along with the shipper and address, to produce the shipping label and record the cost.

These first sections of this chapter have explained the models that are used in object-oriented development to specify the processing aspects of the new system. The use case descriptions, as provided by written narratives or activity diagrams, give the details of the internal steps within each use case. Precondition and postcondition statements help define the context for the use case—that is, what must exist before and after processing. Finally, the SSD describes the inputs and outputs that occur within a use case. Together, these models provide a comprehensive description of the system-processing requirements and give the foundation for system design.

Now that the use cases have been explained, let us find out how to capture important object status information.

The State Machine Diagram—Identifying Object Behavior

Sometimes, it is important for a computer system to maintain information about the status of problem domain objects. For example, a customer might want to know whether a particular sale has been shipped or a manager might want to know if a customer sale has been paid for. Thus, the system needs to be able to track the status of customer sales. When defining requirements, analysts need to identify and document which domain objects require status checking and which business rules determine valid status conditions. Referring back to RMO, an example of a business rule is that a customer sale shouldn't be shipped until it has been paid for.

The status condition for a real-world object is often referred to as the *state* of the object. Defined precisely, a **state** of an object is a condition that occurs during its life when it satisfies some criterion, performs some action, or waits

state a condition during an object's life when it satisfies some criterion, performs some action, or waits for an event

FIGURE **5-10**
SSD for the Ship items *use case*

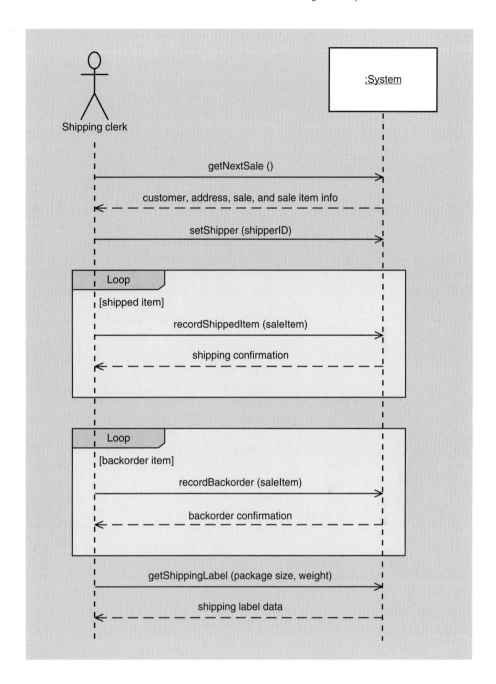

for an event. For real-world objects, we equate the state of an object with its status condition.

The naming convention for status conditions helps identify valid states. A state might have a name of a simple condition, such as *On* or *In repair*. Other states are more active, with names consisting of gerunds or verb phrases, such as *Being shipped* or *Working*. For example, a specific Sale object comes into existence when a customer buys something. Right after it is created, the object might be in a state called *Adding new sale items*, then a state called *Waiting for items to be shipped*, and finally, when all items have been shipped, a state called *Completed*. If you find yourself trying to use a noun to name a state, you probably have an incorrect idea about states or object classes. The name of a state shouldn't be an object (or noun); it should be something that describes the object (or noun).

States are described as semipermanent conditions because external events can interrupt a state and cause the object to go to a new state. An object

transition the movement of an object from one state to another state

state machine diagram a diagram showing the life of an object in states and transitions

pseudostate the starting point of a state machine diagram, indicated by a black dot

destination state for a particular transition, the state to which an object moves after the completion of a transition

origin state for a particular transition, the original state of an object from which the transition occurs

remains in a state until some event causes it to move, or transition, to another state. A **transition**, then, is the movement of an object from one state to another state. Transitioning is the mechanism that causes an object to leave a state and change to a new state. States are semipermanent because transitions interrupt them and cause them to end. Generally, transitions are short in duration—compared with states—and they can't be interrupted. The combination of states and transitions between states provides the mechanisms that analysts use to capture business rules. In our previous RMO example, we would say that a customer sale must be in a *Paid for* state before it can transition to a *Shipped* state. This information is captured and documented in a UML diagram called a **state machine diagram**.

A state machine diagram can be developed for any problem domain classes that have complex behavior or status conditions that need to be tracked. However, not all classes will require a state machine diagram. If an object in the problem domain class doesn't have status conditions that must control the processing for that object, a state machine diagram probably isn't necessary. For example, in the RMO class diagram, a class such as Sale may need a state machine diagram. However, a class such as SaleTransaction probably does not. A sale transaction is created when the payment is made and then just sits there; it doesn't need to track other conditions.

A state machine diagram is composed of ovals representing the states of an object and arrows representing the transitions. **Figure 5-11** illustrates a simple state machine diagram for a printer. Because it is a little easier to learn about state machine diagrams by using tangible items, we start with a few examples of computer hardware. After the basics are explained, we will illustrate modeling of software objects in the problem domain. The starting point of a state machine diagram is a black dot, which is called a **pseudostate**. The first shape after the black dot is the first state of the printer. In this case, the printer begins in the *Off* state. A state is represented by a rectangle with rounded corners (almost like an oval but more squared), with the name of the state placed inside.

As shown in Figure 5-11, the arrow leaving the *Off* state is called a *transition*. The firing of the transition causes the object to leave the *Off* state and make a transition to the *On* state. After a transition begins, it runs to completion by taking the object to the new state, called the **destination state**. A transition begins with an arrow from an **origin state**—the state prior to the transition—to a destination state, and it is labeled with a string to describe the components of the transition.

The transition label consists of the following syntax with three components:

transition-name (parameters, …) [guard-condition] / action-expression

FIGURE **5-11**

Simple state machine diagram for a printer

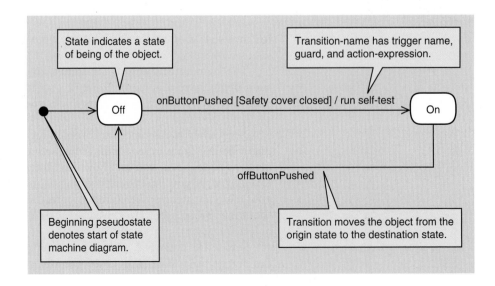

In Figure 5-11, the transition-name is onButtonPushed. The transition is like a trigger that fires or an event that occurs. The name should reflect the action of a triggering event. In Figure 5-11, no parameters are being sent to the printer. The guard-condition is [Safety cover closed]. For the transition to fire, the guard must be true. The forward slash divides the firing mechanism from the actions or processes. **Action-expressions** indicate some process that must occur before the transition is completed and the object arrives in the destination state. In this case, the printer will run a self-test before it goes into the *On* state.

action-expression a description of the activities performed as part of a transition

The transition-name is the name of a message event that triggers the transition and causes the object to leave the origin state. Notice that the format is very similar to a message in an SSD. In fact, you will find that the message event names and transition-names use almost the same syntax. One other relationship exists between the messages and the transitions: Transitions are caused by messages coming to the object. The parameter portion of the message name comes directly from the message parameters.

guard-condition a true/false test to see whether a transition can fire

The **guard-condition** is a qualifier or test on the transition, and it is simply a true/false condition that must be satisfied before the transition can fire. For a transition to fire, first the trigger must occur and then the guard must evaluate to true. Sometimes, a transition has only a guard-condition and no triggering event. In that case, the trigger is constantly firing, and whenever the guard becomes true, the transition occurs.

Recall from the discussion of sequence diagrams that messages have a similar test, which is called a true/false condition. This true/false condition is a test on the sending side of the message, and before a message can be sent, the true/false condition must be true. In contrast, the guard-condition is on the receiving side of the message. The message may be received, but the transition fires only if the guard-condition is also true. This combination of tests, messages, and transitions provides tremendous flexibility in defining complex behavior.

The action-expression is a procedural expression that executes when the transition fires. In other words, it describes the action to be performed. Any of the three components—transition-name, guard-condition, or action-expression—may be empty. If either the transition-name or the guard-condition is empty, it automatically evaluates to true. Either of them may also be complex, with AND and OR connectives.

Composite States and Concurrency

Before teaching you how to develop a state machine diagram, we need to introduce one other type of state: a composite state. In the real world, it is very common for an object to be in multiple states at the same time. For example, when the printer in Figure 5-11 is in the *on* state, it might also be doing other things. Sometimes, it is printing; sometimes, it is just sitting idle; and when it is first turned on, it goes through some self-checking steps. All these conditions occur while the printer is on, and they can be considered simultaneous states. The condition of being in more than one state at a time is called **concurrency**, or **concurrent states**. One way to show this is with a synchronization bar and concurrent paths, as in activity diagrams. Thus, we could split a transition with a synchronization bar so one path goes to the *On* state and the other path goes to the *Idle*, *Printing*, and *Selfcheck* states. A **path** is a sequential set of connected states and transitions.

concurrency or **concurrent state** the condition of being in more than one state at a time

path a sequential set of connected states and transitions

Another way to show concurrent states is to have states nested inside other, higher-level states. These higher-level states are called **composite states**.

composite state a state containing other states and transitions (that is, a path)

A composite state represents a higher level of abstraction and can contain nested states and transition paths. **Figure 5-12**, which is an extension of Figure 5-11, illustrates this idea with respect to a printer. The printer is not only in the *On* state, it is concurrently in either the *Idle* or *Working* state. The rounded rectangle for the *On* state is divided into two compartments. The top compartment

FIGURE **5-12**

Sample composite states for the printer object

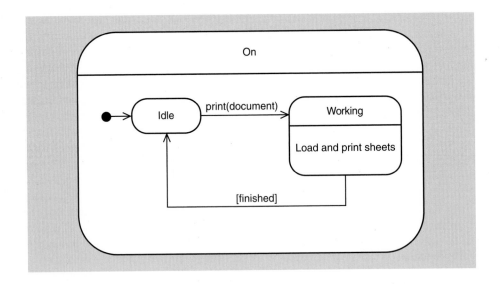

contains the name, and the lower compartment contains the nested states and transition paths.

When the printer enters the *On* state, it automatically begins at the nested black dot and moves to the *Idle* state. Thus, the printer is in the *On* and *Idle* states. When the print message is received, the printer makes the transition to the *Working* state but also remains in the *On* state. Some new notation is also introduced for the *Working* state. In this instance, the lower compartment contains the action-expressions—that is, the activities that occur while the printer is in the *Working* state.

We can extend this idea of composite states and concurrency one step further by allowing multiple paths within a composite state. Perhaps an object has entire sets of states and transitions—multiple paths—that are active concurrently. To document concurrent multiple paths for a single object, we draw a composite state with the lower portion divided into multiple compartments—one for each concurrent path of behavior. For example, imagine a printer that has an input bin to hold the paper. This printer also alternates between two states in its work cycle: *Idle* and *Working*. We may want to describe two separate paths: one representing the state of the input paper tray and the other the state of the printing mechanism. The first path will have the states *Empty*, *Full*, and *Low*. The second path will have the states *Idle* and *Working*. These two paths are independent; the movement between states in one compartment is completely independent of the movement between states in the other compartment.

As before, there are two ways to document this concurrent behavior. First, we could use a synchronization bar with one path becoming three paths. Second, we could use a composite state. **Figure 5-13** extends the printer example from Figure 5-12. In this example, there are two concurrent paths within the composite state. The upper concurrent path represents the paper tray part of the printer. The two paths are completely independent, and the printer moves through the states and transitions in each path independently. When the Off button is pushed, the printer leaves the *On* state. Obviously, when the printer leaves the *On* state, it also leaves all the paths in the nested states. It doesn't matter whether the printer is in a state or in the middle of a transition. When the Off button is pushed, all activity is stopped, and the printer exits the *On* state. Now that you know the basic notation of state machine diagrams, we will explain how to develop a state machine diagram.

FIGURE **5-13**

Concurrent paths for a printer in the On state

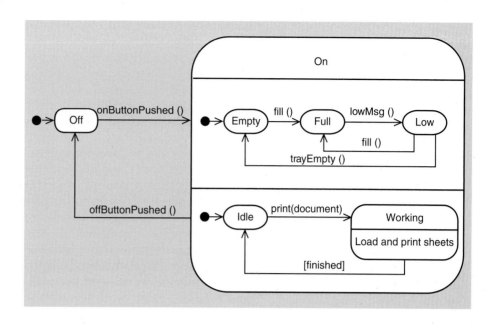

Rules for Developing State Machine Diagrams

State machine diagram development follows a set of rules. The rules help you develop state machine diagrams for classes in the problem domain. Usually, the primary challenge in building a state machine diagram is to identify the right states for the object. It might be helpful to pretend that you are the object itself. It is easy to pretend to be a customer but a little more difficult to say "I am an order" or "I am a shipment. How do I come into existence? What states am I in?" However, if you can begin to think this way, it will help you develop state machine diagrams.

The other major area of difficulty for new analysts is to identify and handle composite states with nested threads. Usually, the primary cause of this difficulty is a lack of experience in thinking about concurrent behavior. The best solution is to remember that developing state machine diagrams is an iterative behavior—more so than developing any other type of diagram. Analysts seldom get a state machine diagram right the first time. They always draw it and then refine it again and again. Also, remember that when you are defining requirements, you are only getting a general idea of the behavior of an object. During design, as you build detailed sequence diagrams, you will have an opportunity to refine and correct important state machine diagrams.

Finally, don't forget to ask about an exception condition—especially when you see the words *verify* or *check*. Usually, there will be two transitions out of states that verify something: one for acceptance and one for rejection.

Here is a list of steps that will help you get started in developing state machine diagrams:

1. **Review the class diagram and select the classes that might require state machine diagrams**—Remember to include only those classes that have multiple status conditions that are important for the system to track. Then, begin with the classes that appear to have the simplest state machine diagrams, such as the SaleItem class for RMO, which is discussed later.
2. **For each selected class in the group, make a list of all the status conditions you can identify**—At this point, simply brainstorm. If you are working on a team, have a brainstorming session with the whole team. Remember that these states must reflect the states for the real-world objects that will be represented in software. Sometimes, it is helpful to think of the physical object, identify states of the physical object, and then translate those that are appropriate into corresponding system states or status conditions. It is

also helpful to think of the life of the object. How does it come into existence in the system? When and how is it deleted from the system? Does it have active states? Does it have inactive states? Does it have states in which it is waiting? Think of activities done to the object or by the object. Often, the object will be in a particular state as these actions are occurring.

3. **Begin building state machine diagram fragments by identifying the transitions that cause an object to leave the identified state**—For example, if a sale is in a state of *Ready to be shipped*, a transition such as beginShipping will cause the sale to leave that state.

4. **Sequence these state-transition combinations in the correct order**—Then, aggregate these combinations into larger fragments. As the fragments are being aggregated into larger paths, it is natural to begin to look for a natural life cycle for the object. Continue to build longer paths in the state machine diagram by combining the fragments.

5. **Review the paths and look for independent, concurrent paths**—When an item can be in two states concurrently, there are two possibilities. The two states may be on independent paths, as in the printer example of *Working* and *Full*. This occurs when the states and paths are independent, and one can change without affecting the other. Alternately, one state may be a composite state, so the two states should be nested. One way to identify a candidate for a composite state is to determine whether it is concurrent with several other states and whether these other states depend on the original state. For example, the *On* state has several other states and paths that can occur while the printer is in the *On* state, and those states depend on the printer being in the *On* state.

6. **Look for additional transitions**—Often, during a first iteration, several of the possible combinations of state-transition-state are missed. One method to identify them is to take every paired combination of states and ask whether there is a valid transition between the states. Test for transitions in both directions.

7. **Expand each transition with the appropriate message event, guard-condition, and action-expression**—Include with each state appropriate action-expressions. Much of this work may have been done as the state machine diagram fragments were being built.

8. **Review and test each state machine diagram**—Review each of your state machine diagrams by doing the following:
   a. Make sure your states are really states of the object in the class. Ensure that the names of states truly describe the object's states rather than the object itself.
   b. Follow the life cycle of an object from its coming into existence to its being deleted from the system. Be sure that all possible combinations are covered and that the paths on the state machine diagram are accurate.
   c. Be sure your diagram covers all exception conditions as well as the normal expected flow of behavior.
   d. Look again for concurrent behavior (multiple paths) and the possibility of nested paths (composite states).

Developing RMO State Machine Diagrams

Let us practice these steps by developing two state machine diagrams for RMO. The first step is to review the domain class diagram and then select the classes that may have status conditions that need to be tracked. In this case, we select the Sale and SaleItem classes. We assume that customers will want to know the status of their sales and the status of individual items on the sale. Other classes that are candidates for state machine diagrams are: InventoryItem, to track in-stock or out-of-stock items; Shipment, to track arrivals; and possibly Customer, to track active and inactive customers.

Developing the SaleItem State Machine Diagram

The first step in developing the SaleItem state machine diagram is to identify the possible status conditions that might be of interest. Some necessary status conditions are *Ready to be shipped*, *On back order*, and *Shipped*. An interesting question comes to mind at this point: Can a sale item be partially shipped? In other words, if the customer bought 10 of a single item but there are only five in inventory, should RMO ship those five and put the other five on back order? You should see the ramifications of this decision. The system and the database would need to be designed to track and monitor detailed information to support this capability. The domain class diagram for RMO indicates that a SaleItem can be associated with either zero (not yet shipped) shipments or one (totally shipped) shipment. Based on the current specification, the definition doesn't allow partial shipments of SaleItems.

This is just another example of the benefit of building models. Had we not been developing the state machine diagram model, this question might never have been asked. The development of detailed models and diagrams is one of the most important activities that a system developer can perform. It forces analysts to ask fundamental questions. Sometimes, new system developers think that model development is a waste of time, especially for small systems. However, truly understanding the users' needs before writing the program always saves time in the long run.

The second step is to identify exit transitions for each of the status conditions. **Figure 5-14** is a table showing the states that have been defined and the exit transitions for each of those states. One additional state has been added to the list—*Newly added*—which covers the condition that occurs when an item has been added to the sale but the sale isn't complete or paid for, so the item isn't ready for shipping.

The third step is to combine the state-transition pairs into fragments and to build a state machine diagram with the states in the correct sequence. **Figure 5-15** illustrates the partially completed state machine diagram. The flow from beginning to end for the SaleItem object is quite obvious. However, at least one transition seems to be missing. There should be some path to allow entry into the *On back order* state so we recognize that this first-cut state machine diagram needs some refinement. We will fix that in a moment. ■

FIGURE **5-14**

States and exit transitions for SaleItem object

State	Transition causing exit from state
Newly added	finishedAdding
Ready to ship	shipItem
On back order	itemArrived
Shipped	No exit transition defined

FIGURE **5-15**

Partial state machine diagram for SaleItem object

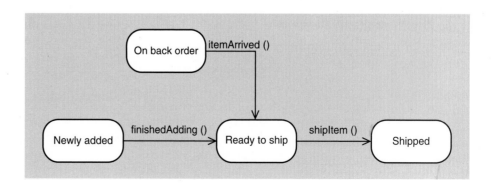

The fourth step is to look for concurrent paths. In this case, it doesn't appear that a SaleItem object can be in any two of the identified states at the same time. Of course, because we chose to begin with a simple state machine diagram, that was expected.

The fifth step is to look for additional transitions. This step is where we flesh out other necessary transitions. The first addition is to have a transition from *Newly added* to *On back order*. To continue, examine every pair of states to see whether there are other possible combinations. In particular, look for backward transitions. For example, can a SaleItem go from *Ready to ship* to *On back order*? This would happen if the shipping clerk found that there weren't enough items in the warehouse, even though the system indicated that there should have been. Other backward loops, such as from *Shipped* to *Ready to ship* or from *On back order* to *Newly added*, don't make sense and aren't included.

The sixth step is to complete all the transitions with correct names, guard-conditions, and action-expressions. Two new transition-names are added. The first is the transition from the beginning black dot to the *Newly added* state. That transition causes the creation—or, in system terms, the instantiation—of a new SaleItem object. It is given the same name as the message into the system that adds it: addItem(). The final transition is the one that causes the order item to be removed from the system. This transition goes from the *Shipped* state to a final circled black dot, which is a final pseudostate. On the assumption that it is archived to a backup tape when it is deleted from the active system, that transition is named archive().

Action-expressions are added to the transitions to indicate any special action that is initiated by the object or on the object. In this case, only one action is required. When an item that was *Ready to ship* moves to *On back order*, the system should initiate a new purchase order to the supplier to buy more items. Thus, on the markBackOrdered() transition, an action-expression is noted to place a purchase order. **Figure 5-16** illustrates the final state machine diagram for SaleItem.

The seventh step—reviewing and testing the state machine diagram—is the quality-review step. It is always tempting to omit this step; however, a good project manager ensures that the systems analysts have time in the schedule to do a quick quality check of their models. A walkthrough (as described in Chapter 2) at this point in the project is very appropriate.

Developing the Sale State Machine Diagram

A Sale object is a little more complex than the SaleItem objects. In this example, you will see some features of state machine diagrams that support more complex objects.

FIGURE **5-16** *Final state machine diagram for SaleItem object*

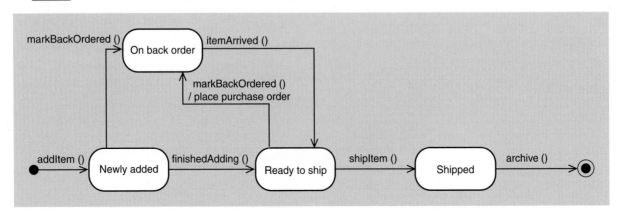

FIGURE **5-17**
States and exit transitions for Sale

State	Exit transition
Open for item adds	completeSale
Ready for shipping	beginShipping
In shipping	shippingComplete
Waiting for back orders	backOrdersArrive
Shipped	paymentCleared
Closed	archive

Figure 5-17 shows the defined states and exit transitions that, on first iteration, appear to be required. Reading from top to bottom, the states describe the life cycle of a sale—for example, the status conditions. First, a Sale comes into existence and is ready to have items added to it—the *Open for item adds* state. The users in RMO indicated that they wanted a Sale to remain in this state for 24 hours in case the customer wants to add more items. After all the items are added, the Sale is *Ready for shipping*. Second, it goes to shipping and is in the *In shipping* state. At this point, it isn't clear how *In shipping* and *Waiting for back orders* relate to each other. That relationship will have to be sorted out as the state machine diagram is being built. Finally, the Sale is *Shipped*, and after the payment clears, it is *Closed*.

In the third step, fragments are built and combined to yield the first-cut state machine diagram (see **Figure 5-18**). The state machine diagram built from the fragments appears to be correct—for the most part. However, we note some problems with the *Waiting for back orders* state.

After some analysis, we decide that *In shipping* and *Waiting for back orders* are concurrent states. And another state is needed, called *Being shipped*, for when the shipping clerk is actively shipping items. One way to show the life of a Sale is to put it in the *In shipping* state when shipping begins. It also enters the *Being shipped* state at that point. The Sale can cycle between *Being shipped* and *Waiting for back orders*. The exit out of the composite state only occurs from the *Being shipped* state, which is inside the *In shipping* state. Obviously, upon leaving the inside state, the order also leaves the composite *In shipping* state.

As we go through the fourth, fifth, and sixth steps, we note that new transitions must be added. The creation transition from the initial pseudostate is required. Also, transitions must be included to show when items are being added and when they are being shipped. Usually, we put these looping activities on transitions that leave a state and return to the same state. In this case, the transition is called addItem(). Note how it leaves the *Open for item adds* state

FIGURE **5-18**
First-cut state machine diagram for Order

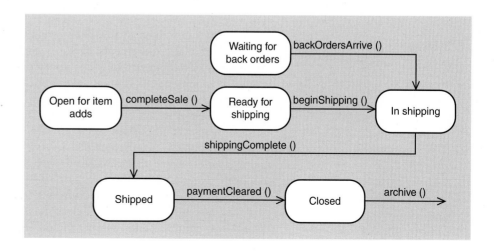

FIGURE **5-19**
Second-cut state machine diagram for Order

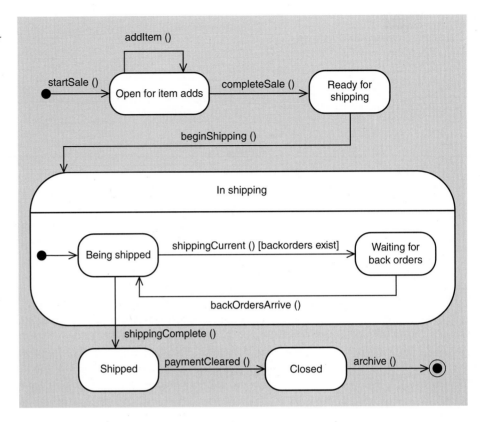

and returns to the same state. **Figure 5-19** takes the state machine diagram to this level of completion.

The benefit of developing a state machine diagram for a problem domain object is that it helps you capture and clarify business rules. From the state machine diagram, we can see that shipping can't commence while the sale is in the *Open for item adds* state, new items can't be added to the sale after it has been placed in the *Ready for shipping* state, and the sale isn't considered shipped until all items are shipped. If the sale has the status of *In shipping*, we know that it is either actively being worked on or waiting for back orders.

As always, the benefits of careful model building help us gain a true understanding of the system requirements. Let us now look at the big picture and see how the different models fit together.

Integrating Requirements Models

The diagrams described in this chapter allow analysts to completely specify the system functional requirements. If you were developing a system using a waterfall systems development life cycle, you would develop the complete set of diagrams to represent all system requirements before continuing with design. However, because you are using an iterative approach, you would only construct the diagrams that are necessary for a given iteration. A complete use case diagram would be important to get an idea of the total scope of the new system. But the supporting details included in use case descriptions, activity diagrams, and system sequence diagrams need only be done for use cases in the specific iteration.

The domain model class diagram is a special case. Much like the entire use case diagram, the domain model class diagram should be as complete as possible for the entire system, as shown for RMO in Chapter 4. The number of

FIGURE **5-20**
Relationships among object-oriented requirements models

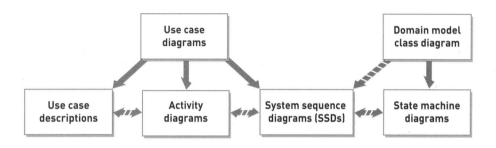

problem domain classes for the system provides an additional indicator of the total scope of the system. Refinement and actual implementation of many classes will wait for later iterations, but the domain model should be fairly complete. The domain model is necessary to identify all the domain classes that are required in the new system. The domain model is also used to design the database.

Throughout this chapter, you have seen how the construction of a diagram depends on information provided by another diagram. You have also seen that the development of a new diagram often helps refine and correct a previous diagram. You should also have noted that the development of detailed diagrams is critical to gaining a thorough understanding of the user requirements. **Figure 5-20** illustrates the primary relationships among the requirements models for object-oriented development. The use case diagram and other diagrams on the left are used to capture the processes of the new system. The class diagram and its dependent diagrams capture information about the classes for the new system. The solid arrows represent major dependencies, and the dashed arrows show minor dependencies. The dependencies generally flow from top to bottom, but some arrows have two heads to illustrate that influence goes in both directions.

Note that the use case diagram and the domain model class diagram are the primary models from which others draw information. You should develop those two diagrams as completely as possible. The detailed descriptions—either in narrative format or in activity diagrams—provide important internal documentation of the use cases and must completely support the use case diagram. Such internal descriptions as preconditions and postconditions use information from the class diagram. These detailed descriptions are also important for development of system sequence diagrams. Thus, the detailed descriptions, activity diagrams, and system sequence diagrams must all be consistent with regard to the steps of a particular use case. As you progress in developing the system and especially as you begin doing detailed system design, you will find that understanding the relationships among these models is an important element in the quality of your models.

Chapter Summary

The object-oriented approach to information systems development has a complete set of diagrams and textual models that together document the user's needs and define the system requirements. These requirements are specified by using domain model class diagrams and state machine diagrams to model the problem domain and use case diagrams, use case descriptions or activity diagrams, and system sequence diagrams (SSDs) to model the use cases.

The internal activities of a use case are first described by an internal flow of activities. It is possible to have several different internal flows, which represent different scenarios of the same use case. Thus, a use case may have several scenarios. These details are documented either in use case descriptions or with activity diagrams.

Another diagram that provides details of the use case's processing requirements is an SSD. An SSD documents the inputs and outputs of the system. The scope of each SSD is usually a use case or a scenario within a use case. The components of an SSD are the actor—the same actor identified in the use case—and the system. The system is treated as a black box in that the internal processing isn't addressed. Messages, which represent the inputs, are sent from the actor to the system. Output messages are returned from the system to the actor. The sequence of messages is indicated from top to bottom.

The domain model class diagram continues to be refined when defining requirements. The behavior of domain objects represented in the class diagram is an aspect of the requirements that is also studied and modeled. The state machine diagram is used to model object states and state transitions that occur in a use case. All the models discussed in this chapter are interrelated, and information in one model explains information in others.

Key Terms

action-expressions 135

alt frame 129

composite states 135

concurrency, or concurrent states 135

destination state 134

guard-condition 135

interaction diagram 126

lifeline, or object lifeline 127

loop frame 129

opt frame 129

origin state 134

path 135

postconditions 122

preconditions 122

pseudostate 134

scenarios or use case instances 121

state 132

state machine diagram 134

system sequence diagram (SSD) 126

transition 134

true/false condition 129

use case description 121

Review Questions

1. What are the models that describe use cases in more detail?
2. What two UML diagrams are used to model domain classes?
3. Which part of a use case description can also be modeled by using an activity diagram?
4. Explain the difference between a use case and a scenario. Give a specific example of a use case with a few possible scenarios.
5. List the parts or compartments of a fully developed use case description.
6. Compare/contrast precondition and postcondition.
7. Compare/contrast postcondition and exception condition.
8. Compare/contrast business process and flow of activities for a use case. Explain how an activity diagram can be used to model both.
9. What is the purpose of an SSD? What symbols are used in an SSD?
10. What are the steps required to develop an SSD?
11. Write a complete SSD message from the actor to the system, with the actor asking the system to begin the process for updating information about a specific product.
12. What is the name of the sequence diagram symbol used to represent the extension of an object throughout the duration of a use case?
13. What are the two ways to show a returned value on a sequence diagram?
14. What are two ways to show repetition on a sequence diagram?
15. What are the three types of frames used on a sequence diagram?
16. What is the symbol for a true/false condition on a sequence diagram?
17. What are the parameters of a message?
18. List the primary steps for developing a SSD.
19. What is an object state?
20. What is a state transition?
21. When considering requirements, states and state transitions are important for understanding which other diagram?
22. What UML diagram is used to show the states and transitions for an object?

23. List the elements that make up a transition description. Which elements are optional?
24. What is a composite state? What is it used for?
25. What is meant by the term *path*?
26. What is the purpose of a guard-condition?
27. Identify the models explained in this chapter and their relationship to one another.

Problems and Exercises

1. After reading the following narrative, do the following:

    i. Develop an activity diagram for each scenario.

    ii. Complete a fully developed use case description for each scenario.

    Quality Building Supply has two kinds of customers: contractors and the general public. Sales to each are slightly different.

    A contractor buys materials by taking them to the checkout desk for contractors. The clerk enters the contractor's name into the system. The system displays the contractor's information, including current credit standing. The clerk then opens up a new ticket (sale) for the contractor. Next, the clerk scans in each item to be purchased. The system finds the price of the item and adds the item to the ticket. At the end of the purchase, the clerk indicates the end of the sale. The system compares the total amount against the contractor's current credit limit and, if it is acceptable, finalizes the sale. The system creates an electronic ticket for the items, and the contractor's credit limit is reduced by the amount of the sale. Some contractors like to keep a record of their purchases, so they request that ticket details be printed. Others aren't interested in a printout.

    A sale to the general public is simply entered into the cash register, and a paper ticket is printed as the items are identified. Payment can be by cash, check, or credit card. The clerk must enter the type of payment to ensure that the cash register balances at the end of the shift. For credit card payments, the system prints a credit card voucher that the customer must sign.

2. Based on the following narrative, develop either an activity diagram or a fully developed description for the use case *Add a new vehicle to an existing policy* in a car insurance system.

    A customer calls a clerk at the insurance company and gives his policy number. The clerk enters this information, and the system displays the basic insurance policy. The clerk then checks the information to make sure the premiums are current and the policy is in force.

    The customer gives the make, model, year, and vehicle identification number (VIN) of the car to be added. The clerk enters this information, and the system ensures that the given data is valid. Next, the customer selects the types of coverage desired and the amount of each. The clerk enters the information, and the system records it and validates the requested amount against the policy limits. After all the coverages have been entered, the system ensures the total coverage against all other ranges, including other cars on the policy. Finally, the customer must identify all the drivers and the percentage of time they drive the car. If a new driver is to be added, then another use case—*Add new driver*—is invoked.

    At the end of the process, the system updates the policy, calculates a new premium amount, and prints the updated policy statement to be mailed to the policy owner.

3. Given the following list of classes and associations for the previous car insurance system, list the preconditions and postconditions for the use case *Add a new vehicle* to an existing policy.

    Classes in the system:

    ■ Policy
    ■ InsuredPerson
    ■ InsuredVehicle
    ■ Coverage
    ■ StandardCoverage (lists standard insurance coverages with prices by rating category)
    ■ StandardVehicle (lists all types of vehicles ever made)

    Relationships in the system:

    ■ Policy has InsuredPersons (one-to-many)
    ■ Policy has InsuredVehicles (one-to-many)
    ■ Vehicle has Coverages (one-to-many)
    ■ Coverage is a type of StandardCoverage
    ■ Vehicle is a StandardVehicle

FIGURE **5-21** *Cellular telephone state machine diagram*

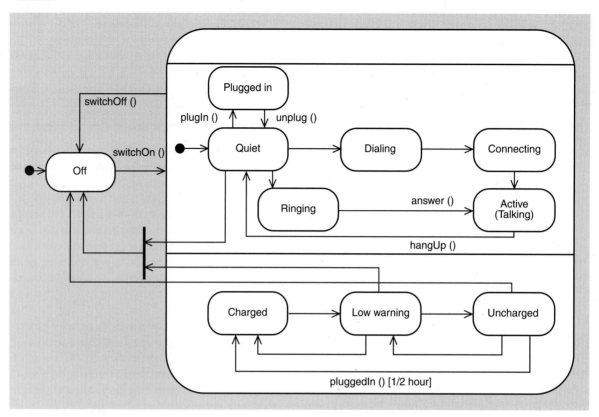

4. Develop an SSD based on the narrative and your activity diagram for problem 1.

5. Develop an SSD based on the narrative or your activity diagram for problem 2.

6. Review the cellular telephone state machine diagram shown in **Figure 5-21** and then answer the following questions. (Note that this telephone has characteristics not found in ordinary telephones. Base your answers only on the state machine diagram.)

   i. What happens to turn on the telephone?

   ii. What states does the telephone go into when it is turned on?

   iii. What are the three ways the telephone can be turned off?

   iv. Can the telephone turn off in the middle of the Active (Talking) state?

   v. How can the telephone get to the Active (Talking) state?

   vi. Can the telephone be plugged in while someone is talking?

   vii. Can the telephone change battery states while someone is talking? Explain which movement is allowed and which isn't allowed.

   viii. What states are concurrent with what other states? Make a two-column table showing the concurrent states.

7. Based on the following description of a shipment made by Union Parcel Shipments, identify all the states and exit transitions and then develop a state machine diagram.

    A shipment is first recognized after it has been picked up from a customer. Once in the system, it is considered active and in transit. Every time it goes through a checkpoint, such as arrival at an intermediate destination, it is scanned and a record is created indicating the time and place of the checkpoint scan. The status changes when it is placed on the delivery truck. It is still active, but now it is also considered to have a status of *delivery pending*. After it is delivered, the status changes again.

From time to time, a shipment has a destination that is outside the area served by Union. In those cases, Union has working relationships with other courier services. After a package is handed off to another courier, it is noted as being handed over. In those instances, a tracking number for the new courier is recorded (if it is provided). Union also asks the new courier to provide a status change notice after the package has been delivered.

Unfortunately, from time to time, a package gets lost. In that case, it remains in an active state for two weeks but is also marked as misplaced. If after two weeks the package hasn't been found, it is considered lost. At that point, the customer can initiate lost-package procedures to recover any damages.

8. Locate a company in your area that develops software. Consulting companies or companies with a large staff of information systems professionals tend to be more rigorous in their approach to systems development. Set up an interview. Determine the development approaches that the company uses. Many companies still use traditional structured techniques combined with some object-oriented development. In other companies, some projects are structured, whereas other projects are object oriented. Find out what kinds of modeling the company does for requirements specification. Compare your findings with the techniques taught in this chapter.

Case Study

TheEyesHaveIt.com Book Exchange

TheEyesHaveIt.com Book Exchange is a type of e-business exchange that does business entirely on the Internet. The company acts as a clearinghouse for buyers and sellers of used books.

To offer books for sale, a person must register with TheEyesHaveIt.com. The person must provide a current physical address and telephone number as well as a current e-mail address. The system then maintains an open account for this person. Access to the system as a seller is through a secure, authenticated portal.

A seller can list books on the system through a special Internet form. The form asks for all the pertinent information about the book: its category, its general condition, and the asking price. A seller may list as many books as desired. The system maintains an index of all books in the system so buyers can use the search engine to search for books. The search engine allows searches by title, author, category, and keyword.

People who want to buy books come to the site and search for the books they want. When they decide to buy, they must open an account with a credit card to pay for the books. The system maintains all this information on secure servers.

When a purchase is made, TheEyesHaveIt.com sends an e-mail notice to the seller of the book that was chosen as well as payment information. It also marks the book as sold. The system maintains an open order until it receives notice that the book has been shipped. After the seller receives notice that a listed book has been sold, the seller must notify the buyer via e-mail within 48 hours that the purchase is noted. Shipment of the order must be made within 24 hours after the seller sends the notification e-mail. The seller sends a notification to the buyer and TheEyesHaveIt.com when the shipment is made.

After receiving the notice of shipment, TheEyesHaveIt.com maintains the order in a shipped status. At the end of each month, a check is mailed to each seller for the book orders that have remained in a shipped status for 30 days. The 30-day waiting period exists to allow the buyer to notify TheEyesHaveIt.com if the shipment doesn't arrive for some reason or if the book isn't in the same condition as advertised.

If they want, buyers can enter a service code for the seller. The service code is an indication of how well the seller is servicing book purchases. Some sellers are very active and use TheEyesHaveIt.com as a major outlet for selling books. Thus, a service code is an important indicator to potential buyers.

For this case, develop these diagrams:

1. A domain model class diagram
2. A list of uses cases and a use case diagram
3. A fully developed description for two use cases: *Add a seller* and *Record a book order*
4. An SSD for each of the two use cases in question 3

Community Board of Realtors

The Multiple Listing Service system has a number of use cases, which you identified in Chapter 3, and three key domain classes, which you identified in Chapter 4: RealEstateOffice, Agent, and Listing.

1. For the use case *Add agent to real estate office*, write a fully developed use case description and draw an SSD. Review the case materials in previous chapters and recall that the system will need to know which real estate office the agent works for before prompting for agent information.

2. For the use case *Create new listing*, write a fully developed use case description and draw an SSD. Recall that the system needs to know which agent made the listing before the system prompts for listing information.

3. Draw a state machine diagram showing the states and transitions for a Listing object.

The Spring Breaks 'R' Us Travel Service

The Spring Breaks 'R' Us Travel Service system has many use cases and domain classes, which you identified in Chapters 3 and 4. Review the domain model class diagram to get a feel for the complexity of some of the use cases.

1. For the use case *Book a reservation*, write a fully developed use case description and draw an SSD. Review the classes that are associated with a reservation in the domain model to understand the flow of activities and repetition involved.

2. For the use case *Add new resort*, write a fully developed use case description and draw an SSD. Review the classes that are associated with a resort in the domain model to understand the flow of activities and repetition involved.

3. Draw an activity diagram to show the flow of activities for the use case *Add a new resort*.

4. Draw a state machine diagram showing the state and transitions for a Reservation object.

On the Spot Courier Services

As On the Spot Courier Services continues to grow, Bill discovers that he can provide much better services to his customers if he utilizes some of the technology that is currently available. For example, it will allow him to maintain frequent communication with his delivery trucks, which could save transportation and labor costs by making the pickup and delivery operations more efficient. This would allow him to serve his customers better. Of course, a more sophisticated system will be needed, but Bill's development consultant has assured him that a straightforward and not-too-complex solution can be developed.

Here is how Bill wants his business to operate. Each truck will have a morning and afternoon delivery and pickup run. Each driver will have a portable digital device with a touch screen. The driver will be able to view his or her scheduled pickups and deliveries for that run. (Note: This process will require a new use case—something the Agile development methodology predicted would happen.) However, because the trucks will maintain frequent contact with the home office via telephony Internet access, the pickup/delivery schedule can be updated in real time—even during a run. Rather than maintain constant contact, Bill decides that it will

be sufficient if the digital device synchronizes with the home office whenever a pickup or delivery is made. At those points in time, the route schedule can be updated with appropriate information.

Previously, customers were able to either call On the Spot and request a package pickup or visit the company's Web site to schedule a pickup. Once customers logged in, they could go to a Web page that allowed them to enter information about each package, including "deliver to" addresses, size and weight category information, and type of service requested. On the Spot provided "three hour," "same day," and "overnight" services. To facilitate customer self-service, On the Spot didn't require exact weights and sizes, but there were predefined size and weight categories from which the customer could choose.

Once the customer entered the information for all the packages, the system would calculate the cost and then print mailing labels and receipts. Depending on the type of service requested and the proximity of a delivery truck, the system would schedule an immediate pickup or one for later that day. It would display this information so the customer would immediately know when to expect the pickup.

(continued on page 149)

(continued from page 148)

Picking up packages was a fairly straightforward process. But there was some variation in what would happen depending on what information was in the system and whether the packages were already labeled. Upon arriving at the scheduled pickup location, the driver would have the system display any package information available for this customer. If the system already had information on the packages, the driver would simply verify that the correct information was already in the system for the packages. The driver could also make such changes as correcting the address, deleting packages, or adding new packages. If this were a cash customer, the driver would collect any money and enter that into the system. Using a portable printer from the van, the driver could print a receipt for the customer as necessary. If there were new packages that weren't in the system, the driver would enter the required information and also print mailing labels with his portable printer.

One other service that customers required was to be able to track the delivery status of their packages. The system needed to track the status of a package from the first time it "knew" about the package until it was delivered. Such statuses as "ready for pickup," "picked up," "arrived at warehouse," "out for delivery," and "delivered" were important. Usually, a package would follow through all the statuses, but due to the sophistication of the scheduling and delivery algorithm, a package would sometimes be picked up and delivered on the same delivery run. Bill also decided to add a status of "cancelled" for those packages that were scheduled to be picked up but ended up not being sent.

1. Based on this description, develop the following for the use case *Request a package pickup* and for the Web customer scenario:

    i. A fully developed use case description

    ii. An activity diagram

    iii. An SSD

    Based on the same description, develop the following for the use case *Pickup a package*:

    i. A fully developed use case description

    ii. An activity diagram

    iii. System sequence diagram

2. Develop a state machine diagram describing all the possible status conditions for a Package object.

Sandia Medical Systems Real-Time Glucose Monitoring

Figure 5-22 shows a set of use cases for the patient and physician actors. Answer the following questions and/or complete the following exercises:

1. Which use cases include which other use cases? Modify the diagram to incorporate included relationships.

FIGURE 5-22 *RTGM system use cases*

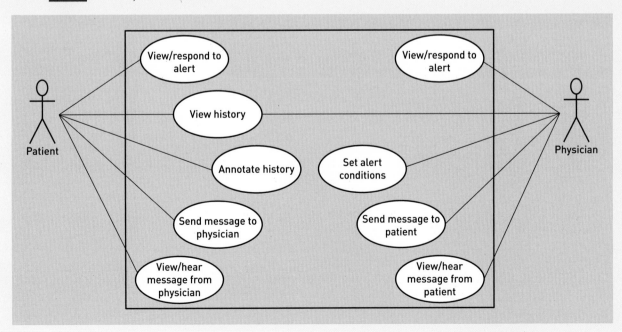

(continued on page 150)

(continued from page 149)

2. Consider the use cases *View/respond to alert* and *View history*. Both actors share the latter, but each has a different version of the former. Why do the actors have different versions of the *View/respond to alert* use case? Would the diagram be incorrect if each actor had his own version of the *View history* use case? Why or why not?

3. Develop an SSD for the *View history* use case. Assume that the system will automatically display the most recent glucose level, which is updated at five-minute intervals by default. Assume further that the user can ask the system to view glucose levels during a user-specified time period and that the levels can be displayed in tabular form or as a graph.

Further Resources

Grady Booch, James Rumbaugh, and Ivar Jacobson, *The Unified Modeling Language User Guide*. Addison-Wesley, 1999.

E. Reed Doke, J. W. Satzinger, and S. R. Williams, *Object-Oriented Application Development Using Java*. Course Technology, 2002.

Hans-Erik Eriksson, Magnus Penker, Brian Lyons, and David Fado, *UML 2 Toolkit*. John Wiley & Sons, 2004.

Martin Fowler, *UML Distilled: A Brief Guide to the Standard Object Modeling Language* (3rd edition). Addison-Wesley, 2004.

Philippe Kruchten, *The Rational Unified Process: An Introduction* (3rd edition). Addison-Wesley, 2005.

Craig Larman, *Applying UML and Patterns: An Introduction to Object-Oriented Analysis and Design and the Unified Process* (3rd edition). Prentice Hall, 2005.

Object Management Group, *UML 2.0 Superstructure Specification*, 2004.

PART 3

Essentials of Systems Design

6

Essentials of Design and the Design Activities

Chapter Outline

- The Elements of Design
- Inputs and Outputs for Systems Design
- Design Activities
- Design the Environment

Learning Objectives

After reading this chapter, you should be able to:

- Describe the difference between systems analysis and systems design
- Explain each major design activity
- Describe the major hardware and network environment options
- Describe the various hosting services available

OPENING CASE

Technology Decisions at Wysotronics, Inc.

As James Schultz walked down the hall toward a meeting with his staff, he thought about his new job. For a year now, Schultz had been the vice president and chief information officer for a medium-sized supplier of electronic components to several large electronics firms, including Samsung and Acer. James's company, Wysotronics, Inc., had been in business for many years but had recently been having some problems with its internal computer systems. James was hired to fix the problems.

Soon after starting his new job, James discovered that quite a few of the systems were functioning properly but that the infrastructure was a hodgepodge of disjointed computers and networks. On the corporate side, there were accounting systems and human resource systems, both of which were desktop client/server systems hosted on a local network computer that resided in the accounting department.

Engineering had its own database and network computers, which hosted several sophisticated engineering systems with intensive computing requirements. The engineers' local desktop systems were the most recent, up-to-date equipment and software. The server was also high capacity, with a large data repository to house all the engineering documents and images.

Marketing and Sales also had their own systems hosted on their own network server, which was also connected to the Internet. The sales staff worked closely with the manufacturing and assembly plants to ensure that deliveries were on time, and they were frequently on the road visiting Wysotronics's clients. It was their job to ensure that clients were satisfied with schedules, deliveries, and quality, and they wanted to be able to access the sales and production databases while they were on the road. Unfortunately, the servers they used weren't very stable and continually had problems.

Perhaps the biggest problem was the supply chain management system. Wysotronics had a large manufacturing plant, an assembly plant, and several suppliers that needed access to the inventory and supply chain system. The current infrastructure didn't have enough capabilities to provide timely information to these facilities and suppliers.

The meeting today was one of many to plan and configure the total infrastructure of corporate systems. As he walked into the room, James was greeted by William Hendricks, who would be making a presentation summarizing past decisions and future directions.

"Hi, Bill," James said. "Will you have some new recommendations for us today? Are there any surprises from your research?"

"No surprises," Bill said. "But you will be pleased to know that our research has validated the decisions you have made recently. We are providing better service to the company than ever before, and we are doing it at less cost than we ever have before. I do have a few recommendations about how to fine-tune our infrastructure to provide even better service, though."

It was obvious that Bill was pretty enthused about the results of his research.

"Before we start, can you give me a brief idea about where the cost savings are coming from?" James asked.

"Sure," Bill said. "As you know, we decided to create a virtual private network using the Internet for all our supply chain and production needs. We moved all the computers to support this system into a colocation facility. We still own the servers, but we have signed a service agreement with the hosting company to manage all the operating system and connection and network maintenance. This has allowed us to focus our efforts on the software itself and not to have to use valuable personnel worrying about the environment or connectivity. And we have not had to invest in additional buildings for a larger data center. Plus, the level of service is incredible. We have had almost 100 percent uptime since the switchover. The people in our plants really are pleased that they can check inventory levels and shipment dates from all their suppliers at any time."

"Wow. That is great news," James said. "And what are you going to recommend for our marketing and sales system?"

"Well, as you know, that is a Web-based system," Bill replied. "It doesn't have the security requirements that our production systems do, but it needs to be widely available. Our research has shown that we can deploy that system through a hosting company that provides 'cloud' computing. We have the option of going with our colocation provider or using another company we have used in the past. I think this other company is going to give us some good price concessions and will still be able to provide excellent service."

"That sounds great!" James said. "I'm interested in hearing about the details. I assume you have also laid out a migration plan to move the systems over?"

"Yes, I have done my homework on this one," Bill said. "I think you will be pleased with the results."

Overview

Previous chapters described the activities and decisions associated with discovering and understanding the major elements of the user's requirements—in other words, the analysis activities. This chapter focuses on the solution system. During analysis, the focus is on *understanding* what the system should do (i.e., the requirements), whereas during design, the focus is on the *solution* (i.e., specifying how the system will be built and what the structural components of the new system will be).

A question new developers often ask is "When are these tasks carried out in a real project?" Unfortunately, there is no single answer. Many projects begin with some of the design decisions having already been made, particularly with regard to the deployment environment when companies already have a strong technology infrastructure in place. For other projects, the new system may be the result of a new thrust for the organization and thus the decisions are wide open. However, it is normal for the project team to start thinking about these issues very early in the project and to begin making preliminary decisions as requirements are being defined. The topics discussed in this and the following chapters are solution-oriented design topics; however, you shouldn't try to come up with a solution until you understand the problem.

This is the first of several chapters that discuss design. Here, we briefly describe all the design activities and discuss the first activity (designing the environment) in more detail. Later chapters explore other design activities and explain in detail the various models and techniques used for systems design.

The Elements of Design

In Chapter 1, we defined systems design as those activities that enable the project team to describe in detail the system that solves the need. Obviously, there are many aspects of a system that need to be designed. The design of any complex artifact requires detailed design documents. For example, think of all the components that must be designed to build an ocean liner or a commercial aircraft. For a commercial aircraft, design ranges from the preliminary shape and size of the aircraft to the major subsystems, such as the mechanical system, the hydraulic system, and the electronic system, on down to the minutest details: the shape of the airfoil, the size and shape of seats, the placement of cockpit display and control devices, and even the external metering devices. Nothing is left to chance. Designing commercial airplanes today is only possible because of many years of experience and an extensive knowledge base regarding how to build airplanes.

The software applications being designed and constructed today are equally complex and are also only possible because so much infrastructure is already in place and there is a massive knowledge base and a set of development tools. Even with all the tools available, designing and constructing software applications is a difficult and complex process.

In this section, we first explore some of the different aspects and levels of design. Next, we take a high-level view of a computer application system to see what things must be included in a design. Later in this section, we identify the types of documents and products that are produced by the design process. Finally, we review what information and documents are available as inputs to design.

What Is Systems Design?

Systems design is really a bridge process. The objective of systems analysis is to thoroughly understand the organization's informational needs or requirements and to document those requirements in a set of specifications. The objective of software construction is to build a system that satisfies those requirements. Systems design, then, is the bridge that takes us from requirements to solution. The objective of systems design is to define, organize, and structure the components of the final solution system that will serve as the blueprint for construction. Another way to think about systems design is that whereas analysis tells us *what* the solution needs to do, design describes *how* the system will be configured and constructed.

Major Components and Levels of Design

Today, information systems are deployed on a range of devices—from individual computers and small mobile digital devices to localized networks of computers to large distributed and Internet-connected computers. We discuss several of these equipment and network configurations in a later section.

The design requirements vary dramatically depending on the targeted environment. Some applications never connect to the Internet, some connect periodically to retrieve specific information, and some need to have a continuous connection in order to execute. For example, you may own a laptop computer with several applications that execute only on your laptop. These might include a spreadsheet program, a word processing program, a tax preparation program, or a program that plays music. Of course, given the connectedness of today's world, even these small programs may have components that from time to time check for updates. However, these kinds of programs don't need to be connected to the Internet or to a network in order to carry out their fundamental purposes.

You may also own a mobile digital device or a smartphone with applications that run in a stand-alone (i.e., not Internet-connected) manner. Maybe you downloaded an application from the Internet, but once downloaded, it executes on its own. The design of these types of programs is usually only moderately complex.

On the other end of the range are applications that run in some type of distributed network environment. The environment can be either a localized private network, such as would be found in medium-sized corporations, or a distributed global network. Almost all large corporations have these types of internal systems that can be used only by employees. For example, a large global corporation may have a human resources system that is accessible by human resources personnel at many company offices throughout the world.

Another ubiquitous type of information system today is Web-based systems deployed on a server (or multiple servers) and accessed entirely through Web pages. Some of these types of systems are small and relatively simple, whereas others are extremely complex, requiring complex database access and sophisticated connectivity and interfaces with multiple external systems. An example of a simple system would be a personal blog hosted on a small server. A more complex system would be a comprehensive catalog and customer sales system, much like the new RMO CSMS.

network diagram a model that shows how the application is deployed across networks and computers

Figure 6-1 is a **network diagram** that illustrates one common configuration for today's information systems. A network diagram is a model that shows how the application is deployed across networks and computers. The system depicted in Figure 6-1 is a network-based system that is also accessible through the Internet. We illustrate this type of comprehensive system to identify the various system components that must be designed and built.

For the entire system, the analysts first identify the overall application deployment environment. This requires defining the hardware and software environments. The hardware environment shown in Figure 6-1 includes the computers, the networks, the firewalls, and so forth. The software environment includes such things as what operating systems, what database management system, and what kind of network protocol will be used.

FIGURE **6-1** *System components requiring systems design*

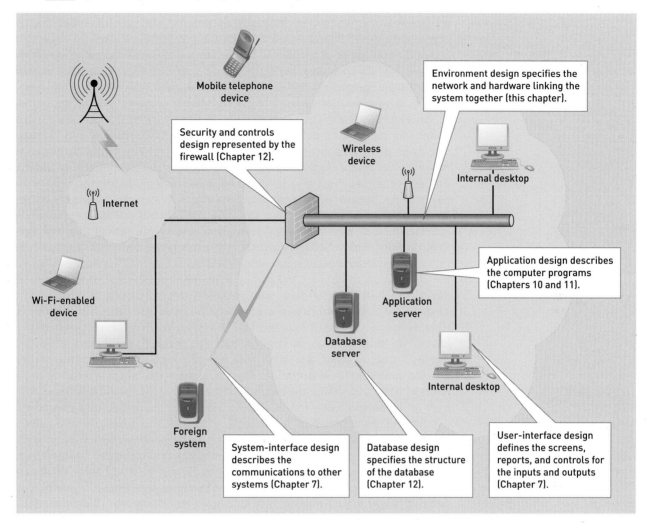

In most situations, the infrastructure for the application information system already exists and the new system must conform to the existing configuration. For example, most companies have an existing infrastructure of computers, networks, and communication devices. Therefore, some parts of the system design may be unnecessary because the new application will be integrated into the existing environment. However, even when there is an existing infrastructure, the new application will at times need to extend it to meet new requirements.

To perform design, analysts first partition the entire system into its major components because an information system is too complex to design all at once. The icons in Figure 6-1 refer to pieces of hardware, and inside the pieces of hardware are their software components. The large cloud on the right side of the figure represents the entire system, and the various icons within it show the parts of the system that must work together to make the system functional. Information systems professionals must make sure to develop a total solution for the users. They haven't done their job if they haven't provided an integrated, complete solution. In other words, design must include the overall infrastructure as well as the various components.

The infrastructure illustrated in Figure 6-1 is a common one in today's computing environments. The servers contain the software applications and the databases. Often, there are local desktop computers that access the application on a local network—either wired or wireless. The local network is isolated from the Internet by a firewall computer. The application is made available to the Internet and to the rest of the world via the firewall computer. The firewall computer is often connected directly to the Internet and provides access for computers and mobile digital devices through Internet Wi-Fi connections. Outside the firewall, connections to telephonic devices can also provide access to the application for mobile telephone devices. Not all applications require this expanded access to the Internet, but more and more companies allow their employees to work at remote locations where they need Internet access to internal applications.

In Figure 6-1, the callout boxes identify the various components requiring design. As mentioned in the previous paragraphs, the network and infrastructure need to be designed or extended at times. The application software that exists on the server needs to be designed, as does the accompanying database. The user interface must also be designed, whether it is desktop screens or Web pages. Some systems have automated interfaces to external systems, and those interfaces must be designed. For example, an internal purchasing system may interface directly with vendors and suppliers so they can submit purchase orders electronically. The following chapters will explain the details of designing these various components of an information system.

In addition to defining the components requiring design, an important idea underlying systems design has to do with the different levels of design. During analysis, we first identified the scope of the problem and described it with a systems vision document. Once we had a clear overview of the problems and issues, we drilled down to understand the detailed requirements. In other words, during analysis, we went from general, high-level information to more detailed and specific information. During design, we take a similar approach. At first, we try to design the overall infrastructure, then the major functions of each of the components, and finally the specific details within each component.

As you begin working in industry, you will find that various names are given to the design at the highest level, including *architectural design*, *general design*, and *conceptual design*. We use the term **architectural design**. During architectural design, you determine the overall structure and form of the solution before trying to design the details. Designing the details is usually called **detail design**. It isn't important at this point to distinguish which activities are architectural design and which are detail design, nor is it important to identify which models or documents belong to architectural design or to detail design. What is important is to recognize that design should proceed in a top-down fashion.

Let us review the implications of this approach for each of the design components identified in Figure 6-1.

For the application software component, the first step is to identify the various subsystems and their relationships to each other. The application also has to be configured to be consistent with the network, the database, and the user-interface components. Part of that early design of the application is the automation system boundary. The system boundary identifies which functions are included within the automated system and which are manual procedures.

For the database component, the first step is to identify the type of database to be used and the database management system. Some details of the tables, the data fields, and indexes might have already been identified, but the final design decisions will depend on the architecture as well as the details of the application.

For the user-interface component, the first step is to identify the general form and structure of the user dialog based on the major inputs and outputs. The project team also describes the relationship of the user-interface elements to the application software and the hardware equipment. In today's world, this can become rather complex because many different types of devices may need to connect to the system. Figure 6-1 illustrates internal computers (wired and wireless), Internet-connected computing devices, and smartphones that connect

architectural design broad design of the overall system structure; also called *general design* or *conceptual design*

detail design low-level design that includes the design of the specific program details

via telephonic protocols. The user interface has to be flexible enough to antici-pate all the various types of connections that will be used. After the overall con-nection and communication protocols have been decided on, detailed screen layouts and report formats can be developed.

Inputs and Outputs for Systems Design

During the analysis activities described in previous chapters, we build docu-ments and models. For object-oriented analysis, we use the event table and developed other models, such as class diagrams, use case diagrams, use case descriptions, activity diagrams, system sequence diagrams, and state machine diagrams. In the optional online chapter, we present the traditional analysis models, such as the event table, data flow diagrams, and entity-relationship diagrams. Regardless of the approach, the input to the design activities consists of those documents and models that are built during earlier activities.

In iterative projects, which we have covered in previous chapters and will explain in more detail in Chapter 8, analysis and design activities are often done concurrently. However, the first focus of any iteration has to be identifying and specifying the requirements (i.e., analysis); determining the solutions (i.e., design) comes later.

During analysis, analysts also build models to represent the real world and to understand the desired business processes and the information used in those processes. Basically, analysis involves decomposition—breaking a complex problem with complicated information requirements into smaller, more under-standable components. Analysts then organize, structure, and document the problem domain knowledge by building requirements models. Analysis and modeling require substantial user involvement to explain the requirements and to verify that the models are accurate.

Design is also a model-building activity. Analysts convert the information gathered during analysis—the requirements models—into models that represent the solution system. Design is much more oriented toward technical issues and therefore requires less user involvement and more involvement by other systems professionals. **Figure 6-2** illustrates this flow from analysis to design, highlight-ing the distinct objectives of analysis and design.

Design involves describing, organizing, and structuring the system solution. The output of the design activities is a set of diagrams and documents that achieves this objective. These diagrams model and document various aspects of the solution system.

The formality of a project also affects design. Formal projects usually require well-developed design documents, which are often reviewed in formal meetings. Developers on informal projects often create their designs with notepads and pencils and then throw away the design once the program is coded. This kind of informal design (used in many Agile projects) is merely a means to an end, which is the actual program code. However, even though outsiders don't see the design documents, the design process must still be followed. A programmer who jumps into code without carefully thinking it through—this if often referred to as *cowboy coding*—ends up with errors, patches, and poorly structured systems.

Figure 6-3 identifies the major models used for analysis and design. These lend themselves to object-oriented development, although they also share many characteristics with traditional development, as explained in online Chapter A. Notice that several analysis models have corresponding design models; for example, class diagram has a design class diagram. As mentioned earlier, devel-opment often flows smoothly from analysis into design, and this approach is further facilitated when the models are quite similar. Take the class diagram for example; here, the analysis model identifies the classes, attributes, and relation-ships, and the design class diagram adds more information, such as type

FIGURE **6-2**
Analysis objectives and design objectives

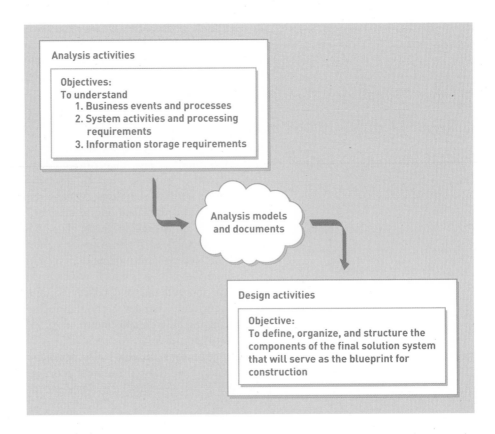

information for the attributes, keys, indexes, and the class methods or functions. It builds and expands on what the analysis-oriented class diagram offers.

You should be familiar with the analysis models shown in Figure 6-3. In the following chapters, you will learn how to create the design models shown on the bottom half of the figure. Two of them—package diagrams and nodes and locations diagrams—are primarily architectural design models. They document the overall structure of the final system. The design class diagram describes the classes in a way that is helpful for database design and application design. The sequence diagrams are an extension of the system sequence diagrams and document the flow of control and execution among the classes. The database schema is of course required to document the database for the application. For most current applications, a relational database approach is used. The user-interface models provide the layout for the computer screens and for the online or printed reports. Communication diagrams are similar to sequence diagrams in that they document the interactions between the classes in the program code. System security and control models aren't formal models; they are documents and notations that identify the important control features of the new application.

In the previous chapters, you learned about analysis activities and you developed skills in using the techniques and tools necessary to create the analysis models shown in the top half of Figure 6-3. In the life of a real project, once the systems analysts have begun to understand the user's business requirements and document those requirements by using the analysis models, they begin their design activities and extend the analysis models into the design models. Next, we discuss those design activities.

Design Activities

Figure 6-4 identifies the activities associated with the following core process that was discussed in Chapter 1: Design the system components that solve the problem or satisfy the need. Each design activity corresponds to the design of

FIGURE **6-3**
Analysis and design models

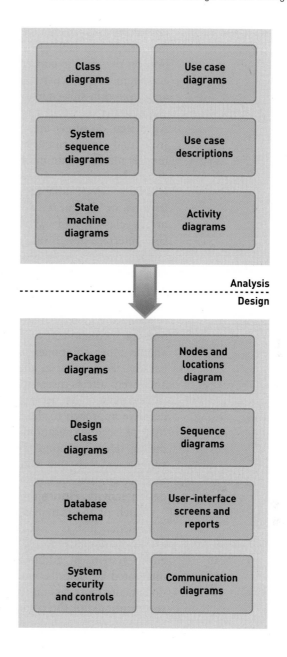

FIGURE **6-4** *Design activities*

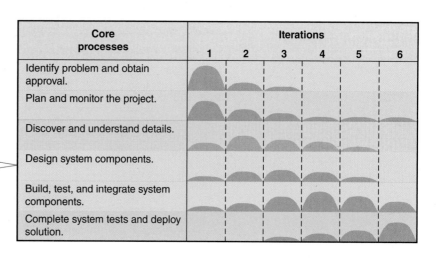

Design activities
Design the environment.
Design application architecture and software.
Design user interfaces.
Design system interfaces.
Design the database.
Design system controls and security.

one of the components identified in Figure 6-1. This section will provide a short introduction to each of these design activities. In-depth explanation and instruction on the specific concepts and skills are given later in the text. Designing the environment is discussed as the last section in this chapter, and the other topics are covered in subsequent chapters.

Systems design is a model-building endeavor, just as systems analysis was. As design decisions are made, especially at the detail level, they are derived from and documented by the building of models. As indicated earlier, the models may be quite informal, but they are the essence of design. For example, in database design, we identify which tables will be required and what fields will be in which table before we begin to build the tables with SQL statements. In software design, we decide which classes are the core classes and which are utility classes and what responsibilities (methods) each class will have. User-interface design often requires storyboards or other visual models to make efficient workflow decisions. All these systems design tasks are model-building tasks.

Systems design involves specifying in detail how a system will work when using a particular technology. Some of the design details will have been developed during systems analysis, but much more detail is often required. In addition, each component of the final solution is heavily influenced by the design of all the other components. Thus, systems design activities are usually done in parallel. For example, the database design is used heavily in software design and even affects user-interface design. Likewise, the application architecture drives many of the decisions for how the network must be configured. When an iterative approach to the SDLC is used, major design decisions are made in the first or second iteration; however, many designed components are revisited during later iterations. To better understand these design activities, you can summarize each one with a question. In fact, systems developers often ask themselves these questions to help them stay focused on the objective of each design activity. **Figure 6-5** presents these questions.

Each of the activities develops a specific portion of the final set of design documents. Just as a set of building blueprints has several different documents, a systems design package consists of several sets of documents that specify the entire system. In addition, just as the blueprints must all be consistent and integrated to describe the same physical building, the various systems design documents must be consistent and integrated to provide a comprehensive set of specifications for the complete system. For example, if an analyst is working on the user interface without consulting the database designer, the analyst could build an interface with the wrong fields or wrong field types and lengths.

FIGURE **6-5**

Design activities and key questions

Design activity	Key question
Design the environment	Have we specified in detail the environment and all the various options in which the software will execute?
Design application architecture and software	Have we specified in detail all the elements of the software and how each use case is executed?
Design system interfaces	Have we specified in detail how the system will communicate with all other systems inside and outside the organization?
Design user interfaces	Have we specified in detail how users will interact with the system to carry out all their tasks (use cases)?
Design the database	Have we specified in detail all the information storage requirements, including all the schema elements?
Design system controls and security	Have we specified in detail all the elements to ensure the system and the data are secure and protected?

Internal consistency is a mandatory element of effective system modeling and design. In what follows, we briefly discuss these design activities to better understand what is involved. In later chapters, you will develop the skills necessary for each of these activities.

Design the Environment

The environment is all the technology required to support the software application that is being developed. For example, in the development of RMO's CSMS, we have focused on the functional and nonfunctional requirements so far. However, the system will need to exist on a set of computer servers, desktop computers, mobile computers, and perhaps additional computing devices. Each of these computing devices will have an operating system, communication capabilities, diverse input and output capabilities, and so forth. Additional software—often referred to as *middleware*—may be needed to facilitate the integration of these diverse computing devices into a comprehensive solution. All these supporting systems—hardware and software—are considered part of the technology architecture, which we discussed in Chapter 2. For the new CSMS to be developed and deployed successfully, the complete environment must be precisely defined. Hence, the first step in the development of a new system is to define this environment.

Every software application must execute in some technology environment. This environment includes the computers and other hardware required for the deployment of the application as well as such things as server computers, desktop computers, mobile computers, firewalls, routers and cabling, fiber optics, and wireless access points. Some applications are simple stand-alone applications that execute on a single computer, laptop, or mobile computing device. Other applications are entirely server based and utilize an application server, a database server, and perhaps some content delivery network, with the users accessing all the application's functions on their computers through a local browser. Other applications are complex distributed applications in which the application itself and the data execute on various computers concurrently. Still other applications may be deployed to remote computing devices, such as smartphones or remote monitoring devices. Today's computing environment has become a world of connected technologies, many of which operate on different protocols and aren't entirely compatible. A big part of designing the environment is identifying and defining all the types of computing devices that will be required. That includes identifying all the locations and communication protocols necessary to integrate computing hardware.

The technology environment includes more than just the hardware. Another important component consists of the operating systems, communication protocols and systems, and other supporting software (i.e., middleware). For example, the deployment of a Web-based system will involve server operating systems as well as the operating systems on the users' computers. The server may also have other systems, such as the Web server, a database management system, a programming-language server, an image and graphics processor, or other specialty software. The design of the supporting software environment is even more complicated when the software systems cannot communicate directly with each other. These incompatibilities must be resolved.

Design the Application Architecture and Software

In designing the application architecture, we include decisions about the structure and configuration of the new system as well as the design of the computer software itself. One of the first steps in this design process is partitioning the software into subsystems. Decisions are also made about the database infrastructure and about the multilayer design in which the user interface is separated from the business logic and database processing. The technology architecture will drive many of these design decisions. For example, which subsystems need to reside on which pieces of equipment? Subsystems may be placed on different server computers based on importance, response time requirements, or privacy and security issues.

Other kinds of processing requirements influence the technology architecture and the application architecture. For example, should users be able to access the new system only at work on their desktops or should they also be able to work from home via an Internet connection? Is it necessary to allow remote wireless devices to connect to the system? What kind of transactions (use cases) and what volume of transactions must the new system be able to handle? These kinds of application decisions will drive the application architecture, the environment, and other hardware requirements. Designing the application architecture is usually a top-down process, with the overall structure defined first and then the detailed design of the various components.

The other part of application design is designing the application software at a detailed level. Detailed design is primarily a model-building activity. Creating models not only enables the design process, but it also provides the documentation necessary for writing code. These models include activity diagrams, sequence diagrams, design class diagrams, and other physical models. For the traditional approach, such models as data flow diagrams are developed. For example, for object-oriented design, one of the primary models is the design class diagram, which identifies the classes, their attributes, and their methods. **Figure 6-6** is a partial design class diagram for RMO's CSMS.

FIGURE **6-6**

Partial design class diagram for RMO's CSMS

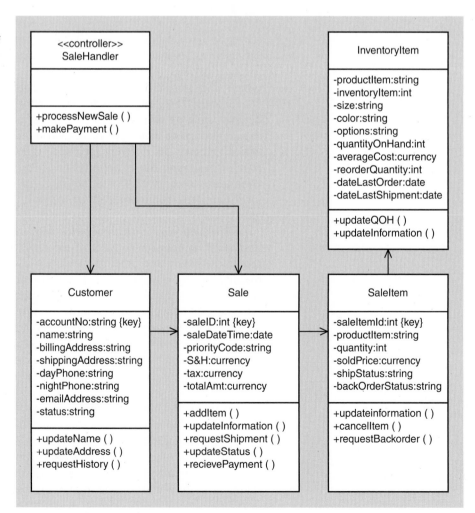

Chapters 10 and 11 will explain the details of how to design the application architecture and software.

Design the User Interfaces

Analysts should remember that to the user of a system, the user interface *is* the system. It is more than just the screens. It is everything the user comes into contact with while using the system—conceptually, perceptually, and physically. Thus, the user interface isn't just an add-on to the system.

New technology has led to many new requirements for the user interface. For example, will users only use computers with large screens or will they also use PDAs and other remote devices with small graphical areas? Will other devices be used for entering information, such as text, verbal commands, pictures, and graphics? These elements and requirements of the user interface need to be considered throughout the development process.

On desktops, laptops, and tablet computers, the interface is a graphical user interface with windows, dialog boxes, and mouse interactions. Increasingly, even these now include sound, video, and voice commands. On mobile devices, such things as touch screens, screen keyboards, voice commands and responses, and movement and positioning inputs and outputs are a standard part of the user interface. As information systems become increasingly interactive and accessible, the user interface is becoming a larger and more important part of the total system.

Designing the user interface can be thought of as an analysis and a design activity. It has elements of analysis in that the developers must understand the user's needs and how the user carries out his or her job. Not only must the user interface carry the right information, but it must also be ergonomically efficient and esthetically attractive. User-interface design is also a design activity in that it requires creativity and conformity to rigorous technology requirements. Many types of models and tools are used to perform user-interface design, including mock-ups, storyboards, graphical layouts, and prototyping with screen-modeling tools. One of the primary difficulties of designing the user interface in today's connected world is that the disparity between desktop screens and smartphone displays often necessitates multiple user interfaces for the same application. Chapter 7 describes many of the tools and techniques used to effectively carry out the user-interface design.

Design the System Interfaces

Few, if any, systems exist in a vacuum in today's connected computing environment. A new information system will affect and utilize many other information systems. Sometimes, one system provides information that is later used by another system, and sometimes, systems exchange information continuously as they run. The component that enables systems to share information is the system interface, and each system interface needs to be designed in detail.

The form of these interfaces will vary dramatically. In some cases, a file is sent from one system to another. In other cases, real-time data exchange is necessary, and live transactions are transferred between systems. In other cases, one system requires a service from another system, and a function call is performed via an application program interface. The format of the interchange can also vary, from binary format to encrypted formats to text-based formats.

From the beginning of a systems design, analysts must ensure that all the systems work together well. In some cases, the new system needs to interface with a system outside the organization—for example, at a supplier's site or a customer's home. Increasingly, organizations are linking systems together across organizational boundaries. For example, at RMO, the new Supply Chain Management System will have information flows from RMO to its key suppliers.

FIGURE **6-7**

Sample system-to-system interface using XML

```
<inventoryRecord>
        <productItem>WS39448-7</productItem>
        <inventoryItem>48763920</inventoryItem>
        <itemCharacteristics>
            <size>large</size>
            <color>blue</color>
            <options>withzippers</options>
        </itemCharacteristics>
        <orderRules>
            <quantityOnHand>54</quantityOnHand>
            <averageCost>38.27</averageCost>
            <reorderQuantity>25</reorderQuantity>
        </orderRules>
        <dates>
            <dateLastOrder>06042012</dateLaseOrder>
            <dateLastShipment>08072012</dateLastShipment>
        </dates>
</inventoryRecord>
```

The new CSMS will also require linkups with the Supply Chain Management system as well as real-time links to banks and other credit verification organizations. One standardized method for defining text-based system interfaces is to use eXtensible Markup Language (XML). Much like HTML, XML uses tags to define the structure of the record. **Figure 6-7** presents an example of an XML record.

Some system interfaces link internal organizational systems, so the analyst may have information available about other systems. Internally at RMO, the Supply Chain Management system will have real-time communication with the Trade Show System that was described in Chapter 1. The Sales subsystem must have access to the supply chain warehouse database in order to know which items are in stock and which aren't available.

System interfaces can become quite complex, particularly with so many types of technology available today. System-interface design is discussed in more detail in Chapter 7.

Design the Database

An integral part of every computer information system is the information itself, with its underlying database. The data model (the domain model) is created early during systems analysis and is then used to create the implementation model of the database. Usually, the first decision is determining the database structure. Sometimes, the database is a collection of traditional computer files. More often, it is a relational database consisting of dozens or even hundreds of tables. Sometimes, files and relational databases are used in the same system. Another decision that needs to be made is whether the database is centralized or distributed. The internal properties of the database must also be designed, including such things as tables, attributes, and links. **Figure 6-8** is an example of an RMO database table definition for inventory items in MYSQL.

Analysts must consider many important technical issues when designing the database. Many of the technical (as opposed to functional) requirements defined during systems analysis concern database performance needs (such as response times). Much of the design work might involve performance tuning to make sure the system actually works fast enough. Security and encryption issues, which are important aspects of information integrity, must be addressed and designed into the solution. Given today's widespread connectivity, a database may need to be replicated or partitioned at various locations around the world. It is also not uncommon to have multiple databases, with distinct database

FIGURE **6-8** *Sample database table definition in MYSQL*

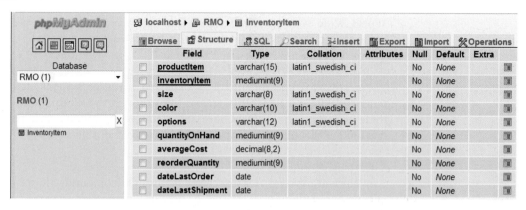

management systems. These databases may be distributed across multiple database servers and may even be located at completely different sites. These highly technical issues often require specialized skills from experts at database design, security, performance, and physical configuration. A final key aspect of database design is making sure the new databases are properly integrated with existing databases. Chapter 12 describes database design in detail.

Design the Security and System Controls

The final design activity is ensuring that the system has adequate safeguards to protect organizational assets—the safeguards referred to as system controls. This activity isn't listed last because it is the least important. On the contrary, especially in today's culture, where outsiders can cause severe damage to a system and its data, designing system controls is a crucial activity. The design of security and system controls should be included in all other design activities: user interface, system interface, application architecture, database, and network design.

User-interface controls limit access to the system to authorized users. System-interface controls ensure that other systems cause no harm to this system. Application controls ensure that transactions are recorded precisely and that other work done by the system is done correctly. Database controls ensure that data is protected from unauthorized access and from accidental loss due to software or hardware failure. Finally—and of increasing importance—network controls ensure that communication through networks is protected. All these controls need to be designed into the system based on the existing technology. Specialists are often brought in to work on controls, and all system controls need to be thoroughly tested. Control issues are addressed in several chapters but most explicitly in Chapter 12.

Design the Environment

The first activity in the list of design activities is designing the environment. This activity is also listed first because it permeates all the other design decisions. For example, a stand-alone, single-desktop system will require very different design decisions for the software, the user interface, the system interfaces, and the database than a complex interconnected and distributed system. Even though all the detailed design decisions concerning the environment might not be completed at the beginning of the project, the major decisions are addressed.

There is an incredible variation in the software systems being deployed today as well as an explosion in the types of devices and configurations that have software applications. Consequently, there is no easy way to organize and discuss the issues that are relevant to designing the environment. In this section, we address issues related to three major industry trends in software deployment: software systems deployed entirely within an organization, software systems built for purely external use (in our case, deployed on the World Wide Web via the Internet), and software systems deployed remotely in a distributed fashion (for internal and external use).

Design for Internal Deployment

There are two types of internally deployed software systems: stand-alone systems and internal network systems. Even though the internal environments for these two types of systems have some common features, each type of system has unique requirements that must be considered during design.

Stand-Alone Software Systems

Any software system that executes on a single computing device without connecting externally via an Internet or network connection is a stand-alone system. Companies still develop stand-alone systems, but most of them are developed by individuals and then sold or delivered to companies or other individuals. For example, the Microsoft Office Suite and the Apple iWork Suite, which allow word documents, spreadsheets, and presentations to be created, execute primarily on single computers. Likewise, many people use such stand-alone software as QuickBooks or H&R Block At Home—programs that are acquired either as packages of CDs or downloaded as executable install files. Another type of stand-alone system are the games that many people download and play on their laptops or desktops.

Design issues for stand-alone systems are usually straightforward. These systems usually read and write data into files without database access. The biggest issue with stand-alone systems is that they often need to be deployed on various pieces of equipment. For example, a tax program may need versions that run on PCs under Windows, PCs under Unix, and Apple computers using the Mac operating system. Other stand-alone programs may need to run in these three configurations as well as be able to run on mobile devices, including tablet computers and smartphones. Each of these environments requires slightly different versions of the interface with the operating system and the functions of the user interface. Usually, different versions of the application will be designed and built so the one best suited for a particular device is deployed.

Internal Network-Based Systems

An internal network-based system is one that is for the exclusive use of the organization that builds it or buys it. It isn't meant to be used by anyone except company employees who are located within the organization's physical facilities. **Figure 6-9** is a network diagram that illustrates a possible hardware configuration for this type of system. Often, such a hardware environment is referred to as a **local area network (LAN)**—a computer network in which the cabling and hardware are confined to a single location, such as a building.

This configuration depicts a simple **client-server architecture** for an internal network system. What distinguishes a client-server architecture from a single-computer architecture is that the individual computers in a client-server architecture must be connected to a server. The computers that users do their work on are called the **client computers**, whereas the main computer is called the **server computer**. The latter "serves" functions and data, which the client computers receive. There are two kinds of systems that can be deployed in a client-server architecture:

■ Desktop application systems
■ Browser-based application systems

The simplest version of a desktop system is a computer program that executes on a client computer. That situation may not even require a server computer. However, many desktop systems access a server computer to retrieve

local area network (LAN) a computer network in which the cabling and hardware are confined to a single location

client-server architecture a computer network configuration with user's computers and central computers that provide common services

client computers the computers at which the users work to perform their computational tasks

server computer the central computer that provides services (such as database access) to the client computers over a network

FIGURE **6-9**

Network diagram for an internal network system

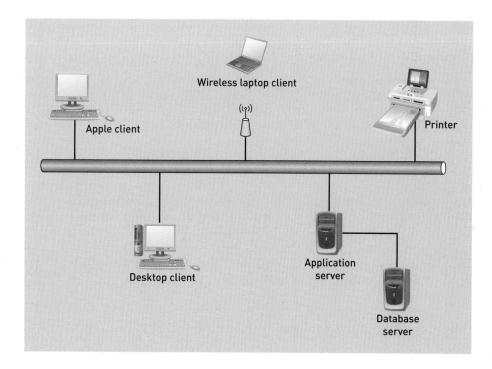

and update data from a database. Even more sophisticated desktop systems may consist of computer programs that communicate together between the client and the server computers. The advantage of this type of system is that the presentation (i.e., the user interface) and the functionality can be customized to the exact requirements of the users. Examples of these types of systems include graphical or engineering systems in which the processing and presentation requirements are very strict and very intensive.

The other type of internal network system is one that is browser based. In a browser-based system, the presentation of screens and reports to the user's computers (i.e., the clients) is handled by an Internet browser, such as Internet Explorer, Firefox, Chrome, or Safari. In this configuration, most of the processing and heavy calculation is done by the server and then passed to the client computers as **Hypertext Markup Language (HTML)** pages. This puts a heavier load on the server computer because it not only has to serve data to all the clients, but it also has to do the processing for all the clients. Therefore, high-speed computers are usually purchased to provide the necessary computing power. Another disadvantage is that the presentation of the user-interface screens and reports must conform to the capabilities provided by the browsers. Often, this isn't a major problem, but at times, it can be limiting. An advantage of using a browser-based design is that the system can easily be extended outside the local LAN and deployed via the Internet. These kind of systems use the same transmission protocol as the Internet: **Transmission Control Protocol/Internet Protocol (TCP/IP)**.

Hypertext Markup Language (HTML) the predominant language for constructing Web pages and which consists of tags and rules about how to display pages

Transmission Control Protocol/ Internet Protocol (TCP/IP) the foundation protocol of the Internet; used to provide reliable delivery of messages between networked computers

three-layer architecture a client/ server architecture that divides an application into view layer, business logic layer, and data layer

view layer the part of the three-layer architecture that contains the user interface

Three-Layer Client-Server Architecture

One effective method of software design is to separate the user-interface routines from the business logic routines and separate the business logic routines from the database access routines. This method of designing the application software is called **three-layer architecture**. Three-layer architecture is used for all types of systems, including desktop applications and browser-based applications. A three-layer architecture divides the application software into three layers:

■ The user interface or **view layer,** which accepts user input and formats and displays processing results

FIGURE **6-10** *Abstract three-layer architecture*

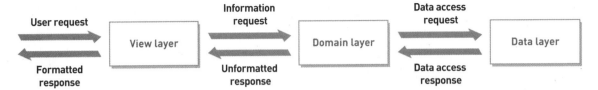

business logic layer or domain
layer   the part of a three-layer architecture
that contains the programs that implement the
business rules and processes

data layer   the part of a three-layer architecture that interacts with the data

- The **business logic** or **domain layer**, which implements the rules and procedures of business processing
- The **data layer**, which manages stored data, usually in one or more databases

Figure 6-10 illustrates in the abstract how these three layers work together to respond to a user request for processing or information. This has proven to be an effective approach to building software. It is effective because the programmers can more easily focus their attention on solving one issue at a time. It is also easier to upgrade and enhance different portions of the system. For example, the user interface can be changed with only minimal impacts on the business logic routines.

One of the advantages of a client-server architecture is that it easily supports—in fact, encourages—software to be developed by using an application program three-layer architecture. **Figure 6-11** illustrates an internally deployed system with a three-layer architecture, and it shows how the three layers might be configured across three separate computing platforms.

The view layer resides on all the client computers as well as on a portion of the application server computer. The HTML is rendered and displayed by the browser on the client computers. The view layer classes that format the HTML are on the application server. The data layer consists of the database server and any application programs on the application server that are necessary to access

FIGURE **6-11** *Internal deployment with three-layer architecture*

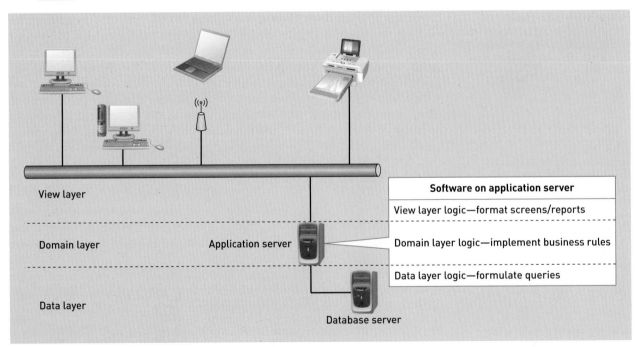

the data. The business logic layer resides on the application server computer and includes all the logic to process the business rules.

A major benefit of using three-layer architecture is its inherent flexibility. Interactions among the layers are always requests or responses, which make the layers relatively independent of one another. It doesn't matter where other layers are implemented or on what type of computer or operating system they execute. The only interlayer dependencies are a common language for requests and responses and a reliable network with sufficient communication capacity.

Multiple layers can execute on the same computer or each layer can operate on a separate computer. Complex layers can be split across two or more computers. System capacity can be increased by splitting layer functions across computers or by load sharing across redundant computers. In the event of a malfunction, redundancy improves system reliability if the server load can be shifted from one computer to another. In sum, three-layer architecture provides the flexibility needed by modern organizations to deploy and redeploy information-processing resources in response to rapidly changing conditions. We will discuss the software aspects of three-layer design in Chapters 10 and 11.

Design for External Deployment

The largest and most rapidly growing arena for new software applications is the deployment of systems that are purely for external use on the Internet. The tremendous increase in broadband connectivity capability and Web-enabled devices has generated incredible opportunities for the creation of purely online businesses. Today, almost all "brick and mortar" businesses have extended their business models to include online purchase of goods and services. Tremendous growth has also occurred in home-based and other small businesses that only do business via the Internet. The software applications that support these business activities are, in most cases, built purely for external use. In other words, there is no need for in-house employees to use these systems. They are to be used by customers who aren't part of the hosting organization. Important issues related to the environment for externally deployed systems include:

- Configuration for Internet deployment
- Hosting alternatives for Internet deployment
- Diversity of client devices with Internet deployment

Configuration for Internet Deployment

Figure 6-12 illustrates a simple Internet deployment configuration. Notice that it is quite similar to the three-layer architecture in the last section. In fact, almost all Internet-deployed applications use a three-layer architecture. The back end (i.e., the application server and the database server) provides the same functionality as an internally deployed client-server system. The view layer architecture has some similarities but also has more complex requirements due to the varied and insecure nature of the Internet. The view layer consists of the HTML pages that are rendered by a browser. It also includes those programs or classes that reside on the server and that format the dynamic HTML.

Internet and Web technologies present an attractive alternative for implementing information systems used by external customers and organization employees. For example, consider the data entry and data access needs of an RMO buyer who purchases items from the company's suppliers. Buyers are typically on the road for several months a year—often for weeks at a time. A traveling buyer therefore needs some means of remotely interacting with RMO's Supply Chain Management (SCM) system to record purchasing agreements and query inventory status.

FIGURE **6-12**

Internet deployment of software applications

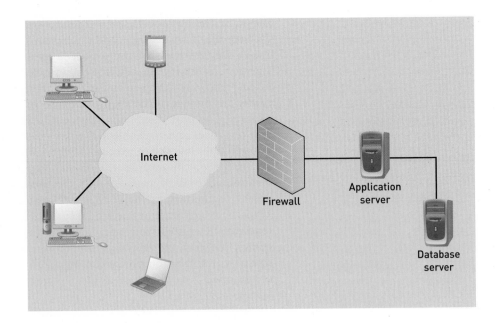

Implementing an application via the Web has a number of advantages over traditional client/server applications, including:

■ Accessibility—Given that Web browsers and Internet connections are nearly ubiquitous, Web-based applications are accessible to a large number of potential users (including customers, suppliers, and off-site employees).

■ Low-cost communication—The high-capacity networks that form the Internet backbone were initially funded primarily by governments. Traffic on the backbone networks travels free of extra charges to the end user. Connections to the Internet can be purchased from a variety of private Internet service providers at relatively low costs.

■ Widely implemented standards—Web standards are well known, and many computing professionals are already trained in their use.

Of course, there are problematic aspects of application delivery via the Internet and Web technologies, including:

■ Security—Web servers are a well-defined target for security breaches because Web standards are open and widely known. Wide-scale interconnection of networks and the use of Internet and Web standards make servers accessible to a global pool of hackers. This is probably the most serious issue that must be addressed with external deployment of applications. Protection must be provided for the home systems, including the data, and for the data as they are transmitted over the Internet.

■ Throughput—When high loads occur, throughput and response time can suffer significantly. The configuration must support not only daily average users but also a peak-load number of users. This is unpredictable and can vary widely.

■ Changing standards—Web standards change rapidly. Client software is updated every few months. Developers of widely used applications are faced with a dilemma: Use the latest standards to increase functionality or use older standards to ensure greater compatibility with older user software.

For RMO, the primary disadvantages of implementing the customer order application or even an RMO buyer-accessing home system via the Internet are security and throughput. If a buyer can access the system via the Web, so can anyone else. Access to sensitive parts of the system can be restricted by a number of means, including user accounts and passwords. But the risk of a security breach will always be present. Protection of data while in transit is important because "sniffer" software may discover user IDs, passwords, and sensitive data during transmission.

Protection of data while in transit is accomplished through **Hypertext Transfer Protocol Secure (HTTPS)**, which is a combination of Hypertext Transfer Protocol (HTTP) and **Transport Layer Security (TLS)** protocol. Web pages that are served through the HTTPS protocol are transmitted in encrypted format, which can be made quite secure. You will learn more about secure transmissions in Chapter 12.

Performance is affected by several factors. First, of course, is the capacity of the server computer and the amount of traffic that it must support. **Figure 6-13** illustrates a simple configuration, with only a Web server, which hosts the software application, and a database server. However, as volumes increase, increased capacity is provided by utilizing larger, more powerful servers and by adding more servers. Conceptually, the configuration is the same as the Internet deployment configuration shown in Figure 6-12: a Web application server and a database server. However, each of two servers becomes many servers that have equipment on the input side to distribute page requests and database requests across these servers. Figure 6-13 illustrates a typical data center configuration that uses multiple application servers and multiple database servers. This configuration adds another layer of complexity to the design of the computing environment.

Many companies that support very high volumes also build server farms, which consist of multiple data centers positioned around the country or even around the world. Each data center houses many individual servers that are linked together with load-balancing hardware. This adds even more complexity as requests for access to the application are routed to the correct data center,

Hypertext Transfer Protocol Secure (HTTPS) an encrypted form of information transfer on the Internet that combines HTTP and TLS

Transport Layer Security (TLS) an advanced version of Secure Sockets Layer (SSL) protocol used to transmit information over the Internet securely

FIGURE **6-13**
Multiple server configuration

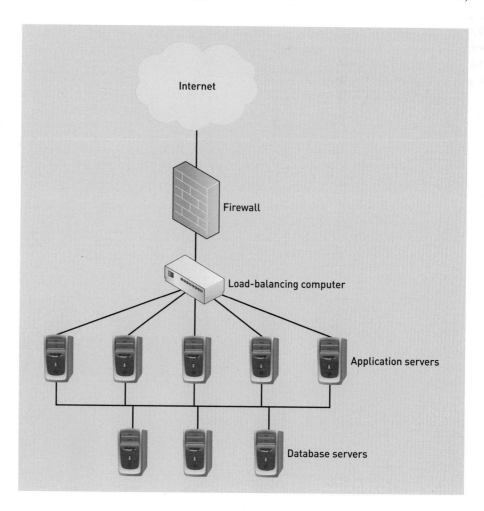

which is frequently the nearest data center but not always so. Sophisticated synchronization algorithms and software are required to keep data current in the various data centers.

Throughput can also be increased by using a **content delivery network (CDN)**. This is an additional set of computers that can be used to deliver static content, such as images or videos. For example, when an RMO customer requests a page from the catalog, the application software determines all the information that is to be returned on the page. The page is dynamically created based on request by using data from the database. However, many of the images that may need to be returned are static images that seldom change. In other words, they aren't dynamically changing. Rather than take up bandwidth going into and out of the data center firewall and load balancer, RMO could use a CDN server to deliver all the images and videos that it uses. **Figure 6-14** is an example of such a configuration, with multiple servers and a separate location for content delivery.

Finally, performance is limited by the RMO user's Internet connection point and the available Internet capacity between that connection and the application server. Unreliable or overloaded local Internet connections can render the application unusable. RMO has no control over the user's connections.

Hosting Alternatives for Internet Deployment

Software applications that are developed for purely external use open up many hosting alternatives. **Hosting** refers to running and maintaining a computer system on someone's behalf where the application software and the database reside. There are several critical issues that must be considered in the

content delivery network (CDN)
a set of server computers, separate from the hosting computers, used to deliver such static content as images or videos

hosting the process of providing physical servers at a secure location and selling those services to other businesses that wish to deploy Web sites

FIGURE **6-14**

Multiple server configuration with content delivery network

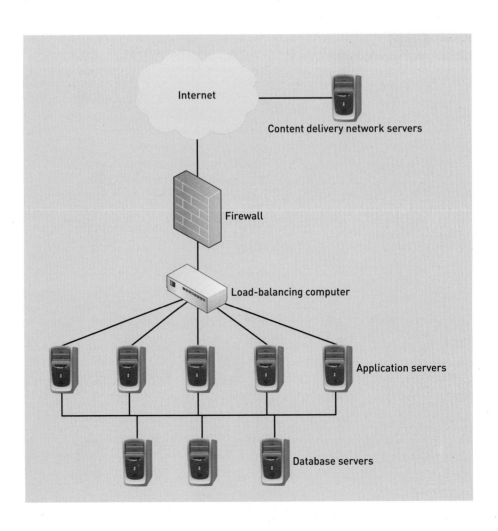

deployment of any system for external use, especially for systems utilized by customers or other outside parties. They include:

■ Reliability—The hardware environment must be completely reliable because customers and other outside parties usually have a very low tolerance for systems that aren't available. This often requires mirroring computers, hard drives, and database records. Backup and recovery must be well established.
■ Security—The systems—hardware and software—must be secure. The current legal regulations for financial and health care data require a very high level of security. Penalties are severe.
■ Physical facilities—To ensure reliability and security, special rooms or even special buildings are often required. In addition, Internet connectivity frequently requires multiple access routes to the Internet backbone. Electrical power must be secure, which often means having backup generators on-site. And air conditioning units must be adequate, with backup facilities, to ensure a constant physical environment.
■ Staff—To ensure reliability and security, a well-qualified technical staff needs to be on-site 24/7.
■ Growth—External systems often grow dramatically as a business expands, increasing the number of servers needed to respond to the traffic. When the number of application servers and database servers increases, there is the need for more sophisticated load balancing. Growth can also leave the physical facilities outstripped, necessitating multiple data centers.

Because of these issues, many companies are outsourcing their hardware environments. In recent years, there has been a large expansion of the services offered to companies to host their applications. A few of the more popular alternatives are now discussed.

colocation a hosting service with a secure location but in which the computers are usually owned by the client businesses

Colocation In a **colocation** arrangement, one company provides a secure data center where other companies (i.e., clients) house their server computers. One feature that comes with the data center includes a securely locked and protected site that meets all regulatory requirements for financial and health care records. The site also often has multiple high-capacity connections to the Internet backbone. And it is often integrated into multiple power grids and has its own emergency power generators. A client can rent rack space to house its own computer servers or it may lease computers from the host provider. Managing the server—its operating system, Internet software, database management software, data backup, and so forth—is done remotely. After the initial setup, the client seldom, if ever, goes to the site. The advantage of this kind of service is that it doesn't incur the costs of a physical, secure, complex data center.

Managed Services A client may want to purchase additional services, such as installing and managing the operating system, the Internet servers, database servers, and load balancing software. The client maintains its own software but doesn't have to hire staff to manage the operating environment. These services are usually called managed services, and almost all hosting companies will provide them. Usually, the client company either owns or leases a specific number of computers for its servers. The advantage of this service is that the client company doesn't have to hire special technical staff to manage the server system software.

virtual server a method to partition the services of a physical Web server so it appears as multiple, independent Internet servers

Virtual Servers In this arrangement, the client company leases a **virtual server** that is configured as a real server, with a certain amount of CPU capacity, internal memory, hard drive memory, and bandwidth to the Internet. How the computer hardware is configured is unknown to the client, which just buys (usually on a monthly or annual lease) a specific server configuration. The client company can purchase virtual servers with or without managed services. The provider company uses special system software to configure a virtual world for that client company. Companies use this type of service for applications with very low volume that don't require the computing power of a whole computer. Prices for these types of services range from only a few dollars a month to one or two thousand dollars a month depending on the size and power of the virtual server. The advantage of this service (in addition to those just listed) is that the client company can start small and add more capacity as needed. Typically, the client purchases a virtual server and increases capacity in a stepwise fashion, adding either a larger virtual server or additional virtual servers as needed.

Cloud Computing   There are two philosophies behind **cloud computing**. First, a client should be able to buy computing capacity much like one purchases such a utility as water or electricity. In other words, the client only purchases as much as is needed and used. Second, the client shouldn't have to be concerned with such environmental issues as how or where this computing capacity is provided, just as an individual doesn't have to worry about how electricity is generated. With cloud computing, the client company purchases computing capacity (with related memory, hard drive storage, and bandwidth) in very small increments for very short time periods. The client company specifies a required environment, such as a Unix operating system with an Apache Web server, but doesn't have any contact with the operating environment. In other words, the client's application software runs "in the cloud." When growth occurs, the cloud automatically provides more capacity. Supposedly, this arrangement saves the client company money because it doesn't have to buy capacity that it doesn't need. At present, this is the ultimate in the separation of application software and the operating environment.

One major selling point with all the hosting companies is the reliability of their equipment and Internet access. Most contracts for these kinds of services include a **Service Level Agreement (SLA)**, which guarantees a specific level of system availability. The volume and activity on a Web site (i.e., a particular software application) makes availability extremely important. For example, how much revenue would Amazon lose if its system weren't available for even a minute during a peak time of day? It isn't unusual to see SLAs that guarantee availability 99.9 percent of the time, with penalties for nonperformance. Some providers even guarantee 100 percent availability. Providers are able to make this guarantee because they have multiple server farms with several layers of redundancy and backup built in.

Figure 6-15 lists hosting options and their various capabilities.

cloud computing   an extension of virtual servers in which the resources available include computing, storage, and Internet access and appear to have unlimited availability

Service Level Agreement (SLA)   part of the contract between a business and a hosting company that guarantees a specific level of system availability

FIGURE **6-15** *Attributes of hosting options*

HOSTING OPTIONS				
Service options	**Colocation**	**Managed services**	**Virtual servers**	**Cloud computing**
Hosting service provides building and infrastructure	Yes	Yes	Yes	Yes
Client owns computer	Yes	Perhaps	No	No
Client manages computer configuration	Yes	No	Possible	No
Scalability	Client adds more computers	Client adds more computers	Client buys larger or more virtual servers	Client adds small increments of computing power
Maintenance	Client provides	Host provides	Host provides	Host provides
Backup and recovery	Client provides	Host provides	Available	Available

Diversity of Client Devices with Internet Deployment

Another critical issue with applications deployed for external use is the extremely wide range of client devices. The problem is that the various devices have different screen sizes, screen display characteristics, Internet browsers, and operating environments. This issue is ameliorated somewhat by the fact that the devices used to view Web pages usually provide some type of browser application as part of their standard software. However, browsers on different devices differ in their capabilities. Designing and implementing the user interface for these browsers is always a challenge.

Client devices fall into three categories by size: full-sized computers, mid-sized tablet computers, and small mobile computing devices.

The full-sized devices include desktop and laptop computers. These normally have full-sized 15-inch or 17-inch screens, although it isn't uncommon to see monitors with 24-inch or 28-inch screens. In addition, these full-sizes devices offer levels of resolution that allow them to display a high level of detail.

The mid-sized tablet devices have more standard display sizes. Most have a screen size of approximately 10 inches, although a few have a screen size of 12 inches. Most tablets can be viewed in either landscape or portrait mode, which the Web designer may need to take into consideration. The resolution is usually lower and less detail can be viewed on these smaller screens.

The number of mobile computing devices has grown enormously in recent times. Not only are these devices viewable in landscape or portrait mode, but there is also a much wider variation of screen size and resolution. Designing the user interface is a big challenge if users are to receive the best possible viewing experience. The small screen size puts a big limitation on the amount of detail that can be presented, but this can be offset somewhat by the devices' zooming capabilities.

It isn't uncommon to build two or three separate view layers so a software application can be viewed on all three types of devices. In fact, most new Internet software applications do have at least two separate view layers to accommodate the disparity in these devices. In some cases, the difference between the view layers is nothing more than the way the HTML is formatted. In other cases, entirely different screens are presented by the various types of devices.

Design for Remote, Distributed Environment

A remote, distributed environment has characteristics of the internal environment and the external, Web-based environment. As with internal configuration, the software applications for a remote, distributed environment are often internal systems used by employees of a business. As with the external, Web-based deployment, the employees aren't constrained to a single location; in fact, they can range throughout the world. Historically, many companies have built their own WANs to service these employees. However, the expense of building and maintaining these individual communication networks has become prohibitive for most companies. Today, almost all these systems are built by using the Internet and are called **Virtual Private Networks (VPNs)**. A VPN is a network built on top of a public network such as the Internet, which offers security and controlled access for a private group.

Virtual Private Networks (VPNs) a closed network with security and closed access built on top of a public network, such as the Internet

Remote Deployment via Virtual Private Network

Earlier, we described an RMO buyer who is on the road and needs access to a home office system. If the buyer just needs access to a few pages, a secure TCP/IP connection with HTTPS is sufficient. However, if the buyer needs access to other secure systems within the home office, a VPN might be a better solution. With a VPN, the buyer can access the home office servers as though he or she were working within the home office building.

FIGURE 6-16
Virtual Private Network using TCP/IP

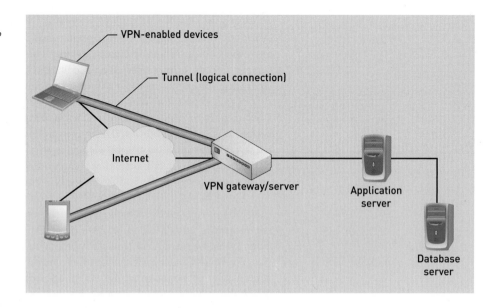

peer-to-peer connection when independent computers communicate and share resources without the need of a centralized server computer

Figure 6-16 illustrates a VPN over the Internet using TCP/IP. As indicated in the figure, there is a secure "pipe" between the remote computers and the home office server. This indicates the presence of a VPN using Internet protocols but with more security and control. In order to implement this type of VPN, special software is used to establish a secure connection and to encrypt all data transmissions. Only computers with the exact software and keys can access the VPN network.

A variation of this configuration can occur when two remote computers need to communicate directly with each other. They can use the home office server to facilitate the connection between them, but after they are connected, they can establish a **peer-to-peer connection** that continues the communication without any other assistance. A peer-to-peer connection is one that goes directly between the remote computers and doesn't require access to the home computer. That configuration can be represented by a pipe directly between the remote computers. The advantage of this configuration is that the speed of the connection can handle high-volume transfers or immediate responses, which is beneficial for, say, online chatting.

An alternative way for a buyer to implement remote access is to construct an application that uses a Web-browser interface. The application executes on a Web server, communicates with a Web browser using HTML, and is accessible from any computer with an Internet connection. Buyers can use a Web browser on their laptop computers and connect to the application via a local Internet service provider. They can also access the application from any other computer with Internet access (e.g., a computer in a vendor's office, a hotel business suite, or a copy center, such as FedEx Kinko's).

An important aspect of all VPNs is that the communication links are always encrypted to maintain security. Because the objective of a VPN is to allow private communications between persons in the same organization, the communication is encrypted. VPN servers and software not only use the secure HTTP protocol (HTTPS://), they include additional authorization, more secure encryption, and a higher level of transport monitoring.

Diversity of Client Devices

In the previous section, we discussed the difficulties of deploying software on a wide range of devices. Software applications deployed remotely have even more complex rendering requirements. Often, specialized equipment is needed to deploy an application to remote employees—for example, when a courier service transmits customer signatures back to the home office upon a package's delivery. Other types of monitoring devices have data capture and data communication requirements unique to a particular software application.

RMO Corporate Technology Architecture

RMO's main offices consist of the corporate headquarters as well as a large retail store, a manufacturing plant, and a large distribution warehouse in Park City, Utah. Park City is where the company got its start and where it opened its first retail store. Salt Lake City is, in many ways, the hub for RMO's daily operations. The primary data center is located in a separate building in Park City. There are two distribution centers: one in Portland, Oregon, and the other in Albuquerque, New Mexico. Additional manufacturing is done in Seattle, Washington. The map shown in **Figure 6-17** indicates where the 10 retail stores are located.

Along with the major data center in Park City, RMO has internal LANs in every office, warehouse, manufacturing plant, and retail store. In addition, the distribution centers and manufacturing plants are all on a VPN that connects these facilities to the central data center and to the corporate offices in Park City. All the retail outlets are connected to the central site, with a separate VPN that connects with the retail applications. Within each building, LANs are

FIGURE **6-17**

*Map showing RMO's warehouses,
manufacturing plants, and retail stores*

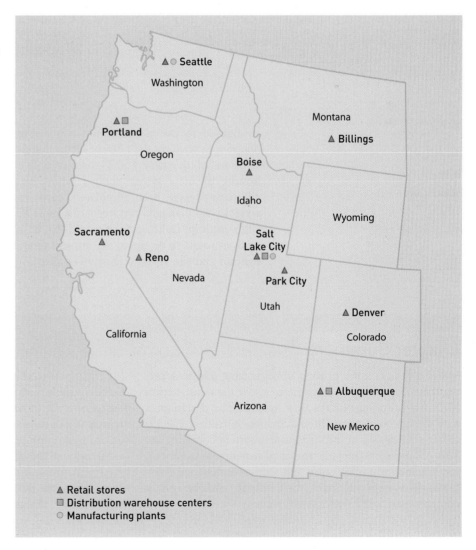

△ Retail stores
▢ Distribution warehouse centers
○ Manufacturing plants

FIGURE **6-18** *RMO's current technology architecture*

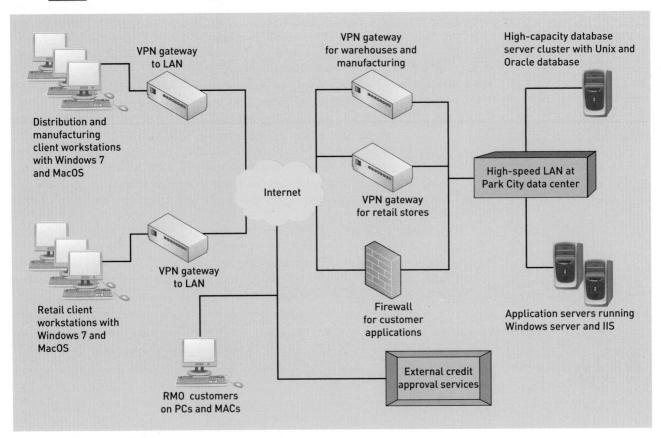

configured to provide common connectivity. **Figure 6-18** provides a network diagram describing the current technology configuration.

Part of the new CSMS project will consist of assessing the feasibility of hosting the new system with a large-capacity provider. Although none of the alternatives have been eliminated, early analysis indicates that utilizing virtual servers with managed services is the most favorable option. Hosting companies in Utah and in California appeared to be the most attractive options. Further analysis needs to be done, and project time is scheduled for more thorough investigation and site visits to the two or three most promising hosting companies.

Chapter Summary

Systems design is the process of organizing and structuring the components of a system to allow the construction (i.e., programming) of the new system. The design of a new system consists of those activities that relate specifically to the design of the various new system components. The components include the deployment environment, the application architecture and software, the user interfaces, the system interfaces, the database, and the system controls associated with system security.

The inputs to the design activities consist of the models that were built during analysis. The outputs of the design consist of a set of diagrams, or models, that describe the architecture of the new system and the detailed logic within various programming components.

Designing the application architecture can be subdivided into architectural and detail design. Detail design often refers to the design of the software programs. Architectural design adapts the application to the deployment environment, including hardware, software, and networks. Modern application software is usually deployed in a distributed multicomputer environment and is organized according to client/server architecture—usually in a three-layer architecture.

In today's widely connected computing environment, the design of the software application must consider client computer characteristics and server environments. Client computing environment options range from simple desktop systems to tablet computers to very small mobile devices. On the server side of the design, there are many options—from in-house equipment to colocation or cloud computing. The specific requirements of the potential client equipment and the server computers impact the final design and the operation of the new system.

Key Terms

architectural design   158

business logic layer   170

client computer   168

client-server architecture   168

cloud computing   176

colocation   175

content delivery network (CDN)   174

data layer   170

detail design   158

domain layer   170

hosting   174

Hypertext Markup Language (HTML)   169

Hypertext Transfer Protocol Secure (HTTPS)   173

local area network (LAN)   168

network diagram   156

peer-to-peer connection   178

server computer   168

Service Level Agreement (SLA)   176

three-layer architecture   169

Transmission Control Protocol/
    Internet Protocol (TCP/IP)   169

Transport Layer Security (TLS)   173

view layer   169

Virtual Private Networks (VPNs)   177

virtual server   175

Review Questions

1. What is the primary objective of systems design?
2. What is the difference between systems analysis and systems design?
3. List the major elements that must be designed for a new software application.
4. List the models that are used for systems analysis.
5. List the models that are used for systems design.
6. What is the difference between user-interface design and system-interface design?
7. On a project that uses iterations to develop the system, in which iteration does systems design begin? Explain why.
8. What is the difference between architectural design and detail design?
9. Designing the security and controls impacts the design of which other elements?
10. Describe what is required for database design.
11. What is a LAN? When would it be used in deploying a new system?
12. What is three-layer design?
13. Describe the contents of each layer in three-layer design.
14. List the different types of client devices in a client/server architecture.
15. What is the difference between HTTPS and HTTP?
16. In the use of software over the Internet, what are the two main security issues that must be considered?
17. Describe the primary factors that affect throughput for Internet systems.
18. List five issues that are important when considering an external hosting company.
19. What is the difference between cloud computing and virtual servers?
20. Why do companies use colocation facilities?
21. Describe the issues to be considered when designing for multiple clients.
22. What is a VPN? Why would a company use a VPN?

Problems and Exercises

1. A financial corporation has desktop applications running in several different offices that are all supported by a centralized application bank of two computers. In addition, there is a centralized database, which requires three servers. Draw a network diagram representing this requirement.

2. A sales organization has an Internet-based customer support system that needs to support every type of client device. The server configuration should be a normal layered application server and database server. Draw a network diagram representing this requirement.

3. A medium-sized engineering firm has three separate engineering offices. In each office, a local LAN supports all the engineers in that office. Due to the requirement for collaboration among the offices, all the computers should be able to view and update the data from any of the three offices. In other words, the data storage server within each LAN should be accessible to all computers, no matter where they are located. Draw a network diagram that will support this configuration.

4. A small start-up company has a Web-based customer sales system that is written by using PHP and JavaScript. The company is deciding whether to host the system on its own servers, contract with a hosting company for a virtual server, or go to Amazon's cloud. Volumes are expected to be low at the beginning, and it is hard to predict a growth pattern, although there is potential for rapid growth. Decide which alternative the company should choose. Defend your decision by giving advantages and disadvantages of each solution based on the characteristics of the start-up company.

5. Describe the differences between HTTPS and a VPN. What kinds of computing and networking situations are better suited to HTTPS? What kind of computing and networking situations are better suited to VPN?

6. Find four separate hosting providers and compare their offerings, including prices. Put your answer in a table showing the results of your research.

7. Compare screen size, resolution, and other important display characteristics of five popular Internet-enabled smartphones. Which would you rate as the best? Defend your answer.

8. Research the issues related to supporting a very large database that must be distributed across multiple servers. Write a list of the issues that need to be addressed and the alternative solutions for a distributed and partitioned database where (a) all servers are colocated in the same data center and (b) the servers are located in separate data centers.

Case Study

County Sheriff Mobile System for Communications (CSMSC)

Law enforcement agencies thrive on information. They need it to respond to emergencies and to anticipate what they will encounter when they arrive on the scene. In previous eras, it was sufficient to receive information through the police dispatch radio. Today, much more than voice-based information is required. Officers often need to check vehicle registrations, personal identities, outstanding warrants, mug shots, maps, and the locations of other officers.

One major difficulty with meeting this need for more information is figuring out how to transmit the data to remote and mobile locations. Local police agencies are sometimes able to restrict their transmission needs to within the city limits. However, county sheriffs and state troopers often have to travel to remote locations that aren't within a metropolitan area's boundaries.

Let us say a local county sheriff's department has received a grant to upgrade its existing communication system and the system must satisfy certain requirements. First, there must be complete coverage throughout the county. This county includes metropolitan areas as well as desert and mountains that are often outside normal radio or wide-area Wi-Fi coverage. About 95 percent of the county is within cell phone coverage, and 5 percent is in uncovered mountainous areas.

Officers normally have access to customized laptops in their vehicles. However, some officers are required to patrol ATV trails and campgrounds. Those officers will still need to be connected with a portable computing device. Hence, another requirement is being able to use smaller portable devices.

Your assignment: Recommend a communication and network solution for the county sheriff's department. It can be any combination of Internet, VPN, Wi-Fi, telephone, and satellite communication. The applications can be custom built, with device-specific or HTML-based user interfaces. Although HTML tends to be more versatile, it has drawbacks regarding security; display can also be an issue on devices that don't have browser support.

As always, the budget is tight, so your solution should be as economical as possible. Develop a network diagram that depicts your proposed solution. Also, explain your solution and justify your design.

RUNNING CASE STUDIES

Community Board of Realtors

The Community Board of Realtors Multiple Listing Service (MLS) will be a Web-based application with extensions to allow wireless smartphone interaction between the agents and their customers. Review the functional and nonfunctional requirements you have developed for previous chapters. Then, for each of the six design activities discussed in this chapter, list some specific tasks to design the environment, application architecture and software, user interfaces, system interfaces, database, and system controls and security. You may want to refer back to the Tradeshow System discussed in Chapter 1 for some design specifics.

The Spring Breaks 'R' Us Travel Service

Let us say that the SBRU information system includes four subsystems: Resort relations, Student booking, Accounting and finance, and Social networking. The first three are purely Web applications, so access to those will be through an Internet connection to a Web server at the SBRU home office. The Social networking subsystem has built-in chat capabilities. It relies on Internet access for the students, as students compare notes before they book their travel reservations and as they chat while traveling. To function properly, the system obviously requires a wireless network at each resort during the trip. SBRU isn't responsible for installing or maintaining the resort wireless network; they only plan to provide some design specifications and guidelines to each resort.

The resort will be responsible for connecting to the Internet and for providing a secure wireless environment for the students.

1. Design the environment for the SBRU information system by drawing a network diagram. Include what might be necessary to support online chatting capabilities.
2. Considering that everything is designed to operate through the Internet with browsers or smartphones, how simple does this architecture appear to be? Can you see why Web and smartphone applications are so appealing?
3. What aspect of design becomes extremely important to protect the integrity of the system?

On the Spot Courier Services

In previous chapters, we have described the technological capabilities that Bill Wiley wants for servicing his customers. One of the problems that Bill has is that his company is very small, so he cannot afford to develop any special-purpose equipment or even sophisticated software.

Given this limitation, Bill's request for advanced technological capabilities is coming at an opportune time. Equipment manufacturers are developing equipment with advanced telecommunications capabilities. And freelance software developers are producing

software applications—many of which provide the capabilities that Bill needs.

The one caveat is that since this will be a live production system, it needs to be reliable, stable, error-free, dependable, and maintainable

Let us review some of the required capabilities of the new system, which has been described in previous chapters:

Customers

■ Customers can request package pickup online via the Internet.

(continued on page 184)

(continued from page 183)

- Customers can check the status of packages online via the Internet.
- Customers can print mailing labels at their offices.

Drivers

- Drivers can view their schedules via a portable digital device while on their routes.
- Drivers can update the status of packages while on their routes.
- Drivers can allow customers to "sign" for packages that are delivered.
- The system "knows" where the driver is on his route and can send updates in real time.
- Drivers can accept payments and record them on the system.

Bill Wiley (management)

- Bill can record package pickups from the warehouse.
- Bill can schedule delivery/pickup runs.
- Bill can do accounting, billing, etc.

- Bill can access the company network from his home.

Given these requirements, do the following:

1. Make a list of the equipment that Bill should purchase to support his new system. Include all equipment that will be needed for the home office, the drivers, and at Bill's residence. Identify and describe actual equipment that can be purchased today. Estimate the cost of the equipment.
2. Describe any special software that may be needed. The software engineer is developing the application software (package scheduling and processing, accounting, etc.), but no special software is required for connecting the devices or communications between them.
3. Develop a network diagram showing how all the equipment will be connected. Identify Internet connections, VPNs, and telephony links as appropriate.

Sandia Medical Devices

As described in previous chapters, the Real-Time Glucose Monitoring (RTGM) system will include processing components on servers and on mobile devices, such as smartphones, with data exchange via 3G and 4G phone networks. Users will include such patient and health care personnel as physicians, nurses, and physician assistants. In the United States, the Health Insurance Portability and Accountability Act of 1996 (HIPAA) mandates certain responsibilities regarding the privacy and security of electronic protected health information (ePHI). The law applies to what are collectively called covered entities—that is, health plans, health care clearinghouses, and any health care providers who transmit health information in electronic form. More information can be obtained from the U.S. Department of Health & Human Services Web site (www.hhs.gov).

In general, covered entities should:

- Ensure the confidentiality, integrity, and availability of all ePHI they create, receive, maintain, or transmit.
- Identify and protect against reasonably anticipated threats to the security or integrity of the information.

- Protect against reasonably anticipated, impermissible uses or disclosures of the information.
- Ensure compliance by their workforces.

Specifically, covered entities should implement policies, procedures, and technologies that:

- Specify the proper use of and access to workstations and electronic media.
- Regard the transfer, removal, disposal, and reuse of electronic media to ensure appropriate protection of ePHI.
- Allow only authorized persons to access ePHI.
- Record and examine access and other activity in information systems that contain or use ePHI.
- Ensure ePHI isn't improperly altered or destroyed.
- Guard against unauthorized access to ePHI that is being transmitted over an electronic network.

Answer these questions in light of HIPPA requirements:

1. Does HIPAA apply to the RTGM system? Why or why not?
2. How should the system ensure data security during transmission between a patient's mobile device(s) and servers?

(continued on page 185)

(continued from page 184)

3. Consider the data storage issues related to a patient's mobile device and the possible ramifications if the device is lost or stolen. What measures should be taken to protect the data against unauthorized access?

4. Consider the issues related to health care professionals accessing server data by using workstations and mobile devices within a health care facility. How will the system meet its duty to record and examine access to ePHI? If a health care professional uses a mobile device outside a health care facility, what protections must be applied to the device and/or any data stored within it or transmitted to it?

5. Consider the issues related to wired and wireless data transmission between servers and workstations within a health care facility. What security duties, if any, apply to transmissions containing ePHI? Does your answer change if the servers are hosted by a third-party provider?

Further Resources

Frederick P. Brooks, *The Design of Design: Essays from a Computer Scientist*, Addison-Wesley, 2010.

Priscilla Oppenheimer, *Top-Down Network Design* (3rd ed.), Cisco, 2010.

Doug Kaye, *Strategies for Web Hosting and Managed Services*, Wiley, 2001.

7

Designing the User and System Interfaces

Chapter Outline

- User and System Interfaces
- Understanding the User Interface
- User-Interface Design Concepts
- The Transition from Analysis to User-Interface Design
- User-Interface Design
- Identifying System Interfaces
- Designing System Inputs
- Designing System Outputs

Learning Objectives

After reading this chapter, you should be able to:

- Describe the difference between user interfaces and system interfaces
- Describe the historical development of the field of human-computer interaction (HCI)
- Discuss how visibility and affordance affect usability
- Describe user-interface guidelines that apply to all types of user-interface types and additional guidelines specific to Web pages and mobile applications
- Create storyboards to show the sequence of forms used in a dialog
- Discuss examples of system interfaces found in information systems
- Define system inputs and outputs based on the requirements of the application program
- Design printed and on-screen reports appropriate for recipients

Interface Design at Aviation Electronics

Bob Crain was admiring the user interface for the manufacturing support system that was recently installed at Aviation Electronics (AE). Bob is the plant manager for AE's Midwest manufacturing facility, which produces aviation devices used in commercial aircraft. These aviation devices provide guidance and control functions for flight crews, and they provide the latest safety and security features that pilots need when flying commercial aircraft.

The manufacturing support system is used for all facets of the manufacturing process, including product planning, purchasing, parts inventory, quality control, finished goods inventory, and distribution. Bob was extensively involved in the development of the system for several years, including the initial planning and development. The system reflected almost everything he knew about manufacturing. The information systems team that developed the system relied extensively on Bob's expertise. That was the easy part for Bob.

What particularly pleased Bob was the final user interface. He had insisted that the development team consider the entire user experience from the very beginning. He didn't want just another cookie-cutter transaction processing system. He wanted a system that acted as a partner in the manufacturing process—much the way that AE's guidance and control system interfaces acted as a pilot's partner.

The first manager assigned to the project placed a low priority on user-interface design. When Bob asked why user-interface design wasn't a key focus of early iterations, the manager replied, "We'll add the user interface later, after we work out the accounting controls." When Bob insisted that the project manager be replaced, the information systems department sent Sara Robinson to lead the project.

Sara had a completely different attitude; she started out by asking about events that affect the manufacturing process and about cases in which users needed support from the system. Although she had a team of analysts working on the accounting transaction details from the beginning, she always focused on how the user would interact with the system. Bob and Sara conducted meetings to involve users in discussions about how they might use the system, even asking users to act out the roles of the user and the system in carrying on a conversation.

At other meetings, Sara presented sketches of screens and asked users to draw on them to indicate the information they wanted to see and options they wanted to be able to select. These sessions produced many ideas. For example, it appeared that many users didn't sit at their desks all day; they needed larger and more graphic displays they could see from across the room. Many users needed to refer to several displays, and they needed to be able to read them simultaneously. Several functions were best performed by using graphical simulations of the manufacturing process. Users made sketches showing how the manufacturing process actually worked, and the team used these sketches to define much of the interface. Sara and her team kept coming back every month or so with more examples to show, asking for more suggestions.

When the system was finally completed and installed, most users already knew how to use it because they had been so involved in its design. Bob knew everything the system could do, but he had his own uses for it. He sat at his desk and clicked the Review Ongoing Processes button on the screen, and the manufacturing support system gave him his morning briefing.

Overview

Information systems interact with people and with other systems. Because few systems operate autonomously or in isolation, designing the interfaces (inputs and outputs) between a system and its users and environment is an important system development task. Poorly designed interfaces with people can result in a system that operates less than optimally or doesn't fulfill its purpose. For example, a human resources system that has a poorly designed user interface may reduce organizational efficiency and be a source of data entry errors. A customer-facing system with a poorly designed user interface might motivate customers to take their business elsewhere. As with user-oriented interfaces, poorly designed interfaces to other automated systems can be a source of errors or inefficiency. Thus, the design of a system's interfaces is an important part of a system development project.

Inputs and outputs of the system are an early concern of any system development project. The project plan lists key inputs and outputs that the analyst identified when defining the scope of the system. During the analysis phase, analysts will have discussed inputs and outputs early and often with system stakeholders to identify users and actors that affect the system and that depend on the information it produces. Requirements models produced during analysis also emphasize inputs and outputs. For example, use case descriptions define inputs and outputs that occur during a use case. The inputs and outputs are further defined as messages and returns in system sequence diagrams (SSDs).

User and System Interfaces

A key step in systems design is to classify the inputs and outputs for each event as either a system interface or a user interface. **System interfaces** are inputs and outputs that require minimal human intervention. They might be inputs captured automatically by such special input devices as scanners, electronic messages to or from another system, or transactions captured by another system. Many outputs are considered system interfaces if they primarily send messages or information to other systems (e.g., a pickup notification to a shipping company) or if they produce reports, statements, or documents for external agents or actors without much human intervention (e.g., end-of-month credit card statements e-mailed to cardholders).

User interfaces are inputs and outputs that more directly involve a system user. User interfaces can be for internal or external users. Their design varies widely depending on such factors as interface purpose, user characteristics, and characteristics of a specific interface device. For example, although all user interfaces should be designed for maximal ease of use, other considerations, such as operational efficiency, may be important for internal users who can be trained to use a specific interface optimized for a specific hardware device (e.g., a keyboard, a mouse, and a large high-resolution display). In contrast, a quite different user interface might be designed for a customer-facing system that assumes a cell phone as the input/output device.

In most system development projects, analysts separate the design of system interfaces from the design of user interfaces because the each requires its own expertise and technology. But as with the design of any system component, considerable coordination is required.

Understanding the User Interface

Many people think the user interface is developed and added to the system near the end of the development process, but the user interface is much more important than that. It is everything that the end user comes in contact with while using the system—physically, perceptually, and conceptually (see **Figure 7-1**).

From a user perspective, the user interface is the entire system. The programs, scripts, databases, and hardware behind the interface are irrelevant. Design techniques that embody this view of user interfaces are collectively called **user-centered design**, which emphasizes three important principles:

■ Focus early on the users and their work.
■ Evaluate designs to ensure usability.
■ Use iterative development.

The early focus on users and their work is consistent with the approach to systems analysis in this text. User-oriented analysis and design tasks are performed as early as possible and are often given higher priority than other tasks. For example, such user-oriented analysis tasks as stakeholder identification and interviews occur early in the project. User interfaces are designed in early iterations, and user-related design decisions drive other design decisions and tasks.

The early focus on users and their work goes beyond issues of task ordering and priority. It embodies an all-encompassing attempt to understand users and answer such questions as: What do they know? How do they learn? How do

system interfaces inputs or outputs that require minimal human intervention

user interfaces system interfaces that directly involve a system user

user-centered design design techniques that embody the view that the user interface is the entire system

FIGURE **7-1**
User-centered design

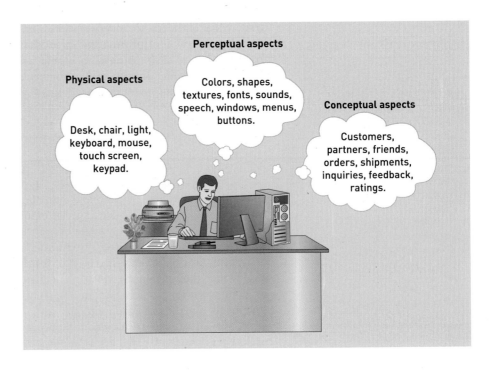

they prefer to work? What motivates them? The extent of user-oriented focus does vary with the type of system being developed. For example, if the system is a shrink-wrapped desktop application marketed directly to end users, the focus on users and their preferences is intense.

The second principle of user-centered design is to evaluate designs to ensure usability. **Usability** refers to the degree to which a system is easy to learn and use. Ensuring usability isn't easy; there are many different types of users with different preferences and skills. Features that are easy for one person to use might be difficult for another. If the system has a variety of end users, how can the designer be sure that the interface will work well for all of them? For example, if it is too flexible, some end users might feel lost. On the other hand, if the interface is too rigid, some users will be frustrated.

Ease of learning and ease of use are often in conflict. For example, menu-based applications with multiple forms, many dialog boxes, and extensive prompts and instructions are easy to learn; indeed, they are self-explanatory. And easy-to-learn interfaces are appropriate for systems that end users use infrequently. But if internal users use the system all day, it is important to make the interface fast and flexible, with shortcuts, hot keys, voice commands, and information-intensive screens. This second interface might be harder to learn, but it will be easier to use after it is learned. Internal users (with the support of their managers) are willing to invest more time learning the system in order to become efficient users.

Developers employ many techniques to evaluate interface designs to ensure usability. User-centered design requires testing all aspects of the user interface. Some usability testing techniques collect objective data that can be statistically analyzed to compare designs. Some techniques collect subjective data about user perceptions and attitudes. To assess user attitudes, developers conduct formal surveys, focus group meetings, design walk-throughs, paper-and-pencil evaluations, expert evaluations, formal laboratory experiments, and informal observation.

The third principle of user-centered design is iterative development—that is, doing some analysis, then some design, then some implementation, and then repeating the processes. After each iteration, the project team evaluates the work on the system to date. Iterative development keeps the focus on the user by

usability degree to which a system is easy to learn and use

continually returning to the user requirements during each iteration and by evaluating the system after each iteration. As with the principle of early focus on users and their work, this principle is reflected throughout this textbook in its approach to system development in general and analysis and design tasks in particular.

Metaphors for Human-Computer Interaction

metaphors analogies between features of the user interface and aspects of physical reality that users are familiar with

Widespread use of visually oriented user interfaces debuted in the mass market with the Apple Macintosh in the 1980s and became widespread with various versions of Microsoft Windows in the 1990s. To make computers easier to use and learn, designers of early visually oriented interfaces adopted **metaphors**, which are analogies between features of the user interface and aspects of physical reality that users are familiar with. Metaphors are still widely applied to user-interface design, as described in **Table 7-1**.

direct manipulation metaphor metaphor in which objects on a display are manipulated to look like physical objects (pictures) or graphic symbols that represent them (icons)

Figure 7-2 is a screen capture of a computer running Windows that illustrates the **direct manipulation, desktop**, and **document metaphors**. The entire display is visually similar to the surface of a physical desktop. Icons and pictures for commonly used tools are located on the left and right sides. The icons can be directly manipulated with a mouse or another pointing device. The windows in the center frame are documents that are visually similar to paper pages laid on the surface of a desk, with a sticky note attached to one of the pages.

desktop metaphor metaphor in which the visual display is organized into distinct regions, with a large empty workspace in the middle and a collection of tool icons around the perimeter

document metaphor metaphor in which data is visually represented as paper pages or forms

dialog metaphor metaphor in which user and computer accomplish a task by engaging in a conversation or dialog via text, voice, or tools such as labeled buttons

The direct manipulation, desktop, and document metaphors emphasize displayed objects with which the user interacts. The **dialog metaphor** emphasizes the communication that occurs between a user and a computer, conceptualized as a conversation. In a conversation or dialog between two people, each person listens and responds to questions and comments from the other person, with the information being exchanged in a sequence. The dialog metaphor is another way of thinking about human-computer interaction because the computer "listens to" and "responds" to user questions or comments, and the user "listens to" and "responds" to the computer's questions and comments. **Figure 7-3** illustrates a conceptual dialog between user and computer.

TABLE **7-1** *Commonly used metaphors for user-interface design*

Metaphor	Description	Example
Direct manipulation	Manipulating objects on a display that look like physical objects (pictures) or that represent them (icons)	The user drags a folder icon to an image of a recycle bin or trash can to delete a collection of files.
Desktop	Organizing visual display into distinct regions, with a large empty workspace in the middle and a collection of tool icons around the perimeter	At computer startup, a Windows user sees a desktop, with icons for a clock, calendar, notepad, inbox and sticky notes (the computer interface version of a physical Post-It note).
Document	Visually representing the data in files as paper pages or forms. These pages can be linked together by references (hyperlinks)	The user fills in a form field for a product he or she owns, and the manufacturer's Web site finds and displays the product's manual as an Adobe Acrobat file, which contains a hyperlinked table of contents and embedded links to related documents.
Dialog	The user and computer accomplishing a task by engaging in a conversation or dialog by using text, voice, or tools, such as labeled buttons	The user clicks a button labeled "troubleshoot" because the printer isn't working. The computer prints questions on the display, and the user responds by typing answers or selecting responses from a printed list.

FIGURE **7-2** *The direct manipulation, desktop, and document metaphors on a typical computer display*

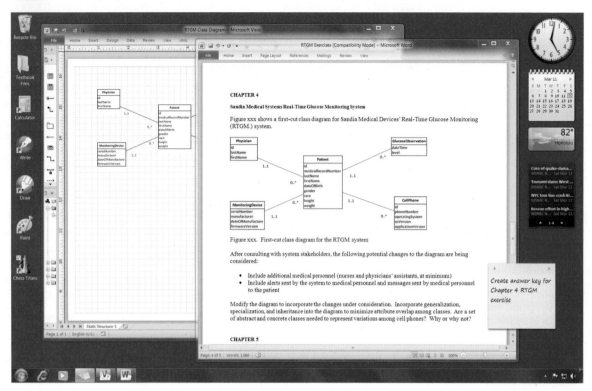

The dialog metaphor can be implemented in various ways in user interfaces. A direct approach uses speech generation and recognition over a voice communication channel, as commonly encountered when calling the customer support number of a large company. A computerized voice asks a series of questions, listens for the answer to each question, and responds to the answers. Another implementation of the dialog metaphor uses questions or instructions displayed by a user through text and responses as well as counter-questions displayed by the computer through text. To minimize the need for user typing, responses to computer questions might be limited to a specific set of possibilities displayed to the user in the form of a list from which the user selects the most appropriate response with a mouse click or by touching a display surface.

Regardless of the specific form of implementation, specifying dialogs between user and computer is one of the more powerful tools used by user-interface designers. Although written and spoken languages vary across the globe, conversation and dialog are fundamental and universal human skills. Modeling the

FIGURE **7-3**

The dialog metaphor for user-computer interaction

interaction between user and computer as a dialog enables the user to incorporate language and related skills that have been honed from an early age.

User-Interface Design Concepts

Many IT researchers and practitioners have published articles, books, and Web sites that offer guidance in user-interface design. Although some guidelines have changed as user-interface technology has changed, many guidelines are universal, having been around for many decades and technology generations. We review some of these universal guidelines in this section and then move on to user-interface development, with specific guidelines for specific interface types, later in this chapter.

Affordance and Visibility

human-computer interaction (HCI)
field of study concerned with the efficiency and effectiveness of user interfaces vis-à-vis computer systems, human-oriented input and output technology, and psychological aspects of user interfaces

Donald Norman is a leading researcher in **human-computer interaction (HCI)**, a field of study concerned with the efficiency and effectiveness of user interfaces to computer systems, human-oriented input and output technology, and psychological aspects of user interfaces. Norman proposes two key principles to ensure good interaction between a person and a machine: affordance and visibility. Both principles apply to user-interface controls, which are elements of a user interface that a user manipulates to perform tasks. Examples of controls include menus, buttons, pull-down lists, sliders, and text entry boxes.

affordance when the appearance of a specific control suggests its function

Affordance means that the appearance of a specific control suggests its function—that is, the purpose for which the control is used. For example, a control that looks like a steering wheel suggests that it is used for turning. Affordance can also be achieved by a user-interface control that the user is familiar with in another context. For example, the media player control icons shown in **Figure 7-4** were first widely used on audiotape and videotape players in the 1970s and have continued to be used in such devices as DVD and portable music players. They are widely incorporated into computer interfaces because so many users are familiar with them.

visibility when a control is visible so that users know it is available, the control providing immediate feedback to indicate that it is responding to the user

Visibility means that a control is visible so users know it is available; it also means that the control provides immediate feedback to indicate that it is responding. For example, the mute button shown in Figure 7-4(a) changes its appearance when the user moves the mouse pointer over it, as shown in Figure 7-4(b). When the user presses the button, it changes its appearance, as shown in Figure 7-4(c).

Visibility and affordance are relatively easy to achieve when the design target is a commonly used platform, such as an iPad, a cell phone running the Android operating system, or a PC running Windows. Such platforms have well-defined user-interface design guidelines and a library of user-interface features and functions that can be reused by application software. When a designer incorporates user-interface objects and styles from these libraries, he or she is tapping into users' experience with similar user interfaces from other applications on those platforms.

Web user-interface design is less standardized because Web browsers are intentionally platform-neutral. Designers can choose from a wide variety of user-

FIGURE **7-4 a-c**

Visibility and affordance in media player controls

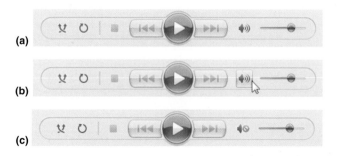

(a)

(b)

(c)

interface libraries, each with its own user-interface objects and styles. Attention to visibility and affordance is especially important in Web interface design because there are no real standards that provide a preexisting framework of user familiarity.

Consistency

User interfaces should be designed for consistency in function and appearance. The way that information is arranged on forms, the names and arrangement of menu items, the size and shape of icons, and the sequence followed to carry out tasks should be consistent throughout the system. Why? People are creatures of habit. After we learn one way of doing things, it is difficult to change. When we operate a computer application, many of our actions become automatic; we don't think about what we are doing.

Figure 7-5 shows the display of Microsoft Word, which illustrates many aspects of consistency among applications that run under Windows and across the various components of the Microsoft Office suite. Icons that appear in the upper-left and upper-right corners of the window frame are standardized across many Windows applications; thus, users know where to look for them and what their functions are at a glance. Similarly, the scrollbar on the right and the zoom in/out slider and resize handles in the lower-right corner are standard across many Windows applications. The top menus and toolbars are similar across Word, PowerPoint, Publisher, and other programs within Microsoft Office. An experienced user of one program learns others more quickly due to these similarities.

Shortcuts

User interfaces and dialogs designed for novices are often an annoyance and impediment to experienced users' productivity. Users who work with an application repeatedly or for long time periods want shortcuts for frequently used functions, which minimize the number of keystrokes, mouse clicks, and menu selections required to complete tasks. Examples include voice commands as well as shortcut keys, such as Windows keyboard sequences Ctrl+C for copy and Ctrl+V for paste. Application designers should use standard shortcuts when available or build their own.

FIGURE **7-5** *Microsoft Word interface features used in various Windows applications*

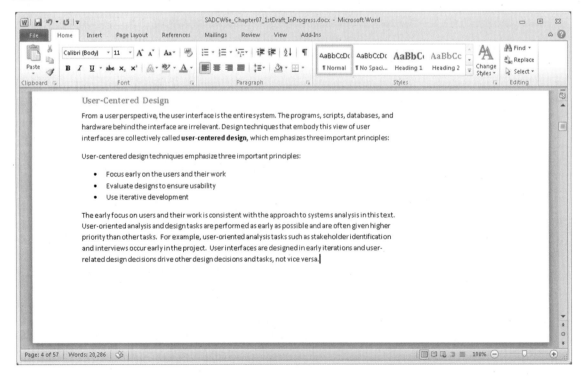

Feedback

Every action a user takes should result in some type of feedback from the computer so the user knows the action was recognized. Feedback can take many forms in a user interface, including:

- Audible feedback, such as clicking sounds when keys are pressed and beeps when on-screen buttons are pressed
- Visible feedback, such as the icon changes shown in Figures 7-4(b) and 7-4(c), or the progress meter shown during the download of a large file

Feedback provides the user with a sense of confirmation and the feeling that a system is responsive and functioning correctly. Lack of feedback leaves the user wondering whether a command or input was recognized or whether the system is malfunctioning. When subsequent processing is delayed by more than a second or two, users may repeatedly press controls or reenter information, resulting in processing errors and user frustration.

Dialogs That Yield Closure

Each dialog should be organized with a clear sequence—a beginning, middle, and end. Any well-defined task has a beginning, middle, and end, so users' tasks on the computer should also feel this way. The user can get lost if it is not clear when a task starts and ends. In addition, users often focus intently on tasks, so when it is confirmed that a task is complete, the user can clear his or her mind and get ready to focus on the next task.

If the system requirements are defined initially as events to which the system responds, each event leads to the processing of one specific, well-defined activity. Each use case can be defined as one or more dialogs, each with a flow of steps and well-defined interactions. Event decomposition sets the stage for dialogs with closure.

Error Handling

User errors are a waste of the time to commit and to correct them. A good user-interface design anticipates common errors and helps the user to avoid them. One way to do this is to limit available options, presenting the user with only valid options for a specific point in a dialog. Adequate feedback, as discussed previously, also helps reduce errors.

When errors do occur, the user interface needs mechanisms to detect then. Validation techniques discussed later in this chapter are useful for catching errors, but the system must also help the user correct the error. When the system does find an error, the error message should state specifically what is wrong and explain how to correct it. Consider this error message that occurs after a user has typed in a full screen of information about a new customer:

The customer information entered is not valid. Try again.

This message doesn't explain what is wrong or what to do next. Furthermore, after this message appears, what if the system cleared the data-entry form and redisplayed it? The user would have to reenter everything previously typed but still have no idea what is wrong. The error message didn't explain it, and now that the typed data has been cleared, the user cannot tell what might have been wrong. A better error message would read more like this:

The date of birth entered is not valid. Check to be sure only numeric characters in appropriate ranges are entered in the Date of Birth field.

The system also should streamline corrective actions. For example, if the user enters an invalid customer ID, the system should tell the user that this has occurred and then place the insertion point in the customer ID text box, with the previously typed number displayed and ready to edit. That way, the user can see the mistake and edit it rather than having to retype the entire ID. The system might also suggest valid values based on past experience or other information that the user has already entered.

Easy Reversal of Actions

Users need to feel that they can explore options and take actions that can be cancelled or reversed without difficulty. This is one way that users learn about the system—that is, by experimenting. It is also a way to prevent errors; as users recognize they have made a mistake, they cancel the action. In the game of checkers, a move isn't final until the player takes his or her fingers off the game piece; it should be the same when a user drags an object on the screen. In addition, designers should be sure to include cancel buttons on all dialog boxes and allow users to go back a step at any time. Finally, when the user deletes something substantial—a file, a record, or a transaction—the system should ask the user to confirm the action and, where possible, delay implementing the action.

A key issue in permitting action reversal is structuring dialogs and corresponding system actions. Novice designers and programmers often assume that user dialog structure and the sequence of corresponding system actions must precisely correspond. For example, a complex transaction may require several discrete steps, each of which accepts data from the user and some of which modify data stored by the system. Although the user dialog should reflect this structure, internal programming doesn't always have to process data as it is received. Instead, it might collect the data as the dialog proceeds and build an internal "to do" list. When the user completes the final step in the dialog, the system can then complete processing for the entire dialog at once. If the user decides near the end of the dialog to cancel the entire sequence, there are no internal changes to undo. Also, user-interface performance may be improved because there are fewer potential processing delays between steps.

Reducing Short-Term Memory Load

People have many limitations, and short-term memory is one of the biggest. Psychologists have demonstrated that people can remember only about seven chunks of information at a time. User-interface designers should avoid requiring the user to remember anything from one form to another or from one dialog box to another during an interaction with the system. If the user has to stop and ask "What was the filename? The customer ID? The product description?" then the design is placing too much of a burden on the user's memory. Memory limitations also apply to steps in a complex process. The interface should help users keep track of where they are in a complex process via visual cues and other aids.

The Transition from Analysis to User-Interface Design

The foundation for user-interface design is laid when use cases are identified and documented, as described in Chapter 3. Use cases that require direct user interaction (i.e., interactive use cases) are the starting point for a dialog, and the corresponding use case, activity, and system sequence diagrams are the initial dialog documentation. Interactive use cases may require the user to input choices and data into the system (such as when making an online order) or may generate outputs in response to a user request (such as when tracking a shipment). During design, dialogs for interactive use cases are further refined by developing menus, forms, and other user-interface elements.

Dialog and user-interface design can proceed in either a top-down or bottom-up fashion. In the top-down approach, menus (groups of related use cases, dialogs, and user interfaces) are defined first, followed by a detailed description of each interactive use case dialog and development of the related user-interface elements. In the bottom-up approach, interactive use cases are prioritized, and related dialogs and user interfaces are developed one at a time. Menus are added later in the project when related sets of fully implemented user interfaces are completed. Neither approach is inherently better; either or both can be a good match for a specific project.

Use Cases and the Menu Hierarchy

Menus are a way of grouping large numbers of related use cases or dialogs within a user interface. In all but the smallest systems, menus are needed to present the user with a tractable number of choices per screen, to group related functions together so users can more easily locate them, and to properly sequence related interfaces for a complex event decomposition. **Figure 7-6** shows two different menu styles. In Figure 7-6(a), the mouse pointer is positioned over the Shop item of the upper menu in a Web page. Clicking Shop would display a second Web page containing another menu. Figure 7-6(b) shows a more complex menu design, with three menu levels displayed.

How does a designer decide which use cases and user interfaces to include in which menus, which menus are required, and how many menu levels are required? These decisions are primarily driven by the number of uses cases or menu choices and the limits of human cognition. Menus usually contain five to 10 choices in order to avoid overloading the user (see Figure 7-6(a)). With careful design, more choices can be provided, especially for experienced users (see Figure 7-6(b)). For a typical business system, dividing the total number of interactive use cases by five provides an initial estimate of the number of menus that include all interactive use cases and allows for additional menu items, such as setting options or preferences. If the number of menus is greater than 10, higher-level menus that contain links to other menus are required. For example, in Figure 7-6(b), the upper menu options, such as File, Home, and Review, each link to a second-level menu formatted as a toolbar. Some choices within the toolbar are links to lower-level menus, such as the displayed Show Markup menu.

FIGURE **7-6** *Two different menu styles*

(a)

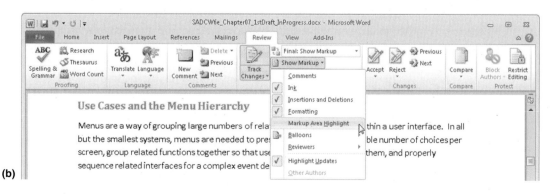

(b)

TABLE **7-2** *RMO use cases grouped by actor and subsystem*

Subsystem	Use Case	Users/Actors
Sales	Search for item	Customer, customer service representative, store sales representative
Sales	View product comments and ratings	Customer, customer service representative, store sales representative
Sales	View accessory combinations	Customer, customer service representative, store sales representative
Sales	Fill shopping cart	Customer
Sales	Empty shopping cart	Customer
Sales	Check out shopping cart	Customer
Sales	Fill reserve cart	Customer
Sales	Empty reserve cart	Customer
Sales	Convert reserve cart	Customer
Sales	Create phone sale	Customer service representative
Sales	Create store sale	Store sales representative
Order fulfillment	Ship items	Shipping
Order fulfillment	Manage shippers	Shipping
Order fulfillment	Create backorder	Shipping
Order fulfillment	Create item return	Shipping, customer
Order fulfillment	Look up order status	Shipping, customer, management
Order fulfillment	Track shipment	Shipping, customer, marketing
Order fulfillment	Rate and comment on product	Customer
Order fulfillment	Provide suggestion	Customer

Use cases with common actors and event decomposition or that implement CRUD actions for a specific domain class are good candidates to be grouped into a single menu or related group of menus. For example, consider the RMO CSMS use cases shown in **Table 7-2**. An initial grouping of these cases by actor and subsystem is a good starting point for menu design.

Table 7-3 shows a grouping of the use cases in Table 7-2 into four menus. Each menu collects uses cases from one subsystem for a customer or internal sales representative. The number of menu choices ranges from four to seven, which won't overload any one menu and may enable multiple menu levels to displayed at one time. A dialog design is created for each menu option. After dialog design proceeds, the designer may redefine the menu options or structure. In fact, designers often discover missing or incomplete use cases during user-interface design, which results in a brief return to analysis activities to complete the documentation.

Menus usually include options that are *not* activities or use cases from the event list. Many options are related to the system controls, such as account maintenance or database backup and recovery, which are discussed later in this chapter. Other added items include help links as well as links to other menus or subsystems.

TABLE `7-3` *RMO CSMS use cases grouped into first-cut menus by similar function and user*

Menu Description	Menu Choices (Use Cases)	Intended User(s)
Shopping cart functions (primary or reserve)	■ Search for item ■ View product comments and ratings ■ View accessory combinations ■ Switch carts (primary to reserve or vice versa) ■ Fill shopping cart ■ Empty shopping cart ■ Check out shopping cart	Customer
Sale creation	■ Search for item ■ View product comments and ratings ■ View accessory combinations ■ Create sale	Customer service and store sales representatives
Order shipment	■ Ship items ■ Manage shippers ■ Create backorder ■ Create item return ■ Look up order status ■ Track shipment	Customer service and store sales representatives
Customer order control	■ Look up order status ■ Track shipment ■ Create item return ■ Rate and comment on product ■ Provide suggestion	Customer

Dialogs and Storyboards

After identifying all required dialogs, the designers must document the dialogs. Many options exist; there are no *de facto* standards. One approach is to list the key steps followed for the dialog with a written description of what the user and computer do at each step. The format for writing these steps can follow the activity diagram described in Chapter 2 or the use case description format in Chapter 4 or it can be more free form.

Designers can also document dialog designs by writing out how the user and system might interact if they were two people engaged in conversation. Sometimes, the designer can go right from the dialog to sketching a series of screens that make the design more visible. Consider the *Check out shopping cart* use case for the RMO CSMS. A dialog between the system and user might follow this pattern:

SYSTEM: What would you like to do?

USER: I'd like to check out.

SYSTEM: Okay. What is your e-mail address or account number?

USER: My e-mail address is nwells22@gmail.com.

SYSTEM: Fine. You are Nancy Wells at 1122 Silicon Avenue. Correct?

USER: Yes.

SYSTEM: All items in your cart are in stock and can be shipped today. Because your order subtotal is over $100, you qualify for free UPS ground shipping (3–5 days). Other shipping options include next day ($35.00), two day ($20.00), and USPS parcel post ($11.70, 5–7 days). How would you like your items to be shipped?

USER:	Free UPS ground shipping.
SYSTEM:	We have a shipping address on file for you (1122 Silicon Avenue). Do you want your order shipped to that address?
USER:	No.
SYSTEM:	To what address would you like the order shipped?
USER:	John Wells, 1612 Jefferson Street NE, Albuquerque, NM 87123.
SYSTEM:	Okay, the total charge is $125.56 ($117.90 item subtotal plus $7.66 sales tax). Shall I charge that amount to your credit card on file (a Visa with an account number ending in 0899)?
USER:	Yes.
SYSTEM:	Your payment has been approved, and your order is being prepared for shipment. A confirmation e-mail has been sent to you, and another will be sent with a shipment tracking number when the order is shipped later today. Can I help you with anything else?
USER:	No.

storyboarding sequence of sketches of the display screen during a dialog

There are many possible implementations of this scenario. Before deciding on a specific implementation, an analyst often uses a technique called **storyboarding**—that is, showing a sequence of sketches of the display screen during a dialog. The sketches don't have to be very detailed to show the basic design concept. Designers can implement a storyboard with a visual programming tool, such as Visual Basic, but using simple sketches drawn with a graphics package can help keep the focus on the fundamental design ideas and avoid biasing the design to the capabilities of one specific application development tool.

Figure 7-7 shows the storyboard for the dialog based on the *Check out shopping cart* use case. The screen formats are primitive though sufficiently detailed to show all the information presented to and entered by the user. The storyboard can be reviewed by users and designers to identify missing or extraneous information and to discuss various options for final implementation, which might be based on a Web page shown on a large display, a traditional Windows dialog, or the user interface for a mobile device app.

User-Interface Design

As computing devices have proliferated, available devices and technologies for user interfaces have grown more powerful and diverse. Displays range from large-format, flat-panel monitors used with desktop computers to much smaller displays on tablets and cell phones. Displays can be supplemented with such simple sounds as clicks and beeps, music, or speech. User input can be captured via speech, touch screen, keyboard and mouse, or digital imaging (scanning).

As the range of user-interface technologies has increased, so has the need to create multiple user interfaces. For example, e-commerce applications typically have one user interface for desktop and laptop computers with large displays, another for cell phones with small displays, and sometimes a third for midsized displays, such as an iPad's. There may even be variations in the user interface among similar devices. For example, a user interface intended for a cell phone might be a generic version designed to run within the phone's Web browser or it might be a customized app with different versions for iPhones and Android-based phones.

Despite user interfaces' wide range of sizes and capabilities, some of their features are used on nearly all computing devices. We begin with a discussion of the common features and then delve into differences.

FIGURE **7-7**

Storyboard for the Check out shopping cart *dialog*

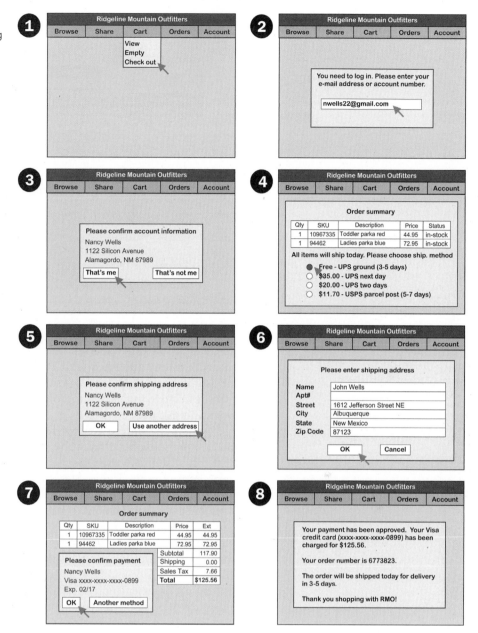

Guidelines for Designing Windows and Forms

After identifying dialogs, menus, and forms with a storyboard or another technique, the system developer can construct the user interface by using one of the many prototyping tools available. Major issues to consider at this stage of interface design include interface layout and formatting, data keying and entry, and navigation controls.

Interface Layout and Formatting

High-quality interfaces are well laid out, with the fields easily identified and understood. One of the best methods to ensure that interfaces are well laid out is to prototype various alternatives and let users test them. Users will let you know which characteristics are helpful and which are distracting. As you design your interfaces, you should think about these:

■ Consistency—All the forms within a system need to have the same look and feel. Consistent use of function keys, shortcuts, control buttons, color, and layout makes a system much more useful and professional looking. If

FIGURE **7-8**
The RMO home page

designing for an OS-supported interface (for example, Windows, iPhone, or Android), follow published guidelines to improve app and form consistency.

■ Labels and headings—Labels should also be easy to identify and read. A clear, descriptive title at the top of the interface helps to minimize confusion about a form's use.

■ Distribution and order—Related fields are usually placed next to each other and can be grouped within a box. Tab order (movement of the cursor or input focus) should follow the users' usual reading order (left to right and top to bottom in the United States and Europe). Blank space should be used so content is easy to distinguish and read.

■ Fonts and colors—Variations in font face and size can help users distinguish different parts of the form, but only a handful of font and size variations should be used for larger screens and as few as possible should be used for small screens. Too much variation is visually distracting and can cause eyestrain. Follow similar guidelines for colors. Avoid using too many colors and ensure that they are complementary. Also, be careful about mixing blue with yellow or mixing red with green, lest you make it difficult for color-blind users.

Figure 7-8 shows the home page displayed when a customer views the RMO Web site. The form includes two menu bars near the top that group related functions within the same part of the page. If the user points to the item Shop for Clothing, a submenu is displayed immediately below it with a similar color scheme and font. If the user points to the item Shop for Gear, Shop for Clothing is displayed in black text against a white background, Shop for Gear is displayed in white text with a blue background, and the latter's submenu items are displayed instead. Labels for the menu items are widely spaced and use an easy-to-read font. Except for the logo and picture, the page uses a small number of complementary colors. The title is positioned near the top of the picture and stands out well from the picture background and other page elements.

Data Entry

Several types of data-entry controls are widely used in user interfaces, including:

text box a rectangular box that accepts text typed on a keyboard or recognized from speech input

list box a text box that contains a list of predefined data values

■ **Text box**—a rectangular box that accepts text typed on a keyboard or recognized from speech input. Example: Product ID in **Figure 7-9**.

■ **List box**—a text box that contains a list of predefined data values. Example: Size in Figure 7-9.

FIGURE **7-9**

The RMO Product Detail form illustrating typical Microsoft Windows data-entry controls

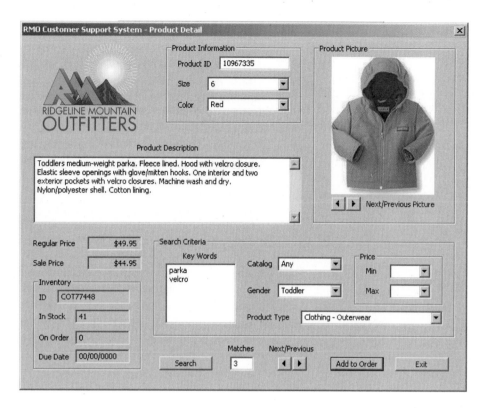

combo box   a text box that contains a predefined list of acceptable entries but permits the user to enter a new value when the list doesn't contain the desired value

radio buttons   a group of choices from which the user selects only one; the system then automatically turns off all other buttons in the group

check boxes   similar to radio buttons, but the user can select multiple items within the group

■ **Combo box**—a text box that contains a predefined list of acceptable entries but permits the user to enter a new value when the list doesn't contain the desired value.

■ **Radio buttons**—a group of choices from which the user selects only one; the system then automatically turns off all other buttons in the group. Example: Shipping Method in screen #4 of Figure 7-7.

■ **Check boxes**—similar to radio buttons, but the user can select multiple items within the group.

These data-entry controls were developed for the Apple Macintosh and later adopted by Windows and other operating systems. Early Web-browser standards had limited support for data-entry controls, but current versions support the controls described previously and many others. iPhone and Android interfaces support similar controls.

The form in Figure 7-9 could be used by an RMO employee to look up information about a product or to modify information in the catalog. Notice how the title and labels make the form easy to read. The natural flow of the form is top to bottom, with related fields placed together. Navigation and close buttons are easily found but aren't in the way of data-entry activity. The form uses standard Microsoft Windows controls, including text, list, and combo boxes. Though not visible in the figure, the form includes features that optimize it for frequent users, including standard Windows keyboard shortcuts, top-to-bottom and left-to-right tab order, and autocompletion of some fields based on database lookups and patches to partially entered text.

Navigation and Support Controls

Standard window interfaces provide several controls for navigation and window manipulation. For Microsoft applications, these controls consist of the Minimize, Maximize, and Close buttons in the upper-right corner, horizontal and vertical scrollbars, and so forth (see Figure 7-5 for specific examples). To maintain consistency across applications, it is a good idea to use built-in or standardized navigation controls whenever possible.

Additional Guidelines for Web Browser User Interfaces

Most user-interface designers first learn to develop Web-based interfaces that operate within a Web browser, such as Internet Explorer, Mozilla, Chrome, or Safari. As Web technologies and standards have matured, the differences in capability between browser-based interfaces (e.g., see Figure 7-11) and those that use operating system support libraries (e.g., see Figure 7-9) have all but disappeared. In many respects, browser-based interfaces have become more powerful. Nonetheless, there are some differences that should be considered when designing Web pages and browser-based forms.

Consistency

Consistency is especially critical within Web sites because most sites contain a large number of pages that serve many different purposes and audiences. For example, a typical corporate Web site provides e-commerce functions (e.g., online ordering), information to investors, directory and public contact information, and such product information as specifications and manuals. In essence, a corporate Web site is the gateway to a large collection of systems serving many different users and tasks. Despite the wide variety of users and tasks, the site as a whole is a single system that should support a single look and feel and should project a consistent, appealing, and desirable image for the corporation as a whole.

Most corporations spend considerable resources developing and maintaining their Web pages and ensuring consistency among them. Thus, user-interface designers of specific parts of the Web site must operate within the constraints of a corporate-wide design. **Cascading style sheets (CSS)** are a Web page encoding standard, and they enable a Web site designer to specify parts of a page that will always look the same and parts that will vary by task or audience. They can also constrain choices within the "variable" parts of a page, including placement and appearance of toolbars and menus, fonts, colors, and background images.

Figures 7-10 and 7-11 show additional pages from the RMO Web site that are displayed as the customer searches for items and completes an order. The menus, the outline surrounding detailed content, and the color and font options are all constrained by CSS. When the user selects menu items or clicks links or controls, the pages that are displayed reuse these elements to ensure consistent appearance and user interaction.

cascading style sheets (CSS) Web page encoding standard that enables a Web site designer to specify parts of a page that will always look the same and parts that will vary by task or audience

FIGURE **7-10**

Product detail page from RMO's Web site

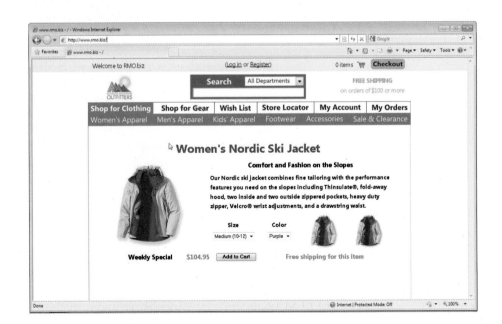

FIGURE **7-11**

Shopping cart page from RMO's Web site

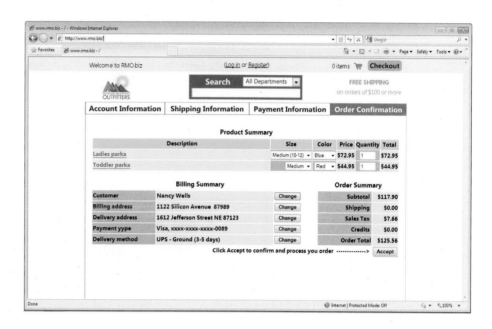

Performance Considerations

Web sites in general and browser-based forms in particular are sensitive to application design and to the quality of the network connections between the user's computing device and the servers that host the site. When a user clicks a hyperlink or a control that acts like a hyperlink, the browser sends information entered by the user (if any) to a Web server, along with a request for a new page. That information traverses multiple networks, is received and processed by a server, and then the response (a new page to be displayed) is sent back over the network. The delay between clicking a hyperlink and the display of the requested page depends on the amount of data to be transmitted, the display and network connection speed of the user computing device, the capacity of the networks that carry the messages, and the number of other users and applications that are competing for that network capacity.

A trade-off exists between the amount of information transmitted between the user's computing device and the server and the time it takes for the page to refresh; the more information that is transmitted, the longer the delay. That trade-off is especially important for communication over the Internet, although it is also a significant issue within corporate networks when user desktops and servers share high-speed connections.

There is also a trade-off between the amount of information and other data contained within a Web page and a Web-based application's performance. Pages with extensive information content or with embedded programming can avoid or postpone page refreshes. For example, a page containing a blank order form may be quite small. But many page refreshes will be required if the browser must interact with the server to validate every input as the user enters it. If the page containing the form also contains embedded validation programs, then many server interactions and page refreshes can be avoided. However, the initial download of the page will take longer because there is more content and the validation programs may be slow if the user computing device isn't very powerful (e.g., a relatively inexpensive cell phone).

Designers of Web-based user interfaces must perform a careful balancing act, providing embedded "intelligence" within a page to avoid refreshes but not overloading page content so as to avoid long delays when the user moves from page to page. Thorough testing is the best way to ensure that the right balance

FIGURE **7-12**

RMO's home page, with three menu levels

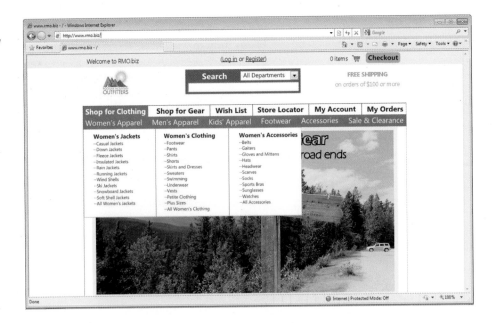

has been found. For example, in **Figure 7-12**, the RMO Web page includes the list of menu items for each of the main menus (e.g., Shop for Clothing) and submenus (e.g., Women's Apparel). As the user points to menu items, submenu content appears automatically without the need for a page refresh. Contrast this to the menu style shown in Figure 7-6(a), where the user must click a menu item and wait for a new page with the next level menu to be downloaded and displayed. The RMO page takes longer to download due to its embedded menu content and related programming, but it avoids page refreshes and their associated delays when displaying submenus.

Pictures, Video, and Sound

Web-based interfaces are often preferred for their inherent ability to mix text, images, and sound. Powerful and compelling interfaces can be constructed, and they are especially important in customer-facing systems. However, heavy use of sound and images exacerbates the previously discussed performance issues and also creates issues of compatibility. The performance implications are most significant for video and high-resolution still images, which consume large amounts of network capacity. For example, the background image in Figure 7-8 would require several hundred kilobytes of downloaded data for high-quality display on a laptop or desktop computer. A smaller image could be substituted for the smaller screen of a cell phone.

Compatibility issues arise for sound and video because there are so many ways in which they can be encoded. Most Web browsers rely on add-on components (sometimes called plug-ins) to play sounds and music and to display video. Unfortunately, all plug-ins don't work with all browsers, particularly older browser versions. Thus, Web site designers must carefully choose which formats and plug-ins will be used. In many cases, designers must create different pages for different plug-ins and write complex programs to query browsers for which plug-ins are present so the proper page can be downloaded.

Users with Disabilities

Designers of all user interfaces must be sensitive to the special needs of persons with disabilities. Because the Web is such a fundamental resource in the modern world, standards have been developed to ensure maximal usability for those who are visually impaired or have limited dexterity. Visually impaired users typically interact with Web pages via text-to-speech software that examines the content of a web page and reads it aloud. Users with limited dexterity often used voice-recognition software to navigate through elements in a page and perform tasks normally done with a keyboard, mouse, or touch screen. Both types of software are examples of a general software class called **assistive technologies**.

The World Wide Web Consortium (W3C) is an organization that sets standards for many aspects of the Web, including compatibility with assistive technologies. As of June 2010, it has published a working draft of the User Agent Accessibility Guidelines (UAAG), version 2.0. Although the draft hasn't received final approval as of this writing, many organizations use its guidelines or those of earlier standards to guide the development of their Web-based user interfaces.

Additional Guidelines for Handheld Devices

Designing Web and app-based user interfaces for handheld devices presents additional design challenges, including:

- Small screen size
- Small keyboards and touch screens
- Limited network capacity
- App design guidelines and toolkits

As of 2012, the size of a typical mobile phone screen is approximately 3.5×2.25 inches and approximately 480×320 pixels. The small screen area provides relatively limited space in which to display content. Thus, designers must pare down the user-interface content to ensure readability and to avoid cluttering the screen. **Figure 7-13** shows a sample mobile RMO Web page. Compared to the larger Web pages shown earlier, the mobile page eliminates many elements, including all images except a scaled-down logo. The remaining textual content is abbreviated, and special attention has been paid to contrast and layout in order to ensure maximal readability.

Small keyboards and touch screens also provide limited capabilities for user input. Mobile device user interfaces must avoid detailed textual input whenever possible and must provide touch controls that are well spaced and easy to locate. On many phones, speech-to-text capabilities provided by the phone's operating system can be used to streamline data entry and navigation, although the state of this technology still makes errors relatively common. Thus, designers must not rely too heavily on speech recognition and must ensure that errors are easily detected and corrected.

As of 2012, most mobile phones include connectivity through cell phone networks and Wi-Fi. Most current cell phone networks, described as third-

assistive technologies software (such as text-to-speech and voice-recognition utilities) that adapts user interfaces to the special needs of persons with disabilities

FIGURE **7-13**

RMO Web page layout for a mobile phone

generation (3G) networks, were originally designed for voice communication, with data communication grafted on as an afterthought. The throughput for 3G networks is usually no more than one-tenth the throughput of a Wi-Fi network. In 2010, deployment of 4G networks began in the United States. 4G networks increase data throughput to approximately that of Wi-Fi, although many users compete for access to that bandwidth.

Because mobile phone data throughput is much more limited than for other computing devices, the performance issues described earlier become much more significant design constraints. Page sizes must be limited to achieve acceptable download and page-refresh rates. High-resolution graphics are used only when absolutely necessary, and bandwidth-consuming video is typically avoided entirely. For RMO's mobile Web site, background graphics are completely avoided and high-resolution images are only used when a customer wants to view product details.

Some organizations deploy custom-developed apps that users can install on their mobile devices. Those apps run within a mobile operating system, such as the iPhone OS, iPad OS, or Google Android OS. Each OS provides a toolkit for user-interface developers and a set of development guidelines that ensure maximal compatibility among apps. Whenever possible, user-interface developers should use these toolkits and guidelines.

Identifying System Interfaces

The user interface includes inputs and outputs that directly involve system users. But there are many other system interfaces that process inputs, interact with other systems in real time, and distribute outputs with minimal human intervention. We define *system interfaces* broadly as any inputs or outputs with minimal or no human intervention. Included in this term are displayed and printed outputs for people, such as billing notices, reports, printed forms, and electronic outputs to other automated systems. Inputs that are automated or come from nonuser-interface devices are also included. For example, inputs from automated scanners, bar-code readers, optical character recognition devices, and other computer systems are included as part of a system interface.

The full range of inputs and outputs in an information system is illustrated in **Figure 7-14** and described here:

■ Inputs from and outputs to other systems—These are direct interfaces with other information systems, normally formatted as network messages. Electronic data interchange (EDI) and many Web-based systems are integrated with other systems through direct messaging. For example, in RMO's integrated supply chain management system and its customer support system, the arrival of inventory items from a supplier might trigger the shipment of a back-ordered item to a customer.

■ Highly automated inputs and outputs—These are captured by devices (such as scanners) or generated by persons who start a process that proceeds without further human intervention. For example, an item in a warehouse might pass a bar-code scanner that records its location as it moves by on a conveyor belt. Also, monthly statements can be printed and mailed through highly automated systems that place the statements within envelopes, apply postage, presort them by ZIP code, and batch them for delivery to the post office.

■ Inputs and outputs to external databases—These can supply input to or accept output from a system. EDI messages are more commonly used, but direct interaction with another system's database may be more efficient. For example, RMO's purchasing system could directly place product orders into a supplier's database.

FIGURE **7-14** *The full range of inputs and outputs in an information system*

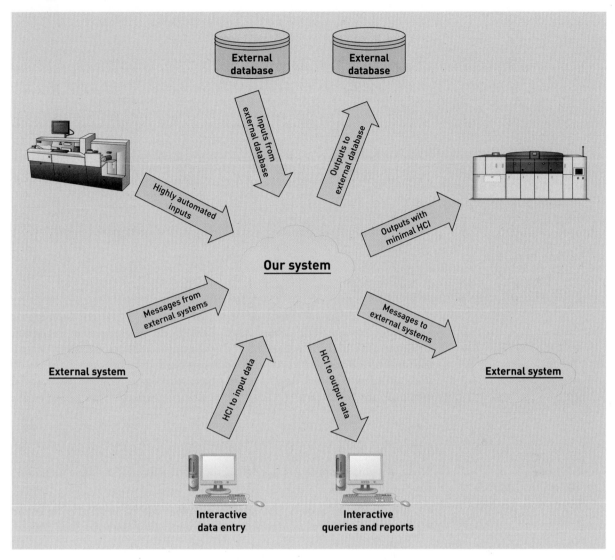

One of the main challenges of EDI is defining the format of the transactions. For example, General Motors—one of the early users of EDI—has thousands of suppliers and thousands of different transaction types, each in a different format. To complicate the situation further, each of these suppliers may be linked via EDI with tens or hundreds of customers, many of whom may also use EDI. So, a single type of transaction may have a dozen or more defined formats. It is easy to see why it is so costly to set up and maintain EDI systems. Even so, EDI is much more efficient and effective than paper transactions, which must be printed and reentered.

Modern EDI messages are generally formatted in **Extensible Markup Language (XML)**. XML is an extension of HTML that embeds self-defining data structures within textual messages. So, a transaction that contains data fields can be sent with XML codes to define the meaning of the data fields. Many newer systems are using XML to provide a common system-to-system interface. **Figure 7-15** illustrates a simple XML transaction that can be used to transfer customer information between systems. Data are surrounded by **XML tags**, such as <name> and </name>, that define the beginning, end, and meaning of the text that appears between them.

Extensible Markup Language (XML) extension of HTML that embeds self-defining data structures within textual messages

XML tags character sequences (such as <name> and </name>) that define the beginning, end, and meaning of the text that appears between them

FIGURE 7-15
Customer information formatted as
an XML message

```
<customer record>
      <accountNumber>RMO10989</accountNumber>
      <name>William Jones</name>
      <billingAddress>
            <street>120 Roundabout Road</street>
            <city>Los Angeles</city>
            <state>CA</state>
            <zip>98115</zip></billingAddress>
      <shippingAddress>
            <street>120 Roundabout Road</street>
            <city>Los Angeles</city>
            <state>CA</state>
            <zip>98115</zip></shippingAddress>
      <dayPhone>215.767.2334</dayPhone>
      <nightPhone>215.899.8763</nightPhone>
</customer record>
```

XML is called an extensible language because users can define any tags they want to use. For XML-based EDI, both systems must recognize the tags, but after a complete set of codes is established, transactions can include many different formats and still be recognized and processed. Many industries and professional organizations have standards committees that define tags used for EDI.

Designing System Inputs

When designing inputs for a system, the system developer must focus on three areas:

■ Identifying the devices and mechanisms that will be used to enter input
■ Identifying all system inputs and developing a list with the data content of each
■ Determining what kinds of controls are necessary for each system input

Automated Input Devices

The primary objective is to enter or update error-free data into the system. The key term here is *error-free*. Several good practices can help reduce input errors:

■ Use electronic devices and automatic entry whenever possible.
■ Avoid human involvement as much as possible.
■ If the information is available in electronic form, use that instead of reentering the information.
■ Validate and correct information at the time and location it is entered.

Automating data entry and avoiding human involvement are essentially different sides of the same coin, although using electronic devices doesn't automatically avoid human involvement. When system developers think carefully about minimizing human input and using electronic input media, they can design a system with fewer electronic input forms and avoid one of the most pervasive sources of input error: typing mistakes by users. Here are some of the more common devices used to avoid human keystroking:

■ Magnetic card strip readers
■ Bar-code readers

- Optical character recognition readers and scanners
- Radio-frequency identification tags
- Touch screens and devices
- Electronic pens and writing surfaces
- Digitizers, such as digital cameras and digital audio devices
- Speech-recognition software

The next principle of error reduction is to reuse the information already captured in automated form whenever possible. For example, consider the automated airline check-in process for customers without checked baggage. The customer swipes a credit card or driver's license, and the system queries its own database or an external database to identify the customer and retrieve the reservation information. The retrieved information is displayed to the customer for confirmation. Because the retrieved information is almost always correct, the task of data entry is reduced to a card swipe and the press of a button on a touch screen, eliminating manual data entry and its associated error rate. If the displayed data is incorrect, the final principle of error reduction is applied by having the customer directly enter corrected data.

Defining the Details of System Inputs

The fundamental approach that analysts use to identify user and system inputs is to search documents developed during analysis activities for information flows that cross the system boundary. The analyst examines system sequence diagrams to identify the incoming and outgoing messages for each activity or use case, and the design class diagrams to identify and describe the data content.

Figure 7-16 is a partial system sequence diagram for an object-oriented version of a payroll system. Various use cases have been combined to consolidate the major inputs on a single diagram. The messages that cross the system boundary identify inputs—system inputs and user-interface inputs. Three inputs cross the system boundary:

- updateEmployee (empID, empInformation)
- updateTaxRate (taxTableID, rateID, rateInformation)
- inputTimeCard (empID, date, hours)

The first input is part of a user interface. The other two inputs are from external systems and don't require user involvement. The information from the tax bureau can be sent as a set of real-time messages or in the form of a downloaded input file. The time card information could come into the system in various formats. Perhaps physical time cards are entered via an electronic card reader. Or an input from a subsystem, such as an electronic employee ID card reader, might send time card information at the end of every workday. These last two input messages need to be precisely defined, including their transmission methods, contents, and formats. The point to note here is that a sequence diagram provides a detailed perspective of the user and system inputs to support the use case and the corresponding business event.

Designing System Outputs

The primary purpose of system outputs is to present information in the right place at the right time to the right people. The tasks in this activity focus on four areas:

- Determining the type of each system output
- Making a list of required system outputs based on application design

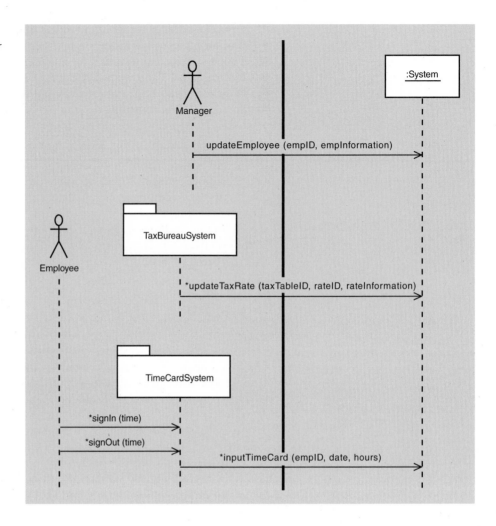

- Specifying any necessary controls to protect the information provided in the output
- Designing and prototyping the output layout

The purpose of the first two tasks is to evaluate the various alternatives and design the most appropriate approach for each needed output. The list of required system outputs is normally specified during the analysis activities as part of modeling system requirements. During design, the task is to coordinate the production of those outputs with the methods that are identified during the application architecture design.

The third task ensures that the designer evaluates the value of the information to the organization and protects it. Frequently, organizations implement controls on the inputs and system access but forget that output reports often have sensitive information.

As with system inputs, outputs are indicated by messages in sequence diagrams that cross the system boundary—originating from an internal system object and flowing to an external actor. Output messages that are based on an individual object (or record) are usually part of the methods of that object class. To report on all objects within a class, a class-level method is used. A class-level method is a method that works on the entire class of objects, not a single object. For example, a customer confirmation of an order is an output message that contains information for a single order object. However, to produce a summary report of all orders for the week, a class-level method looks at all the orders in the Order class and sends output information for each one with an order date within the week's time period.

Designing Reports, Statements, and Turnaround Documents

Modern information systems have made information much more widely available, with a proliferation of all types of reports—paper and electronic. One of the major challenges organizations face today is to organize the overwhelming amount of information in order to support managerial decision-making. One of the most difficult aspects of output design is to decide what information to provide and how to present it to avoid a confusing mass of complex data.

Report Types

There are four types of output reports commonly provided by an information system:

detailed reports reports that contain specific information on business transactions

summary reports reports that summarize detail or recap periodic activity

exception reports reports that provide details or summary information about transactions or operating results that fall outside a predefined normal range of values

executive reports reports used by high-level managers to assess overall organizational health and performance

- **Detailed reports**—These contain specific information on business transactions—for example, a list of all overdue accounts, with each line of the report presenting information about a particular account. A credit manager could use this report to research overdue accounts and determine actions to collect past-due amounts.
- **Summary reports**—These are often used to recap periodic activity. An example of this is a daily or weekly summary of all sales transactions, with a total dollar amount of sales. Managers often use this type of report to track departmental or division performance.
- **Exception reports**—These provide details or summary information about transactions or operating results that fall outside of a predefined normal range of values. When business is progressing normally, no report is needed. For example, a manufacturing organization might produce a report that lists parts that fail quality control tests more than 0.2 percent of the time.
- **Executive reports**—These are used by high-level managers to assess overall organizational health and performance. They thus contain summary information from activities within the company. They might also show comparative performance with industry-wide averages. Using these reports, executives can assess the competitive strengths or weaknesses of their companies.

Internal versus External Outputs

internal outputs reports or other outputs produced for use within the organization

external outputs reports or other outputs produced for use by people outside the organization

turnaround documents external outputs that includes one or more parts intended to be returned with new data or information

Printed outputs are classified as either internal outputs or external outputs. **Internal outputs** are produced for use within the organization. The types of reports just discussed fall under this category. **External outputs** include statements, notices, and other documents that are produced for people outside the organization. Because they are official business documents for an outside audience, they need to be produced with the highest-quality graphics and color. Examples include monthly bank statements, late notices, order confirmation and packing slips (such as those provided to Ridgeline Mountain Outfitters' customers), and legal documents (such as insurance policies). Some external outputs are referred to as **turnaround documents** because they are sent to a customer but include a tear-off portion that is returned for input later, such as a bill that contains a payment stub to be returned with a check. All these printed outputs must be designed with care, but organizations have many more options for printed output. Today's high-speed color laser printers enable all types of reports and other outputs to be produced.

An example of a detailed report for an external output is shown in **Figure 7-17**. When a customer places an order on the Web, the system will be able to print the order information as a confirmation. Of course, a user can always print the Web screen display by using the browser's print capability, but doing so is time consuming because it includes all the graphics and

FIGURE **7-17** *RMO shopping cart order report*

Ridgeline Mountain Outfitters—Shopping Cart Order

Customer Name: Fred Westing
Customer Number: 6747222

Order Number: 4673064
Today's Date: May 18, 2013

Shipping Address:

936 N Swivel Street
Hillville, Ohio 59222

Billing Address:

936 N Swivel Street
Hillville, Ohio 59222

Qty	Product ID	Description	Size	Color	Price	Extended Price
1	458238WL	Jordan Men's Jumpman Team J	12	White/ Light Blue	$119.99	$119.99
1	347827OP	Woolrich Men's Backpacker Shirt	XL	Oatmeal Plaid	$41.99	$41.99
2	8759425SH	Nike D.R.I. – Fit Shirt	M	Black	$30.00	$60.00
1	5858642OR	Puma Hiking Shorts	L	Tan	$15.00	$15.00
					Subtotal	$236.98
					Shipping	$8.50
					Tax	$11.25
					Total	$256.73

Shipping Information:

Shipping Method: Normal 7–10 day
Shipping Company: UPS
Tracking Number: To be sent via email
Email Address: FredW253@aol.com

Payment Information:

American Express ☐ MasterCard ☐ VISA ☒ Discover ☐

Account Number

| X | X | X | X | – | X | X | X | X | – | X | X | X | X | – | 5 | 7 | 8 | 4 |

MO YR

Expiration Date ___ 05 / 15 ___

Thank you for your order. It is a pleasure to serve you.
Check back next week for new weekly specials!!

index links on the page. It is much more user friendly to provide shoppers with a "printer friendly" order confirmation in addition to a Web-based display, as shown in Figure 7-11.

Figure 7-18 is an example of an internal output based on inventory records. The report includes detailed and summary sections, although the figure doesn't show the summary section. A control break is the data item that divides the detailed section into groups. In this example, the control break is on the product item number—called ID on the report. Whenever a new value of the ID is encountered on the input records, the report begins a new control break section. The detailed section lists the transactions of records from the database, and the summary section provides totals and recaps of the information. The report is sorted and presented by product. However, within each product is a list of each inventory item showing the quantity currently on hand.

External outputs can consist of complex, multiple-page documents. A well-known example is the set of reports and statements that you receive with your car insurance statement. This statement is usually a multipage document consisting of detailed automobile insurance information and rates, summary pages, turnaround premium payment cards, and insurance cards for each automobile. Another example is a report of employment benefits, with multiple pages of information customized to the individual employee. Sometimes, the documents are printed in color, with special highlighting or logos. **Figure 7-19** is one page of an example report for survivor protection from an employee benefit booklet. The text is standard wording, and the numbers are customized to the individual employee.

FIGURE **7-18** *RMO inventory report*

Ridgeline Mountain Outfitters — Products and Items

ID	Season	Category	Supplier	Unit Price	Special Price	Discontinued
RMO12587	Spr/Fall	Mens C	8201	$39.00	$34.95	No

Description Outdoor Nylon Jacket with Lining

Size	Color	Style	Units in Stock	Reorder Level	Units on Order
Small	Blue		691	150	
	Green		723	150	
	Red		569	150	
	Yellow		827	150	
Medium	Blue		722	150	
	Green		756	150	
	Red		698	150	
	Yellow		590	150	
Large	Blue		1289	150	
	Green		1455	150	
	Red		1329	150	
	Yellow		1370	150	
Xlarge	Blue		1498	150	
	Green		1248	150	
	Red		1266	150	
	Yellow		1322	150	

ID	Season	Category	Supplier	Unit Price	Special Price	Discontinued
RMO28497	All	Footwe	7993	$49.95	$44.89	No

Description Hiking Walkers with Patterned Tread Durable Uppers

Size	Color	Style	Units in Stock	Reorder Level	Units on Order
7	Brown		389	100	
	Tan		422	100	
8	Brown		597	100	
	Tan		521	100	
9	Brown		633	100	
	Tan		654	100	
10	Brown		836	100	
	Tan		954	100	
11	Brown		862	100	
	Tan		792	100	
12	Brown		754	100	
	Tan		788	100	
13	Brown		830	100	
	Tan		921	100	

Electronic Reports

Organizations use various types of electronic reports, each serving a different purpose and each with its respective strengths and weaknesses. Electronic reports provide great flexibility in the organization and presentation of information. In some instances, screen output is formatted like a printed report but displayed electronically. However, electronic reports can also present information in many other formats. Some have detailed and summary sections, some show data and graphics together, others contain boldface type and highlighting, others can dynamically change their organization and summaries, and still others contain hotlinks to related information. An important benefit of electronic reporting is that it is dynamic; it can change to meet the specific needs of a user in a particular situation. In fact, many systems provide powerful

FIGURE **7-19** *A sample employee benefit report*

Survivor Protection

In the event of your death while working for a participating employer, your designated beneficiaries could receive:

Lump Sum Benefits

$50,000	Basic Life Insurance
$230,000	Supplemental Life Insurance
$148,677	Thrift Plan
$31,686	Tax Sheltered Annuity (TSA) Plan
$255	Social Security for your eligible dependents
$460,618	Total*

You have not elected Universal Life Insurance. If you would like more information on this plan, please call 1-800-555-7772.

*Refer to page 7 for additional information on the amount of coverage needed to provide ongoing replacement income.

Accidental Death Benefits

If your death is due to an accident, your designated beneficiaries will receive the above benefits plus:

$100,000	24-Hour Accidental Death and Dismemberment Insurance
$100,000	Occupational Accidental Death and Dismemberment Insurance, if the accident is work related

Monthly Death Benefits

If you die before receiving the Master Retirement Plan benefits and you are vested and have a surviving spouse, your spouse may be eligible for a Qualified Pre-Retirement Survivor Annuity.

In addition, your family may be eligible for the following estimated monthly benefits from Social Security, not to exceed a maximum of $2,591 based on:

$1,110	for each child under age 18
$1,110	for a spouse with children under age 16; or
$1,058	for a spouse age 60 or older

drill down user-interface design technique that enables a user to select summary information and view supporting detail

ad hoc reporting capabilities so users can design their own reports on the fly. For example, an electronic report can provide links to further information. One technique, called **drill down**, allows the user to activate a "hot spot hyperlink" on the report, which tells the system to display a lower-level report that provides more detailed information. For example, **Figure 7-20** contains a monthly sales summary. The report provides sales totals grouped by product category and season. However, if the user clicks the hotlink for any season, a detailed report pops up with more detailed sales data.

Another variation of this hotlink capability lets the user correlate information from one report to related information in another report. Most people are familiar with hotlinks from using their Internet browsers. In an electronic report, hotlinks can refer to other information that correlates or extends the primary information. This same capability can be very useful in a business report that, for example, links the annual statements of key companies in a certain industry.

Another dynamic aspect of electronic reports is the capability to view the data from different perspectives. For example, it might be beneficial to view sales commission data by region, by sales manager, by product line, or by time period

FIGURE **7-20** *RMO summary report with drill down to the detailed report*

Monthly Sales Summary

Year **2013**   *Month*   **January**

Category	Season Code	Web Sales	Telephone Sales	Mail Sales	Total Sales
Footwear	All	$ 289,323	$ 1,347,878	$ 540,883	$ 2,178,084
Men's Clothing	Spring	$ 1,768,454	$ 2,879,243	$ 437,874	$ 4,691,484
	Summer	213,938	387,121	123,590	724,649
	Fall	142,823	129,873	112,234	384,930
	Winter	2,980,489	6,453,896	675,290	10,109,675
	All	1,839,729	4,897,235	349,234	7,086,198
Totals			747,368	$ 1,698,222	$ 23,391,023
Women's Clothing	Spring				965,610
	Summer				
	Fall				
	Winter				
	All				
Totals					

Monthly Sales Detail

Year **2013**   *Month*   **January**   *Category*   **Men's Clothing**   *Season*   **Winter**

Product ID	Product Description	Web Sales	Telephone Sales	Mail Sales	Total Sales
RMO12987	Winter Parka	$ 1,490,245	$ 3,226,948	$ 337,640	$ 5,054,833
RMO13788	Fur-Lined Gloves	149,022	322,695	33,765	505,482
RMO23788	Wool Sweater	596,097	1,290,775	135,058	2,021,930
RMO12980	Long Underwear	298,050	645,339	68,556	1,003,005
RMO32998	Fleece-Lined Jacket	447,075	1,258,079	100,271	1,805,425
Total		$ 2,980,489	$ 6,743,836	$ 675,290	$ 10,394,615

or to compare the current data with last season's data. Instead of printing all these reports, you can use an electronic format to generate the different views as needed. Sometimes, long or complex reports include a table of contents, with hotlinks to the various sections of the report. Some report-generating programs provide an electronic reporting capability that includes all the functionality found on Web pages, including frames, hotlinks, graphics, and even animation.

Graphical and Multimedia Presentation

The graphical presentation of data is one of the greatest benefits of the information age. Tools that permit data to be presented in charts and graphs have made information reporting much more user friendly for printed and electronic formats. Information is being used more and more for strategic decision-making as businesspeople examine their data for trends and changes. In addition, today's systems frequently maintain massive amounts of data—much more than people can review. The only effective way to use much of this data is by summarizing and presenting it in graphical form. **Figure 7-21** presents a pie chart and a bar graph—two common ways to present summary data.

Multimedia outputs have become available recently as multimedia tool capabilities have increased. Today, it is possible to see a graphical (possibly animated) presentation of the information on a screen and have an audio description of the salient points. Combining visual and audio output is a powerful way to present information. (Of course, video games are pushing the frontier of virtual reality to include visual, audio, tactile, and olfactory outputs.)

FIGURE **7-21** *Sample pie chart and bar graph reports*

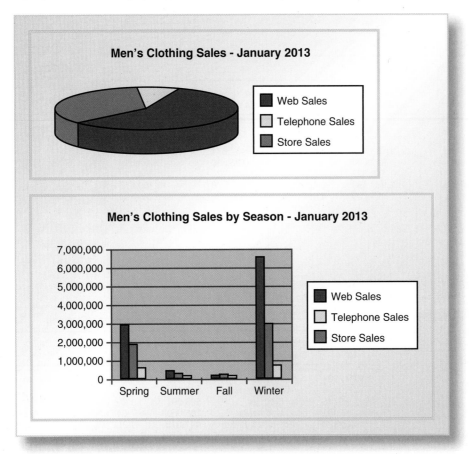

As the design of system outputs progresses, it is beneficial to evaluate the various presentation alternatives. Reporting packages can be integrated into the system to provide a full range of reporting alternatives. Developers should carefully analyze each output report to determine the objective of the output and to select the form of the output that is most appropriate for the information and its use.

Chapter Summary

Inputs and outputs can be classified as system interfaces or user interfaces. The user interface is everything the user comes into contact with while using the system—physically, perceptually, and conceptually. There are many different ways to describe the user interface, including the desktop metaphor, the document metaphor, and the dialog metaphor. Dialog design starts with identifying dialogs based on activities or use cases. A storyboard showing sketches of screens in sequence can be drawn to convey the design for review with users, or prototypes can be created using a tool such as Visual Basic. The object-oriented approach provides UML models that can document dialog designs, including sequence diagrams, activity diagrams, and class diagrams.

Each screen and form used in a dialog needs to be designed, and there are guidelines for the layout, selection of input controls, navigation, and help. These guidelines apply to windows forms and to browser forms used in Web-based systems. Designing a dialog for a Web site is similar to creating any other dialog, except users need more information and more flexibility. Additional Web design guidelines apply to designing for the computer medium, designing the whole site, and designing for the user. In addition, because a Web site reflects the company's image to customers, graphic designers and marketing professionals should be involved.

When designing system inputs, the developer identifies the input devices and identifies all system inputs and lists the data content of each. To develop a list of the inputs to the system, designers use sequence and design class diagrams. The process to design the outputs from the system consists of the same steps as for input design.

Key Terms

Review Questions

1. Why is interface design often referred to as dialog design?

2. What are the three aspects of the system that make up the user interface for a user?

3. What are some examples of the physical, perceptual, and conceptual aspects of the user interface?

4. What are the three metaphors used to describe human-computer interaction?

5. A desktop on the screen is an example of which of the three metaphors used to describe human-computer interaction?

6. What type of document allows the user to click a link and jump to another part of the document?

7. List and briefly describe four guidelines for interface layout and formatting that apply to all types of user display and input devices. What additional guidelines apply to Web sites/pages and user interfaces for mobile computing devices?

8. What is the technique that shows a sequence of sketches of the display screen during a dialog?

9. What UML diagram can be used to show how the interface objects are plugged in between the actor and the problem domain classes during a dialog?

10. What are some of the input controls that can be used to select an item from a list?

11. What two types of input controls are included in groups?

12. What popular analogy is used for direct customer access with a Web site when customers shop online?

13. What does XML stand for? Explain how XML is similar to HTML. Also discuss the differences between XML and HTML.

14. How do you identify the data fields of a system interface by using UML and the object-oriented approach?

15. What are the different considerations for output screen design and output report design?

16. What is meant by drill down? Give an example of how you might use it in a report design.

17. What is the danger from information overload? What solutions can you think of to avoid it?

Problems and Exercises

1. Think of all the software you have used. What are some examples of ease of learning conflicting with ease of use?

2. Visit some Web sites and then identify all the controls used for navigation and input. Are they all obvious? Discuss some differences in visibility and affordance among the controls.

3. A common maxim for designing a man-machine interface is that it is better to change the machine than to try to change the human to accommodate

the machine. Are there machines (or systems) that you use in your daily life that have room for improvement? Are the current generations of Windows PC and Apple Mac as usable as they might be? If not, what improvements can you suggest? Is the World Wide Web as usable as it might be? If not, what improvements can you suggest? Are we just beginning to see some breakthroughs in usability, or have most of the big improvements already been made?

4. Download and install App Inventor from Google. Use it to develop a prototype interface that implements the storyboard in Figure 7-7.

5. Evaluate the course registration system at your university. List the basic steps of a user's dialog with the system. What are some of the problems with the system from the perspective of ease of learning and ease of use? In what ways is the system inflexible? In what ways is needed information not available? Is too much information provided that distracts from the task at hand?

6. Evaluate the online catalog system at your university library. Write a dialog that shows the interaction between the user and the system. Rewrite the dialog to improve it. Create a storyboard to show how your design would look and feel.

7. Find a Web site with direct customer ordering. Browse through some product descriptions and note the design of the dialog and the Web pages. What do you like and dislike about the design? Evaluate the Web site based on visibility and affordance. Does the site achieve an optimal balance between the number of page refreshes and the delay between page refreshes? Would your answer change if you were using a different computing device, a different network, or accessing the system at a different time of day?

Case Study

Video Rental Service

Develop a storyboard that encompasses the following sample dialog. The system supports rental and download of movies or television shows to a computer or digital video recording and playback device. A dialog between the system and user might follow this pattern:

SYSTEM: What would you like to do?

USER: I'd like to rent a movie.

SYSTEM: Okay. What is your e-mail address or account number?

USER: My e-mail address is nwells22@gmail.com.

SYSTEM: Fine. You are Nancy Wells at 1122 Silicon Avenue. Correct?

USER: Yes.

SYSTEM: What movie do you want to rent?

USER: *Titanic.*

SYSTEM: I have two available movies named *Titanic.* The first was released in 1953 and stars Barbara Stanwyck and Clifton Webb. The second was released in 1997 and stars Leonardo DiCaprio and Kate Winslet. Which do you want to rent?

USER: Both.

SYSTEM: Fine. That is *Titanic* (the 1953 release) with a rental charge of $2.99 for seven days and *Titanic* (the 1997 release) with a rental charge of $5.99 for seven days. Would you like to add more rentals?

USER: That's all.

SYSTEM: Okay. The total charge is $8.98. Shall I charge that amount to your credit card file (a Visa with an account number ending in 0899)?

USER: Yes.

SYSTEM: Your movies are downloading to your computer now. They will expire at midnight seven days from now.

USER: Okay. Thanks.

Community Board of Realtors

One of the key use cases for the multiple listing service system is *Create new listing*, where the realtor enters all the important information about a new listing he or she has obtained. Realtors want to be able to create a new listing as soon as possible so other realtors and potential buyers will find the listing online. Some like to enter the information while talking with the owner or while inspecting the property. Realtors are rarely in their offices these days, so being able to create a new listing on a mobile device is a key feature of the multiple listing service system.

Consider the information that must be entered when creating a new listing, and list the dialog steps that are necessary. Keep in mind that when designing for a smartphone, less information can be entered in each step compared to a full screen Web application.

Also keep in mind that typing is error-prone and awkward for many users, so think about opportunities to use check boxes, radio buttons, and list boxes to aid selection. Create a storyboard of this use case for a mobile device, showing each step of the dialog that maximizes the use of check boxes, radio buttons, and list boxes.

On the Spot Courier Services

Review the case description and your solution for the Web scenario of the use case *Request package pickup* from Chapter 5. Then, using a presentation tool, such as Microsoft PowerPoint or Apple Keynote, create a storyboard of the Web pages necessary to support the use case.

The case description in Chapter 5 also identified a new use case, which we can call *View scheduled pickups/deliveries*. Based on current technology, write a dialog showing how this might be supported with a portable digital device. You may use any current technology that you deem applicable, such as GPS tracking, map and directions software, and real-time updates of pickup locations. Consider the possibility that the driver may want to get an overview of his or her stops for the entire run, view the next few stops, or just get directions to the next stop.

The Spring Breaks 'R' Us Travel Service

The Spring Breaks R Us social networking subsystem requires an intuitive and engaging user-interface design for mobile devices. But the social networking subsystem can also play an important role in resort security. For example, each resort could use traveler location, interests, activities, and "likes"—all of which are available through the application—to monitor the well-being of travelers staying at the resort. Most spring break travelers are young, and their parents are concerned about their safety—particularly at isolated resorts in foreign countries. SBRU and the participating resorts could keep track of where travelers are and who they are near, monitor messages about activities and parties, and anticipate crowded conditions or vulnerable travelers wandering off-site. Alerts could notify security if conditions veer away from normal or if messages indicate there are problems. For example, if the pool is overcrowded, some action can be taken. If messages refer to places off-site that are known to be dangerous, security can make an extra patrol. Although many people find this use. of private information objectionable, others—particularly parents—find it essential.

Imagine resort security with a large, wide-screen monitor tracking traveler activities. Design a main screen that includes multiple locations, paths and roads, traveler location and status, messages traveling from traveler to traveler, and other features that security should monitor. Create a storyboard that shows an example of a pop-up alert and a menu of options that security might select after an alert. Should you also show security the staff members' locations and status? How about clicking security staff members to send them a message? How about clicking a location to turn up the lights or to close a security gate? Be creative as you think through the design possibilities. You should include four or five screen layouts for the storyboard.

(continued on page 222)

(continued from page 221)

Sandia Medical Systems Real-Time Glucose Monitoring

Developers have made these choices regarding format and transmission methods between the cell phone app and server-based software components:

- Glucose-level readings within a normal range will be sent to the server once per hour as XML messages via secure HTTP. Because glucose levels are stored every five minutes by the cell phone app, a normal message to the server will contain 12 glucose levels and time stamps. If more than two glucose levels in a row are outside the normal range, the cell phone app will transmit them to the server immediately.
- Physicians and other medical personnel will initiate direct contact with a patient via text message, voice message, or real-time voice call.
- Changes to alert conditions and updates to software will be sent from the server to a patient's cell phone app as XML messages. The XML messages will be sent immediately after the cell phone app has transmitted glucose levels to the server.

No industry group has yet defined XML tags suitable for the RTGM application. Thus, designers must develop appropriate tags for this application.

If you aren't already familiar with XML, do at least 30 minutes of background research on the Web. Refer to the class diagram in the Chapter 4 RTGM case to determine required data content. Then, design XML tags and message formats suitable to transmit time-stamped glucose levels from the cell phone app to the server and to transmit updated alert conditions from the server to the cell phone app.

If medical personnel choose to send a text message to the patient, how will the server transmit that message? To help answer that question, research Short Message Service (SMS), Enhanced Messaging Service (EMS), and SMS gateways.

Further Resources

Randolph G. Bias and Deborah J. Mayhew, *Cost-Justifying Usability: An Update for the Internet Age* (2nd ed). Morgan Kaufmann, 2005.

Paul C. Brown, *Implementing SOA: Total Architecture in Practice*. Addison-Wesley, 2008.

Patrick Carey, *New Perspectives on Creating Web Pages with HTML, XHTML, and XML* (3rd ed.). Cengage Learning, 2010.

Donald Norman, *The Design of Everyday Things*. Basic Books, 2002.

Janice Redish, *Letting Go of the Words: Writing Web Content that Works*. Morgan Kaufmann, 2007.

Ben Shneiderman, Catherine Plaisant, Maxine Cohen, and Steven Jacobs, *Designing the User Interface: Strategies for Effective Human-Computer Interaction* (5th ed.). Addison Wesley, 2009.

Joel Sklar, *Principles of Web Design* (5th ed.). Cengage Learning, 2012.

PART 4

Projects and Project Management

8

Approaches to System Development

Chapter Outline

- The Systems Development Life Cycle
- The Support Phase
- Methodologies, Models, Tools, and Techniques
- Two Approaches to Software Construction and Modeling
- Agile Development

Learning Objectives

After reading this chapter, you should be able to:

- Compare the underlying assumptions and uses of a predictive and an adaptive system development life cycle (SDLC)
- Describe the key activities and tasks of information system support
- Explain what comprises a system development methodology—the SDLC as well as models, tools, and techniques
- Describe the two overall approaches used for software construction and modeling: the structured approach and the object-oriented approach
- Describe the key features of Agile development

Development Approaches at Ajax Corporation, Consolidated Concepts, and Pinnacle Manufacturing

Kim, Mary, and Bob, graduating seniors, were discussing their recent interview visits to companies that recruit computer information system (CIS) majors on their campus. All agreed that they had learned a lot by visiting the companies, but they also all felt somewhat overwhelmed.

"At first, I wasn't sure I knew what they were talking about," Kim said. During her on-campus interview, she had impressed Ajax Corporation with her knowledge of data modeling. And when she visited the company's home office data center for her second interview, the interviewers spent a lot of time describing the company's system development methodology.

"They said to forget everything I learned in school," Kim continued. "That got my attention."

Ajax Corporation had purchased a complete development methodology called IM One from a small consulting firm. Most of its employees agreed that it worked fairly well, having invested a lot of time and money learning and adapting to it. Those who had worked for Ajax for a long time thought IM One was unique, and they were very proud of it.

"Then, they started telling me about their SDLC, iterations, business events, data flow diagrams, and entity-relationship diagrams—things like that," Kim said. She had recognized that many of the key concepts in the IM One methodology were fairly standard models and techniques from the structured approach to system development.

"I know what you mean," said Mary, a very talented programmer who knew just about every popular programming language available. "Consolidated Concepts went on and on about things like OMG, UML, UP, and people named Booch, Rumbaugh, and Jacobson. It turns out they use the object-oriented approach to develop their systems, and they like the fact that I know Java and C# and .NET. There was no problem once I got past all the terminology they used."

Bob, who interviewed with Pinnacle Manufacturing, had a different story to tell. "A few people said analysis and design are no longer a big deal," he said. "And I'm thinking, 'Knowing that would have saved me some time in school.'"

Pinnacle has a small system development group supporting its manufacturing and inventory control. "They said they like to jump in and get to the code as soon as possible," Bob said. "Little documentation and not much of a project plan. They showed me some books on their desks, and it looked like they had been doing a lot of reading about analysis and design. I could see they were using Agile development and Agile modeling techniques and focusing on best practices required for their small projects. It turns out they just organize their work differently by looking at risk and writing user stories while building prototypes. I recognized some sketches of class diagrams and sequence diagrams on the boss's whiteboard, so I felt fairly comfortable."

Kim, Mary, and Bob agreed that there was much to learn in these work environments but also that there are many different ways to describe the key concepts and techniques they learned in school. They were all glad they focused on the fundamentals in their CIS classes and that they had been exposed to a variety of approaches to system development.

Overview

As the experiences of Kim, Mary, and Bob demonstrate, there are many ways to develop an information system, and doing so is very complex. Project managers rely on a variety of aids to help them with every step of the process. So far in this text, you have learned about analysis and design models and techniques, and now you will learn more about the overall system development process. You learned about the systems development life cycle (SDLC) in Chapter 1. That particular SDLC included six core processes and multiple iterations. This chapter discusses the SDLC in more detail, including some variations found in industry. Additionally, an information system requires extensive support after deployment, so the support phase of the SDLC is also discussed.

The entire process of developing an information system requires more than just an SDLC. A system development methodology includes more specific instructions for completing the activities of each core process by using specific models, tools, and techniques. This chapter also reviews two main approaches to defining the information system technology and software development used for business systems: the traditional approach and the object-oriented approach.

The traditional approach refers to structured software development, which describes software as a hierarchy of programs and modules and uses structured analysis, structured design, and structured programming. The object-oriented approach refers to object-oriented software development, which describes software as a set of interacting objects. It uses such models as object class diagrams, sequence diagrams, state charts, and object-oriented programming (OOP). Finally, Agile development is discussed as a philosophy that guides a development project. It focuses on techniques and methods that encourage more user involvement and allow for more flexible projects with changing requirements.

The Systems Development Life Cycle

Chapter 1 demonstrated how analysis and design models and techniques are used to solve business problems by building an information system. For problem-solving work to be productive, it needs to be organized and goal oriented. Analysts achieve these results by organizing the work into projects. As defined in Chapter 1, a project is a planned undertaking, with a beginning and end, that produces a well-defined result or product. The term *information system development project* refers to a planned undertaking that produces a new information system. Some system development projects are very large, requiring thousands of hours of work by many people and spanning several calendar years. In the RMO case study introduced in Chapter 2, the system being developed is a moderately sized computer-based information system requiring a moderately sized project. Many system development projects are smaller, lasting a few months.

For a system development project to be successful, it must be planned and organized. The plan must include a comprehensive set of activities that flow in the proper sequence. Otherwise, activities are omitted or work may need to be done multiple times. The end result, of course, is producing a high-quality information system as measured by its reliability, robustness, efficiency, and fitness for purpose. The systems development life cycle (SDLC), which was introduced in Chapter 1, is a fundamental concept in the success of information system development projects.

The SDLC provides a way to think about the development of a new system as a progressive process, much like a living entity. We can expand this concept and view the information system as having a life itself; in fact, we often refer to the life cycle of a system. During its life cycle, an information system is first conceived, then it is designed, built, and deployed as part of a development project, and, finally, it is put into production and used to support the business. However, even during its productive use, a system is still a dynamic, living entity that is updated, modified, and repaired through smaller projects.

Several projects may be required during the life of a system, first to develop the original system and then to upgrade. In this chapter—and in the rest of this textbook—we will focus on the initial development project, not on the support projects. In other words, our primary concern is with getting the system developed and deployed.

In today's diverse development environment, there are many approaches to developing systems, and they are based on different approaches to the SDLC. Although it is difficult to find a single, comprehensive classification system that encompasses all the approaches, one useful way to categorize them is along a continuum from predictive to adaptive (see **Figure 8-1**).

predictive approach to the SDLC
an approach that assumes the project can be planned in advance and that the new information system can be developed according to the plan

A **predictive approach to the SDLC** assumes that the development project can be planned and organized and that the new information system can be developed according to the plan. Predictive SDLCs are useful for building systems that are well understood and defined. For example, a company may want to convert its old networked client/server system to a newer Web-based system that includes a smartphone app. In this type of project, the staff already understands the requirements very well, and no new processes need to be

FIGURE 8-1

Predictive versus adaptive approaches to the SDLC

The choice of SDLC varies depending on the project

Predictive SDLC	Adaptive SDLC
Requirements well understood and well defined. Low technical risk.	Requirements and needs uncertain. High technical risk.

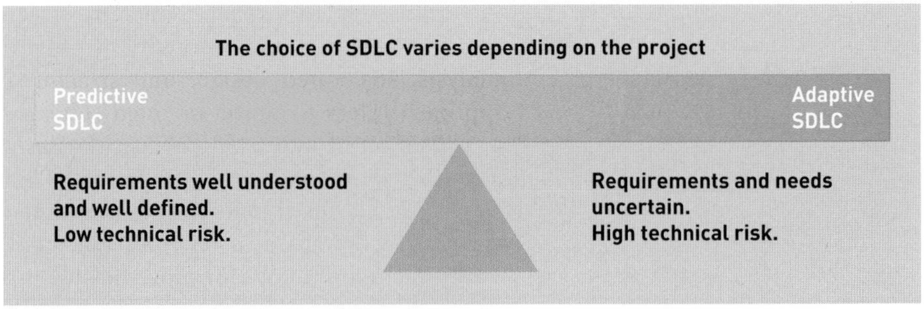

added. Thus, the project can be carefully planned, and the system can be built according to the specifications.

An **adaptive approach to the SDLC** is used when the system's requirements and/or the users' needs aren't well understood. In this situation, the project can't be planned completely. Some system requirements may need to be determined after preliminary development work. Developers should still be able to build the solution, but they need to be flexible and adapt the project as it progresses. Recall that the Tradeshow system described in Chapter 1 used this approach.

In practice, any project could have—and most do have—predictive and adaptive elements. That is why Figure 8-1 shows them as endpoints along a continuum, not as mutually exclusive categories. The predictive approaches are more traditional and were conceived during the 1970s through the 1990s. Many of the newer, adaptive approaches have evolved with object-oriented technology and Web development; they were created during the late-1990s and into the 21st century. Let us look at the more predictive approaches and then examine the newer adaptive approaches.

Traditional Predictive Approaches to the SDLC

The development of a new information system requires a number of different but related sets of activities. In predictive approaches, there is a group of activities that identifies the problem and secures approval to develop a new system; this is called *project initiation*. A second group of activities, called *project planning*, involves planning, organizing, and scheduling the project. These activities map out the project's overall structure. A third group—*analysis*—focuses on discovering and understanding the details of the problem or need. The intent here is to figure out exactly what the system must do to support the business processes. A fourth group—*design*—focuses on configuring and structuring the new system components. These activities use the requirements that were defined earlier to develop the program structure and the algorithms for the new system. A fifth group—*implementation*—includes programming and testing the system. A sixth group—*deployment*—involves installing and putting the system into operation.

These six groups of activities—project initiation, project planning, analysis, design, implementation, and deployment—are sometimes referred to as **phases** of the system development project, and they provide the framework for managing the project. Another phase, called the *support phase*, includes the activities needed to upgrade and maintain the system after it has been deployed. The support phase is part of the overall SDLC, but it isn't normally considered part of the initial development project. **Figure 8-2** illustrates the six phases of a traditional predictive SDLC plus the support phase.

The most predictive SDLC approach (i.e., farthest to the left on the predictive/adaptive scale) is called a **waterfall model**, with the phases of the project flowing down, one after another. As shown in **Figure 8-3**, this model assumes that the phases can be carried out and completed sequentially. First, a detailed plan is developed, then the requirements are thoroughly specified, then the system is designed down to the last algorithm, and then it is programmed, tested, and installed. After a project drops over the waterfall into the next phase,

adaptive approach to the SDLC an approach that assumes the project must be more flexible and adapt to changing needs as the project progresses

phases related groups of development activities, such as planning, analysis, design, implementation, and support

waterfall model an SDLC approach that assumes the phases can be completed sequentially with no overlap

FIGURE **8-2** *Traditional information system development phases*

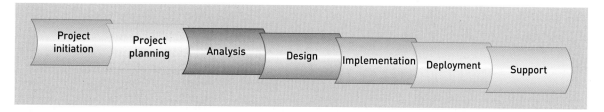

FIGURE **8-3** *Waterfall model of the SDLC*

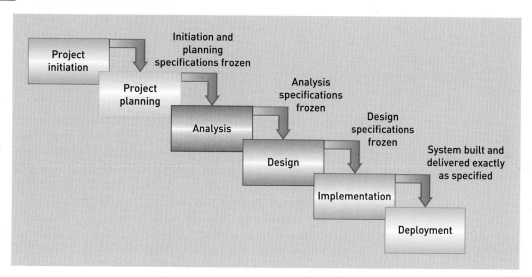

there is no going back. In practice, the waterfall model assumes rigid planning and final decision-making at each step of the development project. As you may have guessed, the waterfall model doesn't always work very well. Being human, developers are rarely able to complete a phase without making mistakes or leaving out important components that had to be added later. However, even though we don't use the waterfall model in its purest form anymore, it still provides a valuable foundation for understanding development. No matter what system is being developed, we need to include initiation, planning, analysis, design, implementation, and deployment activities.

A little farther to the right on the predictive/adaptive scale are modified waterfall models. These are still predictive—that is, they still assume a fairly thorough plan—but there is a recognition that the project's phases must overlap, influencing and depending on each other. Some analysis must be done before the designing can start, but during the design, there is a need for more detail in the requirements or perhaps it is discovered that a requirement cannot be met in the manner originally requested. **Figure 8-4** illustrates how these activities can overlap.

Another reason phases overlap is for efficiency. While the team members are analyzing needs, they may also be thinking about and designing various forms or reports. To help them understand the needs of the users, they may want to design some of the final system. But when they do early design, they will frequently throw some components away and save others for later inclusion in the final system. In addition, many components of a computer system are interdependent, which requires analysts to do analysis and some design at the same time.

Newer Adaptive Approaches to the SDLC

In an adaptive approach, project activities—including plans and models—are adjusted as the project progresses. There are many ways to depict an adaptive SDLC. All include iterations, which were discussed in Chapter 1. Rather

FIGURE **8-4** *Overlap of system development phases*

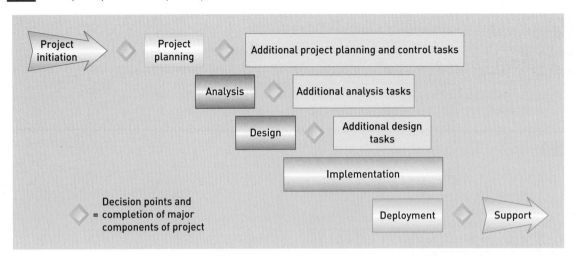

than having the analysis, design, and implementation phases proceed sequentially with some overlap, iterations can be used to create a series of mini-projects that address smaller parts of the application. One of these smaller parts is analyzed, designed, built, and tested during a single iteration; then, based on the results, the next iteration proceeds to analyze, design, build, and test the next smaller part. Using iterations, the project is able to adapt to any changes as it proceeds. Also, parts of the system are available early on for user evaluation and feedback, which helps ensure that the application will meet the needs of the users.

At the far right on the predictive/adaptive scale is the **spiral model**. It contains many adaptive elements, and it is generally considered to be one of the earliest conceptualizations of adapting the project based on the results of each iteration. The life cycle is shown as a spiral, starting in the center and working its way outward, over and over again, until the project is complete (see **Figure 8-5**). This model looks very different from the static waterfall

spiral model an adaptive SDLC approach that cycles over and over again through development activities until completion

FIGURE **8-5**
Spiral life cycle model

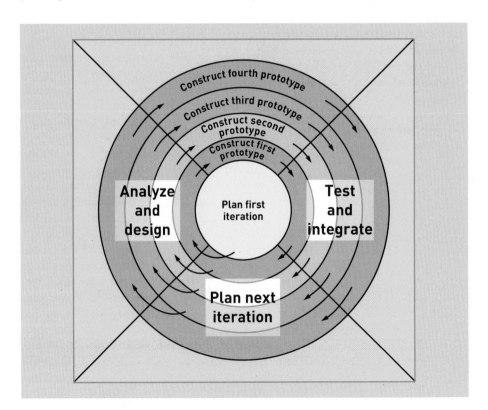

model and sets the tone for the project to be managed differently. For another representation of an adaptive SDLC, which shows several iterations, including analysis, design, and implementation activities, see **Figure 8-6**.

Figure 8-6 illustrates multiple iterations of development activities. Those two dimensions can also be represented as rows and columns of a table. The development activities are placed as rows, and the iterations are defined as columns. You first saw this concept in Figure 1-4, which is repeated here as **Figure 8-7**. The

FIGURE **8-6**

The iteration of system development activities

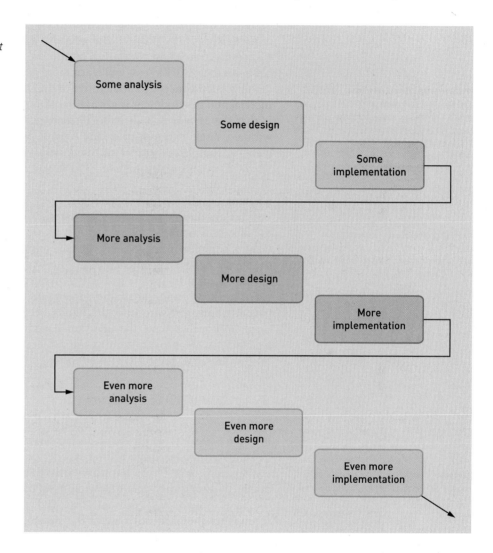

FIGURE **8-7**

Adaptive SDLC with six core processes and multiple iterations

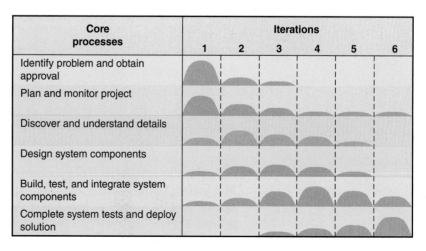

core processes defined in Chapter 1, which are repeated throughout the text, are extensions of the development activities shown in Figure 8-6. The columns, of course, are the multiple iterations of the project. In fact, as we compare the core processes of this adaptive, iterative life cycle with the phases of the waterfall life cycle from Figure 8-3, we see a very close correspondence between the two. The primary difference between these two life cycles is that the waterfall approach attempts to do all planning, all analysis, all design, and so forth, with a single pass. Our iterative approach is adaptive because with each iteration's analysis, design, and implementation, modifications can be made to adapt to the changing requirements of the project. The adaptive approach presented in this textbook is a simplification of and variation on a more formal iterative approach called the Unified Process (UP). You will learn more about the UP in Chapter 14.

A related concept to an iterative SDLC is called **incremental development**. Incremental development is always based on an iterative life cycle. The basic idea is that the system is built in small increments. An increment may be developed within a single iteration or it may require two or three iterations. As each increment is completed, it is integrated with the whole. The system, in effect, is "grown" in an organic fashion. The advantage of this approach is that portions of the system get into the users' hands much sooner so the business can begin accruing benefits as early as possible.

Yet another related concept, which is also based on an iterative approach, is the idea of a **walking skeleton**. A walking skeleton, as the name suggests, provides a complete front-to-back implementation of the new system but with only the "bare bones" of functionality. The walking skeleton is developed in a few iterations early in the project. Later iterations then flesh out the skeleton with more functions and capabilities. It should be obvious that this approach also gets working software into the hands of the users early in the project. Both these approaches have the additional advantage of extensive user testing and feedback to the project team as the project is progressing—another example of how an iterative project is also adaptive.

The Support Phase

The predictive waterfall SDLC explicitly includes a support phase, but adaptive, iterative SDLCs typically don't. In fact, newer adaptive SDLCs consider support to be an entirely separate project worthy of its own support methodology.

The objective of the **support activities** is to keep the system running productively during the years following its initial deployment. They begin only after the new system has been installed and put into production, and they last throughout the productive life of the system. Most business systems are expected to last for years. During the support phase, upgrades or enhancements may be carried out to expand the system's capabilities, and these will require their own development projects. Three major activities occur during support:

■ Maintaining the system
■ Enhancing the system
■ Supporting the users

Every system, especially a new one, contains components that don't function correctly. Software development is complex and difficult, so it is never free of error. Of course, the objective of a well-organized and carefully executed project is to deliver a system that is robust and complete and that gives correct results. However, because of the complexity of software and the impossibility of testing every possible combination of processing requirements, there will always be errors. In addition, business needs and user requirements change over time. Key tasks in maintaining the system include fixing the errors (also known as fixing bugs) and making minor adjustments to processing requirements. Usually, a system support team is assigned responsibility for maintaining the system.

incremental development an SDLC approach that completes portions of the system in small increments across iterations, with each increment being integrated into the whole as it is completed

walking skeleton a development approach in which the complete system structure is built but with bare-bones functionality

support activities the activities in the support phase whose objective is to maintain and enhance the system after it is installed and in use

Most newly hired programmer analysts begin their careers working on system maintenance projects. Tasks typically include changing the information provided in a report, adding an attribute to a table in a database, or changing the design of Windows or browser forms. These changes are requested and approved before the work is assigned, so a change request approval process is always part of the system support phase.

During the productive life of a system, it is also common to make major modifications. At times, government regulations require new data to be maintained or information to be provided. Also, changes in the business environment—new market opportunities, new competition, or new system infrastructure—necessitate major changes to the system. To implement these major modifications, the company must approve and initiate an upgrade development project. An upgrade project often results in a new version of the system. During your career, you may participate in several upgrade projects.

help desk the availability of support staff to assist users with technical or processing problems of the information system

The other major activity during the support phase is providing assistance to the system users. A **help desk**, consisting of knowledgeable technicians, is a popular method for answering users' questions quickly and helping increase their productivity. Training new users and maintaining current documentation are important elements of this activity. As a new systems analyst, you may conduct training sessions or staff the help desk to gain experience with user problems and needs. Many newly hired information systems professionals start their careers working at a help desk for part of their work week.

Methodologies, Models, Tools, and Techniques

Aside from an SDLC, systems developers have a variety of aids at their disposal to help them complete activities and tasks. Among them are methodologies, models, tools, and techniques. The following sections discuss each of these aids.

Methodologies

system development methodology a set of comprehensive guidelines for the SDLC that includes specific models, tools, and techniques

A **system development methodology** provides guidelines for every facet of the systems development life cycle. For example, within a methodology, certain models, such as diagrams, are used to describe and specify the requirements. Related to these models are the techniques for how the project team will do its work. An example of a technique is the guidelines for conducting a user interview that you learned about in Chapter 2. Finally, each project team will use a set of tools—usually computer-based tools—to build models, record information, and write the code. These tools are considered part of the overall methodology used for a given project. **Figure 8-8** illustrates that the techniques,

FIGURE **8-8**

Components of a methodology

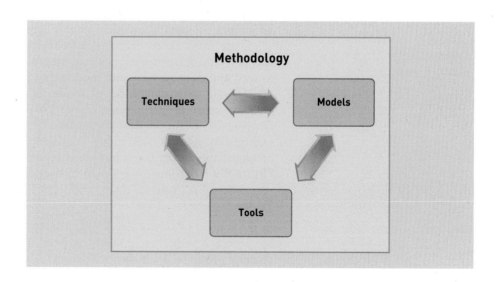

models, and tools support one another to provide a comprehensive, integrated methodology. Some methodologies are developed by systems professionals within the company based on their experience. Others are purchased from consulting firms or vendors.

Some methodologies (whether built in-house or purchased) contain massive written documentation that defines everything the developers may need to produce at any point in the project, including how the documentation itself should look and what reports to management should contain. Other methodologies are much more informal, such as a single document that contains general descriptions of what needs to be done. Sometimes, the methodology a company adopts isn't only informal, it is ad hoc and almost undefined, but such freedom of choice is becoming rare. Management in most IT departments prefers to adopt a flexible methodology so it can be adapted to different types of projects and systems. The methodology used by the organization determines how predictive or adaptive the approach to a system development project should be.

Models

Anytime people need to record or communicate information, it is useful to create a model. As discussed in Chapter 2, a model is a representation of an important aspect of the real world. Sometimes, the term *abstraction* is used because we abstract (separate out) an aspect that is of particular importance to us. For example, consider an airplane model. To talk about the aerodynamics of the airplane, it is useful to have a model that shows the plane's overall shape in three dimensions. Sometimes, a drawing of the cross-sectional details of the plane's wing is what is needed. In other cases, a mathematical formulation of the plane's aerodynamic characteristics might be necessary to understand how it will behave.

Some models are physically similar to the real product. Some are graphical representations of important details. And some are abstract mathematical notations. Each emphasizes a particular type of information. In airplane design, engineers use lots of different models. Learning to be an aerospace engineer involves learning how to create and use all the various models. That is true for an information system developer too, although the models for information systems aren't as standardized or precise as aerospace models. System developers are making progress, but the field is very young, and many senior analysts were self-taught. More importantly, an information system is much less tangible than an airplane; you can't really see, hold, or touch it. Therefore, the models of the information system can seem much less tangible too.

The models used in system development include representations of inputs, outputs, processes, data, objects, object interactions, locations, networks, and devices, among other things. Most of the models are graphical models, which are drawn representations that employ agreed-upon symbols and conventions. These are often called diagrams and charts, and the UML diagrams you have encountered so far in this book are examples. Much of this text describes how to read and create a variety of models that represent an information system.

Another important kind of model is a project-planning model, such as a Gantt chart or net present value (NPV), both of which are discussed in Chapter 9. These models represent the system development project itself, highlighting its tasks and other considerations. Yet another model related to project management is a chart showing all the people assigned to the project. **Figure 8-9** lists some models used in system development.

Tools

tool a software application that assists developers in creating models or other components required for a project

In the context of system development, a **tool** is software support that helps create models or other components required in the project. Tools might be simple drawing programs for creating diagrams. They might also include an application

Some models used in system development

> **Some models of system components**
>
> Flowchart
> Data flow diagram (DFD)
> Entity-relationship diagram (ERD)
> Structure chart
> Use case diagram
> Class diagram
> Sequence diagram
>
> **Some models used to manage the development process**
>
> Gantt chart
> Organizational hierarchy chart
> Financial analysis models - NPV, payback period

integrated development environments (IDEs) a set of tools that work together to provide a comprehensive development and programming environment for software developers

visual modeling tools tools that help analysts create and verify graphical models and may also generate program code

technique guidelines to specify a method for how to carry out a development activity or task

that stores information about the project, such as data definitions, use case descriptions, and other artifacts. A project management software tool, such as Microsoft Project, is another example of a tool used to create models. The project management tool creates a model of the project tasks and task dependencies.

Tools have been specifically designed to help system developers. Programmers should be familiar with **integrated development environments (IDEs)**, which include many tools to help with programming tasks—for example, smart editors, context-sensitive help, and debugging tools. Some tools can generate program code for the developer. Some tools reverse-engineer old programs, generating a model from the code so the developer can determine what the program does if its documentation is missing (or was never done). **Visual modeling tools** are available to systems analysts to help them create and verify important system models. These tools are used to draw such diagrams as class diagrams or activity diagrams. Other visual modeling tools help design the database or even generate program code. **Figure 8-10** lists the types of tools used in system development.

Techniques

You learned several techniques for gathering information in Chapter 2. You learned several techniques for defining functional requirements in Chapters 3, 4, and 5. You learned some user-interface design techniques in Chapter 7. In system development, a **technique** is a collection of guidelines that helps an analyst complete an activity or task. It often includes step-by-step instructions for creating a model, or it might include more general advice on collecting information from system users. Examples include data-modeling techniques, software-testing techniques, user-interviewing techniques, and relational database design techniques.

Sometimes, a technique applies to an entire life cycle phase and helps you create several models and other documents. The modern structured analysis technique (discussed later) is an example of this. **Figure 8-11** lists some techniques commonly used in system development.

FIGURE **8-10**
Types of tools used in system development

> Project management application
> Drawing/graphics application
> Word processor/text editor
> Visual modeling tool
> Integrated development environment (IDE)
> Database management application
> Reverse-engineering tool
> Code generator tool

FIGURE **8-11**
Some techniques used in system development

```
Strategic planning techniques
Project management techniques
User interviewing techniques
Data-modeling techniques
Relational database design techniques
Structured programming technique
Software-testing techniques
Process modeling techniques
Domain modeling techniques
Use case modeling techniques
Object-oriented programming techniques
Architectural design techniques
User-interface design techniques
```

How do methodologies, models, tools, and techniques fit together? A methodology includes a collection of techniques that are used to complete activities within each phase of the systems development life cycle. The activities include the completion of a variety of models as well as other documents and deliverables. Like any other professionals, system developers use software tools to help them complete their activities.

Two Approaches to Software Construction and Modeling

System development is done in many ways, and this diversity can confuse new system developers. Sometimes, it seems as if every company uses its own methodology. In fact, even the various development groups within the same company may use their own methodologies, with each person coming up with his or her own way of developing systems.

Still, there are many common concepts. In virtually all development groups, some variation of the systems development life cycle is used, with phases for project initiation, project planning, analysis, design, implementation, deployment, and support. In addition, virtually every development group uses models, tools, and techniques that make up an overall system development methodology.

All system developers should be familiar with two very general approaches to software construction and modeling because these form the basis of virtually all methodologies: the structured approach and the object-oriented approach. This section reviews the major characteristics of both approaches and provides a bit of history.

The Structured Approach

Earlier in this chapter, we discussed a traditional, predictive approach to the SDLC. Those concepts focused on the phases and activities of the development project itself. This section focuses on the models, including analysis and design models, as well as the programming constructs that are used to develop the software itself. This software construction approach is called structured system development. Sometimes, these two ideas—the predictive approach to the SDLC and the structured approach to software construction—can cause confusion because they are both referred to as the traditional approach. We will be more precise in the terminology in this book, but you should be aware that the industry in general isn't as precise.

Structured System Development

structured approach system development using structured analysis, structured design, and structured programming techniques

Structured analysis, structured design, and structured programming are the three techniques that make up the **structured approach**. Sometimes, these techniques are collectively referred to as the *structured analysis and design technique* (SADT). Developed in the 1960s, the structured programming technique was the first attempt to provide guidelines to improve the quality of computer programs. You certainly learned the basic principles of structured programming in your first programming course. The structured design technique was developed in the 1970s to make it possible to combine separate programs into more complex information systems. The structured analysis technique evolved in the early-1980s to help clarify requirements for a computer system before developers designed the programs.

Structured Programming High-quality programs not only produce the correct outputs each time the program runs, they also make it easy for other programmers to read and modify the program later. And programs need to be modified all the time. **Structured programming** produces a program that has one beginning and one ending, with each step in the program execution consisting of one of three programming constructs:

■ A sequence of program statements
■ A decision point at which one set or another set of statements executes
■ A repetition of a set of statements

structured programming a programming approach where each module has one start point and one end point and uses sequence, decision, and repetition constructs only

Figure 8-12 shows these three structured programming constructs.

A concept related to structured programming is top-down programming. **Top-down programming** divides more complex programs into a hierarchy of program modules (see **Figure 8-13**). One module at the top of the hierarchy controls program execution by "calling" lower-level modules as required. Sometimes, the modules are part of the same program. For example, in COBOL, one main paragraph calls another paragraph by using the Perform keyword. In Visual Basic, a statement in an event procedure can call a general procedure. The programmer writes each program module (paragraph or procedure) by using the

top-down programming the concept of dividing a complex program into a hierarchy of program modules

FIGURE **8-12** *Three structured programming constructs*

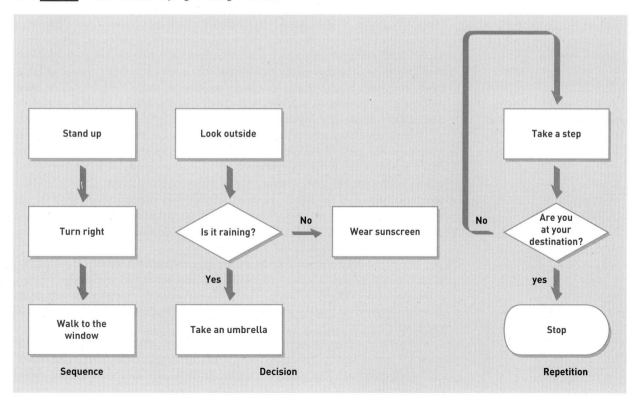

FIGURE **8-13** *Top-down, or modular, programming*

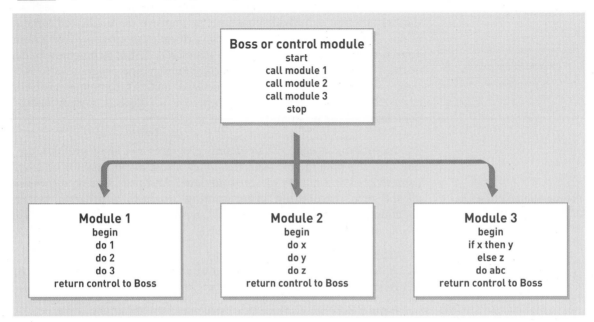

rules of structured programming (one beginning and one end as well as sequence, decision, and repetition constructs).

Sometimes, separate programs are produced that work together as one "system." Each of these programs follows top-down programming and structured programming rules, but the programs themselves are organized into a hierarchy, as with top-down programming. One program calls other programs. When the hierarchy involves multiple programs, such an arrangement is sometimes called modular programming.

Structured Design As information systems became increasingly complex during the 1970s, each system involved many different functions. Each function performed by the system might be made up of dozens of separate programs. The **structured design** technique was developed to provide some guidelines for deciding what the set of programs should be, what each program should accomplish, and how the programs should be organized into a hierarchy. The modules and the arrangement of modules are shown graphically by using a model called a **structure chart** (see **Figure 8-14**).

There are two main principles of structured design: Program modules should be (1) loosely coupled and (2) highly cohesive. *Loosely coupled* means that each module is as independent of the other modules as possible, which allows each module to be designed and later modified without interfering with the performance of the other modules. *Highly cohesive* means that each module accomplishes one clear task. That way, it is easier to understand what each module does and ensure that if changes to the module are required that none will accidentally affect other modules.

The structured design technique defines different degrees of coupling and cohesion and provides a way of evaluating the quality of the design before the programs are actually written. As with structured programming, quality is defined in terms of how easily the design can be understood and modified when the need arises.

Structured design assumes the designer knows what the system needs to do: what the main system functions are, what the required data are, and what the needed outputs are. Designing the system is obviously much more than designing the organization of the program modules. Therefore, it is important to realize

structured design the design process of organizing a program into a set of modules and organizing those modules into a hierarchical structure

structure chart a graphical diagram showing the hierarchical organization of modules

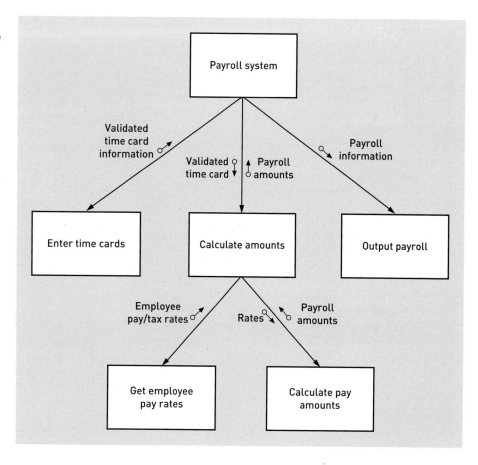

that the structured design technique helps the designer complete only part of the entire design life cycle phase.

By the 1980s, file and database design techniques were used along with structured design. Newer versions of structured design assumed that database management systems are used in the system, and program modules were designed to interact with the database. In addition, because nontechnical people were becoming involved with information systems, user-interface design techniques were developed. For example, menus in an interactive system determine which program in the hierarchy gets called. Therefore, a key aspect of user-interface design is done in conjunction with structured design.

Structured Analysis Because the structured design technique requires the designer to know what the system should do, techniques for defining system requirements were developed. System requirements define in great detail what the system must do but without committing to a specific technology. By deferring decisions about technology, the developers can focus their efforts on what is needed, not on how to do it. If these requirements aren't fully and clearly worked out in advance, the designers cannot possibly know what to design.

The **structured analysis** technique helps the developer define what the system needs to do (the processing requirements), what data the system needs to store and use (data requirements), what inputs and outputs are needed, and how the functions work together to accomplish tasks. The key graphical model of the system requirements that are used with structured analysis is called the **data flow diagram (DFD)**; it shows inputs, processes, storage, and outputs as well as the way these function together (see **Figure 8-15**).

The most recent variation of structured analysis defines systems processing requirements by identifying all the events that will cause the system to react in some way. For example, in an order-entry system, if a customer orders an

structured analysis a technique to determine what processing is required and to organize those requirements by using structured analysis models

data flow diagram (DFD) a structured analysis model showing inputs, processes, storage, and outputs of a system

FIGURE **8-15** *A data flow diagram (DFD) created by using the structured analysis technique*

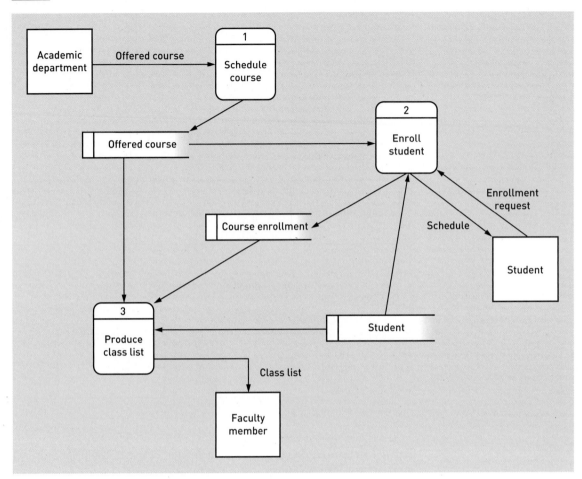

item, the system must process a new order (a major system activity). Each event leads to a different system activity. The analyst takes each of these activities and creates a data flow diagram showing the processing details, including inputs and outputs.

A model of the needed data is also created based on the types of things the system needs to store information (data entities) about. For example, to process a new order, the system needs to know about the customer, the items wanted, and the details about the order. You learned in Chapter 4 that this model is called an entity-relationship diagram (ERD). The data entities from the entity-relationship diagram correspond to the data storage shown on data flow diagrams. **Figure 8-16** shows an example of an entity-relationship diagram. **Figure 8-17** illustrates the sequence followed when developing a system using structured analysis, structured design, and structured programming.

FIGURE **8-16**

An entity-relationship diagram (ERD) created by using the structured analysis technique

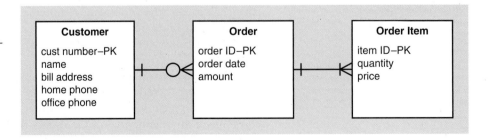

FIGURE **8-17** *How structured analysis leads to structured design and structured programming*

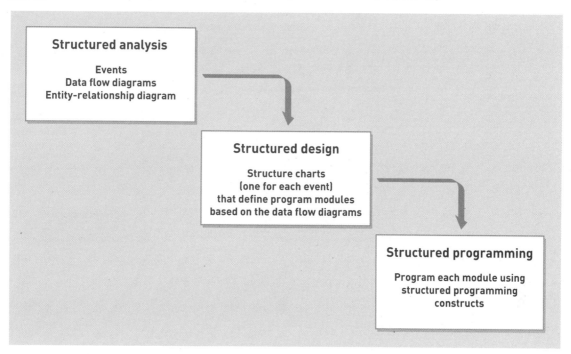

object-oriented approach system development based on the view that a system is a set of interacting objects that work together

object a thing in an information system that responds to messages by executing functions or methods

object-oriented analysis (OOA) the process of identifying and defining the use cases and the sets of objects (classes) in the new system

object-oriented design (OOD) defining all of the types of objects necessary to communicate with people and devices in the system, showing how objects interact to complete tasks, and refining the definition of each type of object so it can be implemented with a specific language or environment

object-oriented programming (OOP) programming using object-oriented languages that support object classes, inheritance, reuse, and encapsulation

The Object-Oriented Approach

An entirely different approach to information systems—the **object-oriented approach**—views an information system as a collection of interacting objects that work together to accomplish tasks (see **Figure 8-18**). Conceptually, there are no processes or programs; there are no data entities or files. The system consists of objects. An **object** is a thing in the computer system that is capable of responding to messages. This radically different view of a computer system requires a different approach to systems analysis, systems design, and programming.

The object-oriented approach began with the development of the Simula programming language in Norway in the 1960s. Simula was used to create computer simulations involving such "objects" as ships, buoys, and tides in fjords. It is very difficult to write procedural programs that simulate ship movement, but a new way of programming simplified the problem. In the 1970s, the Smalltalk language was developed to solve the problem of creating graphical user interfaces (GUIs) that involved such "objects" as pull-down menus, buttons, check boxes, and dialog boxes. More recent object-oriented languages include C++, Java, and C#. These languages focus on writing definitions of the types of objects needed in a system; as a result, all parts of a system can be thought of as objects, not just the graphical user interface.

Given that the object-oriented approach views information systems as collections of interacting objects, **object-oriented analysis (OOA)** defines the objects that do the work and determines what user interactions (called use cases) are required to complete the tasks. **Object-oriented design (OOD)** defines all the additional types of objects that are necessary to communicate with people and devices in the system, it shows how the objects interact to complete tasks, and it refines the definition of each type of object so it can be implemented with a specific language or environment. **Object-oriented programming (OOP)** is the writing of statements in a programming language to define what each type of object does.

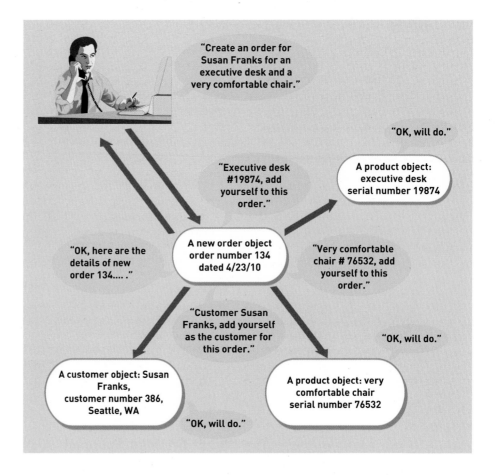

An object is a type of thing. It could be a customer or an employee or it could be a button or a menu. Identifying types of objects means classifying things. Some things, such as customers, exist outside and inside the system. There is the real customer, who is outside the system, and the computer representation of the customer, which is inside the system. A classification or "class" represents a collection of similar objects; therefore, object-oriented development uses a UML class diagram (introduced in Chapter 4) to show all the classes of objects that are in the system (see **Figure 8-19**). For every class, there may be more specialized subclasses. For example, a savings account and a checking account are two special types of accounts (two subclasses of the class Account). Similarly, a pull-down menu and a pop-up menu are two special types of menus. Subclasses exhibit or "inherit" the characteristics of the class above them.

A UML sequence diagram shows how objects interact or collaborate while carrying out a task. **Figure 8-20** shows a use case realization sequence diagram that includes an actor as a stick figure and five other objects that work together by sending messages to complete a use case named *Cancel order*.

The object-oriented approach yields several key benefits, among them naturalness and reusability. The approach is natural—or intuitive—for people because we tend to think about the world in terms of tangible objects. It is less natural to think about complex procedures found in procedural programming languages. Also, because the object-oriented approach involves classes of objects and many systems in the organization use the same objects, these classes can be used over and over again whenever they are needed. For example, almost all systems use menus, dialog boxes, windows, and buttons, but many systems within the same company also use customer, product, and invoice classes. These can also be reused. There is less need to "reinvent the wheel" to create an object.

FIGURE **8-19**

A class diagram created during object-oriented design

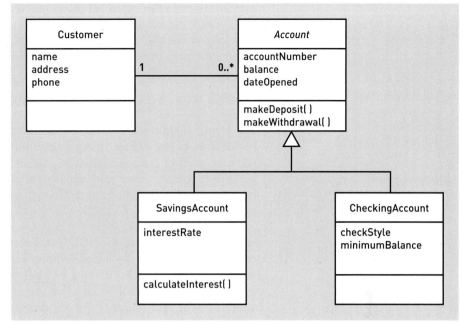

FIGURE **8-20** *A UML sequence diagram showing object interactions for a use case*

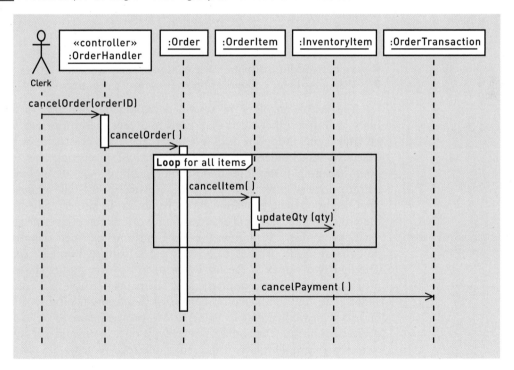

Many systems being developed today combine traditional and object-oriented technology. Some integrated development environments (IDEs) also use traditional and object-oriented technology in the same tool. For example, OOP can be used for the user interface and procedural programming can be used for the rest. Many Web applications are built by using structured programming and modular design. For example, PHP supports both technologies. Many system projects are exclusively traditional in analysis and design, and others are exclusively object-oriented, even within the same information systems department. Everyone should know the basic concepts of each, but your coursework may emphasize one approach over the other. See Online Chapter B for more about structured analysis and structured design.

Agile Development

The highly volatile marketplace has forced businesses to respond rapidly to new opportunities. Sometimes, these opportunities appear in the middle of implementing another business initiative. To survive, businesses must be agile—that is, able to change directions rapidly, even in the middle of a project. **Agile development** is a philosophy and set of guidelines for developing information systems in an unknown, rapidly changing environment, and it can be used with any system development methodology. Usually, Agile development complements adaptive approaches to the SDLC and methodologies that support it. But the emphasis is on taking an adaptive approach and making it agile in all development activities and tasks. Related to Agile development, Agile modeling is a philosophy about how to build models, some of which are formal and detailed, others sketchy and minimal. All the models you have learned how to create in this text can be used with Agile modeling.

Agile Development Philosophy and Values

The "Manifesto for Agile Software Development" (see the "Further Resources" section) identifies four basic values, which represent the core philosophy of Agile development:

- Value responding to change over following a plan
- Value individuals and interactions over processes and tools
- Value working software over comprehensive documentation
- Value customer collaboration over contract negotiation

The people involved in system development—whether as team members, users, or other stakeholders—all need to accept these priorities for a project to be truly agile. Adopting an Agile approach isn't always easy. Managers and executive stakeholders frequently have trouble accepting this less rigid viewpoint, wanting instead to impose more controls on development teams and enforce detailed plans and schedules. However, the Agile philosophy takes the opposite approach, providing more flexibility in project schedules and letting the project teams plan and execute their work as the project progresses.

Some industry leaders in the Agile movement coined the term **chaordic** to describe an Agile project. *Chaordic* comes from two words: *chaos* and *order*. The first two values in the list do seem to be a recipe for chaos, but software projects always have unpredictable elements—hence, a certain amount of chaos. The Agile philosophy recognizes this unpredictability, handling it with increased flexibility and by trusting the project team to develop solutions to project problems. Depending too heavily on a plan and predefined processes exacerbates problems when unpredictable requirements arise. Developers need to accept a certain amount of chaos and mix that with other Agile modeling and development techniques that help to provide order and structure to the project. Chapter 9 will cover many of these Agile project management techniques.

Another important aspect of Agile development is that customers must continually be involved with the project team. They don't just sit down with the project team for a few sessions to develop the specifications and then go their separate ways. They become part of the technical team. Because working software is being developed throughout the project, customers are continually involved in defining requirements and testing components.

Historically, particularly with predictive projects, many systems development efforts attempted to be fixed price endeavors. This was true for both in-house groups and external development teams. However, the approach with Agile development is that systems development projects should be more of a collaborative effort. Hence, contracts take on an entirely different flavor. Fixed prices and fixed deliverables don't make sense. Contracts for Agile projects include other kinds of options for the customer. The approach to the scheduling of activities, the delivery of system components, and the early termination of the project allow the client to maintain control, but it is done with different options than with fixed bid contracts.

Models and modeling are critical to Agile development, so we look next at Agile modeling. Many of the core values are illustrated in the principles and practices of building models.

Agile Modeling Principles

Much of this text teaches techniques for creating models. Your first impression might be that an agile approach means less modeling or maybe even no modeling. **Agile modeling (AM)** isn't about doing less modeling but about doing the right kind of modeling at the right level of detail for the right purposes. AM doesn't dictate which models to build or how formal to make those models. Instead, it helps developers stay on track with their models by using them as a means to an end rather than end deliverables. AM's basic principles express the attitude that developers should have as they develop software. **Figure 8-21** summarizes Agile modeling principles. We discuss those principles next.

agile modeling (AM) a guiding philosophy in which only models that are necessary, with a valid need and at the right level of detail, are created

Develop Software as Your Primary Goal

The primary goal of a software development project is to produce high-quality software. The primary measurement of progress is working software, not intermediate models of system requirements or specifications. Modeling is always a means to an end, not the end itself. Any activity that doesn't directly contribute to the end goal of producing software should be questioned and avoided if it cannot be justified.

Enable the Next Effort as Your Secondary Goal

Focusing only on working software can also be self-defeating, so developers must consider two important objectives. First, requirements models might be necessary to develop design models. So, don't think that if the model cannot be used to write code, it is unnecessary. Sometimes, several intermediate steps are needed before the final code can be written. Second, although high-quality software is the primary goal, long-term use of that code is also important. So, some models might be necessary to support maintenance and enhancement of the system. Yes, the code is the best documentation, but some architectural design

FIGURE **8-21**
Agile modeling principles

Agile Modeling principles

- Develop software as your primary goal.
- Enable the next effort as your secondary goal.
- Minimize your modeling activity—few and simple.
- Embrace change, and change incrementally.
- Model with a purpose.
- Build multiple models.
- Build high-quality models and get feedback rapidly.
- Focus on content rather than representation.
- Learn from each other with open communication.
- Know your models and how to use them.
- Adapt to specific project needs.

decisions might not be easily identified from the code. Look carefully at what other artifacts might be necessary to produce high-quality systems in the long term.

Minimize Your Modeling Activity—Few and Simple

Create only the models that are necessary. Do just enough to get by. This principle isn't a justification for sloppy work or inadequate analysis. The models you create should be clear, correct, and complete. But don't create unnecessary models. Also, keep each model as simple as possible. Normally, the simplest solution is the best solution. Elaborate solutions tend to be difficult to understand and maintain. However, simplicity isn't a justification for being incomplete.

Embrace Change, and Change Incrementally

Because the underlying philosophy of Agile modeling is that developers must be flexible and respond quickly to change, a good Agile developer willingly accepts—even embraces—change. Change is seen as the norm, not the exception. Watch for change, and have procedures ready to integrate changes into the models. The best way to accept change is to develop incrementally. Take small steps and address problems in small bites. Change your model incrementally and then validate it to make sure it is correct. Don't try to accomplish everything in one big release.

Model with a Purpose

We indicated earlier that the two reasons to build models are to understand what you are building and to communicate important aspects of the solution system. Make sure your modeling efforts support those reasons. Sometimes, developers try to justify building models by claiming that (1) the development methodology mandates the development of the model, (2) someone wants a model, even though the person doesn't know why it is important, or (3) a model can replace a face-to-face discussion of issues. Identify a reason and an audience for each model you develop. Then, develop the model in sufficient detail to satisfy the reason and the audience. Incidentally, the audience might be you.

Build Multiple Models

Along with other modeling methodologies, UML has several models to represent different aspects of the problem at hand. To be successful—in understanding the problem or communicating the solution—you need to model the critical aspects of the problem domain or the required solution. Don't develop all of them; be sure to minimize your modeling, but develop enough models to make sure you have addressed all the issues.

Build High-Quality Models and Get Feedback Rapidly

Nobody likes sloppy work. It is based on faulty thinking and introduces errors. One way to avoid error in models is to get feedback rapidly while the work is still fresh. Feedback comes from users as well as from technical team members. Others will have helpful insights and different ways to view a problem and identify a solution.

Focus on Content Rather Than Representation

Sometimes, a project team has access to a sophisticated visual modeling tool. These can be helpful, but at times, they are distracting because developers spend time making the diagrams pretty. Be judicious in the use of tools. Some models need to be well drawn for communication or contractual issues.

Sometimes, it is more productive to build a model with a tool because it is expected that it will be changed frequently, and using a tool is usually more productive than redrawing by hand. In other cases, a hand-drawn diagram might suffice. Sometimes, developers work out a model on a whiteboard in a conference room and take a digital photo to keep a record of the details worked out.

Learn from Each Other with Open Communication

All the adaptive approaches emphasize working in teams. Don't be defensive about your models. Other team members have good suggestions. You can never truly master every aspect of a problem or its models.

Know Your Models and How to Use Them

Being an Agile modeler doesn't mean that you aren't skilled. If anything, you must be *more* skilled to know the strengths and weaknesses of the models, including how and when to use them. An expert modeler applies the previous principles of simplicity, quality, and development of multiple models.

Adapt to Specific Project Needs

Every project is different because it exists in a unique environment, involves different users, stakeholders, and team members, and requires a different development environment and deployment platform. Adapt your models and modeling techniques to fit the needs of the business and the project. Sometimes, models can be informal and simple. For other projects, more formal, complicated models might be required. An Agile modeler is able to adapt to each project.

Chapter Summary

System development projects are organized around a systems development life cycle (SDLC). Some SDLCs are based on a more predictive approach to the project, and other SDLCs are based on a more adaptive approach. The predictive approach to the SDLC includes phases that are completed sequentially or with some overlap. The traditional SDLC phases are project initialization, project planning, analysis, design, implementation, deployment, and support. The adaptive approach to the SDLC is used when the requirements or technology are less certain and it is difficult to plan everything about the project in advance. Adaptive SDLCs use multiple iterations that allow the analysis, design, and implementation of smaller parts of the application to be completed and evaluated. The SDLC used in this text is an example of an adaptive SDLC, and the six core processes correspond to the phases of the traditional predictive SDLC.

All development projects use an SDLC to manage the project, but there is more to system development than the SDLC. Models, techniques, and tools make up a system development methodology that provides guidelines for completing every activity in the SDLC. Most methodologies are based on one of two approaches to software construction and modeling: the traditional approach or the object-oriented approach. The traditional approach uses such models and techniques as use cases, data flow diagrams, entity-relationship diagrams, structure charts, the structured analysis technique, the structured design technique, and structured programming techniques. The object-oriented approach views software as a collection of interacting objects that collaborate to complete tasks. Such models and techniques as use cases, class diagrams, sequence diagrams, package diagrams, state machine diagrams, object-oriented analysis, object-oriented design, and object-oriented programming are used.

Agile development, the leading trend in system development, helps keep system development projects responsive to change. It is a philosophy that values change over following a plan, individuals over process and tools, working software over documentation, and customer collaboration over contract negotiation. Agile modeling describes principles for keeping a project agile.

Key Terms

adaptive approach to the SDLC   228

Agile development   244

Agile modeling (AM)   245

chaordic   244

data flow diagram (DFD)   239

help desk   233

incremental development   232

integrated development
   environments (IDEs)   235

object   241

object-oriented analysis (OOA)   241

object-oriented approach   241

Object-oriented design (OOD)   241

Object-oriented programming (OOP)   241

phases   228

predictive approach to the SDLC   227

spiral model   230

structure chart   238

structured analysis   239

structured approach   236

structured design   238

Structured programming   237

support activities   232

system development methodology   233

technique   235

tool   234

Top-down programming   237

Visual modeling tools   235

walking skeleton   232

waterfall model   228

Review Questions

1. What is a project?

2. What is the range of sizes of an information system development project?

3. What is the system development life cycle (SDLC)?

4. What characteristics of a project call for a predictive approach to the SDLC?

5. What characteristics of a project call for an adaptive approach to the SDLC?

6. What are the seven phases of the traditional predictive SDLC?

7. What are the objectives of the support phase?

8. Explain how the waterfall model of the SDLC controls the changes that occur during a project.

9. Explain the advantages of having the phases of a predictive SDLC overlap.

10. What organizing concept is included in all adaptive SDLCs?

11. What is considered the first adaptive SDCL? Sketch it.

12. For an adaptive SDLC, explain what goes on during each iteration.

13. The SDLC used in this text is based on what adaptive SDLC?

14. What are the core processes in the SDLC used in this book, and what traditional predictive SDLC phase corresponds to each process?

15. What is the iterative approach that involves completing and deploying part of an application over a few iterations and then completing and deploying another part of that application after a few more iterations?

16. Why do adaptive SDLCs not explicitly include the support phase?

17. What are the three activities of the support phase?

18. What is a popular support technique used to answer users' questions and help them increase productivity?

19. What is a system development methodology?

20. What are some examples of models included in a methodology?

21. What are some examples of techniques included in a methodology?

22. What are some examples of tools included in a methodology?

23. What are the two approaches to software construction and modeling?

24. What are the basic characteristics of the traditional approach?

25. What are the basic characteristics of the object-oriented approach?

26. What are the three main structured techniques?

27. What are three diagrams created by the structured approach?

28. What are the main object-oriented techniques?

29. What is Agile development?

30. What are the four "values" reflected in Agile development?

31. What is Agile modeling (AM)?

32. What are the 11 Agile modeling principles?

Problems and Exercises

1. Write a one-page paper that distinguishes among the fundamental purposes of the analysis phase, the design phase, and the implementation phase of the traditional predictive SDLC.

2. Describe an information system project that might have three subsystems. Discuss how three iterations might be used for the project.

3. Why might it make sense to teach analysis and design phases and activities sequentially, like a waterfall, even though iterations are, in practice, used in nearly all development projects?

4. List some of the models that architects create to show different aspects of a house they are designing. Explain why several models are needed.

5. What models might an automotive designer use to show different aspects of a car?

6. Sketch and write a description of the layout of your room at home. Are both the sketch and the written description considered models of your room? Which is more accurate? More detailed? Which would be easier to follow by someone unfamiliar with your room?

7. Describe a technique you use to help you complete the activity "Get to class on time." What are some of the tools you use with this technique?

8. Describe a technique you use to make sure you get assignments done on time. What are some of the tools you use with this technique?

9. What are some other techniques you use to help you complete activities in your life?

10. There are at least two approaches to the SDLC, two approaches to software construction and modeling, and a long list of techniques and models. Discuss the following reasons for this diversity of approaches: The field is young; the technology changes quickly; different organizations have different needs; there are many types of systems; developers have widely different backgrounds.

11. Go to the campus placement office to gather some information on companies that recruit information systems graduates. Try to find any information about the companies' approaches to developing systems. Is their SDLC described? Do any mention an IDE or a visual modeling tool? Visit the companies' Web sites to look for more information.

12. Visit the Web sites of a few leading information systems consulting firms. Try to find information about their approaches to developing systems. Are their SDLCs described? Do the sites mention any tools, models, or techniques?

Case Study

A "College Education Completion" Methodology

Given that you are reading this book, you are probably a college student working on a degree. Think about completing college as a project—a big project lasting many years and costing more than you might want to admit. Some students do a better job managing their college completion projects than others. Many fail entirely (certainly not you), and most complete college late and way over budget (again, certainly not you).

As with any other project, to be successful, you need to follow some sort of "college education completion" methodology—that is, a comprehensive set of guidelines for completing activities and tasks from the beginning of planning for college through to the successful completion.

1. What are the phases that your college education completion life cycle might have?

2. What are some of the activities included with each phase?

3. What are some of the techniques you might use to help complete those activities?

4. What models might you create? Differentiate the models you create to get you through college from those that help you plan and control the process of completing college.

5. What are some of the tools you might use to help you complete the models?

Community Board of Realtors

The Board of Realtors Multiple Listing Service (MLS) system isn't very large in terms of use cases and domain classes. In that respect, the functional requirements are simple and well understood. MLS needs a Web site with public access to the listings, and it also needs to allow agents and brokers to log in to the system to add and update listings. There is very little back-end administrative data maintenance required, except to add or update a real estate office or agent.

1. Compared to the Tradeshow application described in Chapter 1, how long might this project take, and which approach to the SDLC would be most appropriate?
2. If you use a predictive SDLC, how much time might each phase of the project take? How much overlap of phases might you plan for? Be specific about how you would overlap the phases.
3. If you use an adaptive SDLC, how many iterations might you plan to include? What use cases would you analyze, design, and implement in the first

iteration? What use cases would you work on in the second iteration? In additional iterations? Think in terms of getting the core functionality implemented early and then building the supporting functionality.

4. Let us say this project focused on Web access to the MLS. If you also plan to deploy a smartphone application for use by the public and by the agents and brokers, how might this affect your choice of the approach to the SDLC? What are the implications for including the smartphone application in the initial project versus having a separate project for wireless later?
5. Consider using incremental development to include the Web application and the wireless support. Describe what would be included in the first and second deployments of the project. Take into consideration that you might want to work on some initial problem solving for requirements, design, and implementation of the wireless support at the same time you are working on the Web application.

The Spring Breaks 'R' Us Travel Service

Recall from Chapter 2 that SBRU's initial system included four major subsystems: Resort relations, Student booking, Accounting and finance, and Social networking. The project calls for an adaptive approach to the SDLC for several reasons. One, it is relatively large in scope. Two, there is a diverse set of users in several functional areas, internal and external to the company and in several foreign countries. Three, the project needs to use an assortment of newer technologies that can communicate anytime and anywhere.

1. The SBRU information system includes four major subsystems: Resort relations, Student booking, Accounting and finance, and Social networking. Although you have only worked with the domain model class diagram for the Social networking subsystem, list as many of the domain classes that would probably be involved in each of the

subsystems. Note which classes are used by more than one subsystem.

2. Based on the overlapping classes, what domain classes seem to be part of the core functionality for SBRU? Draw a domain model class diagram that shows these classes and their associations.
3. Let us say you plan to implement the basic use cases that create and maintain the classes that are part of the core functionally you just modeled. Describe what domain classes you would focus on in each iteration if you assumed that you would need two iterations for the initial core functionality and two additional iterations to complete each of the subsystems.
4. How might you use incremental development to get some core functionality or some subsystems deployed and put into use before the project is completed?

(continued on page 251)

(continued from page 250)

On The Spot Courier Services

In the On the Spot system, package pickup and delivery are closely integrated with route schedules. However, recall the RMO system, where there is a Sales subsystem, an Order fulfillment subsystem, a Customer account subsystem, and a Marketing subsystem. We could conceive of the On the Spot system as also consisting of four subsystems:

- Customer account subsystem (like customer account)
- Pickup request subsystem (like sales)
- Package delivery subsystem (like order fulfillment)
- Routing and scheduling subsystem

Assuming that On the Spot's system developer approached this new system from this point of view and that the developer also decided to use an adaptive, iterative approach, answer these questions:

1. In what order would you develop the four subsystems? Support your answer.
2. Reviewing your work from Chapter 3, assign each of your use cases to a particular subsystem. Does this change your answer or does it strengthen your original premise? Support your answer.
3. Reviewing your work from Chapter 4, assign each of your classes to a subsystem. (Note: Some classes may be in multiple subsystems. The primary subsystem is the one that "creates" the objects in that class.) Does

this change your answer or does it strengthen your original premise? Support your answer.

4. Considering the Agile modeling principles, discuss each of the following:
   a. In Chapter 3, you developed a list of use cases and a use case diagram. If you follow the Agile modeling philosophy, how much or how little of this model do you think is necessary? Support your answer.
   b. In Chapter 4, you developed a class diagram. If you follow the Agile modeling philosophy, how much or how little of this model do you think is necessary? Support your answer.
   c. In Chapter 5, you developed some use case descriptions, activity diagrams, and system sequence diagrams. If you follow the Agile modeling philosophy, how many or how few of these models do you think are necessary? Support your answer.
   d. In Chapter 6, you developed a network diagram and specified hardware requirements. If you follow the Agile modeling philosophy, how many or how few of these models do you think are necessary? Support your answer.

Save your answers. These questions would be good to review at the end of the semester—after you have learned more about design and implementation.

Sandia Medical Systems Real-Time Glucose Monitoring

Review the original system description in previous chapters and the use case diagram shown in **Figure 8-22** to refamiliarize yourself with the proposed system.
  Consider this additional information:

- Sandia Medical Devices (SMD) and New Mexico Health Systems (NMHS) are developing the system jointly. Project staff will include analysts, designers, and programmers from both organizations. Three technical staff members from each organization have been assigned initially, and the budget includes sufficient funds to add other personnel for short-term assignments as needed. In addition, NMHS will assign a physician and a physician's assistant to the project one day per week.

- It is anticipated that SMD personnel assigned to the project will work primarily at NMHS facilities in office space and with computer equipment dedicated to developing the Real-Time Glucose Monitoring (RTGM) system.
- NMHS anticipates recruiting a handful of its own diabetic employees to provide requirements and to test the prototype RTGM software.
- SMD and NMHS anticipate a six-month development schedule for an initial version of the server software and Android-based client-side software. That will be followed by a three-month period for evaluation and another three-month period for development of improved software versions and

(continued on page 252)

(continued from page 251)

FIGURE **8-22** *Use cases for the patient and physician actors*

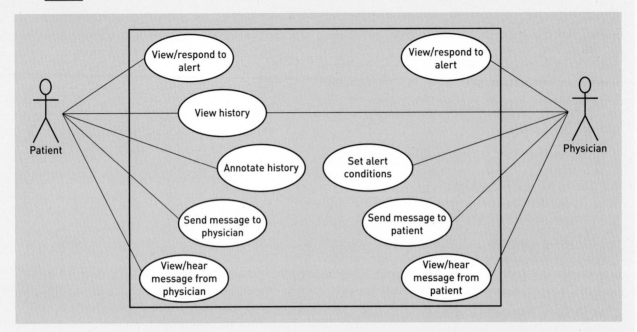

support for a wider range of mobile phone operating systems.

Answer these questions:

1. Given the system goals, requirements, and scope as they are currently understood, is the project schedule reasonable? Why or why not?
2. How well understood are the system requirements at the start of the project? What are the implications of your answer for using a predictive, adaptive, or mixed SDLC? What are the implications of your answer for using Agile techniques?
3. Medical personnel at NMHS have very busy schedules. NMHS's decision to assign two medical practitioners to the project for one day a week represents a significant investment in salary and lost revenue. How should project iterations be structured to ensure rapid progress to completion, high quality, and efficient use of medical practitioner time?

Further Resources

Classic and more recent texts include:

Craig Larman, *Agile and Iterative Development: A Manager's Guide*. Addison-Wesley, 2004.

Scott W. Ambler, *Agile Modeling: Effective Practices for eXtreme Programming and the Unified Process*. Wiley Publishing, 2002.

D. E. Avison and G. Fitzgerald, *Information Systems Development: Methodologies, Techniques, and Tools* (3rd ed.). McGraw-Hill, 2003.

Tom DeMarco, *Structured Analysis and System Specification*. Prentice Hall, 1978.

Ivar Jacobson, Grady Booch, and James Rumbaugh, *The Unified Software Development Process*. Addison-Wesley, 1999.

Steve McConnell, *Rapid Development*. Microsoft Press, 1996.

Meilir Page-Jones, *The Practical Guide to Structured Systems Design* (2nd ed.). Prentice Hall, 1988.

John Satzinger, Robert Jackson, and Stephen Burd, *Object-Oriented Analysis and Design with the Unified Process*. Course Technology, 2005.

9

Project Planning and Project Management

Learning Objectives

After reading this chapter, you should be able to:

- Describe the factors that cause a software development project to succeed or fail
- Describe the responsibilities of a project manager
- Describe the knowledge areas in the project management body of knowledge (PMBOK)
- Describe the Agile approach to the project management knowledge areas
- Explain the activities required to get a project approved (Core Process 1)
- Explain the activities required to plan and monitor a project (Core Process 2)

OPENING CASE

Blue Sky Mutual Funds: A New Development Approach

Jim Williams, vice president of finance for Blue Sky Mutual Funds, spoke first. "There are some things I like about this new approach, but other things worry me," he told Gary Johnson, the company's director of information technology.

"This idea of 'growing' the system through several iterations makes a lot of sense to me. It is always hard for my people to know exactly what they need a new information system to do and what will work best for the company. So, if they can get their hands on the system early, they can begin acceptance testing and try it out to see whether it addresses their needs in the best way.

"Let me see if I understand the big picture, though. Your development team and my investment advisors will decide on a few core processes that the system needs to support and then your team will design and build a system to support those core processes. You will do that in a mini-project that will last about six weeks. Then, you will continue adding more functionality through several other mini-projects until the system is complete and functioning well. Is that right?"

Jim was becoming more enthusiastic about this new approach to system development.

"Yes, that's the basic idea," Gary said. "Your users need to understand that the first few versions of the system won't be complete and may not be completely robust either. But these early versions will give them something to work with and try out. We also need good feedback from their acceptance testing so the system will be thoroughly tested by the time we are through."

"I realize that," Jim said. "My people will like not having to think from the very beginning about everything they need the system to do. They'll like being able to try things out. As I said earlier, I like this approach. However, the part I don't like about this approach is that it will be more difficult for you to give me a firm time schedule and project cost. That worries me. In the past, those have been two of the major tools we used to monitor a project's progress. Are you saying that now we won't have a schedule at all? And you want an open budget?" Jim frowned.

"It's not as bad as it first sounds," Gary said. "This approach is an 'adaptive' approach, by which I mean that because the system is growing, the project is more open ended. The project manager will still create a schedule and estimate the project costs, but she won't even try to identify and lock in all the required functionality for several of the iterations. Because the system's scope is going to continually be refined over the first few iterations, there is the risk of 'scope creep.' That is one of the biggest risks with adaptive approaches. You and I should meet with the project manager fairly frequently to ensure that the scope is controlled and the project doesn't get out of control."

"Okay," Jim said. "You have convinced me to try this new approach. However, let's treat this project as a pilot and see how it works. If it's successful, we will consider using this iterative approach on our other projects." Jim and Gary agreed that a pilot was the best way to get started. Gary then headed off to meet with the project manager and get the project started.

Overview

Chapter 8 introduced you to the SDLC and the various alternatives for organizing software development activities. By now, you may be asking yourself such questions as:

■ "How are all these activities coordinated?"
■ "How do I know which tasks to do first?"
■ "How is the work assigned to the different teams and team members?"
■ "How do I know which parts of the new system should be developed first?"

The purpose of project planning and project management is to bring some order to all these (sometimes seemingly unrelated) tasks. As you will learn in this chapter, the success of any given project highly depends on the skills and abilities of those managing the project. You will also learn that project management skills aren't only for project managers—that all the project team members contribute to the management of the project and thus to its success.

This chapter first discusses the need for project management and the principles associated with it. The rest of this chapter discusses the detailed activities that are associated with the first two core processes of systems development, both of which are primarily project management processes. The purpose of this chapter is to teach you how to plan, organize, and direct a systems development project.

Principles of Project Management

Many of you may have built a Web page with HTML or written a computer program for yourself or a friend. In those cases, where it was just you working, you weren't too concerned about how to organize your work or how to manage the project. However, as soon as two or more developers are working together, the work must be partitioned and organized, with specific assignments for each developer. This is true whether the project uses a predictive approach or an adaptive approach. As discussed in the last chapter, the chosen methodology lays out a complex set of activities and tasks that must be carefully managed. Failing to organize usually causes wasted time and effort as well as confusion and may even cause the project to fail.

Even though every project team designates one person as the project manager, with primary responsibility for the way the team functions, all members contribute to the team's management. The project manager for the RMO CSMS project is Barbara Halifax, but she has a senior systems analyst helping her every step of the way. As the project proceeds, all team members are involved in aspects of managing the project.

As discussed in earlier chapters, a project is a planned undertaking with a beginning and an end, which produces a predetermined result and is usually constrained by a schedule and resources. The development of information systems fits this definition. In addition, it is usually a quite complex project, with many people and tasks that have to be organized and coordinated. Whatever its objective, each project is unique. Different products are produced, different activities are required with varying schedules, and different resources are used. This uniqueness makes information systems projects difficult to control.

The Need for Project Management

Studies suggest that most IT projects are less than successful as measured by three criteria: finishing on time, finishing within budget, and effectively meeting the need as expressed by the original problem definition. Since 1994, the well-known Standish Group has produced an annual CHAOS report, which provides statistics on the outcome of IT development projects for the preceding year. The Standish Group categorizes projects in three ways: (1) successful projects, which are completed on time and within budget while meeting the users' requirements for functionality; (2) challenged projects, which have some combination of being late, over budget, or reduced in scope; and (3) failed projects, which are cancelled or result in the system never being used. The numbers vary somewhat year by year, with more recent years showing a slight improvement. In 2009, the results indicated that 32 percent were successful, 44 percent were challenged, and 24 percent were failed projects (see **Figure 9-1**). Billions of dollars are spent on projects that don't meet their objectives.

The Standish Group's report doesn't just indicate the rate of IT project failure or success; it also identifies the reasons for each. Here are some of the reasons for failure:

- Undefined project management practices
- Poor IT management and poor IT procedures

FIGURE **9-1**
Project completion results as reported by the Standish Group

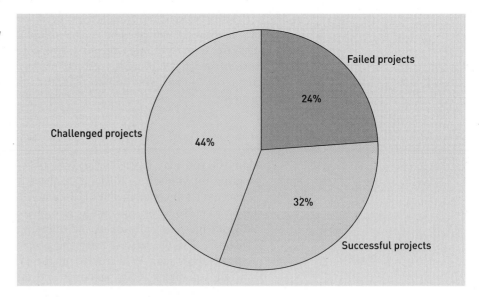

- Inadequate executive support for the project
- Inexperienced project managers
- Unclear business needs and project objectives
- Inadequate user involvement

It is notable that the primary reasons projects fail are a lack of executive involvement and a lack of management skills. The other major reason is lack of involvement by the user community. In other words, projects don't tend to fail for lack of programming skills or enthusiastic developers.

For an IT project to be successful, strong IT management and business direction need to be present. The other major element in all project success is sound project management procedures as well as experienced and competent project managers. In fact, good project managers always ensure that they have received clear directives from business executives and committed user involvement with the requirements for the new system.

The Role of the Project Manager

project management organizing and directing other people to achieve a planned result within a predetermined schedule and budget

Project management is organizing and directing other people to achieve a planned result within a predetermined schedule and budget. At the beginning of a project, a plan is developed that specifies the activities that must take place, the deliverables that must be produced, and the resources that are needed. Thus, project management can also be defined as the processes used to plan the project and then to monitor and control it.

One of the most exciting careers for IT-oriented people is being a project manager. As projects become more complex because of shorter time frames, distributed project teams (including off-shore and cross-cultural teams), rapidly changing technology, and more sophisticated requirements, highly qualified project managers are sought after and paid well. Many universities are adding project management courses to their curricula to respond to the needs of industry. There is a strong need and a high demand for people who are capable project managers. As your career progresses, you should develop your management skills. You may even want to become active in the Project Management Institute (PMI), which is the most well-known professional organization for project managers.

Overall, project managers must be effective internally (managing people and resources) and externally (conducting public relations). Internally, the project manager serves as locus of control for the project team and all its activities. He or she establishes the team's structure so work can be accomplished. This list identifies a few of these internal responsibilities:

■ Developing the project schedule
■ Recruiting and training team members
■ Assigning work to teams and team members
■ Assessing project risks
■ Monitoring and controlling project deliverables and milestones

Externally, the project manager is the main contact for the project. He or she must represent the team to the outside world and communicate the team members' needs. Major external responsibilities include:

■ Reporting the project's status and progress
■ Working directly with the client (the project's sponsor) and other stakeholders
■ Identifying resource needs and obtaining resources

client the person or group that funds the project

A project manager works with several groups of people. First of all, there is the **client** (i.e., the customer), who pays for the development of the new system. Project approval and the release of funds come from the client. For in-house developments, the client may be an executive committee or a vice president. The client approves and oversees the project, along with its funding.

oversight committee clients and key managers who review the progress and direct the project

For large, mission-critical projects, an **oversight committee** (sometimes called the steering committee) may be formed. This consists of clients and other key executives who have a vision of the organization's strategic direction and a strong interest in the project's success. On the other hand, the

users the person or group of people who will use the new system

users are the people who will actually use the new system. The user typically provides information about the detailed functions and operations needed in the new system.

Communication with the client and oversight committee is an important part of the project manager's external responsibilities. Similarly, working with the team leaders, team members, and any subcontractors is an important part of a project manager's internal responsibilities. The project manager must ensure that all internal and external communication is flowing properly. **Figure 9-2** depicts the various groups of people involved in a development project.

Project Management and Ceremony

Another dimension that has a heavy impact on project management is the level of formality, sometimes called ceremony, required for a given project.

ceremony the level of formality of a project; the rigor of holding meetings and producing documentation

Ceremony is a measure of the amount of documentation generated, the traceability of specifications, and the formality of the project's decision-making processes. Some projects, particularly small ones, are conducted with very low ceremony. Meetings occur in the hallway or around the water cooler. Written documentation, formal specifications, and detailed models are kept to a minimum. Developers and users usually work closely together on a daily basis to define requirements and develop the system. Other projects, usually larger, more complex ones, are executed with high ceremony. Meetings are often held on a predefined schedule, with specific participants, agendas, minutes, and follow-through. Specifications are formally documented with an abundance of diagrams and documentation and are frequently verified through formal review meetings between developers and users.

A project's ceremony isn't the same as whether its approach is predictive or adaptive. However, even though the approach and ceremony are different, large predictive projects often tend to have high ceremony, with lots of meetings and documentation. Unfortunately, the extensive documentation tended to increase

FIGURE **9-2**

Participants in a system development project

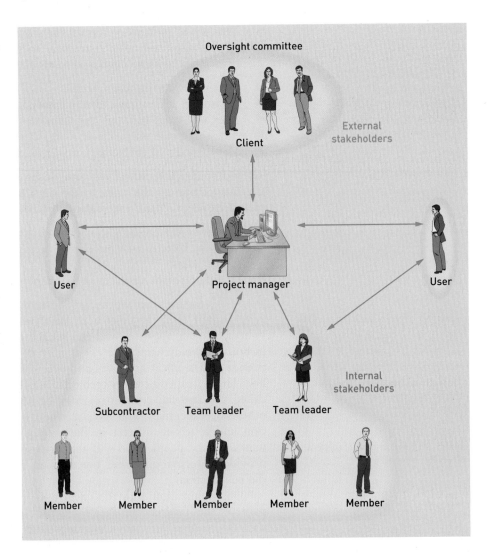

the length of the project and sometimes contributed to cost overruns. Techniques such as rapid application development (RAD) were utilized to help manage large predictive projects with less formality. This approach required less documentation and fewer status and review meetings. Of course, many smaller projects were often managed with less ceremony.

Adaptive projects can also be more or less formal in the way they are managed. The Unified Process, which will be explained in Chapter 14, is quite formal, with high ceremony. Each iteration is precisely defined, with such specific outcomes as specifications, diagrams, prototypes, and deliverables. However, adaptive, iterative approaches also lend themselves to being managed with much less formality. The inherent characteristics of an iterative approach, with its "just in time" project plans, easily adjust to less documentation, fewer diagrams for specifications, and less formal status reporting. The Agile approach, discussed in several chapters, is typically a low-ceremony approach.

Project Management Body of Knowledge (PMBOK)

The Project Management Institute (PMI) is a professional organization that promotes project management, primarily within the United States but also throughout the world. In addition, professional organizations in other countries promote project management. The PMI has a well-respected and rigorous certification program, and many corporations encourage their project managers to become certified.

As part of its mission, the PMI has defined a body of knowledge for project management. This body of knowledge, referred to as the **project management body of knowledge (PMBOK)**, is a widely accepted foundation of information that every project manager should know. The PMBOK is organized into these nine knowledge areas:

project management body of knowledge (PMBOK) a project management guide and standard of fundamental project management principles

- Project Scope Management—Defining and controlling the functions that are to be included in the system as well as the scope of the work to be done by the project team
- Project Time Management—Creating a detailed schedule of all project tasks and then monitoring the progress of the project against defined milestones
- Project Cost Management—Calculating the initial cost/benefit analysis and its later updates and monitoring expenditures as the project progresses
- Project Quality Management—Establishing a comprehensive plan for ensuring quality, which includes quality control activities for every phase of a project
- Project Human Resource Management—Recruiting and hiring project team members; training, motivating, and team building; and implementing related activities to ensure a happy, productive team
- Project Communications Management—Identifying all stakeholders and the key communications to each; also establishing all communications mechanisms and schedules
- Project Risk Management—Identifying and reviewing throughout the project all potential risks for failure and developing plans to reduce these risks
- Project Procurement Management—Developing requests for proposals, evaluating bids, writing contracts, and then monitoring vendor performance
- Project Integration Management—Integrating all the other knowledge areas into one seamless whole

As you progress in your career, you would be wise to keep a record of the project management skills you observe in others as well as those you learn from your own experiences. One place to start is with the set of skills a systems analyst needs, as described in earlier chapters. A good project manager knows how to develop a plan, execute it, anticipate problems, and make adjustments. Project management skills *can* be learned.

Agile Project Management (APM)

In the last chapter, you learned about the Agile approach to developing systems and the four values of Agile development, which tended to prefer flexibility over plans and defined procedures. Obviously, these values have a large impact on the way a project is managed. However, one of the concerns with them is that they imply a working environment that has no controls or plans—one that can turn into pure chaos. In Chapter 8, we introduced a term, *chaordic*, that describes a project that expects and allows chaos while remaining controlled or ordered.

Agile project management is still a young discipline, and the IT industry is still learning how best to balance the flexibility and chaos of an Agile team with the order and control needed for a project. More than anything else, Agile project management is a way of balancing these two conflicting requirements: how to be agile and flexible while maintaining control of the project schedule, budget, and deliverables.

To help you understand Agile project management better, we will now go through five of the nine knowledge areas of the PMBOK and discuss the issues involved in implementing them by using Agile principles.

Agile Scope Management

Scope management refers to the scope of the new system and the scope of the project. In traditional predictive projects, the project manager and the team attempted to define the scope in both areas at the beginning of the project, during the planning phase. Unfortunately, for most new systems, there were so many unknowns that the scope was almost never defined accurately. The Agile philosophy accepts the fact that the scope isn't well understood and that there will be many changes, updates, and refinements to the requirements as the project progresses. However, uncontrolled scope can result in a project that never finishes, even if it is an Agile project. The project manager must have a process and mechanisms in place to control the scope of the project. How can this be done?

Let us assume that one of the major outcomes of the planning iteration was the decision to develop a prioritized list of business requirements that the new system needs to support. **Figure 9-3** represents this list, with the higher-priority items toward the top and the lower ones toward the bottom. These requirements can be prioritized by using several criteria, including importance to the business, risk, complexity, size, and other dependencies. In most projects, some combination of these criteria is used to prioritize the requirements. Figure 9-3 also indicates that the project team has made a preliminary assignment of these requirements to iterations. As new requirements are defined, they are prioritized, inserted into the stack, and assigned to an iteration.

Controlling the scope is a decision made by the client, with input provided by the project team and the users. With an iterative project, a deliverable is usually provided at the end of each iteration. Because the system is growing throughout the project, with the highest priority requirements implemented first, the client is able to shut down the project when he or she feels that the

FIGURE **9-3**

Scope management with changing requirements

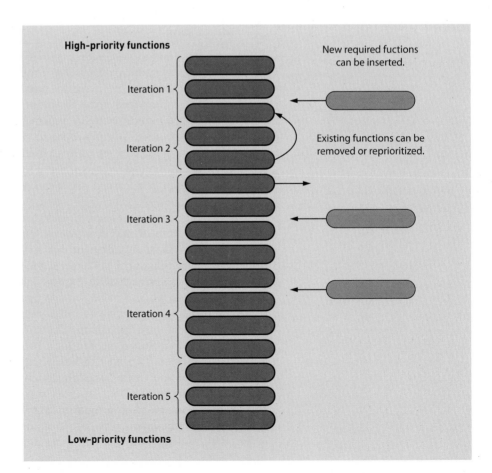

High-priority functions

Iteration 1

Iteration 2

Iteration 3

Iteration 4

Iteration 5

Low-priority functions

New required fuctions can be inserted.

Existing functions can be removed or reprioritized.

system is complete enough to satisfy the business need. Most projects usually require one or two more iterations to do final integration and testing to ensure that the system will scale for high volume and that it meets all the "hardening" requirements for security purposes.

Agile Time Management

Traditional time management is primarily concerned with scheduling tasks: creating the schedule, assigning work according to the schedule, and monitoring progress against the schedule. In predictive projects, the schedule is created during the initial planning phase and entered into a project scheduling system, such as Microsoft Project.

In an Agile project, because the requirements are always changing, it can be very difficult to create and maintain a meaningful project schedule. The initial planning effort will usually include the beginning set of requirements and divide the project into iterations, with a preliminary assignment of requirements to iterations. However, it is expected that the number of iterations and the assignments will change as new requirements are discovered and put on the prioritized stack.

Within an iteration, which often lasts from two to four weeks, a more detailed schedule can be developed. The Agile philosophy includes the idea that only for small work projects, in which the tasks are performed at nearly the same time (i.e., within one iteration), can a meaningful schedule be developed. In addition, the project team, not the project manager or team leader, will schedule its own work. Thus, for an Agile project, each iteration is usually planned as the first task within the iteration. The tasks are identified, estimates of the effort are developed, and work is assigned by the project team members. Because there are so many iterations in a project, the project team gets lots of practice and quickly becomes proficient at estimating and scheduling the work.

Agile Cost Management

It is normal for the client stakeholder to ask "How long will it take and how much will it cost for this new system to be developed?" These questions are hard to answer. For predictive projects, the project manager gives estimates, but as we saw earlier, these are usually incorrect. Agile project managers admit more readily that time and cost estimates are difficult to make, especially with a project in which the requirements are expected to change throughout. Hence, estimating the project's cost isn't as important as controlling the cost during the life of the project. The project manager's responsibility to control costs is just as important for an Agile project as it is for a traditional predictive project.

Agile Risk Management

In most adaptive, iterative projects, including Agile projects, close attention is given to project risks, particularly technical risks. Iterative projects are often risk-driven, meaning that early iterations focus specifically on addressing the most critical project risks. Although a similar emphasis on risk can be included in a predictive project, it is more difficult to integrate specific risk-reducing activities into the project schedule. The major difference between the two types of projects is that in predictive projects, separate prototypes are built, whereas in adaptive projects, the high-risk portions of the new system are built first.

Agile Quality Management

Usually, quality management has to do with the quality of the deliverable from the project. However, in an Agile project, we also consider the quality of the process. How well is the project working, and how well do the internal procedures promote project success?

In a predictive project, the final set of tasks consists of the system test, the integration test, and the user acceptance test. However, scheduling these extensive tests at the end of the project renders it very difficult and expensive to make the necessary changes. An alternative is to deploy the system with minimal testing, which helps the budget but can cause many problems for the company.

In an Agile project, each iteration has a deliverable. Often, each iteration also integrates a new piece into the growing total system. Within each iteration, the new pieces are tested by themselves and as integrated with the rest of the system. The users also get involved in testing the system's ability to meet their business needs. Hence, testing and quality control are spread across the entire project and usually provide a better-tested and more robust system.

Another kind of quality control that should be done as part of an Agile project is a process evaluation at the end of each iteration. In other words, the project team does a self-evaluation to figure out how well it did and what could be done to improve the next iteration.

Activities of Core Process 1: Identify the Problem and Obtain Approval

The adaptive SDLC used in this text includes six core processes. Chapter 2 outlined the activities of Core Process 3 ("Discover and understand details"), and Chapter 6 outlined the activities of Core Process 4 ("Design system components"). In this chapter, we discuss the activities of Core Processes 1 and 2.

Core Process 1 is probably the most critical process for project success. As was noted in the Standish report, establishing such things as strong executive support, clear business case and direction, and effective planning is critical to project success. These important factors are identified and resolved during the activities of Core Process 1. **Figure 9-4** highlights the four activities associated with Core Process 1.

Identify the Problem

Information system development projects are initiated for various reasons, including: (1) to respond to an opportunity, (2) to resolve a problem, and (3) to respond to an external directive.

Most companies are continually looking for ways to increase their market shares or open up new markets. One way they create opportunities is with strategic plans—short term and long term. In many ways, planning is the optimal way to identify new projects. As the strategic plans are developed, projects are identified, prioritized, and scheduled.

Projects are also initiated to resolve immediate business problems. Such projects can be initiated as part of a strategic plan, but they are more commonly

FIGURE **9-4** *Activities of Core Process 1*

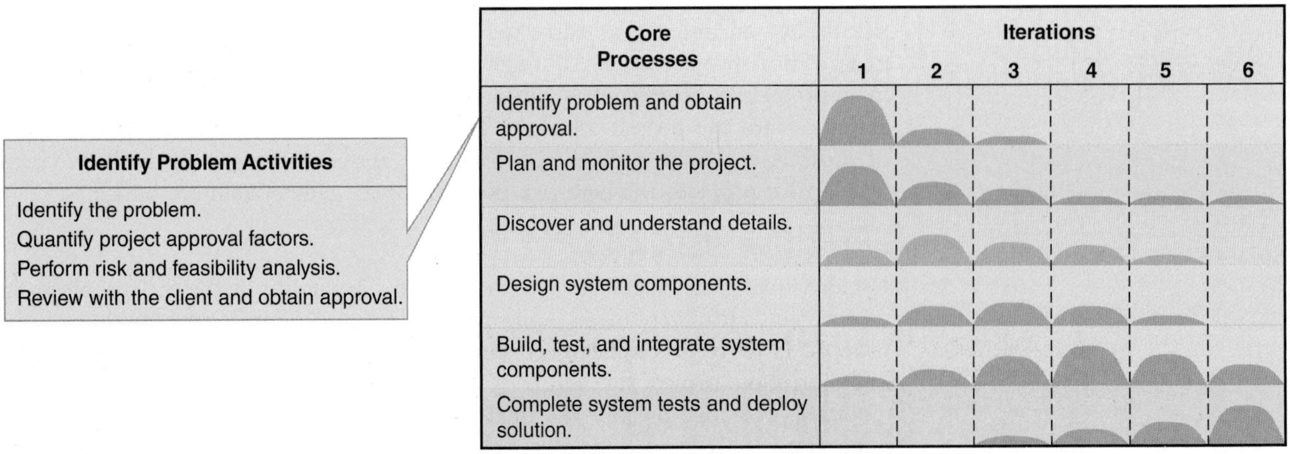

requested by middle managers who want to take care of some difficulty in the company's operations. Sometimes, these needs are so critical that they are brought to the attention of the strategic planning committee and integrated into the overall business strategy. At other times, an immediate need can't wait, such as a new sales commission schedule or a new report needed to assess productivity. In these cases, managers of business functions will request the initiation of individual development projects.

Finally, projects are initiated to respond to outside directives. One common version of this is legislative changes that require new information gathering and reporting—for example, changes in tax laws and labor laws. Legislative changes can also expand or contract the range of services and products that an organization can offer in a market. The regulatory changes in the telecommunications industry have opened doors for cable TV and telephone companies, which are vying for opportunities to provide cellular services, Internet access, and personalized entertainment.

Identifying and carefully defining the problem is a critical activity for a successful project. The objective is to ensure that the new system actually meets the business need. The purpose is to precisely define the business problem and determine the scope of the new system. This activity defines the target you want to hit. If the target is ill defined, all subsequent activities will lack focus. For example, a request might be made for a system that would "keep track of salesperson commissions." Without knowing more about the context surrounding this request, a system could be built that only records the commissions, ignoring the complexities of tax reporting, internal-versus-outside salespersons, deferred commissions, complex relationships, shared commissions, and so forth. Thus, even though all the specifications may not be defined in this initial activity, enough defining needs to be done to understand most of the implications of the required solution.

System Vision Document a document to help define the scope of a new system

An effective way to define the problem is to develop a **System Vision Document**, which was introduced in Chapter 1. There are three components to this document: the problem description, the anticipated business benefits, and the system capabilities.

The first task in developing a System Vision Document is to review the business needs that initiated the project. If the project was initiated as part of the strategic plan, then the planning documents need to be reviewed. If the project originated from departmental needs, then key users need to be consulted to help the project team understand the business need. From this task, a brief problem description is developed. As these needs are identified, the team also develops a detailed list of the expected business benefits. The list of **business benefits** contains the results that the organization anticipates it will accrue from a new system. Business benefits are normally described in terms of the specific results that can change the financial statements, either by decreasing costs or increasing revenues.

business benefits the benefits that accrue to the organization; usually measured in dollars

As the business benefits are being identified, the project team will identify the new system's specific capabilities to support the realization of these benefits. The objective of this task is to define the scope of the problem in terms of the requirements for the information system. This scoping statement, as defined by a list of **system capabilities**, helps identify the size and complexity of the new system and the project that will be required.

system capabilities the required capabilities of a new system; part of a System Vision Document

Members of the development team, working with the users and the client, combine these three components—the problem description, the business benefits, and the system capabilities—into a System Vision Document. **Figure 9-5** presents RMO's System Vision Document. Note the differences between the business benefits and the system capabilities. The business benefits focus on the financial benefit to the company. The system capabilities focus on the system itself. The benefits are achieved through the capabilities provided by the system.

FIGURE **9-5** *System Vision Document for RMO's CSMS*

Consolidated Sales and Marketing System
System Vision Document

Problem Description

Sales and marketing on the Web has changed drastically since the CSS system was built. Customers are more sophisticated, and they are used to catalog and sales systems that are easy to use and provide many services, such as one-click ordering, deferred-purchase tracking, simplified searches, and comparison shopping. In addition, research has shown that sales increase dramatically when social media marketing tools are combined with basic sales functionality. Hence, the new CSMS is needed not only to respond to today's competition but to launch RMO into today's world of social media and mobile computing. The longer RMO delays in starting this project, the more opportunities it misses.

System Capabilities

This document identifies the required system capabilities at a high level. Later documents will specify the detailed requirements. These capabilities are required:

- Provide a shopping cart capability.
  - Support customer sales with high automation (one-click, etc.).
  - Recommend related product purchases and comparison shopping.
  - Allow customer ratings and recommendations.
  - Include "friend" network capability.
- Include comprehensive order fulfillment.
  - Support multiple and split-order shipping and tracking.
  - Support back-ordering and tracking.
  - Allow customer comments and feedback.
- Provide customer account and billing capability.
  - Provide individualized customer accounting.
  - Support electronic billing and many electronic payment methods.
  - Accumulate customer "points" and allow transfer and sharing.
- Include marketing functions for promotions and specials.
  - Provide flexible promotions and sales.
  - Accumulate and track "points" from suppliers directly to customers.
  - Interface with social marketing media for advertising and social marketing activities.
  - Support mobile devices for social marketing and sales.

Business Benefits

The primary business benefit of these capabilities will be to increase sales by connecting with customers and improving the customer experience. The specific benefits include:

- Increasing the size of customer purchases
- Increasing the frequency of customer purchases
- Increasing customer satisfaction
- Increasing product recommendations from customers to friends
- Attracting new customers through recommendations and social marketing
- Building customer loyalty with recommendations and service
- Increasing speed of product availability
- Eliminating shipping delays and outages

RMO's existing CSS system was built under a tight deadline, and the company recognized that it would have a fairly short life. There were still many things to learn about Web marketing, but the existing CSS system will help the company define the requirements for its CSMS system.

Quantify Project Approval Factors

The first activity produced a high-level overview document that identified the need for a new system. However, that document alone may not be adequate to receive approval and funding. During this second activity, the project team, working with the users, will attempt to define more precisely the scope and impact of the project.

The objective is to provide sufficient justification so funds will be released and the project can start. Sometimes, the need is so great or so obvious that project approval is almost automatic. In other situations, it may be necessary to prepare a thorough cost-benefit analysis. These criteria must frequently be considered to obtain project approval:

■ The estimated time for project completion
■ The estimated cost for the project and system
■ The anticipated benefits from the deployment of the new system

These are rough estimates. In the traditional predictive approach to systems development, estimates were often made with a considerable amount of detail. However, the estimates were usually far off the mark. The problem was, of course, that with most new systems, the team was venturing into unknown needs, requirements, and technologies. With the more adaptive approaches, the stakeholders recognize that the requirements are unknown and that it is more important to monitor and control scope, cost, and schedule than to try to make estimates.

The Estimated Time for Project Completion

During Core Process 2 ("Plan and monitor the project"), a more detailed project schedule is created. During project initiation, there usually isn't enough known about the project to create a schedule. But there is nevertheless a need to estimate the project's completion date, even though this is one of the hardest things to do.

Sometimes, there are business constraints that dictate the completion of the project. For example, new legislative requirements may affect the deployment date. A window of opportunity may also provide a powerful motivation to complete a project at a specific time. These considerations should be made manifest and considered in the project approval and project planning processes.

The major inputs toward estimating the project completion date are the scoping document and the amount of effort required to develop the listed requirements. As indicated earlier, it is difficult to make an estimate with any degree of accuracy. At this early point in the project, gross estimates of team size and time frame are usually the best that can be achieved. For a predictive approach, the list of requirements can serve as the starting point for estimating the effort required to define and develop a particular function. For an adaptive approach, the same information can be used to estimate the number of iterations required and the size and number of teams working on the various subsystems.

Figure 9-6 shows an example of a time estimate document for RMO.

For RMO, the development of the time estimate was a one-day exercise. Because the project didn't yet have approval or funding, neither a project manager nor any systems analysts had been assigned to the project. However, a project manager had been assigned to obtain approval, and two systems analysts were assigned to help him. These three experienced technical people

FIGURE **9-6**
Project completion date estimate for the CSMS project

Time Estimate for the New CSMS Project			
Subsystem	Functional requirements	Iterations required	Estimated time
Sales subsystem*	15	5	20 weeks
Order Fulfillment subsystem*	12	5	20 weeks
Customer Account subsystem**	10	4	15 weeks
Marketing subsystem**	6	3	13 weeks
Reporting subsystem**	7	3	12 weeks
Total development time (2 teams)			40 weeks
Final hardening and acceptance testing		2	8 weeks
Total project time			48 weeks

*Assigned to Tiger team
**Assigned to Cougar team

met for four hours with the key users from the various RMO departments. The object of these meetings was to build a comprehensive list of all the functional requirements from each department. After the meetings, the group met again to organize this list of requirements into groupings that could be assigned to various iterations for the development of the software.

An assumption that was made by the director of new development was that there would be two subteams of four people each allocated to this project. As indicated in Figure 9-6, the time estimate for this project is 48 weeks from the date it begins.

The Estimated Cost for the Project and System

The estimated costs of developing the new CSMS are shown in **Figure 9-7**. By far, the largest cost item in the project's budget is the salaries of the project team. Other cost elements include the cost of the new computers, training for the users, offices, facilities, and utilities for the project team, travel expenses for the project team to do site visits, and software licenses. As you can see, this estimate is a little over $1.5 million.

After the system has been put into production, there will be annual operating costs, as shown in **Figure 9-8**. The largest cost is for a hosting service to provide some of the equipment, the connection to the Internet, and server administration services. These estimated costs were based on RMO using a hosting service to provide the equipment, the connection to the Internet, and server

FIGURE **9-7**
Summary of development costs for CSMS

Summary of Development Costs for CSMS	
Expense category	Amount
Salaries/wages (includes benefits costs) (1 PM, 8 analysts, 1 support)	$936,000.00
Equipment/installation	$308,000.00
Training	$78,000.00
Facilities	$57,000.00
Utilities	$97,000.00
Travel/miscellaneous	$87,000.00
Licenses	$18,000.00
Total	**$1,581,000.00**

FIGURE **9-8**
Summary of estimated annual operating costs for CSMS

Summary of Estimated Annual Operating Costs for CSMS	
Recurring expense	**Amount**
Connectivity/hosting	$156,000.00
Programming	$75,000.00
Help desk	$90,000.00
Total	**$321,000.00**

administration. The project team estimated about $13,000 a month for those expenses, which is enough for 15 very large managed servers. This appeared to be more than adequate, depending on the traffic volume. Other costs were for one full-time programmer and two help desk personnel.

The Anticipated Benefits from the Deployment of the New System

The System Vision Document identifies the anticipated business benefits of the new system. In this task, we analyze those business benefits and provide an estimate of their value to the organization. This value becomes part of the total decision criteria. Obviously, the dollar amount associated with these savings or revenues must be estimated by the client. It isn't the project manager's job to predict the value of business benefits. However, the project manager can help the client identify categories of potential benefits. Typical areas of increased revenue or cost reduction benefits include:

■ Opening up new markets with new services, products, or locations
■ Increasing market share in existing markets
■ Enhancing cross-sales capabilities with existing customers
■ Reducing staff by automating manual functions or increasing efficiency
■ Decreasing operating expenses, such as shipping charges for "emergency shipments"
■ Reducing error rates through automated editing or validation
■ Reducing bad accounts or bad credit losses
■ Reducing inventory or merchandise losses through tighter controls
■ Collecting receivables (accounts receivable) more rapidly

The project team at RMO worked with the vice president of sales and marketing to identify benefit areas and estimate a value for each one. This size of an investment and ongoing expense was going to require board approval within RMO. The board will want to know what the benefits of the new system will be and what the return on the investment will be. One of the difficulties for RMO is to determine how to assign a value to a benefit. A typical question might be "Do we assign the value of all our sales given that this system is needed to stay competitive in the marketplace? Or do we assign only the value of the increased sales we expect to get from marketing and higher volume?" If sales will drop because RMO becomes less competitive in the marketplace, the total sales value could be used. However, if the existing system is good enough to maintain a good client base, then only the increased sales should be used. These kinds of decisions are made by the client, not the project team. In this case, the vice president of sales and marketing at RMO decided to use a more conservative estimate. **Figure 9-9** summarizes the estimates he generated.

Many organizations like to compare the estimated costs with the anticipated benefits to calculate whether the benefits outweigh the costs. This process is called a **cost/benefit analysis**. Companies use a combination of methods to measure the overall benefit of the new system. One popular approach is to determine the **net present value (NPV)** of the new system. The two concepts behind net present value are (1) that all benefits and costs are calculated in terms of today's dollars (present value) and (2) that benefits and costs are

cost/benefit analysis process of comparing costs and benefits to see whether investing in a new system will be beneficial

net present value (NPV) the present value of dollar benefits and dollar costs of a particular investment

FIGURE **9-9**
Estimated annual benefits for CSMS

Estimated Annual Benefits for CSMS	
Benefit or cost saving	**Amount**
Recapture/prevention of lost sales	$200,000.00
Increase sales to existing customers	$300,000.00
Sales to new customers	$350,000.00**
Increased efficiency in order processing	$50,000.00
Reduction of data center and equipment costs because of hosting	$146,000.00
Total	**$1,046,000.00**

**plus 8% annual growth

combined to give a net value. The future stream of benefits and costs are netted together and then discounted by a factor for each year in the future. The discount factor is the rate used to bring future values back to current values.

Figure 9-10 shows a copy of the NPV calculation done for RMO's new CSMS. There are various techniques for calculating the NPV of a given investment. In this example, Year 0 represents the development period prior to the deployment of the system. The annual benefits for each year are extended across the top row. The development costs are shown on the second row. Annual expenses are shown on the third. Those three rows are combined in the fourth row to give the net benefits and costs. The fifth row shows the discount value, given a 6 percent discount rate. The sixth row is the product of the fourth and fifth rows and represents the net value in terms of today's dollars (i.e., the NPV). The seventh row shows a cumulative total of annual NPVs.

In Figure 9-10, the numbers in the seventh row eventually change from negative to positive. The point in time when that happens is called the **break-even point**. The length of time before the break-even point is reached is called the **payback period**. The payback period occurs in the year that the cumulative value goes positive. To calculate it, first take the last year that the cumulative value is negative—in this case, Year 2. Add to that year the number of days in the following year (in this case, Year 3) that it takes for the cumulative value to go positive. The method for doing that is to take absolute values of the ending value in Year 2 divided by the sum of the absolute values for the end of Year 2 and Year 3—in this case, 226,865 divided by (226,865 + 430,743). Here, that calculation indicates that the cumulative value goes positive after 35 percent of the year has passed. Multiply .35 times the 365 days in the year to get 128 days into Year 3. Many companies require a payback period of two to three years on new software.

The previous cost/benefit calculation depends on an organization's ability to quantify the costs and benefits. If it can indeed estimate a dollar value for a benefit or a cost, the organization treats that value as a **tangible benefit** or

break-even point   the point in time at which dollar benefits offset dollar costs

payback period   the time period after which the dollar benefits have offset the dollar costs

tangible benefit   a benefit that can be measured or estimated in terms of dollars

FIGURE **9-10** *Five-year cost/benefit analysis for CSMS*

	A	B	C	D	E	F	G	H
1		RMO Cost/Benefit Analysis for CSMS						
2		Category	Year 0	Year 1	Year 2	Year 3	Year 4	Year 5
3	1	Value of benefits		$1,046,000	$1,074,000	$1,104,240	$1,136,899	$1,172,171
4	2	Development costs	-$1,581,000					
5	3	Annual expenses		-$321,000	-$321,000	-$321,000	-$321,000	-$321,000
6	4	Net benefit/costs	-$1,581,000	$725,000	$753,000	$783,240	$815,899	$851,171
7	5	Discount factor	1.0000	0.9434	0.8900	0.8396	0.7921	0.7473
8	6	Net present value	-$1,581,000	$683,965	$670,170	$657,608	$646,274	$636,080
9	7	Cumulative NPV	-$1,581,000	-$897,035	-$226,865	$430,743	$1,077,017	$1,713,097
10	8	Payback period	2 years +	226865 / (226865+430743) = .35			or 2 years + 128 days (.35*365)	

cost. However, in many instances, an organization can't measure some of the costs and benefits to determine a value. Never discount the importance of ascertaining the "behind the scenes" reasons for a project. There may be political reasons for or against the project that override all other feasibility analyses. If there is no reliable method for estimating or measuring the value, it is considered an **intangible benefit**. In some instances, the importance of the intangible benefits far exceeds the tangible costs—at least in the opinion of the client, who pursues developing the system even though the dollar numbers don't indicate a good investment.

intangible benefit a benefit that accrues to an organization but that can't be measured quantitatively or estimated accurately

Examples of intangible benefits include:

■ Increased levels of service (in ways that can't be measured in dollars)
■ Increased customer satisfaction (not measurable in dollars)
■ Survival
■ Need to develop in-house expertise (such as a pilot program with new technology)

Examples of intangible costs include:

■ Reduced employee morale
■ Lost productivity (the organization may not be able to estimate it)
■ Lost customers or sales (during some unknown period of time)

Determining Project Risk and Feasibility

Project risk and feasibility analysis verifies whether a project can be started and completed successfully. Because each project is a unique endeavor, every project has unique challenges that affect its potential success.

The objective of this activity is to identify and assess the potential risks to project success and to take steps to eliminate or at least ameliorate these risks. They should be identified during the project approval process so all stakeholders are aware of the potential for failure. The team can also establish plans and procedures to ensure that those risks don't interfere with the success of the project. Generally, the team assigns itself these tasks when confirming a project's feasibility:

■ Determine the organizational risks and feasibility.
■ Evaluate the technological risks and feasibility.
■ Assess the resource risks and feasibility.
■ Identify the schedule risks and feasibility.

Determine Organizational Risks and Feasibility

Each company has its own culture, and any new system must be accommodated to that culture. There is always the risk that a new system departs so dramatically from existing norms that it can't be successfully deployed. The analysts involved with feasibility analysis should evaluate organizational and cultural issues to identify potential risks for the new system. Such issues might include:

■ Substantial computer phobia
■ A perceived loss of control on the part of staff or management
■ Potential shifting of political and organizational power due to the new system
■ Fear of change of job responsibilities
■ Fear of loss of employment due to increased automation
■ Reversal of long-standing work procedures

It isn't possible to enumerate all the potential organizational and cultural risks that exist. The project management team needs to be very sensitive to the reluctance within the organization to identify and resolve these risks.

After identifying the risks, the project management team can take positive steps to counter them. For example, the team can hold additional training sessions to teach new procedures and provide increased computer skills. Higher levels of user involvement in developing the new system will tend to increase user enthusiasm and commitment.

Evaluate Technological Risks and Feasibility

Generally, a new system brings new technology into the company, even state-of-the-art technology. Other projects use existing technology but combine it into new, untested configurations. If an outside vendor is providing a capability in a certain area, the client organization usually assumes the vendor is expert in that area. However, even an outside vendor may find the requested level of technology too complicated.

The project management team needs to carefully assess the proposed technological requirements and available expertise. When these risks are identified, the solutions are usually straightforward. The solutions to technological risks include providing additional training, hiring consultants, or hiring more experienced employees. In some cases, the scope and approach of the project may need to be changed to ameliorate technological risk. The important point is that a realistic assessment will identify technological risks early, making it possible to implement corrective measures.

Assess Resource Risks and Feasibility

The project management team must also assess the availability of resources for the project. The primary resource consists of team members. Development projects require the involvement of systems analysts, system technicians, and users. Required people may not be available to the team at the necessary times. An additional risk is that people assigned to the team may not have the necessary skills for the project. After the team is functioning, members may have to leave the team. This threat can come either from staff who are transferred within the organization if other special projects arise or from qualified team members who are hired by other organizations. Although the project manager usually doesn't like to think about these possibilities, skilled people are in short supply and sometimes do leave projects.

The other resources required for a successful project include adequate computer resources, physical facilities, and support staff. Generally, these resources can be made available, but the schedule can be affected by delays in the availability of these resources.

Identify Schedule Risks and Feasibility

The development of a project schedule always involves high risk. Every schedule requires many assumptions and estimates without adequate information. For example, the needs (and, hence, the scope) of the new system aren't well known. Also, the time needed to research and finalize requirements has to be estimated. The availability and capability of team members aren't completely known.

Another frequent risk in developing the schedule occurs when upper management decides that the new system must be deployed within a certain time. Sometimes, there is an important business reason for setting a fixed deadline, such as RMO's need to complete the CSS in time for online ordering over the holidays. Similarly, universities require the completion of new systems before key dates in the university schedule. For example, if a new admissions system isn't completed before the admissions season, then it might as well wait another full year. In cases like these, schedule feasibility can be the most important feasibility factor to consider.

If the deadline appears arbitrary, the tendency is to create the schedule to show that it can be done. Unfortunately, this practice usually spells disaster. The project team should create the schedule without any preconceived notion of required completion dates. After the schedule is completed, comparisons can be done to see whether timetables coincide. If not, the team can take corrective measures, such as reducing the scope of the project, to increase the probability of the project's on-time completion.

One objective of defining milestones and iterations during the project schedule is to permit the project manager to assess the ongoing risk of schedule slippage. If the team begins to miss milestones, the manager can possibly implement corrective measures early. Contingency plans can be developed and carried out to reduce the risk of further slippage.

Review with Client and Obtain Approval

As mentioned earlier, the amount of expenditure for the RMO project required board approval. However, before a presentation could be given to the board, RMO's executive committee needed to understand and agree to the project. A project this size has major impacts on all areas of the company. The departments, such as sales and marketing, will be directly impacted. They will have to allocate staff and resources to help in defining the requirements, developing test cases, and testing the new system as it is developed. In other words, the people in this department will have extra duties for the next 12 months or so. Even departments not directly involved will need to support this heavy development activity, perhaps tightening their budgets. In any event, it is always good policy to get the approval and support of the entire company. This process starts by making presentations to the senior executives of RMO. Often, a project manager will be asked to make the presentation or at least be present to answer questions.

After the executive committee approves the project, it goes to the board. After board approval, the IT department begins to assign full-time resources to the project. It is also a good idea at this point to have a company-wide memo or meeting to mark the beginning of this major activity. If the entire company knows that all the executives are supporting it and requesting cooperation, the project will proceed much more smoothly.

Activities of Core Process 2: Plan and Monitor the Project

This core process lasts throughout the entire project. A major planning effort occurs immediately after the project is approved. Ongoing planning and project monitoring continue during all project iterations. Not only must each iteration be planned as it starts, but progress must continually be monitored and corrective actions may be required. **Figure 9-11** illustrates by the height of the effort

FIGURE **9-11** *Activities of Core Process 2*

Plan and Monitor Activities
Establish the project environment.
Schedule the work.
Staff and allocate resources.
Evaluate work processes.
Monitor progress and make corrections.

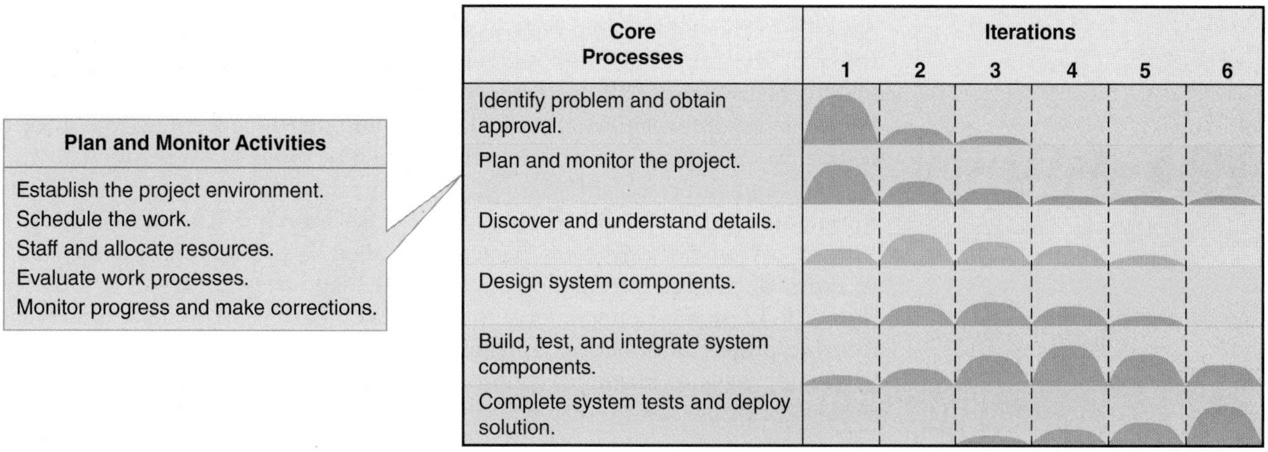

curve in each iteration that planning and monitoring activities must be an integral part of every project iteration. The specific activities associated with this core process are also listed in Figure 9-11. We will discuss each of these activities individually.

Establish the Project Environment

So far in this text, we have discussed different types of projects, such as predictive and adaptive projects, as well as tools, techniques, and methodologies to use with these different types of projects. We have also discussed such concepts as ceremony, project reporting, stakeholders, user participation, and the project team work environment. All these elements must be put in place as the project gets under way. Some of these decisions will already have been made based on the organization's standard policies and procedures. Others will be decided during the approval process. In any case, the project manager must ensure that the project's parameters and the work environment are finalized so the work of the project can proceed without roadblocks or delays. There are important project structure considerations that must be addressed as the project gets under way. For example, what kind of communication processes will be needed to keep the team and external stakeholders informed about what is going on? In addition, the members of the project team all need computers and IDEs and other tools to do their work. Of course, specific procedures about how the project team meets with the users, how they write code, and how they submit code for acceptance also need to be finalized. We will discuss three important considerations:

- Recording and communicating—internal/external
- Work environment—support/facilities/tools
- Processes and procedures

Recording and Communicating—Internal/External

The project manager and project team members will be involved in all types of meetings where decisions will be made and information developed. Determining what is important and how to record this information needs to be set out in specific project procedures. The other critical issue with information is what, how, how frequently, and to whom this information needs to be disseminated. One of the first tasks for a project manager on a new project is to establish the procedures and guidelines for how to handle the project's information.

A critical success factor for IT projects is to have the support of the organization's executives and other key stakeholders. A good project manager understands this need and structures his or her project so he or she communicates frequently, with the appropriate detail, to each of his or her stakeholders. Figure 9-2 identified the various stakeholders and participants in a project. Some of these stakeholders will be integrally involved with the project. Other stakeholders will be only marginally involved, receiving periodic status reports. The client stakeholders (the ones paying the project costs) will need to be kept aware of the project's status and of any difficulties or delays. A stakeholder analysis helps identify all those persons who have an interest in the project and defines what information they will want and need concerning the project. Generally, we refer to this as external reporting of project information.

Maintaining project information can be done via electronic means. Schedule information can be published to a Web site so everyone can view it. Another type of project-tracking tool, sometimes called a *project dashboard*, allows all types of project information to be posted and viewed by Web browsers. **Figure 9-12** is an example of a project dashboard system that allows easy access to project information. Spreadsheets, e-mails, newsletters, and list servers all provide ways to maintain, collect, and distribute information. Once the electronic systems are set up, they will often take care of themselves.

FIGURE **9-12** *Sample dashboard showing project information and status*

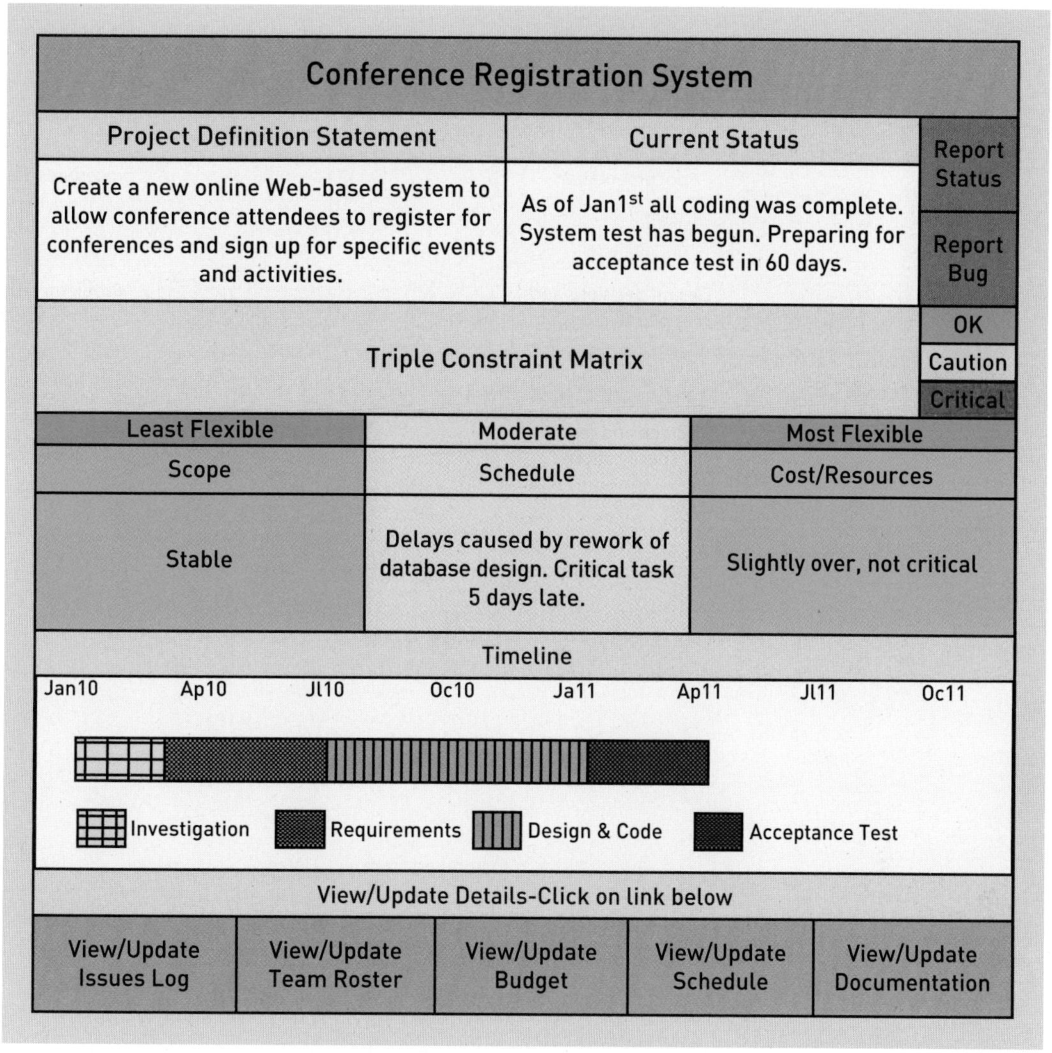

Conference Registration System

Project Definition Statement	Current Status	Report Status
Create a new online Web-based system to allow conference attendees to register for conferences and sign up for specific events and activities.	As of Jan1st all coding was complete. System test has begun. Preparing for acceptance test in 60 days.	Report Bug

| OK |
| Caution |
| Critical |

Triple Constraint Matrix

Least Flexible	Moderate	Most Flexible
Scope	Schedule	Cost/Resources
Stable	Delays caused by rework of database design. Critical task 5 days late.	Slightly over, not critical

Timeline

Jan10   Ap10   Jl10   Oc10   Ja11   Ap11   Jl11   Oc11

Investigation   Requirements   Design & Code   Acceptance Test

View/Update Details–Click on link below

View/Update Issues Log	View/Update Team Roster	View/Update Budget	View/Update Schedule	View/Update Documentation

The members of the project team also need to have mechanisms in place to communicate among themselves and document project decisions. This is an entirely different type of information—information about the system under development. For example, during analysis activities, the project team documents the results of user meetings by using various means, such as writing use case descriptions. During design, information also needs to be recorded and distributed among members of the team as appropriate. During testing, when errors are found, they must be documented and assigned to programmers to be fixed. Finally, the entire recording and communication requirement is often made more critical by members of the project team (as well as the users) being located at various sites around the globe. **Figure 9-13** illustrates some of the information that may need to be captured and maintained. The data repository in the figure usually consists of many different types of data structures and storage techniques—from wikis to databases to issue-tracking systems.

There is one caveat related to recording and communicating. With traditional predictive projects, the tendency was to create reams and reams of documentation. As you learned in the previous chapter, adaptive projects that use the Agile philosophy emphasize code over documentation. A novice project manager may interpret that to mean that no documentation is required. However, even with an Agile approach, the basic user definitions need to be

FIGURE **9-13**

System information stored in data repositories

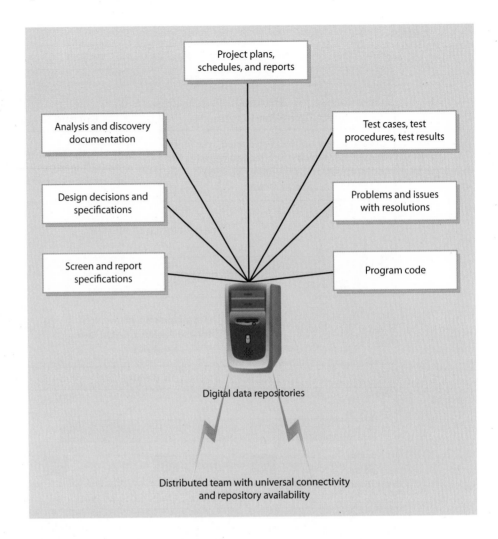

Project plans, schedules, and reports

Analysis and discovery documentation

Test cases, test procedures, test results

Design decisions and specifications

Problems and issues with resolutions

Screen and report specifications

Program code

Digital data repositories

Distributed team with universal connectivity and repository availability

documented for later verification. It isn't uncommon during programming for a programmer to have to refer back to notes and models to remember the exact details and decisions of a particular requirement. An experienced project manager knows the right amount of documentation so the project isn't overloaded with overhead but critical decisions are recorded.

It should be obvious that a comprehensive recording and communication scheme needs to be put in place. Fortunately, in today's connected world, there are many tools available so external and internal communication can be done easily. With so many electronic tools, all project information should be available online and accessible to all stakeholders. In fact, with the use of wikis, it is now common to allow many team members and even users to assist in the recording and updating of critical project information.

The CSMS team wanted to maintain its project information in digital format and have it available to all stakeholders, including team members, users, the client, and the members of the steering committee. RMO is a very open shop guided by the philosophy that information should be widely distributed. **Figure 9-14** shows all the tools that the CSMS project team uses to communicate and capture information. The core team members had previously worked on several Agile projects, so they had learned that there is a correct balance of documentation—not too much but enough to be able to trace key decisions and requirements. Barbara Halifax, the project manager, wanted to ensure the tools were in place so it was easy to record information when it was prudent to do so.

User documents, such as sample invoices, were scanned and placed in a document repository. User functional definitions were recorded in a

FIGURE **9-14**

Electronic digital repositories of information for CSMS

Electronic Digital Repositories		
Information captured	**Electronic tools**	**Who can update/view**
User definitions and functions User documents	Forum software Document server Scanners	Analysts, users/all
Screens and reports layouts	Web design tools Visio PowerPoint/Keynote	Analysts, users/all
Design specifications and diagrams	Wiki software Visio	Analysts/all
Issues and outstanding problems	Issue-tracking software	Analysts, users/all
Program code	Apache subversion (SVN)	Analysts/analysts
Project schedule	MS project	Analysts/all
Project status and information	Forum software	Analysts, users/all
Daily team coordination meeting	Video laptop conferencing	Project team
Distributed team communication	IM chat with video	Project team
Project update newsletter	Blog software	Project manager/all

forum system. Using a forum allowed team members and users to update it when key issues were discussed and needed to be remembered. Sample screen and report layouts were either sketched out or drawn with Visio or Keynote. Hand sketches were often scanned and saved. Most design decisions and specifications went right into the program code and weren't documented. However, some decisions were global, and those were captured in a wiki.

Each day, the project team had a "stand up" meeting—a short coordination meeting. Most of the team members were in the Park City Center, but some users were assigned to the team from other locations. Sometimes, team members were visiting user sites and therefore not available, and there were some team members who worked in the Salt Lake City office. Therefore, the daily meeting was conducted as a video conference call, with each person using his or her webcam and personal computer. The meeting normally lasted about 15 minutes.

Finally, there was some discussion of sending out a biweekly newsletter about the progress on the project. Barbara felt that it was important for the entire company to stay informed about the project in order to encourage their enthusiasm and support. However, instead of a printed newsletter, she opted to do it in the form of a blog. All users were invited to sign up with an RSS feed to keep informed about the project's progress.

Work Environment

Although the work environment may relate more to the work processes of the project team, the project manager must ensure that it is adequate to allow the project team to work productively. There are five major components of the work environment:

■ Personal computer(s) and/or workstation(s)
■ Personal development software and tools
■ Development server with repositories, sandboxes, and communication tools

- Office space, conference rooms, and equipment, including printers, scanners, and projectors
- Support staff

Most importantly, of course, is the computer equipment and other hardware that the team will need. Obviously, each developer will need his or her own computing configuration, which may consist of multiple computers or monitors. Other important hardware includes the development servers, printers, and internal development network. If the team is distributed, video cameras and projectors may be necessary to conduct distributed team meetings. Along with the hardware, resources must be made available to administer things such as the development server.

Related to the hardware is the computer software and other tools. Software tools can get quite elaborate—from stand-alone Integrated Development Environment (IDE) tools to modeling software to code repository software. The development server, with its environment and software, must also be configured and deployed. The server may be set up as a virtual server or as a stand-alone computer. Applications include such things as code repository, issue-tracking application, testing system, and the project dashboard.

Along with the hardware and software, a work configuration must be provided for each developer, with log-on permissions, sandbox environments, repository access, and so forth. The final two components are the office space and other facilities that may be needed. This will include access to conference rooms, presentation equipment, and maybe even transportation vehicles. Finally, the productivity of the team members is always enhanced when adequate support staff is available to take care of myriad details that always accompany an active project.

Processes and Procedures

The final major set of decisions has to do with the project's internal processes and procedures. Earlier, we discussed a project's level of formality. Larger projects require more elaborate reporting processes and meeting schedules. When there are many people involved, coordination of activities becomes critical. Procedures include:

- Reporting and documentation—What is done? How is it done? Who does it?
- Programming—Single or pair programming? How is work assigned? By whom?
- Testing—Programmer tests or user tests? How to mark items ready for testing?
- Deliverables—What are they? How and when are they handed over to users? How are they accepted?
- Code and version control—How is the code controlled to prevent conflicts? How to coordinate bug fixing with new development? How and when are deliverables released?

Schedule the Work

Scheduling the work is necessary for any size or type of project. However, the techniques used can vary widely depending on the type of project. For predictive, highly controlled projects, a detailed and complete schedule that covers the entire project is usually built. Again, these kinds of schedules only work because the software to be built is well understood. However, even in those projects with detailed and comprehensive schedules, accommodation is required as things change during the life of the project. At the other end of the spectrum, small Agile projects sometimes don't even have a project schedule, with the team members being responsible for scheduling their own work. Coordination is accomplished by talking and keeping each other informed of what each person is working on. This is what is meant by *chaordic*.

Scheduling the work for many of today's projects lies somewhere between these two extremes. Large projects may have several independent teams of developers working on various subsystems. Even though the work between the teams is fairly independent, coordination is still required. Adaptive projects also anticipate additional requests and changes to the original scheduled tasks.

For adaptive types of projects, creating the project schedule is done throughout the life of the project. During the initial planning phase, the initial list of use cases or user stores are developed for each subsystem. The use cases are divided up and tentatively assigned to the iterations. Let us call this the **project iteration schedule**. As each iteration is begun, a detailed schedule of tasks and work to be done is developed. You saw an example of creating an iteration schedule in Chapter 1. Let us call this schedule a **detailed work schedule**, meaning that it schedules the work within an iteration. Sometime during each iteration—often as one iteration is finishing and before the next iteration begins—the project manager, with assistance from the team leaders and key users, will review and rework the project iteration schedule. During this process, the changes and any new requirements are prioritized and placed on the schedule.

Creating the project iteration schedule must take into account the total size and configuration of the solution system and the number of teams available to work on the project. Separate lists of requirements are made by subsystem, and a project iteration schedule can then be made for each subsystem. Some tasks, such as designing the database, may go across all subsystems and need to be scheduled separately or be included in every subsystem list. **Figure 9-15** shows a sample project iteration schedule for the CSMS Sales subsystem. As you can see, the length of each iteration is fairly constant at around four weeks. All the identified tasks, which represent the requirements, have been assigned to iterations. In this case, we have identified five iterations.

Developing a detailed work schedule for a single iteration is a three-step process:

- Develop a work breakdown structure.
- Estimate effort and identify dependencies.
- Create a schedule by using a Gantt chart.

project iteration schedule the list of iterations and use cases or user stories assigned to each iteration

detailed work schedule the schedule that lists, organizes, and describes the dependencies of the detailed work tasks

FIGURE **9-15**

Project iteration schedule for the CSMS Sales subsystem

Project Iteration Schedule for the CSMS Sales Subsystem		
Iteration	**Time estimate**	**Use cases assigned to iteration**
1	4 weeks	1. Search for item. 2. View detailed descriptions. 3. View rotating (3-D) images. 4. Compare item characteristics.
2	4 weeks	5. View comments and ratings. 6. Search comments and ratings for friends. 7. View accessory combinations (images). 8. Save item + accessories as "combo."
3	5 weeks	9. Add item (or combo) to shopping cart. 10. Remove item (or combo) from shopping cart. 11. Add item (or combo) to "on reserve" cart. 12. Remove item (or combo) from "on reserve" cart.
4	4 weeks	13. Check out active cart. 14. Create and process store sale. 15. Create and process phone sale.
5	3 weeks	16. Clean up, final test, harden site, tune database, etc.
Total	20 weeks	

work breakdown structure (WBS)
the list or hierarchy of activities and tasks of a project; used to estimate the work to be done and to create a detailed work schedule

A **work breakdown structure (WBS)** is a list of all the required individual activities and tasks for the project. There are two general approaches for creating a WBS: by deliverable or by a timeline. The first approach identifies all the deliverables that must be completed for a given iteration. Then, the WBS identifies every task that is necessary to create each deliverable. The second approach works through the normal sequence of activities that are required for the final deliverable. Experienced developers who have worked on Agile projects understand the steps and tasks that are required to create a particular deliverable. Of course, each iteration is slightly different depending on the particular functions and deliverables that are included.

Figure 9-16 is a sample handwritten WBS for the first iteration of the Sales subsystem. The tasks have been partitioned according to the core processes

FIGURE **9-16** *Work breakdown structure for first iteration*

Work Breakdown Structure
Iteration 1 of Sales Subsystem

I. Project planning
   Develop WBS and build schedule and then plan the work — 1/2 day.

II. Analysis tasks
   Meet with sales department — 1 day.
   Meet with marketing department — 1 day.
   Define required information and data elements (share with
      Cougar team) — 1 day.
   Model user activities — 1 day.

III. Design tasks
   Design database schema (work with Cougar team) — 1 day.
   Design screen layouts and cross links — 2 days.
   Identify program classes and methods — 1 day.

IV. Build tasks
   Build database (coordinate with Cougar team) — 1/2 day.
   Write program code — 4 days.
   Integrate 3-D imaging code — 2 days.
   Build test data — 2 days.
   Set up user "simulated live environment" — 1/2 day.
   Perform acceptance tests with users — 2 days.
   Release accepted version — 1/2 day.
   Perform team introspection — 1/2 day.

Note: The use cases in this iteration require data and a database schema that is also used in building the catalog. Building the catalog will be done in the first iteration of the Marketing subsystem. The two iterations will be done in parallel, and the two teams (Tiger team and Cougar team) will coordinate and share information.

Note: This is a four-person team. The work pattern is (1) all together for fact finding and design tasks and (2) pairs for programming and testing tasks.

Note: The four use cases that will be developed during this iteration are:
   1. Search for item.
   2. View detailed information for an item.
   3. View rotating (3-D) image of item.
   4. Compare characteristics and prices across multiple items.

Planning, Analysis, Design, and Building. In the figure, each task also has an estimate of the time required. Sometimes, two estimates are provided: the effort required and the expected duration. The effort required is given in person-days of work, and the duration is a measure of lapsed calendar time. Of course, these are related depending on the number of people working on the specific task. In Figure 9-16, only duration is shown; however, the time estimates assume a project team of four people.

When developing a WBS, new analysts frequently ask "How detailed should the individual tasks be?" A few guidelines can help answer that question:

■ There should be a way to recognize when the task is complete.
■ The definition of the task should be clear enough so one can estimate the amount of effort required.
■ As a general rule for software projects, the effort should take one to five working days.

The second step in developing a detailed work schedule for a single iteration is to determine the dependencies between the tasks and the amount of effort required for each. The most common way to relate tasks is to consider the order in which they are completed; that is, as one task finishes, the next one starts. This is called a finish-start relationship. Other ways to relate tasks include start-start relationships, in which tasks start at the same time, and finish-finish relationships, in which tasks must finish at the same time. The effort required should be the actual amount of work required to complete the task. As with the identification of the tasks in the WBS, the dependencies and effort estimates should be done by the developers who are going to actually do the work.

The third step in developing a detailed work schedule is to actually create the iteration schedule. In Figure 1-7 of Chapter 1, we presented a graph of the tasks involved in the first iteration of the Tradeshow system, their sequence, and the estimated calendar time to complete them. The graph was, in actuality, a simplified PERT/CPM chart. We provide more information about PERT charts in Online Chapter C. The other form for presenting a schedule is a bar chart that shows the activities as bars on a horizontal time line; this is called a **Gantt chart**. A widely used tool for building Gantt charts is Microsoft Project. New versions of MS Project are network enabled and provide a powerful tool to not only create schedules but to also distribute schedule information across the organization by using the HTML protocol so it can be viewed in a browser. The benefit of using a tool such as MS Project is that the project manager can update progress easily and make that information widely available.

Figure 9-17 shows an iteration schedule from the RMO CSMS project formatted as a bar chart. In the figure, the tasks from the work breakdown structure are listed in the Task Name column and the durations are listed in the Duration column. The Predecessor column identifies dependencies between tasks. As you can see, every task except the first has at least one predecessor task, and every task except the last is a predecessor to one or more other tasks. There are various ways to document dependencies. The most common way is to show the finish of one task occurring before the start of another (FS). Other common ways are finish-finish (FF), where both must finish at the same time, and start-start (SS), where both start at the same time. Any dependency can have a lag time, such as that shown on line 11 of Figure 9-17. The final column documents what resources have been assigned to each task. In this example, the Tiger Team is divided into two subteams of two people each: TT1 and TT2.

The bars in Figure 9-17 illustrate the duration of each task superimposed on a calendar. The red bars indicate a critical path on the schedule. The **critical path** is defined as those tasks that must stay on schedule. If any of the critical path tasks cause a schedule slip, then the entire project is delayed. The blue bars are those tasks that aren't on the critical path. Obviously,

Gantt chart a bar chart that portrays the schedule by the length of horizontal bars superimposed on a calendar

critical path a sequence of tasks that can't be delayed without causing the entire project to be delayed

FIGURE **9-17** *An iteration schedule for the first iteration of the Shopping Cart subsystem*

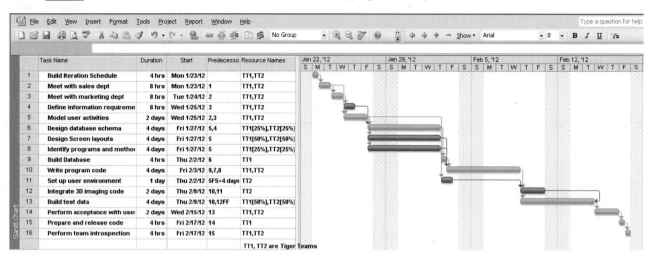

a project manager will monitor critical path tasks quite closely. Online Chapter C gives more detailed explanations and instructions on how to use MS Project to create Gantt chart schedules.

Staff and Allocating Resources

In an Agile project, the various teams are self-organizing. They decide how they are going to work together and assign the tasks to be done among themselves. However, the job of identifying what expertise is needed for the project and getting those people assigned to the project falls on the shoulders of the project manager. This includes finding the right people with the correct skills and then organizing and managing them throughout the project. The staffing activity consists of five tasks:

- Developing a resource plan for the project
- Identifying and requesting specific technical staff
- Identifying and requesting specific user staff
- Organizing the project team into work groups
- Conducting preliminary training and team-building exercises

Based on the tasks identified in the project schedule, the project manager can develop a detailed resource plan. In fact, the schedule and the resource requirements are usually developed concurrently. In developing the plan, the project manager recognizes that (1) resources usually aren't available as soon as requested and (2) a period of time is needed for a person to become acquainted with the project. After developing the plan, the project manager can then identify specific people and request that they become part of the team.

On small projects, members of the project team may all work together. However, a project team that is larger than four or five members is usually divided into smaller work groups. Each group will have a group leader who coordinates the tasks assigned to the group. The project manager is responsible for dividing the team into groups and assigning group leaders.

Finally, training and team-building exercises are conducted. Training may be done for the project team as a whole when such new technology as a new database or a new programming language is used. In other cases, team members who are unfamiliar with the tools and techniques being used may require individual training. The team should conduct appropriate training for technical people and users. Team-building exercises are especially important when members haven't worked together before. The integration of users with technical people is an important consideration in developing effective teams and workgroups.

Evaluate Work Processes (How Are We Doing?)

Although evaluating how the project team performed is sometimes done on predictive projects, it isn't a common practice. However, on iterative projects, many companies require an "end of iteration" review of how well the team performed and worked together. One of the advantages of an iterative project is that the same team often stays together for a number of iterations. After each iteration, team members can evaluate how well they worked together and how they can improve their effectiveness and performance as a team. In an Agile project, this is referred to as a **retrospective**. Here are the kinds of questions the team might want to ask:

retrospective a meeting held by the team at the end of an iteration to determine what was successful and what can be improved

- Are our communication procedures adequate? How can they be improved?
- Are our working relationships with the user effective?
- Did we hit our deadlines? Why or why not?
- Did we miss any major issues? How can we avoid this in the future?
- What things went especially well? How can we ensure it continues?
- What were the bottlenecks or problem areas? How can we eliminate them?

Monitor Project Progress and Make Corrections

In theory, executing and controlling the project plan sounds easy, but in fact, it is quite complicated. To execute any project, you need some type of project plan. How a team builds and executes project plans will vary depending on whether the project structure is based on a predictive approach or an adaptive approach. In the predictive approach, the project plans are quite large and complex. The adaptive approach is less daunting because the detailed project plan is done for each iteration. Because the piece of work is smaller and often better understood, these plans tend to be smaller and less complex.

Figure 9-18 is a high-level process chart that illustrates the basic process for monitoring and controlling the project. The first box—*Assign work to person or team*—refers to a task that is complex all by itself due to the fact that teams are made up of people with varying skill levels and experiences.

The task for the second box—*Collect status*—is less complex. When collecting status information, you should adhere to certain guidelines. First, providing

FIGURE **9-18** *Process to monitor and control project execution*

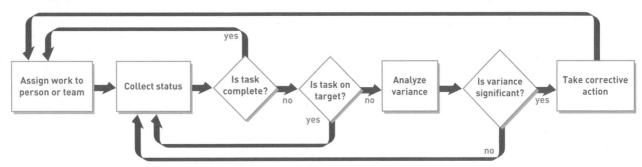

status information should be a standard process for all team members. Second, status information should be collected and posted electronically for all to see. Status information can be reported at milestones as complete or not complete.

The task for the third box—*Analyze variance*—requires the project manager to try to determine why the task isn't on target and how significant the delay is with regard to the impact on the total project.

The task for the fourth box—*Take corrective action*—can be complex. Experienced project managers have a whole set of tools they can use to try to correct the variance. Sometimes, the correction is as simple as reassigning team members, or maybe it just requires some extra hours of overtime. At other times, tasks may have to be rearranged. In more serious instances, the entire schedule may have to be reworked or more team members may need to be recruited for the team. The objective of corrective action is to get the project back to a known and predictable schedule.

Every development project—whether it follows a predictive or adaptive approach—has lots of questions that need answers and many decisions that need to be made. In many cases, these issues are quickly resolved and the project moves rapidly forward. However, in other instances, the answer to a question or the resolution of an open issue will require additional research. For example, a set of rules for sales commissions might include when and how commissions are calculated, what happens to commissions on returned merchandise, when commissions are paid, how the commission schedule varies to encourage sales of high-margin items and sale items, and so on. If management is still making decisions about these rules, you will need to track these issues until they are resolved.

The monitoring and control of open issues and risks for a project is usually no more complex than building various tracking logs. These logs can be built in a simple spreadsheet and posted on the project's Web site or central repository. It is a good idea to make these logs available to all team members. **Figure 9-19** presents an example of a tracking log. The column headings will vary depending on the type of log you use. The tracking log in Figure 9-19 shows issues that need to be resolved by a certain date and the persons responsible for resolving those issues.

FIGURE 9-19 *Sample issue-tracking log*

	A	B	C	D	E	F	G	H	I
1	Issue log#	Issue date	Issue description	Priority	Issue impact	Person responsible	Target fix date	Resolution description	Actual fix date
2		1/18/2012	Commission structure for sales promotion is undefined	Urgent	Database structure may need to be modified	William Henry	2/1/2012		
3									
4									
5									

Chapter Summary

This chapter focused on the principles and activities related to planning and managing a systems development project. It covered three major themes: (1) the principles of project management, (2) the activities to get a project initiated and approved, and (3) the activities to plan the project and monitor its progress.

Project management is the organizing and directing of other people to achieve a planned result. Historically, software projects haven't had a very good track record. Strong project management is seen as one factor that improves success rates of software development projects. Other factors, such as the adaptive approach to the SDLC, can also contribute to project success.

In this chapter's first section, many important skills, techniques, and concepts that relate to project management were discussed. The project management body of knowledge (PMBOK) provides an extensive conceptual foundation for learning about project management. Agile project management requires the same foundation concepts and skills as the PMBOK, although many of the specific techniques may be different.

This chapter's second major section focused on the specific activities of Core Process 1, the objective of which is to identify the business need and get the project initiated. These activities include:

- Identifying the problem
- Extending the project approval factors
- Performing risk and feasibility analysis
- Reviewing with the client and obtaining approval

This chapter's third major section focused on those activities that are necessary to get the project planned, scheduled, and started. These activities include:

- Establishing the project environment
- Scheduling the work
- Staffing and allocating resources
- Evaluating work processes
- Monitoring progress and making corrections

Key Terms

break-even point   268

business benefits   263

ceremony   257

client   257

cost/benefit analysis   267

critical path   279

detailed work schedule   277

Gantt chart   279

intangible benefit   269

net present value (NPV)   267

oversight committee   257

payback period   268

project iteration schedule   277

project management   256

project management body of
   knowledge (PMBOK)   259

retrospective   281

system capabilities   263

System Vision Document   263

tangible benefit   268

users   257

work breakdown structure (WBS)   278

Review Questions

1. List the six major reasons that projects fail.
2. List six critical factors that contribute to project success.
3. Define project management.
4. List five internal responsibilities of a project manager.
5. What is the difference between the client and the user?
6. What is meant by an organic approach?
7. What is the importance of "ceremony"?
8. List the nine areas of the PMBOK.
9. What is meant by Agile project management?
10. How is scope management accomplished with Agile project management?
11. What are the four activities of Core Process 1?
12. What are three reasons that projects are initiated?
13. What is the difference between system capabilities and business benefits?

14. What factors are usually considered when approving a project?

15. List 10 types of benefits that may be considered when approving a project.

16. Explain how net present value (NPV) is calculated.

17. What is the difference between tangible benefits and intangible benefits?

18. What are some factors to consider when assessing organizational feasibility?

19. What are the five activities of Core Process 2?

20. List seven types of information that should be captured during a project.

21. What is the difference between the project iteration schedule and the detailed work schedule?

22. What is a work breakdown structure used for?

23. What is the benefit of an iteration review and retrospective?

Problems and Exercises

1. Read the following description and then make a list of expected business benefits that the company might derive from a new system:

   Especially for You Jewelers is a small jewelry company in a college town. Over the last couple of years, it has experienced a tremendous increase in its business. However, its financial performance hasn't kept pace with its growth. The current system, which is partly manual and partly automated, doesn't track accounts receivables sufficiently, and the company is finding it difficult to determine why the receivables are so high. It runs frequent specials to attract customers, but it has no idea whether these are profitable or if the benefit—if there is one—comes from associated sales. Especially for You wants to increase repeat sales to its existing customers, thus it needs to develop a customer database. It also wants to install a new direct sales and accounting system to help solve these problems.

2. Read this narrative and then make a list of system capabilities for the company:

   The new direct sales and accounting system for Especially for You Jewelers will be an important element in the growth and success of the jewelry company. The direct sales portion needs to track every sale and be able to link to the inventory system for cost data to provide a daily profit and loss report. The customer database needs to be able to produce purchase histories to assist management in preparing special mailings and special sales to existing customers. Detailed credit balances and Aged accounts for each customer would help solve the problem with the high balance of accounts receivables. Special notice letters and credit history reports would help management reduce accounts receivable.

3. Develop a System Vision Document for Especially for You Jewelers based on the work you did for Problem 1 and Problem 2.

4. Develop a work breakdown structure (WBS) based on the following narrative. It should cover all aspects of the move—from the beginning of the project (now) to the end, when all employees are moved into their new offices. Format your solution in tabular form with the following column headings: Task ID No, Task Description, Estimated Effort, Predecessor Task ID. For your solution, follow these guidelines:

   - Include dependencies.
   - Include effort (work) estimates.
   - Have 30 to 40 detailed tasks.
   - Cover a period of at least two months to a maximum of six months.

   You are an employee of a small company that has outgrown its facility. It is a Web development and hosting company, so you have technical network administrators, developers, and a couple people handling marketing and sales. There are 10 employees.

   The president of your company has purchased a nearby single-story building, and the company is going to move into it. The building will need some internal modifications to make it suitable. The president has asked you to take charge of the move. Your assignment is to (1) get the building ready, (2) arrange for the move, and (3) carry out the move.

   The building is nearly finished, so the job shouldn't be too difficult (no construction is necessary—just some refurbishing). The building has several offices as well as a larger area that needs to be set up with cubicles.

   You and the president are walking through the building, and she tells you what she wants:

   "Let's use the offices as they are," she says. "We will need a reception desk for visiting customers. The office in the back corner should be okay for our computer servers. Let's put the salespeople in these offices along the east wall. We are short a few offices, so let's put up a few cubicles in the large room for our junior developers.

"Of course, we will need to get everybody connected to our system, and I think Ethernet would be faster than wireless for us. And we all need to have phones.

"Let's plan the move for a long weekend, like a Thursday, Friday, and Saturday. Of course, we need to be careful not to shut down the clients we are already hosting.

"Will you put together a schedule for the move for our employees and set up instructions for all the employees so they know how they are supposed to get ready for the move? Thanks."

5. Enter your WBS from Problem 4 into MS Project. First, enter the tasks, dependencies, and durations. Write a paragraph on your experience using MS Project.

6. Develop a six-year NPV spreadsheet similar to the one shown in Figure 9-10. Use the table of benefits, costs, and discount factors shown in **Figure 9-20**. The development costs for the system were $225,000.

7. Using MS Project, Build a Gantt chart based on the table shown in **Figure 9-21**. Enter the tasks, dependencies, and durations. Print out the PERT chart (Network chart) and the Gantt chart.

Figure 9-21 presents a list of tasks for a student who wants to have an international experience by attending a university abroad. Assume that all predecessor tasks must finish before the succeeding task can begin (the simplest version). Also, insert a few overview tasks, such as Application tasks, Preparation tasks, Travel tasks, and Arrival tasks. Be sure to state your assumption.

FIGURE **9-20**
Benefits, costs, and discount factors for calculating NPV

Year	Annual benefits	Annual operating costs	6% discount factor
1	$55,000	$5,000	0.9524
2	$60,000	$5,000	0.9070
3	$70,000	$5,500	0.8638
4	$75,000	$5,500	0.8227
5	$80,000	$7,000	0.7835
6	$80,000	$8,000	0.7462

FIGURE **9-21** *WBS task list for attending a university abroad*

Task Id	Description	Duration (days)	Predecessor task
1	Obtain forms from the international exchange office.	1	None
2	Fill out and send in the foreign university application.	3	1
3	Receive approval from the foreign university.	21	2
4	Apply for the scholarship.	3	2
5	Revive notice of approval for the scholarship.	30	4
6	Arrange financing.	5	3, 5
7	Arrange for housing in a dormitory.	25	6
8	Obtain a passport and the required visa.	35	6
9	Send preregistration forms to the university.	2	8
10	Make travel arrangements.	1	7, 9
11	Determine clothing requirements and go shopping.	10	10
12	Pack and make final arrangements to leave.	3	11
13	Travel.	1	12
14	Move into the dormitory.	1	13
15	Finalize registration for classes and other university paperwork.	2	14
16	Begin classes.	1	15

8. The state university wants to implement a better system to keep track of all the computer equipment it owns and needs to maintain. The university purchases a tremendous number of computers and software that are distributed throughout the campus and are used by faculty, staff, departments, and colleges. Currently, the university has very sparse records of its equipment and almost no records about maintenance or the software that has been purchased. A list of use cases has been defined; it will serve as the starting point to develop this system.

Take the following list of use cases to create a project iteration schedule. You should try to arrange the use cases so that similar ones are developed together. Also, the most important use cases should be developed first. State your assumptions, and explain your reasons for your solution.

Note: For brevity, we use the word *computer* to refer to any type of computing equipment, such as a desktop computer, laptop computer, server computer, printer, monitor, projector, wireless access point, and so forth.

1. Buy a computer.
2. Sell a computer.
3. Put a computer in service.
4. Take a computer out of service (surplus).
5. Assign a computer to a person.
6. Record the location of a computer.
7. Repair a computer (in house).
8. Return a computer for repair.
9. Identify computers ready for replacement.
10. Search for a specific computer by various options.
11. Buy a software license.
12. Renew a software license.
13. Install software on a computer.
14. Remove software from a computer.
15. Record a warranty for a computer.
16. Purchase a warranty for a computer.
17. Search for multiple computers by various options.
18. Search for software on computers by various options.
19. Assign a computer to a department or college.

Case Study

Custom Load Trucking

It was time for Stewart Stockton's annual performance review. As Monica Gibbons, an assistant vice president of information systems, prepared for the interview, she reviewed Stewart's assignments over the last year and his performance. Stewart was one of the "up and coming" systems analysts in the company, and she wanted to be sure to give him solid advice on how to advance his career. For example, she knew that he had a strong desire to become a project manager and accept increasing levels of responsibility. His desire was certainly in agreement with the needs of the company.

Custom Load Trucking (CLT) is a nationwide trucking firm that specializes in the movement of high-tech equipment. With the rapid growth of the communications and computer industries, CLT was feeling more and more pressure from its clients to move its loads more rapidly and precisely. Several new information systems were planned that would enable CLT to schedule and track shipments and truck locations almost to the minute. However, trucking wasn't necessarily a high-interest industry for information systems experts. With the shortage in the job market, CLT had decided not to try to hire project managers for these new projects but to build strong project managers from within the organization.

As Monica reviewed Stewart's record, she found that he had done an excellent job as a team leader on his last project, where he was a combination team leader/ systems analyst on a four-person team. He had been involved in systems analysis, design, and programming, and he had also managed the work of the other three team members. He had assisted in the development of the project schedule and had been able to keep his team right on schedule. It also appeared that the quality of his team's work was as good as, if not better than, other teams on the project. Monica wondered what advice she should give him to help him advance his career. She was also wondering if now was the time to give him his own project.

1. Do you think the decision by CLT to build project managers from its existing employee base is a good one? What advice would you give CLT to make sure it has strong project management skills in the company?
2. What kind of criteria would you develop for Monica to use to measure whether Stewart (or any other potential project manager) is ready for project management responsibility?
3. How would you structure the job for new project managers to ensure or at least increase the possibility of a high level of success?
4. If you were Monica, what kind of advice would you give Stewart about managing his career and attaining his immediate goal of becoming a project manager?

Community Board of Realtors

The Board of Realtors Multiple Listing Service (MLS) system is a fairly focused system. In Chapter 3, you identified a use case diagram for the customer users. In Chapter 8, you extended the functions to include aspects of the system that would be required for the real estate agents to enter their information. You also made some preliminary estimates of iterations and time to complete. Let us expand and refine those answers to include concepts from this chapter.

1. Given the total vision of this system, develop a System Vision Document. Focus primarily on finding the benefits to the community board, the real estate agents, and home buyers.
2. Including the use cases and functions identified in Chapters 3 and 8, make a list of all the use cases that must be developed. Divide them into subsystems as appropriate. You should have at least two subsystems: one for viewing data and one for updating data. Add any additional use cases (and subsystems) that might be important to the Community Board of Realtors itself. [Hint: Think about user goals and CRUD.]
3. Decide on a work sequence and develop a project iteration schedule.
4. Estimate the development cost and the time required.
5. Develop a work breakdown structure (WBS) for the project's first iteration.
6. Enter your WBS into MS Project to create a detailed work schedule. (Instructions on how to use MS Project are given in Online Chapter C, which you can find on the Cengage Web site.)

The Spring Breaks 'R' Us Travel Service

Let us assume that you are a project manager for Spring Breaks and have been asked to prepare the necessary documents to get this project approved and planned. You have been told that four programmers will be available to work on this project and that it will have the highest priority within the company. In other words, the company would like to have this application up and running as soon as possible. The travel season is fast approaching, and Spring Breaks would like to be able to use the system for this very next season.

As part of the approval and planning activities, you decide that the most important items to develop will be a System Vision Document and a project iteration plan. Given those elements, you can make an estimate of the completion date and the development cost for the system.

1. Based on the answers you gave to the Chapter 2 running case questions, develop a System Vision Document.
2. Based on the functional descriptions you provided for the Chapter 2 running case and the use cases you defined in Chapter 3, finish identifying a complete list of use cases for each of the four subsystems. One important decision you will have to make is which subsystems to develop first. In other words, can the subsystems be deployed independently and, if so, which should be deployed first? Defend your answer.
3. A related decision is whether to organize your programmers into one larger team or multiple smaller teams and how many programmers you can use on this project. Make that decision and then defend your answer.
4. Once those decisions are made, develop a project iteration plan. If you have multiple independent teams, your project iteration plan will have parallel paths.
5. Based on your previous answers, develop an estimate for the total project cost and the time required to complete the project.
6. Assuming an annual revenue increase of $250,000 per year (benefit) and an annual operating cost of $75,000, develop a five-year NPV worksheet by using your estimates for developing the system. Use a 6 percent discount factor.

On the Spot Courier Services

In this chapter, we have discussed the first two core processes of a systems development project and the activities within these core processes. Obviously, for a normal project, these first two core processes are done at the beginning, when the project manager is still learning about the needs of the business and trying to

(continued on page 288)

(continued from page 287)

develop a vision of the solution. However, in this case, you have already gathered a lot of information about On the Spot from the previous chapters. Let us go ahead and use this information to develop some project management planning documents.

In Chapter 2, you developed a list of use cases. In Chapter 8, you identified four required subsystems in the total solution. Chapter 6 provided a good review of the essential system capabilities. Using the discussions of On the Spot from these chapters and the items you have produced from your previous work, produce these items:

1. Create a System Vision Document.
2. Review all the use cases that you identified in Chapter 2 and then enhance the list to achieve a complete solution. Assign each use case to one of these four subsystems from Chapter 8:

   ■ Customer account subsystem (like customer account)

   ■ Pickup request subsystem (like sales)
   ■ Package delivery subsystem (like order fulfillment)
   ■ Routing and scheduling subsystem

3. Create a project iteration schedule for each subsystem. The project consultant is planning to assign one team of two people to this project, and the subsystems will be built consecutively. Based on the answers you provided in Chapter 8, combine your four individual schedules into a total project iteration schedule.
4. Create a work breakdown structure (WBS) for the first iteration of the project as you have outlined it. Estimate the effort required for each task in the WBS.
5. Enter the WBS into MS Project to create a detailed work schedule. (Instructions on how to use MS Project are given in Online Chapter C on the Cengage Web site.)

Sandia Medical Devices

Use cases were identified for the RTGM system in Chapter 5 (see **Figure 9-22**). Additional descriptions of the system requirements are found in Chapters 3, 4, and 8. You might want to review those to refresh your memory of the needs for this system.

Complete these tasks:

1. Based on the use case diagram and other project information, develop a list of software components (subsystems) that must be acquired

FIGURE **9-22** *RTGM system use cases*

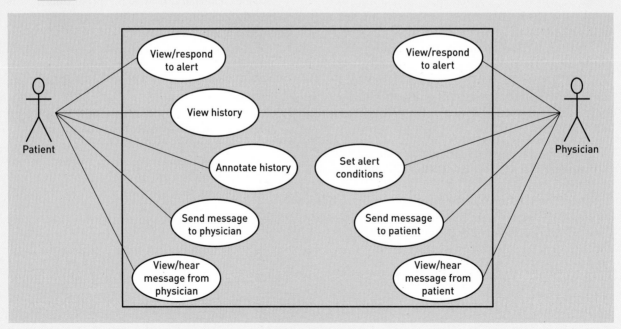

(continued on page 289)

(continued from page 288)

or developed. Describe the function(s) of each component in detail. Be sure to consider components that aren't directly tied to use cases, such as the software interface between the glucose monitoring wristband and the cell phone.

2. Prioritize the list of software components based on risk.

3. Prepare a project iteration schedule based on iterations that last between two and four weeks.

The schedule should include all the tasks needed to develop a complete version of the system, which will then be subjected to live testing and evaluation by real users for three months.

4. Prepare a detailed work schedule for the first iteration. If you have access to project management software, prepare the schedule and a Gantt chart by using the software.

Further Resources

Scott W. Ambler, *Agile Modeling: Effective Practices for eXtreme Programming and the Unified Process*. John Wiley and Sons, 2002.

Charles G. Cobb, *Making Sense of Agile Project Management: Balancing Control and Agility*. John Wiley and Sons, 2011.

Jim Highsmith, *Agile Project Management: Creating Innovative Products*. John Wiley and Sons, 2009.

Gopal K. Kapur, *Project Management for Information, Technology, Business, and Certification*. Prentice-Hall, 2005.

Craig Larman and Bas Vodde, *Scaling Lean and Agine Development: Thinking and Organizational Tools for Large-Scale Scrum*. Addison-Wesley, 2009.

Jack R. Meredith and Samuel J. Mantel Jr., *Project Management: A Managerial Approach* (6th ed.). John Wiley and Sons, 2004.

Project Management Institute, *A Guide to the Project Management Body of Knowledge* (4th ed.). Project Management Institute, 2008.

Kathy Schwalbe, *Information Technology Project Management*, (6th ed.). Course Technology, 2009.

Robert K. Wysocki, *Effective Project Management: Traditional, Agile, Extreme*. John Wiley and Sons, 2009.

PART 5

Advanced Design and Deployment Concepts

10

Object-Oriented Design: Principles

Chapter Outline

- Object-Oriented Design: Bridging from Analysis to Implementation

- Object-Oriented Architectural Design

- Fundamental Principles of Object-Oriented Detailed Design

- Design Classes and the Design Class Diagram

- Detailed Design with CRC Cards

- Fundamental Detailed Design Principles

Learning Objectives

After reading this chapter, you should be able to:

- Explain the purpose and objectives of object-oriented design

- Develop UML component diagrams

- Develop design class diagrams

- Use CRC cards to define class responsibilities and collaborations

- Explain some of the fundamental principles of object-oriented design

OPENING CASE

New Capital Bank: Part 1

Bill Santora, the project leader responsible for developing an integrated customer account system at New Capital Bank, just met with the review committee and finished a technical review of the new system's first-cut design. This first-cut design focused on four core use cases that were chosen as fundamental and would be implemented in the first development iteration.

New Capital Bank had been using object-oriented techniques for quite a while, but it had been slower to adopt some of the newer Agile approaches. Bill had been involved in some early pilot projects that had utilized Unified Modeling Language (UML) to develop systems using object-oriented techniques. However, this development project was his first large-scale project that would be entirely Agile.

As Bill was collecting his presentation materials, his supervisor, Mary Garcia, spoke to him.

"Your technical review went very well, Bill," she said. "The committee found only a few minor items that need to be fixed. And even though I am not completely current on the new approach, it was easy for me to understand how these core functions will work. I still find it hard to believe that you will have these four pieces implemented in the next few weeks, though."

"Wait a minute," Bill said, laughing. "It won't be ready for the users then. Getting these four core functions coded and running doesn't mean that we are almost done. This project is still going to take a year to complete."

"Yes, I know," Mary said. "But it is nice that we will have something to show after only one month. Not only do I feel more confident in this project, but the users love to see things developing."

"I know," Bill said. "Remember how much grief I got when I originally laid out this plan based on an iteration approach? It was difficult to detail the project schedule for the later iterations, so I had a hard time convincing everybody that the project schedule wasn't too risky. The upside is that because each iteration is only four weeks long, we have something to show right at the beginning. You don't know how relieved I am that the design passed the review! The team has done a lot of work to make sure the design is solid, and we all felt confident."

"Well, building it incrementally makes a lot of sense and certainly seems to be working," Mary said. "I especially liked the diagrams you showed. It was terrific how you showed that the three-layer architectural design supported each use case. Even though I don't consider myself an advanced object-oriented technician, I could understand how the object-oriented design fit into the architecture. I think you wowed everybody when you demonstrated how you could use the same basic design to support our internal bank tellers and a Web portal for our customers. Congratulations."

Bill picked up on Mary's enthusiasm.

"How about the design class diagrams?" he asked. "Don't they give a nice overview of the classes and the methods? We use them extensively as a focus for discussion on the team. They really help the programmers write good, solid code."

"By the way, have you scheduled a review with the users?" Mary asked.

"No, not yet," Bill replied. "The architectural design is mostly technical stuff, and we aren't quite ready to meet with the users. The users will help us by verifying our understanding of the information availability, but much of what we do is too technical for them to follow."

"I am excited to see the first pieces run," Mary said. "It just makes so much sense to be able to test these core functions during the rest of the project. Let me congratulate you again." Then, they headed off to lunch together.

Overview

In Chapters 3, 4, and 5, you learned how to do object-oriented analysis by developing functional requirements models. You learned that analysis consists of two parts: discovery and understanding. Understanding is taking the information gleaned from user interviews and constructing a set of interrelated and comprehensive models. Model building is an essential part of understanding the user needs and how they influence the proposed system. However, remember that the objective of analysis models isn't to describe the new system but to understand—in precise terms—the requirements.

This chapter and the next focus on how to develop object-oriented design models based on the requirements models, which are then used by the programmers to code the system. This chapter focuses on two levels of design introduced in Chapter 6: architectural design—often referred to as high-level design—and detailed design—where the design of each use case is specified. This chapter starts by teaching the models and processes required to develop an overall architectural structure for the new system. The primary UML diagram that is discussed is a component diagram, which is based on the ideas you learned in Chapter 6 about the deployment environment.

In the latter part of this chapter, you will begin learning the process of detailed design. We first explain design class diagrams, which are an extension of the problem domain class diagram, with design information added. Next, we explain class responsibility collaboration (CRC) cards to begin teaching the details of use case–centered, object-oriented design.

This chapter ends with an important discussion of design principles for good object-oriented design. Throughout this chapter and the next, we discuss not only the basics of object-oriented design but also the teaching principles underlying it so that the systems you build are well structured and maintainable. The design principles will provide you with a solid foundation for designing systems correctly.

Object-Oriented Design: Bridging from Analysis to Implementation

So, what is object-oriented design? It is a process by which a set of detailed object-oriented design models are built and then used by the programmers to write and test the new system. Systems design is the bridge between user requirements and programming the new system. One strength of the object-oriented approach is that the design models are often just extensions of the requirements models. Obviously, it is much easier to extend an existing model than to create entirely new design models. However, it is a good practice to create design models and not just jump into coding. Just as a builder doesn't build something larger than a doghouse or a shed without a set of blueprints, a system developer would never try to develop a large system without a set of design models.

One tenet of Agile, adaptive approaches to development is to create models only if they have meaning and are necessary. Sometimes, new developers misinterpret this guideline to mean that they don't need to develop design models at all. The design models may not be formalized into a comprehensive set of documents and diagrams, but they are certainly necessary. Developing a system without doing design is comparable to writing a research paper without an outline. You could just sit down and start writing; however, if you want a paper that is cohesive, complete, and comprehensive, you should write an outline first. You could write a complex paper without an outline, but in all probability, it would be disjointed, hard to follow, and missing important points—and it would earn a low grade! The outline can be jotted down on paper, but the process of thinking it through and writing it down allows the writer to ensure that it is cohesive. Systems design provides the same type of framework.

One important point about adaptive approaches is that requirements, design, and programming are done in parallel within iterations. Thus, a complete set of design documents isn't developed at one time. The requirements for a particular use case or several use cases may be developed and then the design documents are developed for that use case. Immediately following the design of the solution, the programming can be done. Some people call this "just in time" systems design.

Overview of Object-Oriented Programs

Let us quickly review how an object-oriented program works. Then, we will discuss the design models and how they must be structured to support object-oriented programming (OOP).

An object-oriented program consists of a set of program objects that cooperate to accomplish a result. Each program object has program logic and any necessary attributes encapsulated into a single unit. These objects work together

FIGURE **10-1**
Object-oriented event-driven program flow

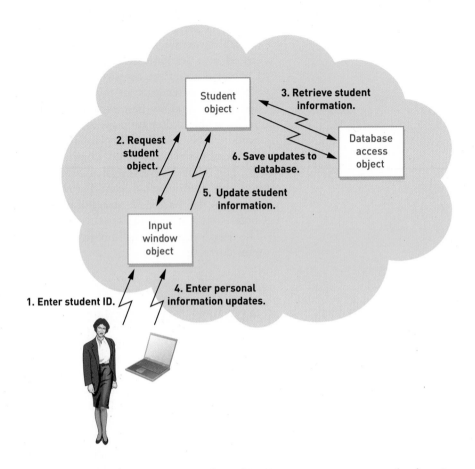

by sending each other messages and working in concert to support the functions of the main program.

Figure 10-1 depicts how an object-oriented program works. The program includes a window object that displays a form in which to enter student ID and other information. After the student ID is entered, the window object sends a message (number 2) to the Student class to tell it to create a new student object (instance) in the program and to go to the database, get the student information, and put it in the object (message 3). Next, the student object sends the information back to the window object to display it on the screen. The student then enters the updates to her personal information (message 4), and another sequence of messages is sent to update the student object in the program and the student information in the database.

An object-oriented system consists of sets of computing objects. Each object has data and program logic encapsulated within itself. Analysts define the structure of the program logic and data fields by defining a class. The class definition describes the structure or a template of what an executing object looks like. The object itself doesn't come into existence until the program begins to execute. This is called an **instantiation** of the class—that is, making an instance (an object) based on the template provided by the class definition.

instantiation creation of an object based on the template provided by the class definition

Figure 10-1 illustrates three objects in this simple program execution. Each object also represents a structure of three-layer architecture. The three objects don't need to exist on the same machine. In fact, in a multilayer architecture, the three classes of objects will generally exist on three separate machines. You learned about multilayer architectures in Chapter 6.

Object-Oriented Design Models and Processes

The objective of object-oriented design is to identify and specify all the objects that must work together to carry out each use case. As shown in Figure 10-1, these objects include user-interface objects, problem domain objects, and data access objects. In addition to simply identifying the objects, you need to specify the detail methods and attributes within the objects so a programmer can understand how the objects collaborate to execute a use case.

Figure 10-2 illustrates which requirements models are directly used to develop which design models. The requirements models in the left column were developed during analysis, and the design models in the right column are the ones we will develop during design. As you might infer from the number of arrows pointing to them, interaction diagrams are the core diagrams used for detailed design. They are either sequence diagrams (as illustrated in the figure) or communication diagrams.

At this point, you should be familiar with the requirements, but let us take a minute to review them. The domain model class diagram identifies all the classes, or "things," that are important in the problem domain. The use case diagrams identified the elementary business processes that the system needs to support—in other words, all the ways users want to use the system to carry out processing goals. The activity diagrams and use case descriptions document the internal workflow of each use case. An activity diagram shows the steps

FIGURE **10-2**

Design models with their respective input requirements models

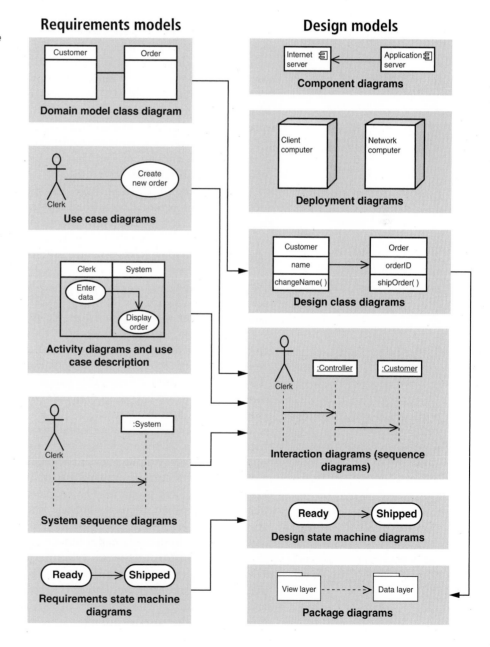

necessary to carry out a particular use case. The system sequence diagrams are closely related to activity diagrams, except that they show the messages or data that are sent back and forth between the user and the system during the steps of the use case. Finally, state machine diagrams keep track of all status condition requirements for one particular class. They also show the business rules that control the changing of one state (status condition) to another.

The right side of Figure 10-2 shows the design models. Architectural design is one of the first steps in systems design, inasmuch as it provides the big picture and overall structure of the new system. At the top of the right column are component diagrams and deployment diagrams. You learned a bit about the deployment environment in Chapter 6—but without using official UML notation. This chapter explains how to draw UML component diagrams. You also learned a little about multilayer software design in Chapter 6. This chapter explains how to use these diagrams to document the architectural design of the software system.

Moving down the right column of Figure 10-2, we next see design class diagrams, or DCDs, which are an expansion of the domain model class diagram. They are used to document the design elements of software classes, as you will learn later in this chapter. Recall that the classes in the domain model class diagram aren't software classes, but the classes in the DCD are indeed software classes. Domain classes are the basis for designing software classes. Next are interaction diagrams, which can be either UML sequence diagrams or UML communication diagrams. You learned how to build system sequence diagrams (SSDs) in Chapter 5. The design version of sequence diagrams is much more detailed and is used to carry out much of the detailed design activity. You will learn how to create interaction diagrams in Chapter 11. State machine diagrams are also used by programmers to develop the detailed class methods. You already learned about those in Chapter 5. The design version is similar to the version used during analysis.

Finally, there are UML package diagrams, which are simply a way to group design elements together. Package diagrams can be used to group any type of design elements, but we will use them primarily to group design classes in Chapter 11.

We begin systems design by thinking about the overall structure of the new solution system—that is, the architecture.

Object-Oriented Architectural Design

Usually, the first step in systems design is architectural design. In most cases, developers begin to think about how the system will be deployed and what the overall structure will look like during the early steps of requirements gathering and documentation. At the beginning of a project, it is normal to say "This is a Web-based system" or "This will only be used internally on our network and desktops." Those comments are the beginning of the architecture design of the solution system.

Software systems are generally divided into two types: single-user systems and enterprise-level systems. Single-user systems are found on a single desktop or execute from a server without sharing resources. Typical examples are a spreadsheet program, an engineering drawing program, a simple accounting program, or even an e-mail client program. The architectural design of a single-user system is usually simple. Often, there is only one layer, which runs on a single computer. However, even for a single-user system, it is wise to develop the system as a multilayer program so the boundaries between the various levels are well defined.

The term **enterprise-level system** can mean many different things. We define it as a system that has shared components among multiple people or groups in an organization. Enterprise-level systems almost always use client-server architectures with multiple layers. You learned about client-server and three-layer architectures in Chapter 6. A typical example of this architecture is

enterprise-level system a system that has shared resources among multiple people or groups in an organization

an internal-networked client-server environment in which the client computers contain the view and domain layer programs and the data access layer is on a central server. Characteristic of enterprise-level systems, the database and data access are on a central server because it is a shared resource throughout the organization. Because the central database is shared across the enterprise, it is placed on a central server that all users of the application program can share. Because local client computers are often powerful workstations, the view layer and domain logic can be executed locally.

Our definition of an enterprise-level system is a broad one. Two major categories of systems fit this definition in relation to systems design: (1) client-server network-based systems and (2) Internet-based systems.

These two methods of implementing enterprise-level systems have many similar properties. Both require a network, both have central servers, and both have the view layer on the client machines. However, some fundamental differences also exist in the design and implementation of these two approaches. The primary difference is in how the view layer interacts with the domain and data access layers. As developers, we must be able to distinguish between these two types of systems because we must consider important design issues.

Figure 10-3 identifies three fundamental differences that affect the architectural design of the system: state, client configuration, and server configuration. The concept of state relates to the permanence of the connection between the client view layer and the server domain layer. If the connection is permanent, as in a client/server system, values in variables can be passed back and forth and are remembered by each component in the system. The view layer has direct access to the data fields in the domain layer.

In a stateless system, such as the Internet, the client view layer doesn't have a permanent connection to the server domain layer. The Internet was designed so that when a client requests a screen via a URL address typed in the browser, the server sends the appropriate document and then the two disconnect. In other words, the client doesn't know the state of the server, and the server doesn't remember the state of the client. This transient connection makes it difficult to implement such things as a sale in a shopping cart. To add more permanence to the stateless environment, Web designers have developed other techniques, such as cookies, session variables, and XML data transmission. As a systems designer, you must consider these additional components when designing an Internet enterprise-level system.

Concerning client configuration, the client side of a network-based system contains the view layer classes and often the domain layer classes. Formatting, displaying, and event processing within the screens are all directly controlled by the view layer and domain layer program logic. There is great flexibility in the design and programming of these electronic screens. The view layer classes and domain layer classes can communicate directly with each other.

FIGURE **10-3**

Differences between client/server and Internet systems

Design Issue	Client/Server Network System	Internet System
State	"Stateful" or state-based system—e.g., client/server connection is long term.	Stateless system—e.g., client/server connection is not long term and has no inherent memory.
Client configuration	Screens and forms that are programmed are displayed directly. Domain layer is often on the client or split between client and server machines.	Screens and forms are displayed only through a browser. They must conform to browser technology.
Server configuration	Application or data server directly connects to client tier.	Client tier connects indirectly to the application server through a Web server.

In an Internet-based system, all information is displayed by a browser. The formatting, displaying, and event processing all must conform to the capabilities of the browser being used. Special techniques and tools, such as scripting languages, applets, and style sheets, have been developed to simulate the network-based capability.

The server configuration in a network-based system consists of data access layer classes and sometimes domain layer classes. These classes collaborate through direct communication and access to each other's public methods. In an Internet-based system, all communications from the client tier must go through the HTTP server. Communication isn't direct, and methods and program logic are invoked indirectly through passed parameters. This indirect technique of accessing domain layer logic is more complex and requires additional care in designing the system.

Component Diagrams and Architectural Design

component diagram a type of implementation diagram that shows the overall system architecture and the logical components within it

The **component diagram** identifies the logical, reusable, and transportable system components that define the system architecture. The essential element of a component diagram is the component element with its interfaces.

application program interface (API) the set of public methods that are available to the outside world

A component is an executable module or program, and it consists of all the classes that are compiled into a single entity. It has well-defined interfaces, or public methods, that can be accessed by other programs or external devices. The set of all these public methods that are available to the outside world is called the **application program interface (API)**. **Figure 10-4** illustrates the UML notation for a component and its interfaces. It isn't necessary to list all the interfaces on a single component. Only those that are pertinent to the context of the diagram are listed. There are two ways to represent a component: as a general class or as a specific instance. The same rules apply in this situation that applied to class and object notation: A general class uses the name of the component class, and an instance name is underlined. The name of the component is written inside.

FIGURE **10-4** *UML component diagram notation*

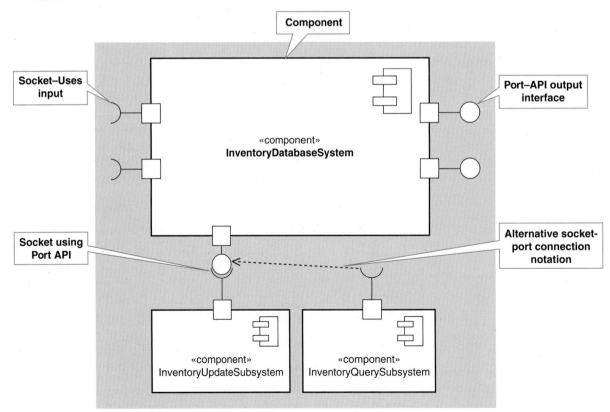

The top rectangle in Figure 10-4 illustrates the notation for a component, along with its interfaces. The icon in the upper-right corner is a small rectangle with two plugs extending from the left side, signifying that it is a moveable, executable component and is possibly reusable and pluggable.

The figure also shows two types of interfaces: an output port and an input socket. The output port is similar to a programming interface; it defines the method names (e.g., a portion of the API) that can be used by other components to access the component's functions. The input socket represents the services that the component needs from other components. Notice the ball and socket notation. They go together so the input of one component precisely fits the output of another component.

The bottom portion of the figure shows how the port interfaces and sockets can be used in a component diagram. The InventoryDatabaseSystem component at the top presents an interface to the world, as denoted by the port API interface along its bottom edge. The InventoryUpdateSubsystem component uses that interface to access the methods of the InventoryDatabaseSystem. In this figure, we show via a dashed arrow that the InventoryQuerySubsystem component accesses the same interface.

In our design examples, we would like to show how to do multilayer design for a Web-based system, and we would like to illustrate the locations of various Web pages. In other words, we want to have some notation to show where Web pages reside and are deployed. Because UML notation doesn't have standard notation for a window, we extend the notation to include it. UML does have rules for stereotyping a symbol and for extending the language. **Figure 10-5** shows a Web page notation.

The figure simply displays a class notation with a stereotype notation, such as ≪GUI≫ and ≪frameset≫, along with a small window icon in the top-right corner. This notation will serve for either a desktop system window or a Web system frameset. A frameset is a high-level object that can hold items to be displayed by a browser. We will use a frameset notation and stereotype to indicate a Web page. You can think of a frameset as the window in a browser that can display frames or a set of frames.

Next, let us use the component and window notation to do an architectural design of some straightforward Internet systems.

Two-Layer Architectural Design of Internet Systems

Many colleges have courses in Web development. Most of those courses fall into one of two categories: Web site design or Web programming languages. Both classes are important and beneficial for your education as a system developer.

Web programming courses teach you various programming languages and the ways to insert program logic into Web pages. You learn JavaScript, VBScript, PHP, and ASP (Active Server Pages). You may learn how to use advanced database tools, such as Cold Fusion, to access databases from your pages. You may also learn how the browser and server work together to serve

FIGURE **10-5**

Extension notation for GUI window and Web page

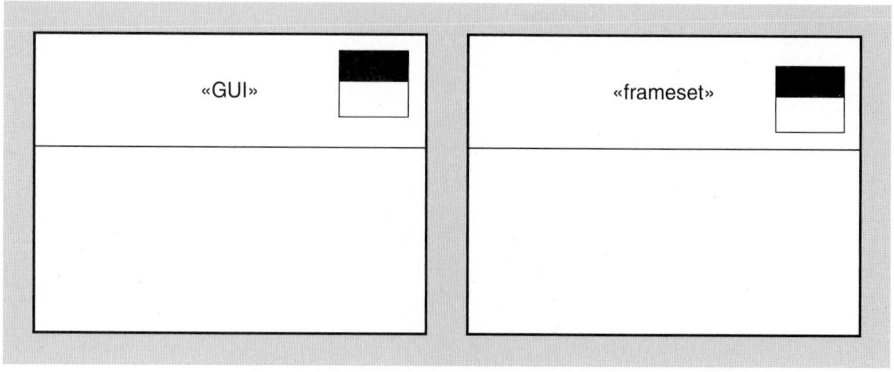

up pages that have sufficient programming logic to support the business application. Advanced versions of this course even teach you the Java or .NET environments so you can configure an entire application.

We don't intend for this short section to replace that course. Instead, we introduce the architecture of these Web-based systems and provide a few principles of good design that you can apply as you develop skills in other courses. In Chapter 6, we explained three-layer design as one effective approach to developing robust, easily maintainable systems. But how can designers implement a three-layer design in a Web-based architecture? This question is particularly important if an organization wants to use the same problem domain logic for both types of enterprise systems: a client/server system and an Internet-based system.

Figure 10-6 illustrates a simple, generic Internet architecture. Remember, we are doing logical design at this point and aren't yet concerned with the physical computer configuration. We will discuss the computer configuration when we learn more about deployment diagrams in the next section. Of course, because there is an Internet cloud between two components, we can naturally assume they are in different physical locations.

Figure 10-6 includes four recognizable components. The browser is an executable component whose purpose is to format, display, and execute active code, such as JavaScript or ActiveX Controls. The Internet Server is an executable component whose purpose is to retrieve pages and invoke other components. The diagram in Figure 10-6 shows two additional examples of components: executable programs in the Common Gateway Interface (CGI) and the Application Server.

In the interest of brevity, the ports and sockets have been omitted from this diagram. The interfaces between these components are industry standard, and

FIGURE **10-6** *Component diagram showing two-layer Internet architecture*

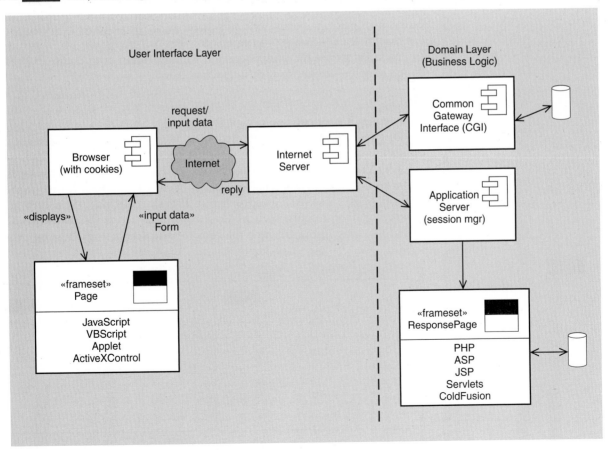

the unique port/socket combination doesn't need to be emphasized here. However, every two-headed arrow does represent two port/socket pairs.

As indicated in Chapter 6, many simple business systems can be designed as two-layer systems. These systems primarily capture information from the user and update a database. No complex domain layer logic is required. In those instances, the domain layer and data access layer are usually combined. The business logic in the domain layer frequently relates only to data formatting and to deciding which database table to update. Many business applications fall into this category. For example, a simple address book system could be easily designed as a two-layer system.

The CGI was the original way to process input data. The CGI directory contains compiled programs that are available to receive input data from the server. The programs in the CGI directory can be written in any compiled language, such as C++. This technique is effective and usually has quick response and processing times. The only downside is that these programs can be quite complex and difficult to write. They process the input data, access any required database, and format a response page in HTML, as indicated by the ResponsePage in the diagram.

The other potential direction for input data is directly to a URL for a Web page with embedded program code. The extension shown on the ResponsePage indicates the type of program code embedded in the page: ASP for Active Server Pages, PHP for PHP Hypertext Preprocessor, JSP for Java Server Pages, CFM for Cold Fusion Pages, and so forth. Depending on the type of extension, an application server—which is the language processor—is invoked to process the embedded code. The embedded code, via the Application Server, can process the code, including reading and writing to a database. The Application Server, working with cookies on the browser, can manage sessions with the user. Session variables are set up to maintain information about the user across multiple page requests and forms. The Application Server also formats the ResponsePage based on the HTML statements and the code and forwards it back to the Internet Server.

Even though we have referred to this design as two-layer architecture, the user interface classes often contain the business logic and data access. Due to the structure of Web servers, the program (defined as object-oriented classes) that processes the input forms also outputs the HTML code that is sent back to the client browser. For example, in Java-based systems, Java servlets receive the data from input on a Web form, process the data, and format the output HTML page. To process the data, the servlets usually include any required business logic and data access logic. The .NET environment is similar. For every Web form, there is a code-behind class written in Visual Basic, C#, J#, or some similar language. Another popular language combination is PHP, which executes on the server, and JavaScript, which executes in a browser. The code-behind object receives the data from the Web form, processes the data, and formats the output HTML page. Thus, it often isn't clear whether the architecture is one layer or two layers. However, we refer to this architecture as two-layer to emphasize that it is dynamic and that the HTML response pages are built dynamically.

This architecture works well for two-layer applications that aren't too complex—for example, when the response pages already have most of the HTML written. The embedded code performs such functions as validating the data and storing it in the database. Note that the business logic is minimal, so mixing it with the data access logic still provides a maintainable solution.

Several integrated development environments (IDEs) for developing Web-based systems are based on a two- or three-layer architecture called Model-View-Controller (MVC). You will learn more about design patterns and MVC in Chapter 11. The basic idea is that even a Web-based system can be developed in multiple layers of the view—that is, the user interface, the model (which is the business logic and database access), and the controller (which provides a link between the view and the model).

As noted in Chapter 2, RMO's new CSMS needs to support a Web user interface and an internal desktop interface. It is critically important for the same back end—business logic and database access—to link with either user interface. Consequently, the design team must specify the architectural design in enough detail to ensure the programmers implement a system that can support both user interfaces.

Once the architectural design is determined, it is time to drill down to a lower level of abstraction. In other words, you stop treating the logical components as black boxes and start to look inside. Each component is an executable program and is made up of classes. So, the next step in the application design is to begin defining the interaction of design classes for each use case. Designing at this level is usually called detailed design.

Fundamental Principles of Object-Oriented Detailed Design

Now that we have learned about architectural design, we can turn to detailed design. If you refer back to Figure 10-2 and proceed down the right column, the next two diagrams to discuss are the design class diagram and the interaction diagrams (sequence diagrams and communication diagrams). These two diagrams are the most important for detailed design.

The objective of object-oriented detailed design is to identify and specify all the objects that must work together to carry out each use case. As shown in Figure 10-1, there are user-interface objects, problem domain objects, and database access objects. As you may suppose, a major responsibility of detailed design is to identify and describe each set of objects within each layer and to identify and describe the interactions or messages that are sent between these objects.

The most important model in object-oriented design is a sequence diagram—or its first cousin, a communication diagram. In Chapter 5, you learned to develop system sequence diagrams (SSDs) to model input and output requirements for a use case. The full sequence diagram is used for design and is a type of interaction diagram. A communication diagram is also a type of interaction diagram. During design, developers extend the SSD by modifying the single :System object to include all the interacting user interface, problem domain, and database access objects. In other words, they look inside the :System object to see what is happening inside the system. We will spend a good deal of time in Chapter 11 learning how to develop these detailed sequence diagrams. **Figure 10-7** shows a simple sequence diagram based on Figure 10-1, which

FIGURE **10-7**

Sequence diagram for updating student name

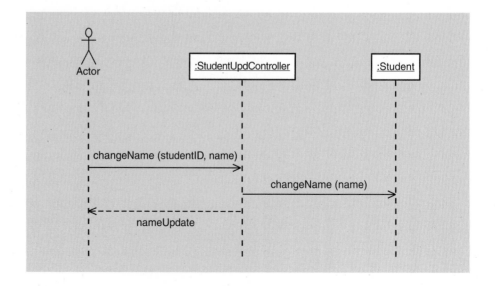

FIGURE **10-8** *Student class examples for the domain class and the design class diagrams*

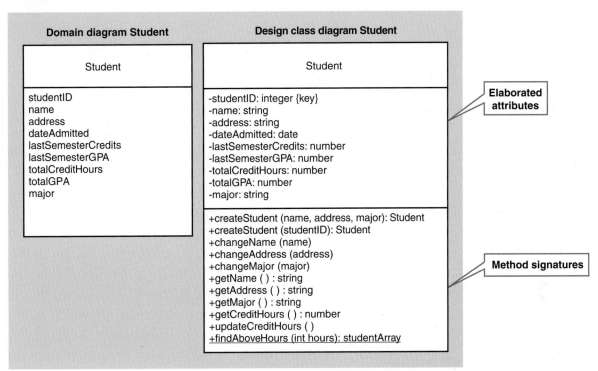

updates student information. A sequence diagram uses the same notation as an SSD, which you learned to develop in Chapter 5.

The other major design model, which you will learn to develop later in this chapter, is the design class diagram. Its main purpose is to document and describe the programming classes that will be built for the new system. It describes the set of object-oriented classes needed for programming, navigation between the classes, attribute names and properties, and method names and properties. A design class diagram is a summary of the final design that was developed by using the detailed sequence diagrams, and it is used directly when developing the programming code. **Figure 10-8** shows the original domain model that was developed during analysis and the design class diagram version of that class. The design class version has a new compartment at the bottom that specifies the method signatures for the class. The attributes have also been enhanced. We explain the details of this notation in the next section. Detailed design is the process that takes the domain model to the design class model.

As an object-oriented systems designer, you must provide enough detail so a programmer can write the initial class definitions, including the method code. For example, a design class specification helps define an object's attributes and methods. **Figure 10-9(a)** illustrates some sample code, written in Java, for the Student class. **Figure 10-9(b)** shows the same sample code written in Visual Basic .NET. Referring back to Figure 10-8, you should be able to see how the design class provides the input to write the code for Figure 10-9. Notice that the class name, the attributes, and the method names are derived from the design class notation.

Object-Oriented Design Process

Object-oriented design is model-driven and use case–driven. As you saw in Figure 10-2, the design process takes the requirements models as input and produces the design models as output. Obviously, we need a process for organizing this activity, and it is focused around use cases. In other words, we develop the design models use case by use case. For example, a design sequence diagram is developed for each use case. After a group of them has been designed, the

FIGURE **10-9a** *Example class definition in Java for Student class*

```java
public class Student
{
        //attributes
        private int studentID;
        private String firstName;
        private String lastName;
        private String street;
        private String city;
        private String state;
        private String zipCode;
        private Date dateAdmitted;
        private float numberCredits;
        private String lastActiveSemester;
        private float lastActiveSemesterGPA;
        private float gradePointAverage;
        private String major;

        //constructors
        public Student (String inFirstName, String inLastName, String inStreet,
            String inCity, String inState, String inZip, Date inDate)
        {
            firstName = inFirstName;
            lastName = inLastName;
            ...
        }
        public Student (int inStudentID)
        {
            //read database to get values
        }

        //get and set methods
        public String getFullName ( )
        {
            return firstName + " " + lastName;
        }
        public void setFirstName (String inFirstName)
        {
            firstName = inFirstName;
        }
        public float getGPA ( )
        {
            return gradePointAverage;
        }
        //and so on

        //processing methods
        public void updateGPA ( )
        {
            //access course records and update lastActiveSemester and
            //to-date credits and GPA
        }
}
```

FIGURE **10-9b** *Example class definition in VB .NET for Student class*

```vb
Public Class Student

    'attributes
    Private studentID As Integer
    Private firstName As String
    Private lastName As String
    Private street As String
    Private city As String
    Private state As String
    Private zipCode As String
    Private dateAdmitted As Date
    Private numberCredits As Single
    Private lastActiveSemester As String
    Private lastActiveSemesterGPA As Single
    Private gradePointAverage As Single
    Private major As String

    'constructor methods
    Public Sub New(ByVal inFirstName As String, ByVal inLastName As String,
         ByVal inStreet As String, ByVal inCity As String, ByVal inState As String,
         ByVal inZip As String, ByVal inDate As Date)
      firstName = inFirstName
      lastName = inLastName
      ...
    End Sub

    Public Sub New(ByVal inStudentID)
       'read database to get values
    End Sub

    'get and set accessor methods
    Public Function GetFullName() As String
       Dim info As String
       info = firstName & " " & lastName
       Return info
    End Function

    Public Property firstName()
       Get
           Return firstName
       End Get
       Set(ByVal Value)
       firstName = Value
       End Set
    End Property

    Public ReadOnly Property GPA()
       Get
           Return gradePointAverage
       End Get
    End Property

    'Processing Methods
    Public Function UpdateGPA()
       'read the database and update last semester
       'and to date credits and GPA
    End Function

End Class
```

FIGURE **10-10**
Object-oriented detailed design steps

Design Step	Chapter
1. Develop the first-cut design class diagram showing navigation visibility.	10
2. Determine class responsibilities and class collaborations for each use case using class-responsibility-collaboration (CRC) cards.	10
3. Develop detailed sequence diagrams for each use case. (a) Develop the first-cut sequence diagrams. (b) Develop the multilayer sequence diagrams.	11
4. Update the design class diagram by adding method signatures and navigation information using CRC cards and/or sequence diagrams.	11
5. Partition the solution into packages as appropriate.	11

design class diagram is completed for that entire group of use cases. We can divide the process of object-oriented design into five major steps, as summarized in **Figure 10-10**, which also shows the chapter that each step is discussed in.

First, a preliminary version, or first-cut model, of the design class diagram is created. Some basic information, such as attribute names, must be listed in the first-cut model to develop the sequence diagrams. This step provides a foundation for the second and third steps.

The second step often used by developers is to develop a set of CRC cards for each use case. These help provide an overall understanding of the internal steps required for the system to support the use case. CRC cards provide a simple method for identifying all the objects involved in a particular use case and their responsibilities. The results of a CRC activity will be sets of cards that can be used to help develop a sequence diagram; if a use case is simple enough, the cards can be used to program the use case. CRC cards are explained in detail later in this chapter.

Design Classes and the Design Class Diagram

As shown in Figure 10-2, the design class diagrams and the detailed sequence diagrams work together. A first iteration of the design class diagram is created based on the domain model and on software design principles. The preliminary design class diagram is then used to help develop sequence diagrams. As design decisions are made during development of the sequence diagrams, the results are used to refine the design class diagram.

The domain model class diagram shows a set of problem domain classes and their associations. During analysis, because it is a discovery process, analysts generally don't worry much about the details of the attributes. However, in OOP, the attributes of a class must be declared as public or private, and each attribute must also be defined by its type, such as character or numeric. During detailed design, it is important to elaborate on these details as well as to define the methods and parameters that are passed to the methods and the return values from methods. Sometimes, developers also define the internal logic of each method at this point. We complete the design class diagram by integrating information from interaction diagrams and other models.

As developers build the design class diagrams, they add many more classes than were originally defined in the domain model. Referring to Figure 10-1, the Input window objects and Database access objects are examples of additional classes that must be defined. The classes in a system can be partitioned into distinct categories, such as user-interface classes. At times, designers may also develop distinct class diagrams by subsystem. We now turn to design class diagram notation and discuss the design principles used in developing the first iteration of the design class diagram.

Design Class Symbols

UML doesn't specifically distinguish between design class notation and domain model notation. However, practical differences occur simply because the objective of design modeling is distinct from that of domain modeling. Domain modeling shows things in the users' work environment and the naturally occurring associations among them. At that point, the classes aren't specifically software classes. After we start a design class diagram, though, we are specifically defining software classes. Because many different types of design classes are identified during the design process, UML has a special notation—called a *stereotype*—that allows designers to designate a special type of class. A **stereotype** is simply a way to categorize a model element as a certain type. A stereotype extends the basic definition of a model element by indicating that it has some special characteristic we want to highlight. The notation for a stereotype is the name of the type placed within printer's guillemets, like this: «control».

Four types of design classes are considered standard: an entity class, a control class, a boundary class or view class, and a data access class. **Figure 10-11** shows the notation used to identify these four stereotypes.

An **entity class** is the design identifier for a problem domain class. It is also usually a persistent class. A **persistent class** is one with objects that exist after the program quits. Obviously, the way to make data persistent is to write it to a file or database.

A **boundary class,** or **view class**, is specifically designed to live on the system's automation boundary. In a desktop system, these classes would be the windows classes and all the other classes associated with the user interface.

A **control class** mediates between the boundary classes and the entity classes. In other words, its responsibility is to catch the messages from the boundary class objects and send them to the correct entity class objects. It acts as a kind of switchboard, or controller, between the view layer and the domain layer.

A **data access class** is used to retrieve data from and send data to a database. Rather than insert database access logic, including SQL statements, into the entity class methods, a separate layer of classes to access the database is often included in the design.

stereotype a way of categorizing a model element by its characteristics, indicated by guillemets (« »)

entity class a design identifier for a problem domain class

persistent class an entity class whose objects exist after a system is shut down

boundary class, or **view class** a class that exists on a system's automation boundary, such as an input window

control class a class that mediates between boundary classes and entity classes, acting as a switchboard between the view layer and domain layer

data access class a class that is used to retrieve data from and send data to a database

FIGURE **10-11**

Standard stereotypes found in UML design models

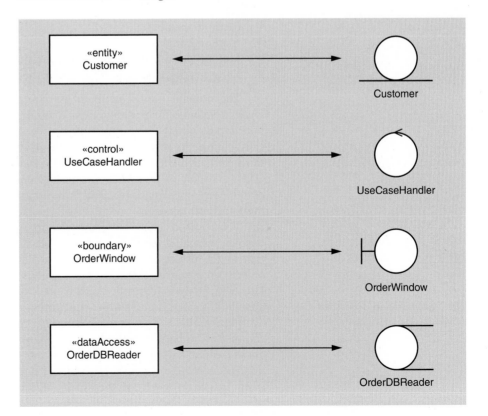

FIGURE **10-12**
Notation used to define a design class

FIGURE **10-12**

Notation used to define a design class

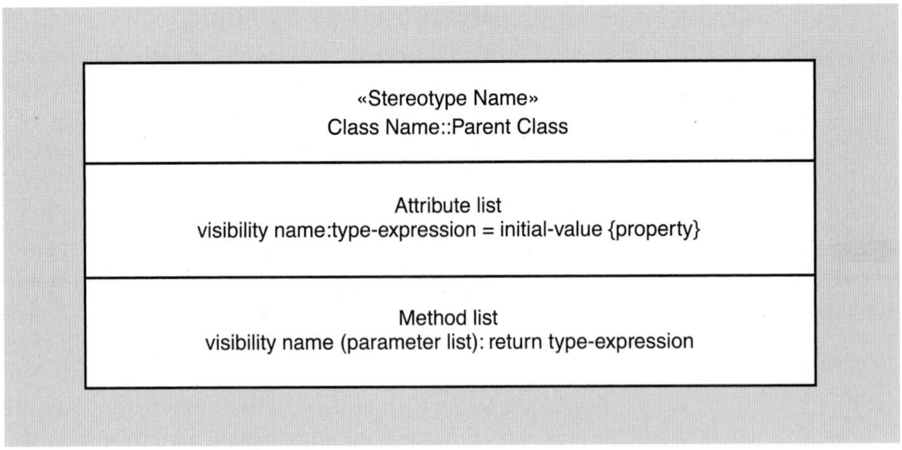

Design Class Notation

Figure 10-12 shows the details within a design class symbol, as you first saw in the design class shown in Figure 10-8. The name compartment includes the class name and the stereotype information and parent class. The lower two compartments contain more details about the attributes and the methods.

The format that analysts use to define each attribute includes:

visibility a notation that indicates (by plus or minus sign) whether an attribute can be directly accessed by another object

- Visibility—**Visibility** denotes whether other objects can directly access the attribute. (A plus sign indicates that an attribute is visible, or public; a minus sign indicates that it isn't visible, or private.)
- Attribute name
- Type-expression (such as character, string, integer, number, currency, or date)
- Initial-value, if applicable
- Property (within curly braces), such as {key}, if applicable

method signature a notation that shows all the information needed to invoke, or call, the method

The third compartment contains the method signature information. A **method signature** shows all the information needed to invoke (or call) the method. It shows the format of the message that must be sent, which consists of these:

- Method visibility
- Method name
- Method parameter list (incoming arguments)
- Return type-expression (the type of the return parameter from the method)

The domain model attribute list contains all attributes discovered during analysis activities. The design class diagram includes more information on attribute types, initial values, and properties. It can also include a stereotype for clarification. As shown in Figure 10-8 in the Student design class diagram, the third compartment contains the method signatures for the class. Remember that UML is meant to be a general object-oriented notation technique and not specific to any one language. Thus, the notation won't be the same as programming method notation.

The method called findAboveHours (int hours): studentArray, which is denoted with an underline in Figure 10-8, is a special kind of method. Remember that in the object-oriented approach, a class is a template to create individual objects or instances. Most of the methods apply to one instance of the class. However, analysts frequently need to look through all the instances at once. Such a method is called a **class-level method** and is denoted by an underline.

class-level method a method that is associated with a class instead of with objects of the class

Figure 10-13 is an example of design classes with attributes and methods; it shows how inheritance works for design classes. In Chapter 4, you learned about generalization/specialization. In the problem domain model, generalization/specialization becomes inheritance in the design model and in a programming language. Each of the three subclasses inherits all the attributes and

FIGURE **10-13** *Sale class (abstract) with three concrete subclasses showing inheritance*

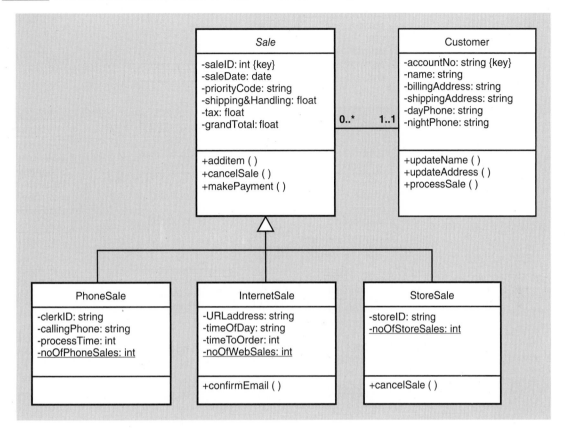

methods of the parent Sale class. Hence, each subclass has a saleID, a saleDate, and so forth. In this example, each subclass also has additional attributes that are unique to its own specific class. Each of the subclasses also has a unique attribute that is underlined, such as noOfPhoneSales. Underlined attributes are **class-level attributes** and have the same characteristics as class-level methods. A class-level attribute is a static variable, and it contains the same value in all instantiated objects of the same type.

Not only methods and attributes are inherited by the subclasses; associations are also inherited. In Figure 10-13, the Sale object must be associated with exactly one customer. Each subclass inherits the same association and must be associated with exactly one customer. Finally, notice that the title of the Sale class is italicized. An italicized class name indicates that it is an **abstract class**— a class that can never be instantiated. In other words, there are never any Sale class objects. All orders in the system must be instantiated as one of the three subclasses. Every order in the system will be either a PhoneSale, an InternetSale, or a StoreSale. Each of the three subclasses is considered a **concrete class** because it can be instantiated; in other words, objects can be created. The purpose of an abstract class is illustrated by the figure. It provides a central holding place for all the attributes and methods that each of the three subclasses will need. This example demonstrates one way that OOP implements reuse. The methods and attributes in the abstract class only need to be written once in order to be reused by each of the subclasses.

class-level attribute an attribute that contains the same value for all objects in the system

abstract class a class that can never be instantiated (no objects can be created of this type)

concrete class a class that can be instantiated (objects can be created of this type)

Developing the First-Cut Design Class Diagram

To illustrate how to start the design process, we develop a first-cut design class diagram based on the domain model. **Figure 10-14** is a partial RMO domain model class diagram, as developed in Chapter 4.

FIGURE **10-14**

Partial RMO Sales subsystem domain model class diagram

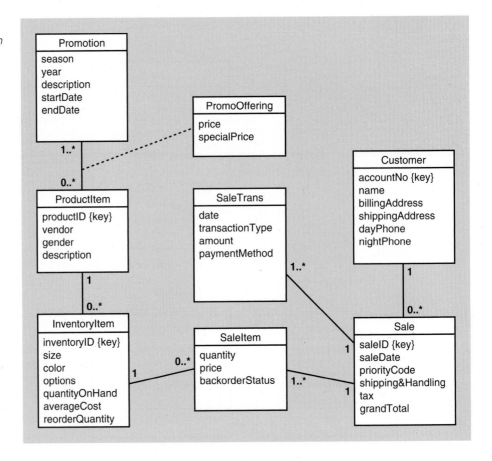

The first-cut design class diagram is developed by extending the domain model class diagram. It requires two steps: (1) elaborating on the attributes with type and initial value information and (2) adding navigation visibility arrows. As indicated earlier, object-oriented design is use case–driven. So, let us choose a use case to start with and focus only on classes involved in that use case.

Elaboration of Attributes

The elaboration of the attributes is fairly straightforward. The type information is determined by the designer based on his or her expertise. In most instances, all attributes are kept invisible or private and are indicated with minus signs before them. We also need to add a new compartment to each class for the addition of method signatures.

Navigation Visibility

As stated earlier, an object-oriented system is a set of interacting objects. The sequence diagrams document what interactions occur between which objects. However, for one object to interact with another by sending a message, the first object must be visible to the second object. In this context, **navigation visibility** refers to the ability of one object to view and interact with another object. We use two types of navigation visibility during design: attribute navigation visibility and parameter navigation visibility. Attribute navigation visibility occurs when a class has an attribute that references another object. Visibility is obtained through the attribute reference. Parameter navigation visibility occurs when a class is passed a parameter that references another object. A parameter is usually passed through a method call. Sometimes, developers refer to navigation visibility as just *navigation* or *visibility*. However, we prefer the term *navigation visibility* to distinguish the concept from public and private visibility on attributes and methods.

navigation visibility a design principle in which one object has a reference to another object and thus can interact with it

FIGURE **10-15**

Attribute navigation visibility between Customer and Sale

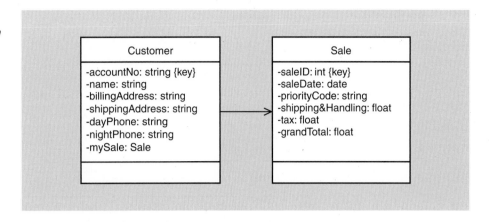

Figure 10-15 shows one-way attribute navigation visibility between the Customer class and the Sale class. Notice the variable called mySale in the Customer class. This variable holds a value to refer to a Sale instance. The navigation arrow indicates that a Sale object must be visible to the Customer object. We have included the mySale attribute in the example to emphasize this concept.

Now let us think about adding navigation visibility to the RMO design class diagram. Remember that we are designing just the first-cut class diagram, so we might need to modify the navigation arrows as the design progresses. We ask the following basic question when building navigation visibility: Which classes need to have references to or be able to access which other classes? Here are a few general guidelines:

■ One-to-many associations that indicate a superior/subordinate relationship are usually navigated from the superior to the subordinate—for example, from Sale to SaleItem. Sometimes, these relationships form hierarchies of navigation chains—for example, from Promotion to ProductItem to InventoryItem.

■ Mandatory associations, in which objects in one class can't exist without objects of another class, are usually navigated from the more independent class to the dependent class—for example, from Customer to Sale.

■ When an object needs information from another object, a navigation arrow might be required, pointing either to the object itself or to its parent in a hierarchy.

■ Navigation arrows may be bidirectional.

Figure 10-16 is a first-cut design class diagram for the use case *Create phone sale* based on the two steps described earlier in this section. The first step is to elaborate on the attributes with type information and visibility. The second step is to identify which classes may be involved and which classes require navigation visibility to other classes. We identify the classes that appear to be necessary to carry out the use case. We determine what other classes are necessary based on what information is needed. For example, price information is in the PromoOffering class and description information is in the ProductItem class. One thing to remember about visibility is that the classes are programming classes, not database classes. So, we aren't thinking about foreign keys in a relational database. We are thinking about object references in a programming language.

Figure 10-16 has one additional design class in the diagram for this use case—SaleHandler—which is stereotyped as a controller class. As mentioned previously, a controller class, or use case controller, is a utility class that helps in the processing of a use case. Notice that it has visibility at the top of the visibility hierarchy.

FIGURE **10-16**
First-cut RMO design class diagram for the Create phone sale *use case*

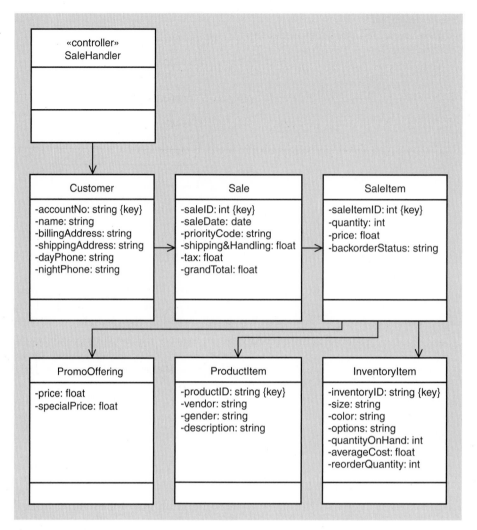

Three points are important to note. First, as detailed design proceeds use case by use case, we need to ensure that the sequence diagrams support and implement the navigation that was initially defined. Second, the navigation arrows need to be updated as design progresses to be consistent with the design details. Finally, method signatures will be added to each class based on the design decisions made when creating the sequence diagrams for the use cases.

As a preliminary step before developing sequence diagrams, many developers like to use CRC cards in brainstorming sessions to help identify the sets of classes involved in each use case. The next section explains how CRC cards can help with detailed object-oriented design.

Detailed Design with CRC Cards

CRC cards *a brainstorming technique for designing interactions in use cases by assigning responsibilities and collaborations for classes*

CRC cards are a brainstorming technique that is quite popular among object-oriented developers. Here, "CRC" is an acronym for Class Responsibility Collaboration. Developers use this technique during design to help identify responsibilities of the class and the sets of classes that collaborate for a particular use case.

A CRC card is simply a 3 × 5 or 4 × 6 index card with lines that partition it into three areas: class name, responsibility, and collaboration classes. **Figure 10-17** illustrates the two sides of a CRC card from the RMO CSMS. The card is partially filled out. Along the top of the card is the name of the class. The left partition lists the responsibilities for objects in this class. Responsibilities

FIGURE **10-17** *Example CRC card (front and back)*

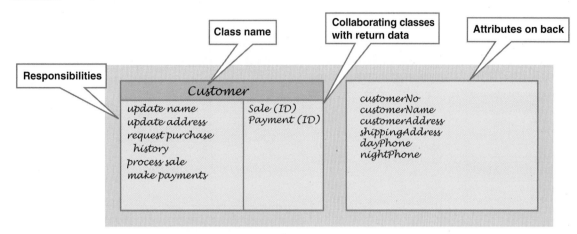

include information that the class maintains and actions that the class carries out in support of some use case. The right partition lists other classes with which this class collaborates in support of a particular use case. The information within parentheses is return information from the collaborating class to the main class. On the back of the card, you have the option of listing important attributes that are required for a particular use case.

The process of developing a CRC model is usually done in a brainstorming session. A design session using CRC cards already has substantial information from which to begin. Before starting the design session, each team member should have a copy of the domain model class diagram. Of course, the use case diagram or list of use cases also needs to be available. Other detailed information, such as activity diagrams, system sequence diagrams, and use case descriptions, should be provided, along with a stack of blank CRC-formatted index cards. For each use case you need to design, these steps are done iteratively:

■ Selecting a use case—Because the process is to design, or realize, a single use case, start with a set of unused CRC cards. Because we are doing multilayer design, make up one card as the use case controller card.

■ Identifying the problem domain class that has responsibility for this use case—This object will receive the first message from the use case controller. Using the domain model that was developed during analysis, select one class to take responsibility. Focus only on the problem domain classes. On the left side of the card, write the object's responsibility. For example, a Customer object may take responsibility to make a new sale, so one responsibility may be *Create phone sale*.

■ Identifying other classes that must collaborate with the primary object class to complete the use case—Other classes will have required information. For our example of creating a sale, we will need a Sale class card and a SaleItem class card; pricing information will probably come from the PromoOffering class; and the InventoryItem class will have to be checked for stock on hand. List these classes on the primary problem domain card. As you identify the other classes, write their responsibilities on their cards. Also, on the backs of all cards, write the pertinent information or attribute of each object class.

At the end of this process, you will have a small set of CRC cards that collaborate to support the use case. This process can be enhanced with several other activities. First, the CRC cards can be arranged on the table in the order they are executed or called. In other words, the calling order can be determined

FIGURE **10-18** *CRC cards model for* Create phone sale *use case*

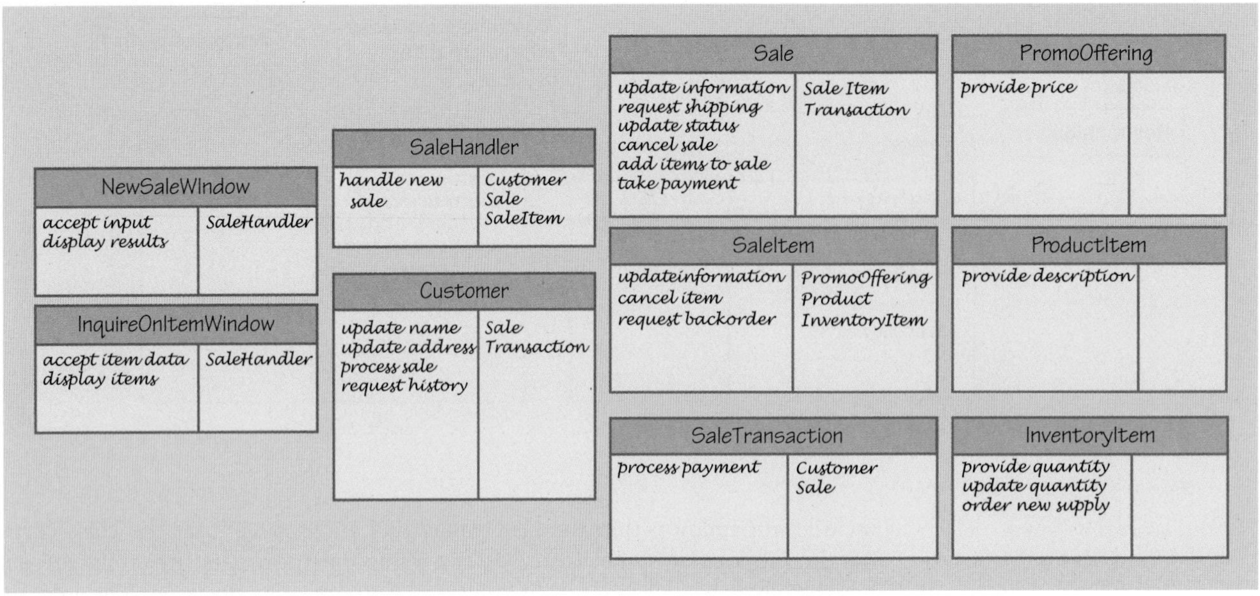

at this time. For example, the Customer object creates a Sale object, the Sale object creates SaleItem objects, and SaleItem objects access ProductItem and InventoryItem objects to get required information. **Figure 10-18** illustrates a solution set of CRC cards for the use case *Create phone sale*.

Another helpful step is to include the user-interface classes. If a user is part of the team and if some preliminary work has been done on the user interface requirements, it could be effective to add CRC cards for all user-interface window classes that are required for the use case. By including user-interface classes, all the input and output forms can be included in the design. Obviously, this is a much more complete design.

Any other required utility classes can also be added to the solution. For example, for a three-layer design, data access objects will be part of the solution. Each persistent class will have a data access class to read and write to the database.

At the end of the design for one use case, two other important tasks remain. Because the CRC cards only have data for a single use case, the information can be transferred to the design class diagram. The design class diagram then becomes a central repository for all information about every class in the new system.

A second task is to put an elastic band around the set of CRC cards for the next step. If the use case is a simple one, the CRC cards can be taken by a programmer and the use case can be implemented.

To finish the example, let us go back to the DCD and update it based on the design information created during the brainstorming session. **Figure 10-19** shows an updated DCD, with several methods added and updates in the visibility. We first note that a new class has been added. Evidently, we overlooked the OrderTransaction class in the first-cut DCD. We also note that the OrderHandler class needs visibility to the Sale class to process a payment. Compare the responsibilities identified on the CRC cards and the method names described in each class. Note the close correlation.

Often, when developers begin using CRC cards, they list many different responsibilities for a given class. For example, developers might say the SaleItem class should get the price. In reality, the PromoOffering class provides the price and the SaleItem only uses it. In other words, it usually helps to think of responsibilities as being similar to methods—requests to do something rather than random actions that need to occur.

FIGURE **10-19**

Updated DCD for the Create phone sale *use case*

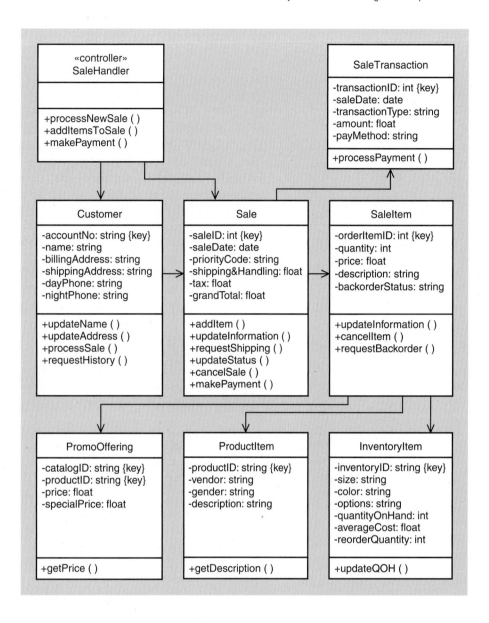

Fundamental Detailed Design Principles

Now that you understand how an object-oriented program works and you know the notation for a design class, let us review several basic principles that will guide design decisions. We used these principles throughout this chapter as we discussed the steps of object-oriented design because they are important to all parts of the process.

Coupling

In the previous example, in which Customer had navigation visibility to Sale, we could also say that Customer and Sale are coupled, or linked. **Coupling** is a qualitative measure of how closely the classes in a design class diagram are linked. A simple way to think about coupling is as the number of navigation arrows on the design class diagram. Low coupling is usually better for a system than high coupling. In other words, fewer navigation visibility arrows indicate that a system is easier to understand and maintain.

We say that coupling is a qualitative measure because no specific number measures coupling in a system. A designer must develop a feel for coupling—that is, recognize when there is too much or too little. Coupling is evaluated as

coupling a qualitative measure of how closely the classes in a design class diagram are linked

a design progresses—use case by use case. Generally, if each use case design has a reasonable level of coupling, the entire system will too.

Refer back to Figure 10-1 to observe the flow of messages between the objects. Obviously, objects that send messages to each other must have navigation visibility and thus are coupled. For the Input window object to send a message to the Student object, it must have navigation visibility to it. Thus, the Input window object is coupled to the Student object. But notice that the Input window object isn't connected to the Database access object, so those objects aren't coupled. If we designed the system so the Input window object accessed the Database access object, the overall coupling for this use case would increase; that is, there would be more connections. Is that good or bad? In this simple example, it might not be a problem. But for a system with 10 or more classes, too many connections with navigation visibility can cause high levels of coupling, making the system more complex and therefore harder to maintain.

So, why is high coupling bad? The main reason is that a change in one class ripples through the entire system. Therefore, experienced analysts make every effort to simplify coupling and reduce ripple effects in the design of a new system.

Cohesion

Cohesion refers to the consistency of the functions within a single class. **Cohesion** is a qualitative measure of the focus or unity of purpose within a single class. For example, in Figure 10-1, you would expect the Student class to have methods—that is, functions—to enter student information, such as identification number or name. That would represent a unity of purpose and a highly cohesive class. But what if that same object also had methods to make classroom assignments or assign professors to courses? The cohesiveness of the class would be reduced.

cohesion a qualitative measure of the focus or unity of purpose within a single class

Classes with low cohesion have several negative effects. First, they are hard to maintain. Because they perform many different functions, they tend to be overly sensitive to changes within the system, suffering from ripple effects. Second, it is hard to reuse such classes. Because they have many different—and often unrelated—functions, it usually doesn't make sense to reuse them in other contexts. For example, a button class that processes button clicks can easily be reused. However, a button class that processes button clicks and user log-ins has limited reusability. A final drawback is that classes with low cohesion are usually difficult to understand. Frequently, their functions are intertwined and their logic is complex.

Although there is no firm metric to measure cohesiveness, we can think about classes as having very low, low, medium, or high cohesion. Remember, high cohesion is the most desirable. An example of very low cohesion is a class that has responsibility for services in different functional areas, such as a class that accesses the Internet and a database. These two types of activities are different and accomplish different purposes. To put them together in one class causes very low cohesion.

An example of low cohesion is a class that has different responsibilities but in related functional areas—for example, one that does all database access for every table in the database. It would be better to have different classes to access customer information, order information, and inventory information. Although the functions are the same—that is, they access the database—the types of data passed and retrieved are very different. Thus, a class that is connected to the entire database isn't as reusable as one that is only connected to the customer table.

An example of medium cohesion is a class that has closely related responsibilities, such as a single class that maintains customer information and customer account information. Two highly cohesive classes could be defined: one for customer information, such as names and addresses, and another class or set of classes for customer accounts, such as balances, payments, credit information, and all financial activity. If the customer information and the account information are limited, they could be combined into a single class with medium cohesiveness. Either medium or highly cohesive classes can be acceptable in systems design.

Protection from Variations

protection from variations a design principle in which parts of a system that are unlikely to change are segregated from those that will

One of the underlying principles of good design is **protection from variations**—the idea that the parts of a system that are unlikely to change should be segregated (or protected) from those that will change. As you design systems, you should try to isolate the parts that will change from those that are more stable.

Protection from variations is a principle that drives the multilayer design pattern. Designers could mix all the user-interface logic and business logic together in the same classes. In fact, in early user-oriented, event-driven systems, such as those built with early versions of Visual Basic and PowerBuilder, the business logic was included in the view layer classes—often in the windows input forms. Many Web applications also combine HTML and business logic. The problem with this design was that when an interface needed to be updated, all the business logic had to be rewritten. A better approach is to decouple the user-interface logic from the business logic. Then, the user interface can be rewritten without affecting the business logic. In other words, the business logic—being more stable—is protected from variations in the user interface.

Also, what if updates to the business logic require the addition of new classes and new methods? If the user-interface classes are tightly coupled to the business classes, there could be a ripple effect of changes throughout the user-interface classes. However, because the user interface can simply send all its input messages to the use case controller class, changes to the methods or classes in the business logic and domain layer are isolated to the controller class. You will find that protection from variations affects almost every design decision, so you should watch for and recognize the application of this principle in all design activities.

Indirection

indirection a design principle in which an intermediate class is placed between two classes to decouple them but still link them

Indirection is the principle of decoupling two classes or other system components by placing an intermediate class between them to serve as a link. In other words, instructions don't go directly from A to B; they are sent through C first. Or in message terminology, don't send a message from A to B; let A send the message to C and then let C forward it to B.

Although there are many ways to implement protection from variations, indirection is frequently used for that purpose. Inserting an intermediate object allows any variations in one system to be isolated in that intermediate object. Indirection is also useful for many corporate security systems. For example, many companies have firewalls and proxy servers that receive and send messages between an internal network and the Internet. A proxy server appears as a real server—ready to receive such messages as e-mail and HTML page requests. However, it is a fake server, which catches all the messages and redistributes them to the recipients. This indirection step allows security controls to be put in place and protect the system.

Object Responsibility

object responsibility a design principle in which objects are responsible for carrying out system processing

One of the most fundamental principles of object-oriented development is the idea of **object responsibility**; that is, objects are responsible for carrying out the system processing. These responsibilities are categorized in two major areas: knowing and doing. In other words, what is an object expected to know, and what is an object expected to do or to initiate?

"Knowing" includes an object's responsibilities for knowing about its own data and knowing about other classes with which it must collaborate to carry out use cases. Obviously, a class should know about its own data, what attributes exist, and how to maintain the information in those attributes. It should also know where to go to get information when required. For example, during the initiation of an object, data that aren't passed as parameters may be required. An object should know about or have navigation visibility to other objects that can provide the required information. In Figure 10-8, the first constructor method for the Student class doesn't receive a studentID value as a parameter. Instead, the Student class takes responsibility for creating a new studentID value based on some rules it knows.

"Doing" includes all the activities an object does to assist in executing a use case. Some of those activities include receiving and processing messages. Another activity is to instantiate, or create, new objects that may be required for completion of a use case. Classes must collaborate to carry out a use case, and some classes are responsible for coordinating the collaboration. For example, for the use case *Create phone sale*, the Sale class has responsibility to create SaleItem objects. Another class, such as InventoryItem, is only responsible for providing information about itself.

Chapter Summary

The primary creative activity of system developers is to write computer software that solves the business problem. So far, this textbook has focused on two major activities: understanding the user's requirements (the business problem) and figuring out and visualizing a solution system. This chapter focuses on how to configure and develop the solution system—that is, design the system. Systems design is the bridge that puts business requirements in terms that the programmers can use to write the software that becomes the solution system.

Architectural design is the first step in configuring the new system. Its purpose is to determine the structure and configuration of the new system's various components. Component diagrams show the various executable components of the new system and how they relate to one another. Many new systems are enterprise-level systems, in that they are used at locations throughout the organization. They also share resources, such as a common database of information.

Once the architectural design is known, detailed design can begin. The objective of detailed design is to determine the objects and methods within individual classes to support the use cases. The process of detailed design is use case–driven, in that it is done for each use case separately.

The process of detailed design can be divided into two major areas: developing a design class diagram (DCD) and developing the set of interacting classes and their methods for each use case via a sequence diagram. The DCD is usually developed in two steps. A first-cut DCD is created based on the domain model class diagram, but then it is refined and expanded as the sequence diagrams are developed. A preliminary idea of how the objects collaborate can be created using class responsibility collaboration (CRC) cards. For simple use cases, a set of CRC cards may be sufficient to write code. For more complex use cases, the CRC cards serve as the beginning point for developing sequence diagrams.

One reason that we suggest a more formal system of design rather than just starting to write code is that the final system is much more robust and maintainable. Doing design as a rigorous activity builds better systems. Some fundamental principles should be considered as a system is developed; two critical ideas are coupling and cohesion. A good system has low coupling between the classes, and each of the classes has high cohesion. Another important principle is "protection from variations," meaning that some parts of the system should be protected from and not tightly coupled to other parts of the system that are less stable and subject to change. Being a good developer entails learning and following the principles of good design.

Key Terms

abstract class   311

application program
  interface (API)   300

boundary class, or view class   309

class-level attributes   311

class-level method   310

cohesion   318

component diagram   300

concrete class   311

control class   309

coupling   317

CRC cards   314

data access class   309

enterprise-level system   298

entity class   309

indirection   319

instantiation   296

method signature   310

navigation visibility   312

object responsibility   319

persistent class   309

protection from variations   319

stereotype   309

visibility   310

Review Questions

1. Describe in your own words how an object-oriented program works.

2. What is instantiation?

3. List the models that are used for object-oriented systems design.

4. What UML diagram is used to model architectural design?

5. Explain how domain classes are different from design classes.

6. What is an enterprise-level system? Why is it an important consideration in design?

7. What are some of the differences between a client-server network system and an Internet system?

8. What is an API? Why is it important?

9. What notation is used to identify the interface of a component?

10. What is the difference between the notation for problem domain classes and design classes?

11. In your own words, list the steps for doing object-oriented detailed design.

12. What do we mean by use-case driven design, and what is use case realization?

13. What are a) persistent classes, b) entity classes, c) boundary classes, d) control classes, and e) data access classes?

14. What are class-level methods and class-level attributes?

15. What are attribute and method visibility, and what are the types of visibility shown on a design class diagram?

16. What is a method signature?

17. Compare and contrast abstract and concrete classes. Give an example of each.

18. Describe navigation visibility. Why is it important in detailed design?

19. List some typical conditions that dictate in which direction navigation visibility occurs.

20. What information is maintained on CRC cards?

21. What is the objective of a CRC card design session?

22. Compare and contrast the ideas of coupling and cohesion.

23. What is protection from variations, and why is it important in detailed design?

24. What is meant by object responsibility, and why is it important in detailed design?

Problems and Exercises

1. Given the following system description, develop a component diagram for a desktop-operated internal network system (i.e., Internet access not required).

   The new Benefits for Employees, Spouses, and Dependents (BESD) system will be used primarily by the human resources department and will contain confidential information. Consequently, it will be built as a totally in-house system, without any Internet elements. The database for the system is the human resource employee database (HRED), which is shared by several other systems within the company.

   There are two types of screens from a systems design viewpoint: simple inquiry screens and complex inquiry/update screens. The simple inquiry screens just access the database, with no business logic required. The complex screens usually do fairly complicated calculations based on sophisticated business rules. These programs often have to access other data tables from other databases in the company.

   The database will always remain on a central database server. The application program itself will be installed on each desktop that is allowed access. However, authentication is a centralized process, and it will control which screens and program functions can be accessed by which users.

2. Develop a component diagram for the following description of a Facebook application.

   The Facebook platform is available for entrepreneurs to develop applications for use among all Facebook users. A new application is being written that allows Facebook users to send gifts and greeting cards to their friends. (These are real gifts and greeting cards, not just electronic images.) The application running within Facebook is on its own server and has its own database of information, which includes a list of gifts and cards that have been sent or received. The actual retail store of gifts and cards to send must be located on a different server because it is part of a regular Internet sales storefront. This storefront maintains the database of inventory items to sell and collects credit card payment information.

3. In this chapter, we developed a first-cut DCD, a set of CRC cards, and a final DCD for the *Create phone sale* use case for RMO. Create the same three drawings for the *Look up item availability* use case.

4. Find a company that does object-oriented design by using CRC cards. The information systems unit at your university often uses object-oriented techniques. Sit in on a CRC design brainstorming session. Interview some of the developers about their feelings regarding the effectiveness of doing CRC design. Find out what documentation remains after the sessions and how it is used.

5. Find a company that has an enterprise system. (If you are working for a company, see what systems they use.) Analyze the system and then develop a component diagram and a deployment diagram.

6. Find a system that was developed by using Java. If possible, find one that has an Internet user interface and a network-based user interface. Is it multilayered—three layers or two layers? Can you identify the view layer classes, the domain layer classes, and the data access layer classes?

7. Find a system that was developed by using Visual Studio .NET (Visual Basic or C#). If possible, find one that has an Internet user interface and a network-based user interface. Is it multilayered? Where is the business logic? Can you identify the view layer classes, the domain layer classes, and the data access layer classes?

8. Pick an OOP language with which you are familiar. Find a programming IDE tool that supports that language. Test its reverse-engineering capabilities to generate UML class diagrams from existing code. Evaluate how well it does and how easy the models are to use. Does it have any capability to input UML diagrams and generate skeletal class definitions? Write a report on how it works and what UML models it can generate.

9. Draw a UML design class that shows the following information.

   The class name is Boat, and it is a concrete entity class. All three attributes are private strings with initial null values. The attribute boat identifier has the property of "key." The other attributes are the manufacturer of the boat and the model of the boat. There is also an integer class-level attribute containing the total count of all boat objects that have been instantiated. Boat methods include creating a new instance, updating the manufacturer, updating the model, and getting the boat identifier, manufacturer, and model year. There is a class-level method for getting the count of all boats.

The State Patrol Ticket-Processing System (Revisited)

In Chapter 3, you identified use cases and considered the domain classes for the State Patrol Ticket Processing System. Review the descriptions in Chapter 3 for the use case *Record a traffic ticket*. Recall that the domain classes included Driver, Officer, Ticket, and Court.

1. Draw a domain class diagram for the ticket-processing system based on the four classes just listed and include attributes, association, and multiplicity.
2. List the classes that would be involved in the use cases and decide which class should be responsible

for collaborating with the other classes for the use case *Record a traffic ticket*. Consider some possibilities: 1) a driver object should be responsible for recording his/her ticket; 2) the officer object should be responsible for recording the ticket that he or she writes; and 3) a ticket object should be responsible for recording itself.

3. Create a set of CRC cards showing these classes, responsibilities, and collaborations for the use case.
4. Draw a first-cut design class diagram (DCD) based on your CRC cards.

RUNNING CASES

Community Board of Realtors

In Chapter 3 and Chapter 5, you identified and then modeled use cases for the Multiple Listing Service (MLS) application. You also identified and modeled domain classes. Use your solutions from these chapters to do the following:

1. Draw a basic component diagram showing the architectural design for the system, assuming that it is a two-layer Internet architecture.

2. Use the CRC cards technique to identify the classes that are involved in the *Create new listing* use case. Recall that creating a new listing involves an agent, a real estate office, and a listing. Decide which class should have the primary responsibility for collaborating with the other classes and then complete the CRC cards for the use case.
3. Draw a first-cut design class diagram (DCD) based on the CRD cards for this use case.

The Spring Breaks 'R' Us Travel Service

In Chapter 3, you identified use cases for the Spring Breaks 'R' Us Travel Service. In Chapter 5, you elaborated those use cases. In Chapter 4, you identified the classes associated with these use cases. Using your solutions from these chapters, do the following:

1. Draw a basic component diagram showing the architectural design for the system, assuming it is a two-layer Internet architecture.

2. Use the CRC cards technique to identify the classes that are involved in the *Book a reservation* use case. Recall that creating a booking involves at least a student group, a resort, a week, and a room type. Decide which class should have the primary responsibility for collaborating with the other classes and then complete the CRC cards for the use case.
3. Draw a first-cut design class diagram (DCD) based on the CRD cards for this use case.

On the Spot Courier Services

In Chapter 6, you considered the issues relevant to the specification of the hardware equipment and networking requirements. The case description in Chapter 6

also reviewed the three primary types of users for the system and many of their respective system-supported activities.

(continued on page 324)

(continued from page 323)

Consistent with the network that you recommended in Chapter 6, develop a component diagram. Show which parts of the system may use general Internet access and which parts may use VPN capabilities.

In Chapter 5, you developed activity diagrams and system sequence diagrams for two use cases: *Request a package pickup* and *Pickup a package*. In Chapter 4, you developed a domain model class diagram for the system.

For each of the two use cases, develop a first-cut design class diagram and a set of CRC cards. The design class diagram should elaborate the attributes and show navigation visibility. You may also need to add more classes from your solution in Chapter 4. It isn't uncommon for developers to enhance early models as they begin to understand system requirements better. The CRC cards should include classes for the Controller class and any classes for screens you identified in Chapter 7.

Sandia Medical Devices

Review the original system description in Chapter 2, additional project information in Chapters 3, 4, 6, 8, and 9, and the use case diagram shown in **Figure 10-20** to refamiliarize yourself with the proposed system.

Complete these tasks:

1. Develop a deployment diagram that shows the equipment specified in Chapter 6 and the list of software components you developed while answering question 1 in Chapter 9.

2. For the moment, assume that the database will store two glucose levels for each patient—normal minimum and normal maximum—and that an alert will be generated if three or more consecutive glucose readings are above or below those levels. Expand the domain class diagram in Chapter 4 to include this information and then develop a first-cut design class diagram to support the patient use case *View/respond to alert*.

FIGURE **10-20** *Use cases for the patient and physician actors*

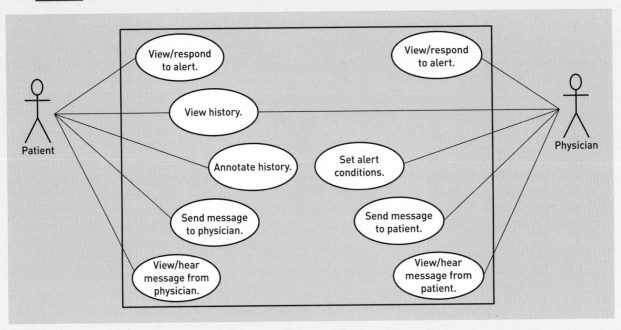

Further Resources

Grady Booch, James Rumbaugh, and Ivar Jacobson, *The Unified Modeling Language User Guide*. Addison-Wesley, 1999.

Grady Booch, et al., *Object-Oriented Analysis and Design with Applications* (3rd edition). Addison-Wesley, 2007.

E. Reed Doke, J. W. Satzinger, and S. R. Williams, *Object-Oriented Application Development Using Java*. Course Technology, 2002.

E. Reed Doke, J. W. Satzinger, and S. R. Williams, *Object-Oriented Application Development Using Microsoft Visual Basic .NET*. Course Technology, 2003.

Hans-Erik Eriksson, Magnus Penker, Brian Lyons, and David Fado, *UML 2 Toolkit*. John Wiley and Sons, 2004.

Martin Fowler, *UML Distilled: A Brief Guide to the Standard Object Modeling Language* (3rd edition). Addison-Wesley, 2004.

Ivar Jacobson, Grady Booch, and James Rumbaugh, *The Unified Software Development Process*. Addison-Wesley, 1999.

Philippe Kruchten, *The Rational Unified Process, An Introduction*. Addison-Wesley, 2000.

Craig Larman, *Applying UML and Patterns: An Introduction to Object-Oriented Analysis and Design and the Unified Process* (3rd edition). Prentice Hall, 2004.

Jeffrey Putz, *Maximizing ASP.NET Real World, Object-Oriented Development*. Addison-Wesley, 2005.

James Rumbaugh, Ivar Jacobson, and Grady Booch, *The Unified Modeling Language Reference Manual*. Addison-Wesley, 1999.

11

Object-Oriented Design: Use Case Realizations

Chapter Outline

- Detailed Design of Multilayer Systems
- Use Case Realization with Sequence Diagrams
- Designing with Communication Diagrams
- Updating and Packaging the Design Classes
- Design Patterns

Learning Objectives

After reading this chapter, you should be able to:

- Explain the different types of objects and layers in a design
- Develop sequence diagrams for use case realization
- Develop communication diagrams for detailed design
- Develop updated design class diagrams
- Develop multilayer subsystem packages
- Explain design patterns and recognize various specific patterns

New Capital Bank: Part 2

The integrated customer account system project for New Capital Bank was now two months old. The first development iteration had gone pretty well, although there were a few snags because the team was still learning the ins and outs of iterative development projects. Bill Santora, the project leader, was discussing some of the system's technical details with Charlie Hensen, one of his team leaders, in preparation for the iteration retrospective.

"How is the team feeling about doing detailed design?" Bill asked Charlie, who was one of the early critics of doing more formal design. "I know some of the programmers wanted to just start coding from the use case descriptions that were developed with the users. They weren't very happy about taking the time to design. Is that still a problem?"

"It really has worked out quite well," Charlie said. "As you know, I was skeptical at first and thought it would waste a lot of time. Instead, it has allowed us to work together better because we know what the other team members are doing. I also think the system is much more solid. We're all using the same approach, and we've discovered that there are quite a few classes we share. Of course, we don't waste a lot of time making fancy drawings. We do document our designs with some quick drawings, but that is about as far as we take it."

"What would you say were the strengths and weaknesses of our approach?" Bill asked. "Or are there ways you think we could do it better in this next iteration?"

"I really like the approach of first doing a rough design using CRC cards," Charlie replied. "It's nice to have a couple of users there with us to verify that our collaborations are correct. For the simple use cases, we can work with the users to lay out the user interface. Between the CRC cards and the user-interface specifications, we should have enough to program from, especially now that we have the basic structure set up. Then, for the more complex use cases, we can go ahead and do a detailed design with sequence diagrams. The nice thing about the sequence diagrams is that they're detailed enough for us to hand the designs over to some of the junior programmers. It makes them much more effective in their team contributions."

"So, would you change our approach, or do you think it's working the right way?" Bill asked, still looking for ways to improve the process.

"Well, it really is working pretty well right now," Charlie said. "One thing I really like about it is that we have a common DCD that everyone can access and review. That really helps when you're ready to insert some code into a class to check and see what is already there. The central repository for all our code and for the diagrams we do formalize is a great tool. I wonder if there is a way to get more use out of that tool. Other than that, I would say let's stick with this approach for another iteration and see if it needs changing after that."

Overview

Chapter 10 explained the design concepts and models used for multilayer systems and their architectural design. The latter portion of that chapter introduced the concepts associated with object-oriented detailed design. You also learned how to begin the detailed design process by using CRC cards and design class diagrams to identify which classes collaborate to carry out use cases. Simple use cases can frequently be programmed from the design information developed with these two steps.

This chapter pursues object-oriented detailed design in more depth and formality. Detailed design is a subject that can be addressed at multiple levels. For the beginner, a fairly straightforward yet complete process can be defined. This chapter focuses on the foundation principles, which are based on the concepts of use case realization by using sequence diagrams and design patterns. Once you become proficient with these two subjects, you can consider yourself an object-oriented designer. Several good books are available on design patterns and design methods.

use case realization the process of elaborating the detailed design with interaction diagrams for a particular use case

The method used to extend the process of detailed design is called **use case realization**. In use case realization, each use case is taken individually to determine all the classes that collaborate on it. As part of that process, any other utility or support classes are identified. Care is taken during this process to define the classes so the integrity of the multilayer architectural design is maintained.

As the details of the classes are designed—use case by use case—the design class diagram is also updated as necessary.

The last section of this chapter is a brief introduction to design patterns. As with any engineering discipline, certain procedures have become tried and proven solutions. Even though object-oriented development is a relatively young engineering discipline, it offers standard ways to design use cases that lead to solid, well-constructed solutions. You will learn a few of those standard designs or patterns.

Detailed Design of Multilayer Systems

The discussion of CRC cards in Chapter 10 introduced the idea of collaborating objects to execute various use cases. However, design sessions using CRC cards focus on the problem domain classes, with very little attention given to multi-layer issues as they affect the detailed design. This chapter describes in depth the detailed design of all layers of a multilayer system.

Looking back at Figure 10-1 in Chapter 10, there are three objects representing the three layers of a system. Each object has certain responsibilities. The input window object has the primary responsibility of formatting and presenting student information on the screen. It also has the responsibility of accepting input data—either student ID or changed information—and forwarding it to the system. The object probably also does some editing of the input data. Where does this object come from? What are this object's attributes and methods? Identifying and defining the window objects are part of the application design and the user interface design.

The student object represents the middle layer, or business logic layer, for the use case. A CRC design session will help you design the structure of the objects in this layer. However, you probably noticed that CRC cards are quite informal, especially when trying to ascertain class methods from object responsibilities. CRC cards provide little direction in defining method signatures with appropriate input and output parameters. This chapter formalizes the process of precisely identifying methods and defining method signatures.

Again referring to Figure 10-1, the database access object represents the third layer in the multilayer design. It is responsible for connecting to the database, reading the student information, and sending it back to the student object. It is also responsible for writing the student information back to the database when necessary. This object doesn't come from a problem domain class; it is a utility object created by the designer.

Several questions should come to mind as you review a detailed systems design. First, how do all these objects get created in memory? For example, how and when does the student object get created? How about the database access object? Other questions include: Will other objects be necessary? What object represents authentication? What is the life span of each object? Maybe the student object should go away after the update—but what about the database access object?

Patterns and the Use Case Controller

Patterns, also called *templates*, are used repeatedly in everyday life. A chef uses a recipe, which is just another word for a pattern, to combine ingredients into a flavorful dish. A tailor uses a pattern to cut fabric for a great-fitting suit. Engineers take standard components and combine them into established configurations, or set patterns, to build buildings, sound systems, and thousands of other products. Patterns are created to solve problems. Over time and with many attempts, people who work on a particular problem develop a set solution to the problem. The solution is general enough that it can be applied over and over again. As time passes, the solution is documented and published, and eventually it becomes accepted as the standard.

design patterns standard design techniques and templates that are widely recognized as good practice

Standard design templates have become popular among software developers because they can speed object-oriented design work. The formal name for these templates is **design patterns**. Design patterns became a widely accepted object-oriented design technique in 1996 with the publication of *Elements of Reusable Object-Oriented Software* by Erich Gamma, Richard Helm, Ralph Johnson, and John Vlissides. These four authors are now referred to as the Gang of Four (GoF). As you learn more about design patterns, you will often see references to a particular design pattern as a GoF pattern. In their book, the authors identified 23 basic design patterns. Today, scores of patterns have been defined—from low-level programming patterns to mid-level architectural patterns to high-level enterprise patterns. Two important enterprise platforms—Java and .NET—have sets of enterprise patterns, which are described in various books and publications.

In Chapter 10, you were introduced to the concept of a use case controller. Let us formalize the concept of a use case controller and explain its importance as a design pattern. For any particular use case, messages come from the external actor to a windows class (that is, an electronic input form) and then to a problem domain class. One issue in systems design is the question of which problem domain class should receive input messages to reduce coupling, maintain highly cohesive domain classes, and maintain independence between the user interface and the domain layer. Designers often define intermediary classes that act as buffers between the user interface and the domain classes. We call these classes *use case controllers*. For example, the use case *Fill shopping cart* might have a controller class named CartHandler.

Figure 11-1 provides a more formal specification for the use case controller pattern. Note that this specification has five main elements:

- Pattern name
- Problem that requires a solution
- Solution to or explanation of the pattern
- Example of the pattern
- Benefits and consequences of the pattern

You should read this specification to understand the important principles of the controller pattern, what problem it solves, how it works, and its benefits. This same template will be used later in this chapter with the other design patterns.

A use case controller acts as a switchboard, taking input messages and routing them to the correct domain class. In effect, the use case controller acts as an intermediary between the outside world and the internal system. What if a particular window object needs to send messages to several problem domain objects? Without the use case controller, the input window would need references to all these domain objects. The coupling between the input window object and the internal system would be very high; there would be many connections. The coupling between the user interface objects and the problem domain objects could be reduced by making a single use case controller object to handle all the input messages. A use case controller also contains logic that controls the flow of execution for the use case. In this way, domain layer design classes can remain more cohesive by focusing only on the precise functions that truly belong to that domain object.

In the examples that follow, we define a controller class for each use case. This is a common practice, and many development environments (such as Java Struts) automatically define a controller class for each use case. Of course, this creates many artifact objects in a system. If there are 100 use cases, there would be 100 use case controller artifact objects. To reduce the number of controllers, developers sometimes combine the control of several closely related use cases into a single use case controller. Either approach, if done judiciously, provides a good solution.

FIGURE **11-1** *Pattern specification for the controller pattern*

Name:	Controller
Problem:	Domain classes have the responsibility of processing use cases. However, since there can be many domain classes, which one(s) should be responsible for receiving the input messages? User-interface classes become very complex if they have visibility to all of the domain classes. How can the coupling between the user-interface classes and the domain classes be reduced?
Solution:	Assign the responsibility for receiving input messages to a class that receives all input messages and acts as a switchboard to forward them to the correct domain class. There are several ways to implement this solution: (a) Have a single class that represents the entire system, or (b) Have a class for each use case or related group of use cases to act as a use case handler.
Example:	The RMO Customer account subsystem accepts inputs from a :CustomerForm window. These input messages are passed to the :CustomerHandler, which acts as the switchboard to forward the message to the correct problem domain class. 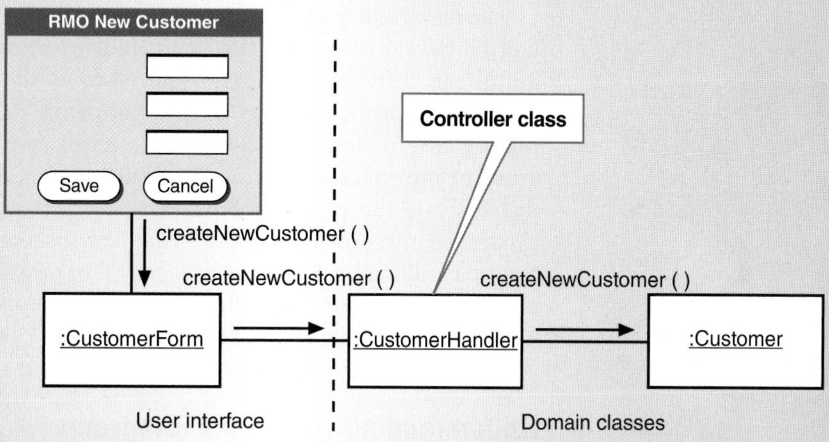 **RMO New Customer** Save Cancel createNewCustomer ( ) Controller class createNewCustomer ( ) createNewCustomer ( ) :CustomerForm → :CustomerHandler → :Customer User interface Domain classes Other cases of the controller pattern will be used for each RMO use case.
Benefits and Consequences:	Coupling between the view layer and the domain layer is reduced. The controller provides a layer of indirection. The controller is closely coupled to many domain classes. If care is not taken, controller classes can become incoherent, with too many unrelated functions. If care is not taken, business logic will be inserted into the controller class.

A use case controller is a completely artificial class created by the person doing the system design. Sometimes, such classes are called *artifacts* or *artifact objects*. As we get deeper into the explanation of design, you will see the need to create many kinds of service classes as artifacts—classes that are needed to execute the use case but aren't based on any domain model classes.

The process illustrated in the preceding paragraphs and in Chapter 10—that is, balancing the design principles of coupling, cohesion, class responsibility, indirection, and protection from variations—is precisely the process of systems design. As you read this chapter, you will see the importance of the design principles discussed in Chapter 10.

Use Case Realization with Sequence Diagrams

Developing interaction diagrams is at the heart of object-oriented detailed design. The realization of a use case—determining what objects collaborate and the messages they send to each other to carry out the use case—is done through the development of an interaction diagram. Two types of interaction diagrams can be used during design: **sequence diagrams** or **communication diagrams**. This section shows you how to design with sequence diagrams; then, in the next section, we briefly explain how communication diagrams are used in systems design.

Adaptive projects that use iteration and Agile modeling techniques minimize the formality of design diagrams. For these kinds of projects, the same developers will sometimes do the analysis, the design, and the programming. In those instances, the design diagrams are often done on a whiteboard or scratch paper and then discarded after the programming is done. However, it isn't uncommon to have distributed teams in which the analysts and designers are in one location and the programmers are in another location—often offshore. In that case, design diagrams are helpful in communicating design decisions throughout the team.

In this chapter (and possibly for your homework assignments), the diagrams will be developed using MS Visio. However, in real projects, hand-drawn diagrams can be scanned and transmitted to all the team members just as easily.

The following sections explain in detail the steps and techniques required for use case realization. In the first section, we provide a partial sequence diagram to introduce the terms and composition of a sequence diagram. We then demonstrate the process of use case realization by using the problem domain classes for the use cases *Create customer account* and *Fill shopping cart*. These examples illustrate the core process of organizing and structuring the problem domain classes into the solution for the use case. The final examples explain how to add the data access layer classes and the view layer classes. Each layer is illustrated with a detailed example using the same use case.

Understanding Sequence Diagrams

You first heard about sequence diagrams in Chapter 5 when you learned how to develop a system sequence diagram (SSD). By now, you should feel comfortable reading, interpreting, and developing an SSD. Remember that an SSD is used to document the inputs to and outputs from the system for a single use case or scenario. The system itself is treated as a single object named :System. The inputs to the system are messages from the actor, and the outputs are usually return variables showing the data being returned. **Figure 11-2** reviews the elements of an SSD by showing a partial example for the *Create customer account* use case. As shown in the figure, each message has a source and a destination.

Remember that the syntax of an input message, as discussed in Chapter 5, is:

* [true/false condition] return-value := message-name (parameter-list)

The starting point for the detailed design of a use case is always its SSD. Remember that the SSD only has two lifelines—one for the actor and one for the system. The most important information on an SSD is the sequence of messages between the actor and the system. Frequently, there are input and output messages. There may be a single input message or many. The input messages may have data parameters or not. There also may be Loop frames, Alt frames, and Opt frames as well as repeating inputs and outputs. A Loop frame denotes a set of messages within a loop. An Alt frame is similar to an if-then-else

sequence diagrams type of interaction diagram that emphasizes the sequence of messages sent between objects for a specific use case

communication diagrams type of interaction diagram that emphasizes the objects that send and receive messages for a specific use case

FIGURE **11-2**

SSD for the Create customer account *use case*

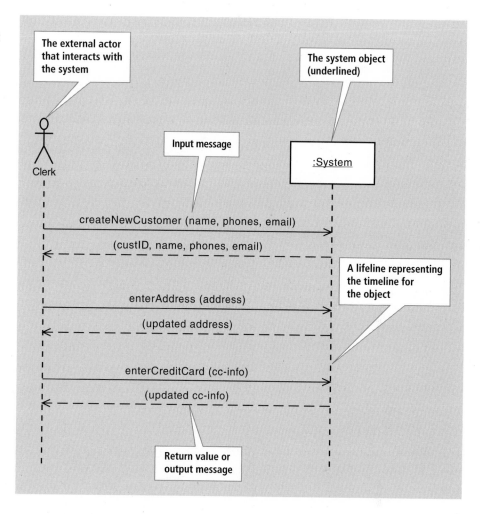

statement or switch statement, which allows the firing of different sets of messages. An Opt frame is an optional invoking of a set of messages. We will see examples of these later in this chapter.

The three input and output messages in Figure 11-2 characterize the sequence of actions required to add a customer. The first message sends the basic information about customer name, telephone numbers, and e-mail addresses. The same information is returned with a customer ID number. The return information occurs after the system has created a new record in the database. The next two input messages simply add more information to the account: the customer address information and credit card information. The return messages are essentially a reformat of the input data for the customer to verify. Additional complexity might be added in real life for such things as verifying and correcting the entered information.

Figure 11-3 illustrates a two-level detailed design for this use case. Callouts show the classes associated with each of the two layers. This use case has one view layer object: the :CustomerForm object. Notice that the input messages from the external actor always go to the view layer objects. The purpose of the design process is to take each input message and determine what the system must do to respond to the message. For the first message—createNewCustomer()— the system simply uses the :CustomerForm screen to accept and possibly edit the input values. At this point in the design, we won't worry about the required editing. The primary objective of the sequence diagram is to identify which classes collaborate and what messages they must send to each other.

After the :CustomerForm object receives the createNewCustomer() message, it sends a message to the :CustomerHandler object, which in turn sends a

FIGURE **11-3** *Partial sequence diagram for the* Create customer account *use case*

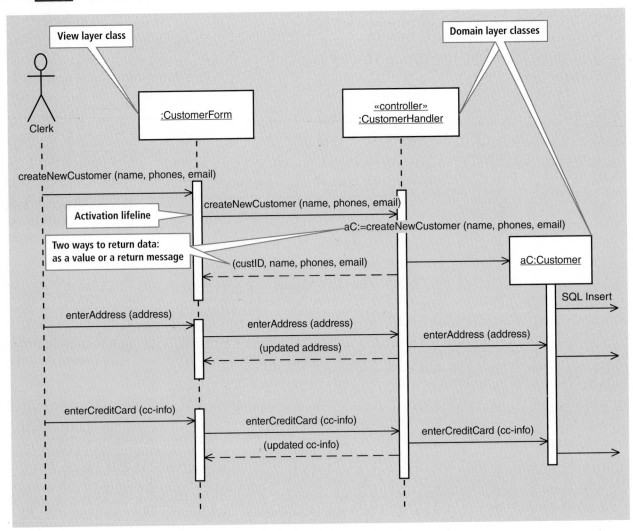

message to the Customer class, asking it to instantiate a new customer object. This message is sent directly to the object rectangle, which is an optional but preferred notation to indicate a message that invokes a constructor. The label in the rectangle—aC:Customer—indicates that the box represents a customer object with a reference variable name of aC. Notice on the create message that the object reference is returned to the controller, which gives it visibility to the customer object. It will need this reference later to send other messages.

After the customer object is created, it has the responsibility of saving itself to the database. Because this is a fairly simple use case, we have limited it to a two-layer design. In other words, the customer object includes customer logic and database access statements to write itself to the database.

Once the customer object is created and saved to the database, the controller receives a pointer to the object that it can use to access the custID field as well as any other data for the newly created customer object: Ac. Once the data is returned to the controller, it is then returned to the :CustomerForm screen, which is visible to the external actor. Optionally, we could have added dashed lines going from the :CustomerForm screen to the external actor to denote the user looking at the screen. Note that there are two ways to document the returning data: either as a return value on the function or as a dashed line returning specific values.

One thing should be evident at this point: When a message is sent from an originating object to a destination object, in programming terms, it means that the originating object is invoking a method on the destination object. Thus, by defining the messages to various internal objects, we are actually identifying the methods of that object. The data that is passed by the messages corresponds to the input parameters of the methods. The return data on a message is the return value from a method. Hence, once a use case is realized with this detailed design process, the set of classes and required methods can be extracted so programming can be completed. These method names become part of the documentation within a design class diagram.

Figure 11-3 contains a new notation called the **activation lifeline**, which is represented by small vertical rectangles. Because a message invokes a method on the destination object, one valuable piece of information might be the duration of that method's execution (i.e., the time during which a method is active). The activation lifeline represents that information. That is why the input message is normally at the top of the rectangle and the return message at the bottom. Notice that the customer object has the constructor method attached to the bottom of the object. It remains active until all the data is saved, even while other get methods are invoked.

activation lifeline a representation of the period during which a method of an object is alive and executing

Design Process for Use Case Realization

Before we jump into the examples, let us review the final outcomes and the required steps to get there. As indicated in Chapter 10, the purpose of detailed design is to identify the classes required for the new system and the methods for each of those classes. Therefore, one outcome is a comprehensive design class diagram (DCD) with the attributes elaborated and the method signatures specified. This DCD may be modeled as one large diagram or as several subsystem diagrams. The other outcome is either a detailed sequence diagram or a set of CRC cards for each use case or each use case scenario. These two models are the primary input that are used to write the program code.

Figure 11-4, which also appeared as Figure 10-10, lists the steps for doing detailed design and the chapters where each step is discussed. We continue to follow those steps in this chapter. In Chapter 10, you learned about the first-cut DCD and the CRC cards. In this section, you learn how to develop the detailed sequence diagrams and update the first-cut DCD with additional information. As you learned in Chapter 10, to develop the first-cut sequence diagram, each input message is reviewed—one at a time—to determine what other internal messages and problem domain classes are required to fully process the input request. This enables us to develop the first-cut sequence diagram, which includes only the problem domain classes. Once the processing with the problem domain classes is known, the data access layer and the view layer classes and messages are added to the diagram. Finally, the DCD is updated with method signatures from the details generated

FIGURE **11-4**
Object-oriented detailed design steps

Design Step	Chapter
1. Develop the first-cut design class diagram showing navigation visibility.	10
2. Determine the class responsibilities and class collaborations for each use case using CRC cards.	10
3. Develop detailed sequence diagrams for each use case. (a) Develop the first-cut sequence diagrams. (b) Develop the multilayer sequence diagrams.	11
4. Update the DCD by adding method signatures and navigation information.	11
5. Partition the solution into packages, as appropriate.	11

during use case realization. In the next two sections, we realize two use cases: *Create customer account* and *Fill shopping cart.*

First-Cut Sequence Diagram: *Create customer account* Use Case

A detailed sequence diagram uses all the elements that an SSD uses. The difference is that the :System object is replaced by all the internal objects and messages within the system. We will identify the internal objects that collaborate and the messages they send to each other to carry out the use case or the use case scenario. Figure 11-2 illustrates the SSD for this use case.

The next step is to look at the problem domain classes and determine which classes are required for this use case. **Figure 11-5** is the class diagram for the Customer account subsystem of the RMO CSMS, as first presented in Figure 4-22. Obviously, the Customer class is needed. Of the other classes attached to the Customer class, the Address class and the Account class will probably be created within this use case. The other classes will be created with other use cases.

To create the first-cut DCD, we will elaborate the attributes with type information. Then, we will make some logical decisions about navigation visibility. Using the guidelines from Chapter 10, we decide that the Customer class would

FIGURE **11-5** *Class diagram for the Customer account subsystem*

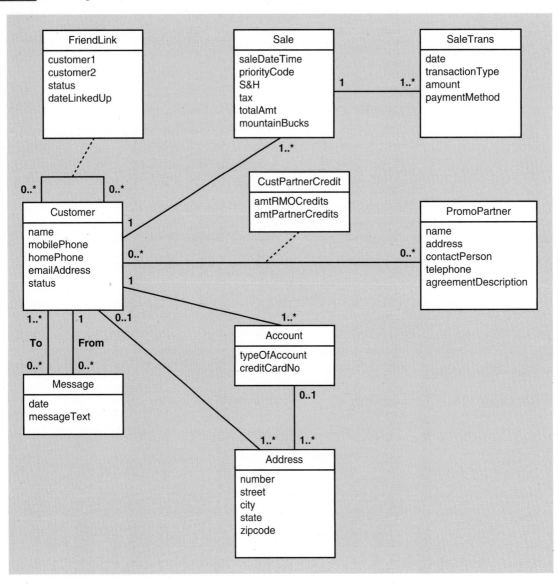

FIGURE **11-6**

First-cut DCD for the Create customer account *use case*

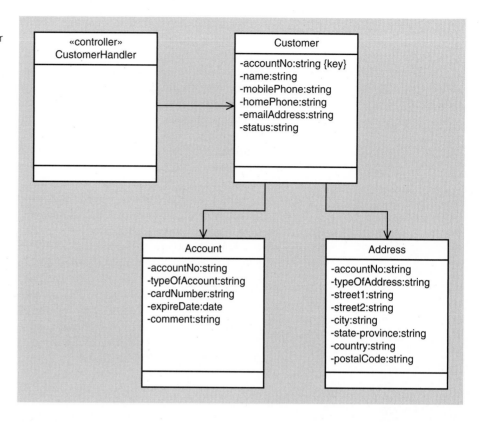

have visibility to the other two classes: Account and Address. **Figure 11-6** provides the first-cut DCD for this use case.

Based on Figure 11-2, which is the SSD for this use case, and Figure 11-6, which is the DCD, we proceed with the detailed design of the *Create customer account* use case. The first step in expanding an SSD is to place the problem domain objects in the diagram, along with the input messages from the SSD. **Figure 11-7** shows this first step in the detailed design.

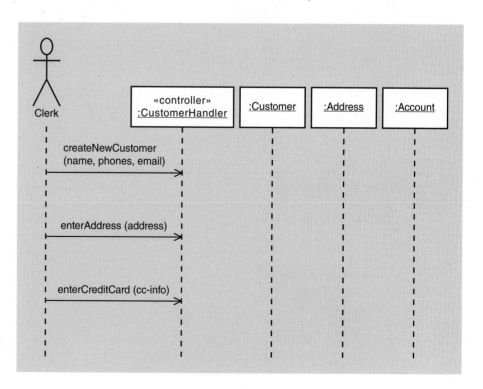

The next step is to determine the internal messages that must be sent between the objects, including which object should be the source and destination of each message. Decisions about what messages are required and which objects are involved are based on the design principles described earlier: coupling, cohesion, object responsibility, and controllers. **Figure 11-8** is the completed first-cut sequence diagram for the design of the *Create customer account* use case. The :CustomerHandler controller receives the input messages, searches for the correct order object, and forwards the createNewCustomer message to the correct :Customer object. The :Customer object takes responsibility to save itself to the database based on the createNewCustomer input message. For the other messages—enterAddress and enterCreditCard—it also takes responsibility for creating these new objects. And it does this because it "owns" them in the sense that they can't exist without a customer object. These newly created objects then save themselves to the database.

Note that there are a few differences between Figure 11-8 and Figure 11-3. Figure 11-3 represents only a partial sequence diagram because it doesn't identify all the domain classes. However, it does include the view-layer object :CustomerForm, which was added to demonstrate a more complete sequence diagram. Figure 11-8 focuses only on the domain classes. It is important for the first-cut sequence diagram to identify all the domain classes and the required internal messages between them. In a later step, the view layer classes and data access layer classes will be added.

FIGURE **11-8** *First-cut sequence diagram for the* Create customer account *use case*

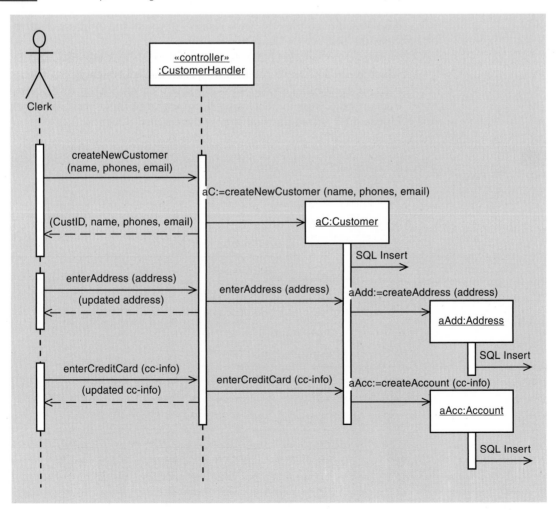

When identifying and creating messages, we must first determine the origin and destination objects for the message. The origin object is the one that needs information or help in carrying out a responsibility so it will initiate a message. The destination object is the one that has the information to help in the solution and will receive the message and process it. After determining the origin and destination objects, we must name the message. Because a message is requesting a service from the destination object, the message name should reflect the requested service. For example, when a quantity needs to be updated in the destination object, the message name should indicate the requested process to update the quantity. Notice also that the input parameters provide the information that the destination object needs in order to provide the service.

Let us analyze this solution based on some of the principles of good design that we discussed in Chapter 10 and earlier in this chapter: coupling, cohesion, object responsibility, and use case controllers.

The use case controller provides the link between the internal objects and the external environment. The problem domain objects are coupled to the use case controller, and the use case controller is coupled to the external view layer. Thus, the problem domain objects aren't coupled directly with the view layer. The responsibilities assigned to :CustomerHandler are to catch incoming messages, distribute them to the correct internal domain objects, and return the required information to the external environment.

The responsibility assigned to :Customer is to be in charge of creating itself and to control all the other required updates. The :Address and :Account objects create themselves and save themselves to the database. Coupling is straightforward, being basically vertical on the hierarchy. Thus, the assignment of responsibilities and corresponding messages conforms to good design principles. Other issues will need to be addressed as the design expands to include three layers.

First-Cut Sequence Diagram: *Fill Shopping Cart* Use Case

Before moving ahead to multilayer design, let us work through a slightly more complex example of a first-cut diagram. **Figure 11-9** is an activity diagram for the *Fill shopping cart* use case. You will remember from Figure 3-16 that this use case "included" three other use cases, as shown in Figure 11-9. By designing the use case in this manner, with other use cases included, our solution will only have to focus on those functions that actually add items to the shopping cart.

The SSD for this use case is quite simple. **Figure 11-10** shows that there are only two input messages to the system: adding an item and adding an accessory item. As you analyze the SSD, notice that adding an item to the shopping cart and adding an accessory to the cart are the same operation. The only difference is that adding accessories requires a loop for any additional accessories added for that same item. Because this is the only difference, we can simplify the diagram by limiting the solution to the first message. (Note: An additional class—AccessoryPackage—is required for the use case *Search and view accessories*, but because we aren't designing that use case, it isn't required for this solution.)

We begin this design by developing the first-cut DCD. Refer back to Figure 4-21, which presented the class diagram for the CSMS Sales subsystem. Using that diagram, we can identify the classes that are required for this use case. The Customer, Cart, and CartItem classes are necessary because the use case will be adding items for this customer to the customer's cart. To create a cart item, the system will need to know what product it is, if there are items in stock, and the price for the item. Therefore, other classes that are required are InventoryItem, ProductItem, and PromoOffering. As we develop the solution, we may have to add classes, but this appears to be sufficient for now. Navigation visibility between these classes will be from the controller to the

FIGURE **11-9**
Activity diagram for the Fill shopping
cart *use case*

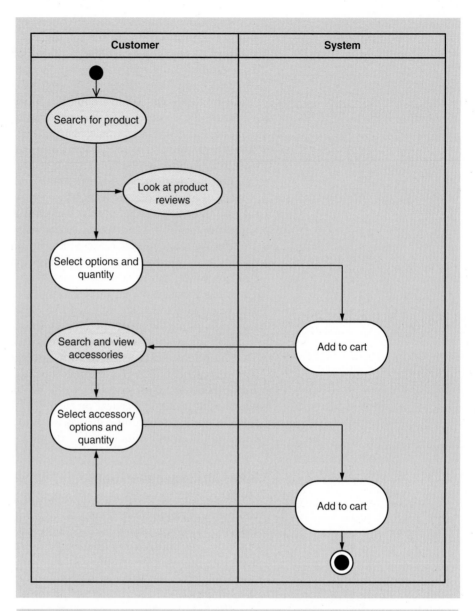

FIGURE **11-10**
System sequence diagram for the
Fill shopping cart *use case*

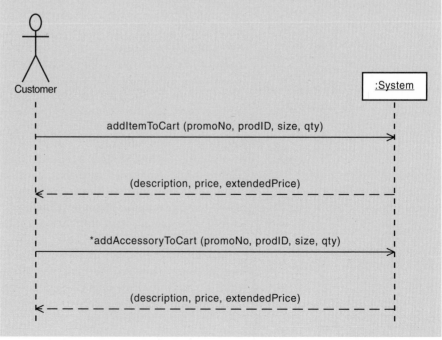

Customer class and to the Cart class once it has been created. The Cart class will be able to access the CartItem class. The CartItem class should have visibility to the other classes, such as ProductItem and InventoryItem, that contain the necessary information. **Figure 11-11** shows the first-cut DCD.

Figure 11-12 is the first-cut sequence diagram for the *Fill shopping cart* use case. Even though there are a lot of messages on it, notice that the bottom half of the diagram is simply a duplicate of the top half with a couple of minor changes. Along the left side of the diagram appear the same four input and output messages from the SSD that are shown in Figure 11-10. Along the top are the seven objects from the seven classes in the first-cut DCD shown in Figure 11-11. The other messages on the diagram are the result of our design activities.

The input addItemToCart message originates from the :Customer object and is directed to the :CartHandler controller object. The controller object determines if this is the very first item for this customer and, if so, sends a message to the :Customer object to create a new online cart. In UML, when a create message is sent to an object, it is often drawn directly to the object's box and not to the lifeline. The reference to the cart—aCrt—is then passed

FIGURE **11-11** *First-cut DCD for the* Fill shopping cart *use case*

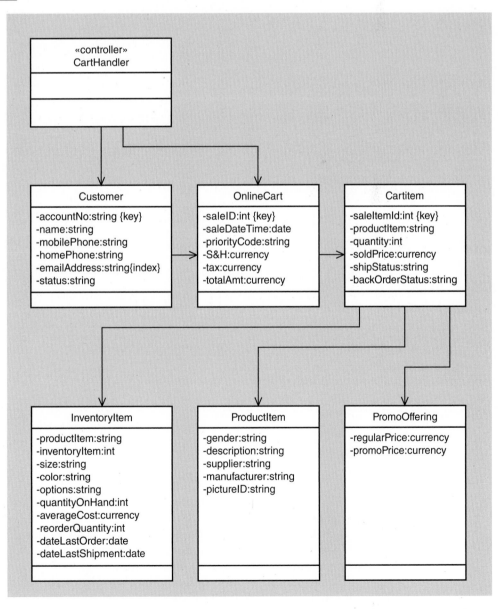

FIGURE **11-12** *First-cut sequence diagram for the* Fill shopping cart *use case*

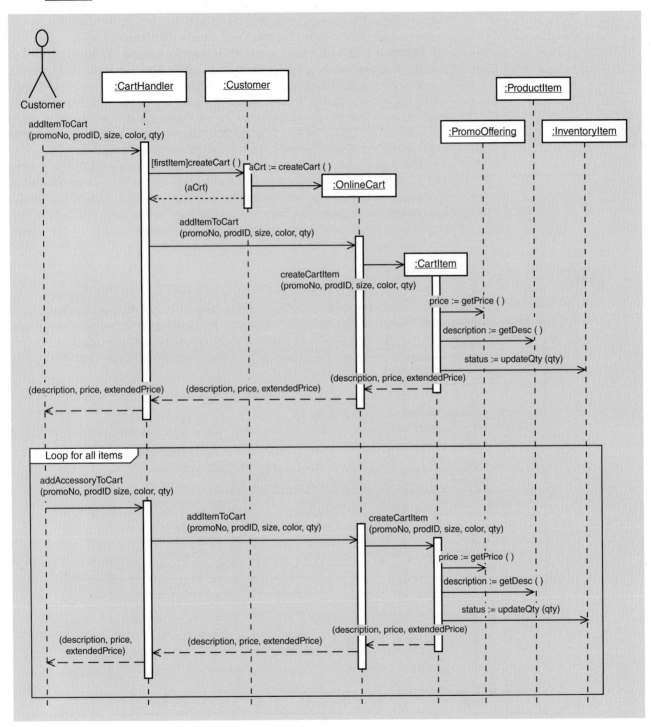

back to the controller, which uses it to send the addItemToCart message to the :OnlineCart object. The :OnlineCart object processes that message by creating a :CartItem object. The :CartItem object then takes responsibility to get its price and description and see that the inventory is updated. It does this by sending appropriate messages to those objects.

The bottom part of the sequence diagram is similar except that adding accessory items may require multiple accessories to be added for each primary item. Hence, it is shown with the rectangular box for the loop notation. Notice that the source for this loop is the asterisk in that message, as shown in the SSD. It is exploded into a loop box to indicate that all the enclosed messages participate in the loop.

The relevant design principle is that objects that "own" other objects are responsible for creating those objects; for example, a customer creates an online cart and then the online cart creates items in the cart. This approach ensures that cart objects aren't created in the absence of an owning customer object. :CartItem has responsibility to get its own information. One interesting design decision was whether :CartHandler should send the addItemToCart message directly to :OnlineCart or should send it to :Customer and allow the :Customer object to forward it to the :OnlineCart object. As noted, once a cart is created, the controller will send messages directly to the cart. This decision promotes a simpler solution with fewer messages. This design is a solid design based on good design principles.

As we identify the specific messages, along with source and destination and the passed parameters, we need to consider some critical issues. As before, an important question is: Which object is the source or initiator of a message? If the message is a query message, the source is the object that needs information. If the message is an update or create message, the source is the object that controls the other object or that has the information necessary for its creation.

Another important consideration is navigation visibility. To send a message to the correct destination object, the source object must have visibility to the destination object. Remember that the purpose of doing design is to prepare for programming. As a designer, you must think about how the program will work and consider programming issues. Given these two considerations and the source considerations discussed in the previous paragraph, we have determined that the following internal messages will be required. For each message, a source object and a destination object have been identified.

- createCart()—The :CartHandler object will know whether it has received earlier messages to add items. If it hasn't, it knows it must tell the :Customer object to create a cart.
- createCart()—The :Customer object owns the :OnlineCart object.
- addItemToCart()—A forwarded version of the input message from :CartHandler to :OnlineCart. Because :CartItem objects are dependent on a cart, :OnlineCart is the logical object to create :CartItem objects. The controller has visibility to :OnlineCart from the previous return message, when aCrt was returned.
- createCartItem()—The internal message from :Cart to :CartItem. Because :CartItem will be responsible for obtaining the data for its attributes, it needs visibility to :PromoOffering, :ProductItem, and :InventoryItem. As a result, those keys are sent as parameters.
- getPrice()—The message to get the price from the :PromoOffering object. The :CartItem initiates the message. It has visibility because it has the key values.
- getDescription()—The message initiated by :CartItem to get the description from :ProductItem.
- updateQty(qty)—The message that checks for sufficient quantity on hand. This message also initiates updates of the quantity on hand. The :CartItem object initiates the message.

By focusing only on the domain classes, we could design the core processing for the use case without having to worry about the user interface or the database. Figure 11-12 is rather complex, even though it only contains domain objects. However, this design provides a solid base for programming. Working with design models enables the designer to think through all the requirements to process a use case without having to worry about code. More importantly, it enables the designer to modify and correct a design without having to throw away code and write new code. In the next section, we will add the view layer and data access layer objects to the telephone order scenario.

Guidelines and Assumptions for First-Cut Sequence Diagram Development

From the two previous examples, we can distill several guidelines that can help you develop a design for a use case or scenario using sequence diagrams. Several assumptions are also implicit in this process. (Note that the tasks aren't done sequentially but only when necessary to build the sequence diagram. We identify them here as separate tasks simply to ensure that all three are completed.)

Guidelines

Designing a use case or scenario by using sequence diagrams involves performing these tasks:

- Take each input message and determine all the internal messages that result from that input. For each message, determine its objective. Determine what information is needed, what class needs it (the destination), and what class provides it (the source). Determine whether any objects are created as a result of the input.

- As you work with each input message, identify the complete set of classes that will be affected by the message. In other words, select all the objects from the domain class diagram that need to be involved. In Chapter 5, you learned about use case preconditions and postconditions. Any classes that are listed in either the preconditions or postconditions should be included in the design. Other classes to include are those that are created, classes that are the creators of objects for the use case, classes updated during the use case, and those that provide information used in the use case.

- Flesh out the components for each message; that is, add iteration, true/false conditions, return values, and passed parameters. The passed parameters should be based on the attributes found in the domain class diagram. Return values and passed parameters can be attributes, but they may also be objects from classes.

These three steps will produce the preliminary design. Refinements and modifications may be necessary; again, we are focusing only on the problem domain classes involved in the use case.

Assumptions

The development of the first-cut sequence diagram is based on several assumptions, including:

- Perfect technology assumption—We first encountered this assumption in Chapter 3, when we identified business events. The assumption continues here. We don't include messages such as the user having to log on.

- Perfect memory assumption—You might have noticed our assumption that the necessary objects were in memory and available for the use case. We didn't ask whether those objects were created in memory. We will change this assumption when we get to multilayer design. In multilayer design, we do include the steps necessary to create objects in memory.

- Perfect solution assumption—The first-cut sequence diagram assumes that there are no exception conditions. No logic is included to handle a situation in which the requested catalog or product isn't found. More serious exception conditions, such as the failure of a credit check, might also be encountered. Many developers design the basic processing steps first and then add the other messages and processes to handle the exception conditions later. We do the same here.

Developing a Multilayer Design

The development of the first-cut sequence diagram focuses only on the classes in the problem domain layer. In many instances, this may be sufficient documentation to program the solution—either by yourself or with another programmer. Once you have a solid design for the problem domain classes, adding the view layer and the data access layer is a straightforward process. Conforming to the principles of Agile modeling, we don't want to create diagrams unless there is real benefit. We also don't normally keep the design diagrams as documentation because over time the system will be modified and the diagrams will become obsolete. As Agile modeling suggests, be prudent in the development of models. However, there are times when it is important to see the total picture and identify the need and use of the view layer classes and the data access layer classes. A systems developer needs to know how to do complete design for those instances when it is necessary.

Every system will need view layer classes to represent the input and output screens for the application. Data access layer classes aren't always required. The data access layer is required when the business logic is fairly complex and should be isolated from the SQL statements that access the database. In the sequence diagram shown in Figure 11-8, which illustrates the creation of a new customer account, the business logic and data access logic are combined. Each domain layer object also contains the SQL insert statements to write itself to the database. We can do that because there really is no business logic required. Hence, Figure 11-8 shows a two-layer design after the view layer is added. In this section, we show a complete three-layer design for the *Fill shopping cart* use case.

Designing the Data Access Layer

The principle of **separation of responsibilities** is the motivating factor behind the design of the data access layer. On large, complex systems, designers create three-layer designs, including classes whose sole responsibility is executing database SQL statements, getting the results of the query, and providing the information to the domain layer. As hardware and networks became more sophisticated, multilayer design was used to support multitier networks in which the database server was on one machine, the business logic was on another server, and the user interface was on several desktop client machines. This way of designing systems creates more robust and more flexible systems.

In most cases, problem domain classes are also **persistent classes**, which means that their data values must be stored by the system even when the application isn't executing. The whole purpose of a relational database is to provide this ability to make problem domain objects persistent. Executing SQL statements on a database enables a program to access a record or a set of records from the database. One of the problems with object-oriented programs that use relational databases is that there is a slight mismatch between programming languages and database SQL statements. For example, in a database, tables are linked through the use of foreign keys (see Chapter 13), such as a cart having a CustomerID as a column so the order can be joined with the customer in a relational join. However, in object-oriented programming languages, the navigation is often in the opposite direction (i.e., the Customer class may have an array of references that point to the OnlineCart objects, which are in computer memory and are being processed by the system). In other words, design classes don't have foreign keys.

In this chapter, we take a somewhat simplified design approach in order to teach the basic ideas without getting embroiled in the complexities of database access. Let us assume that every domain object has a table in a relational database. (More complex situations exist in which tables must be combined to provide the correct set of objects in memory.) There are several techniques (that provide different designs) for linking the domain layer to the data access layer. One way is to have the constructor method of each problem domain object invoke the data access object to get the necessary information to complete the instantiation of the new object. Another way is to send a message to the data

separation of responsibilities a design principle that recommends segregating classes into separate components based on the primary focus of the classes

persistent classes problem domain classes that must be remembered between program executions (i.e., require storage in a database)

access layer object and have it read the database and then instantiate a new problem domain object. This second technique is better when a set of objects needs to be created from a database access that returns an unknown number of rows. However, both techniques are good solutions.

The Data Access Layer for the Fill shopping cart Use Case

To design the data access layer, we no longer assume that the objects are automatically in memory when we need them; that is, we disregard the perfect memory assumption. The design of the use case now requires additional messages to save data to the database and to retrieve data to instantiate classes. Inasmuch as there are two sets of almost identical messages, we will only illustrate the upper set of messages from Figure 11-12.

Figure 11-13 is the sequence diagram that includes the domain classes and the data access classes. Notice that all the original internal messages from Figure 11-12

FIGURE **11-13** *Sequence diagram for the* Fill shopping cart *use case with data access layer*

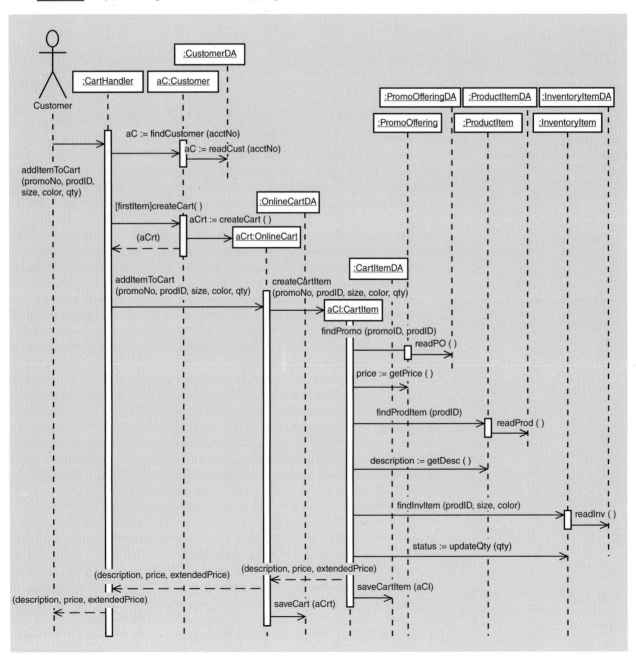

are still there. However, associated with each problem domain object is a data access class, and associated with each original internal message are additional messages to retrieve data from or save data to the database via the data access object.

It is important during this process to ensure that source objects have navigation visibility to destination objects so messages can be sent. We assume but don't show that the data access objects have global visibility. (In your programming class, you will learn that factory or singleton classes are often designed with global methods.) After the appropriate problem domain object is created, a reference to it is returned to the object that needs visibility. As you look closely at Figure 11-13, note that every object that sends a message to another object must have navigation visibility to that object. Remember this important design point as you develop your design solutions.

An effective method for understanding what is going on in Figure 11-13 is to begin with the internal messages from the first-cut sequence diagram in Figure 11-12. Let us review each one and see what changes are required.

■ [firstTime]createCart—The cart handler is going to send a message to a customer object to create a cart. First, it needs to ensure that there is a customer object in memory. It sends a findCustomer message to the aC:Customer object to find and create itself from the database. It does so by sending a message to the :CustomerDA object to read the database and return the appropriate customer object. Only then can it send the createCart message to aC:Customer. Also, note that at the end of this execution, the aCrt:OnlineCart object sends a message to the data access object to save the data to the database.

■ addItemToCart—This message is initially the same in both figures. After aCrt:CartItem has been created and populated with data, a message is sent to the data access object to save the data to the database.

■ getPrice, getDesc, updateQty—These three messages all access or update the database. Therefore, each also requires a previous message to find the appropriate data from the database, which is stored in a domain object in memory.

Even though Figure 11-13 appears rather crowded, looking at each internal message to a problem domain class makes the figure easier to understand. The primary thing to remember is that data access objects are necessary to retrieve data and thereby provide navigation visibility to the required object.

Designing the View Layer

The final step in the multilayer design of a particular use case is to add the view layer. User interface design is a complex process, as you learned in Chapter 7. It often precedes the realization of the use cases with sequence diagrams because it is more complex than can be documented with sequence diagrams. However, documenting the user interface design with the view layer on a sequence diagram often helps to envision the integration of the user interface with other system object classes.

User interface design and the integration of the view layer into a sequence diagram are made even more complex by the fact that many systems require a Web-based interface and an internal, network-based interface. Fortunately, browsers are becoming more sophisticated, so many new systems can now be designed for only one type of interface. Designing a system with multiple user interfaces is a difficult endeavor.

Although adding the user interface classes may sound simple, it must be done in conjunction with the detailed design of the user interface forms, as described in Chapter 7. **Figure 11-14** is a partial sequence diagram showing only the view layer classes and the controller class. There are two sources of inputs for the design of the view layer. First, of course, is the user-interface components that were designed during user-interface design. The second source is either the first-cut sequence diagram or the sequence diagram with the data access classes identified.

FIGURE **11-14** *Partial sequence diagram for the* Fill shopping cart *use case with view layer*

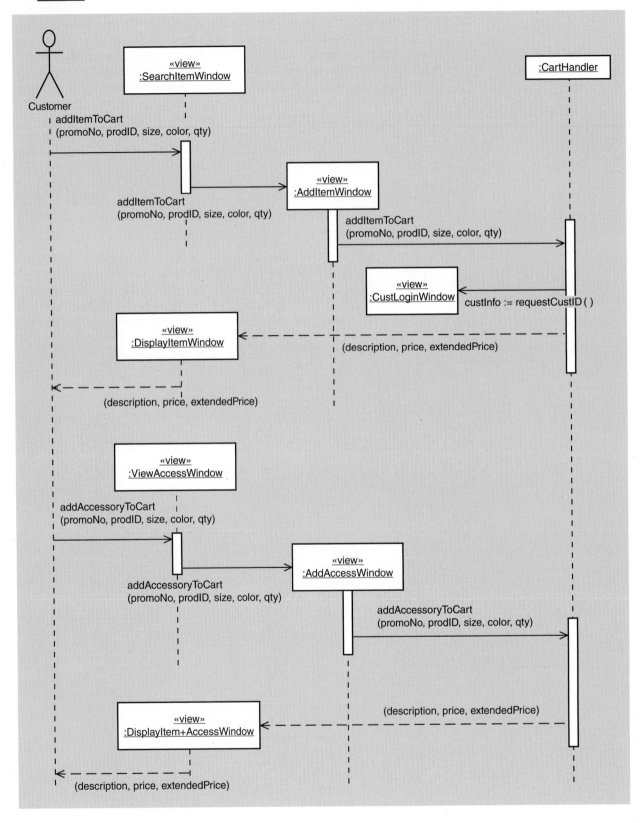

Remember that the *Fill shopping cart* use case included other use cases for *Search for item* and *View accessory combinations*. Obviously, all those use cases go together for a rich and efficient user experience. In Figure 11-14, we have added the two view layer objects for searching items and viewing accessories. The first input message—addItemToCart—will go through the :SearchItemWindow object. In other words, when the customer finds something he or she likes, he or she will initiate adding it to his or her cart from that window. The message then causes a detailed :AddItemWindow object to display and show the details to verify the addition to the cart. This later window will forward the message on to the :CartHandler object.

In Figure 11-13 the :CartHandler object required that there be an identified user. Therefore, in our example, we have shown a login window where the customer can log in to the system. Remember from the perfect technology assumption that we often omit this step. Either way is acceptable. Eventually, it does need to be added to the solution, so we included it this time.

Once the item has been added to the cart, another window displays that shows the results of adding this new item. Depending on the design of the user interface, this window might show the single newly added item or it could also be the total shopping cart.

The next three view layer objects—:ViewAccessWindow, :AddAccessWindow, and :DisplayItem+AccessWindow—function in a manner similar to the other view layer objects. The only difference is that the data includes the item and the accessories that have been added to the online cart.

Adding the view layer to your design is a good way to verify that the user interface that was developed with the users is consistent with the application design. All the input messages that were identified and documented on SSDs must be handled by the user interface. If there are messages without input windows or windows without messages, you will know that part of the design is incomplete and that more definition is required.

Designing with Communication Diagrams

Communication diagrams and sequence diagrams are interaction diagrams, and they capture the same information. The process of designing is the same whether you use communication diagrams or sequence diagrams. Which model you use is primarily your personal preference. Many designers prefer to use sequence diagrams because use case descriptions and dialog designs follow a sequence of steps. Communication diagrams are useful for showing a different view of the use case—one that emphasizes coupling. Communication diagrams are also easier to use to sketch design ideas in a meeting, as they are easier to change and rearrange on the fly.

A communication diagram uses the same symbols for actors, objects, and messages as a sequence diagram. The lifeline and activation lifeline symbols aren't used. However, a different symbol—the link symbol—is used. **Figure 11-15** illustrates the four symbols used in most communication diagrams.

The format of the message descriptor for a communication diagram differs slightly from that for a sequence diagram. Because no lifeline shows the passage of time during a scenario, each message is numbered sequentially to indicate the order of the messages. The syntax of the message descriptor in a communication diagram is as follows:

[true/false condition] sequence-number: return-value := message-name (parameter-list)

As you can see in Figure 11-15, a colon always directly follows the sequence number.

The connecting lines between the objects or between actors and objects represent links. In a communication diagram, a link shows that two items share a message—that one sends a message and the other receives it. The connecting

FIGURE **11-15** *Symbols used in a communication diagram*

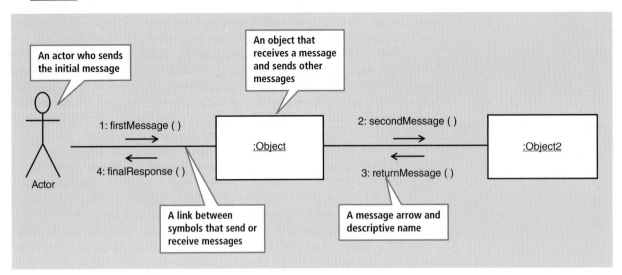

lines are essentially used only to carry the messages, so you can think of them as the wires used to transmit the messages. The numbers on the messages indicate the sequence in which the messages are sent. The hierarchical dot numbering scheme is used when messages depend on previous messages.

Figure 11-16 presents a communication diagram for the RMO *Fill shopping cart* use case. The diagram contains only domain model objects and not the view layer or the data access layer. However, multilayer design can be done just as effectively with communication diagrams as with sequence diagrams.

When you compare the communication diagrams with the sequence diagrams, it should be evident that the focus of a communication diagram is on the objects themselves. Drawing a communication diagram is an effective way to get a quick overview of the objects that work together. However, as you look at the diagrams, you should see that it is more difficult to visualize the sequence of the messages. You have to hunt to find the numbers to see the sequence of the messages.

Many designers use communication diagrams to sketch out a solution. If the use case is small and not too complex, a simple communication diagram may suffice. However, for more complex situations, a sequence diagram may be required to allow you to visualize the flow and sequence of the messages. It isn't unusual to find a mix within the same set of specifications—some use cases described by communication diagrams and others shown with sequence diagrams.

FIGURE **11-16** *Communication diagram for the* Fill shopping cart *use case*

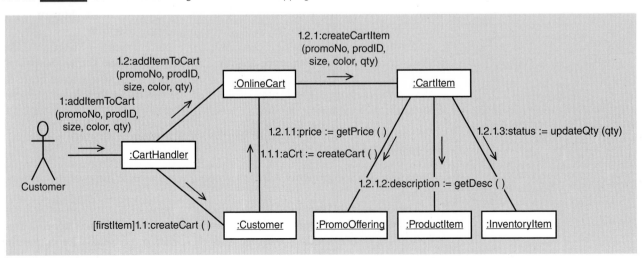

Updating and Packaging the Design Classes

Design class diagrams can now be developed for each layer. In the view layer and the data access layer, several new classes must be specified. The domain layer also has some new classes added for the use case controllers.

In Figure 11-5, we developed a partial first-cut design class diagram for the domain layer based on the *Create customer account* use case. At that point in the development, no method signatures had been developed. Now that several sequence diagrams have been created, method information can be added to the classes. We also mentioned that the navigation arrows may need updating from the decisions that were made during sequence diagram development. In Chapter 10, we briefly introduced the idea of creating method names in the classes based on responsibilities identified on the CRC cards. However, at that point, we didn't have enough information to rigorously define method signatures with names, return types, and parameter lists. Use case realization with sequence diagrams generates enough information to be rigorous in defining methods.

The first step in updating the DCD is to add the method signatures based on the information from the sequence diagrams. Three types of methods are found in most classes: (1) constructor methods, (2) data-get and data-set methods, and (3) use case-specific methods. Constructor methods create new instances of objects. Get and set methods retrieve and update attribute values. To avoid information overload, most developers don't include the get and set methods in the DCD. The third type of method—use case-specific methods—must be included in the design class diagram.

As in sequence diagrams, every message has a source object and a destination object. When a message is sent to an object, the object must be prepared to accept the message and initiate some activity. This process is nothing more than invoking or calling a method on an object. In other words, every message that appears in a sequence diagram requires a method in the destination object. In fact, the syntax for a message looks very much like the syntax for a method. Thus, the process of adding method signatures to a design class is to go through every sequence diagram and find the messages sent to that class. Each message indicates a method.

Let us work through an example based on the InventoryItem class. In Figure 11-14, two messages are sent to InventoryItem. The first is a constructor, and the other is an update message: updateQty(qty). The update message return can be void or it can return a success status value as a string. We need to add a method signature that corresponds to this message. Adding this method to the InventoryItem class is shown in **Figure 11-17**.

This process is continued for every class in the domain layer, including the added use case controller classes. **Figure 11-18** contains the completed design class diagram for the domain layer classes for the two use cases illustrated in this

FIGURE **11-17**

DCD for the InventoryItem class with method signature

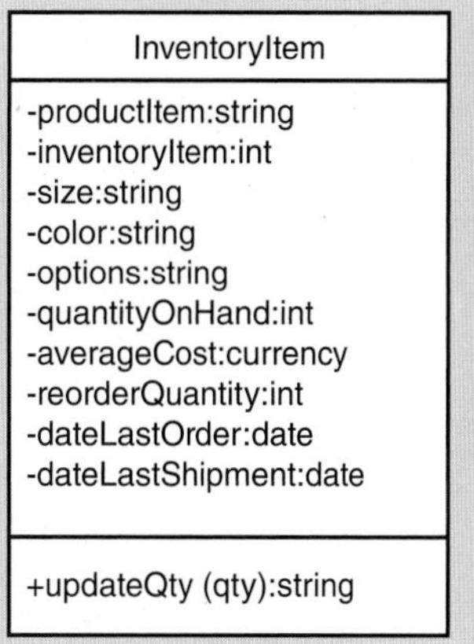

FIGURE **11-18** *Updated partial DCD for the domain layer*

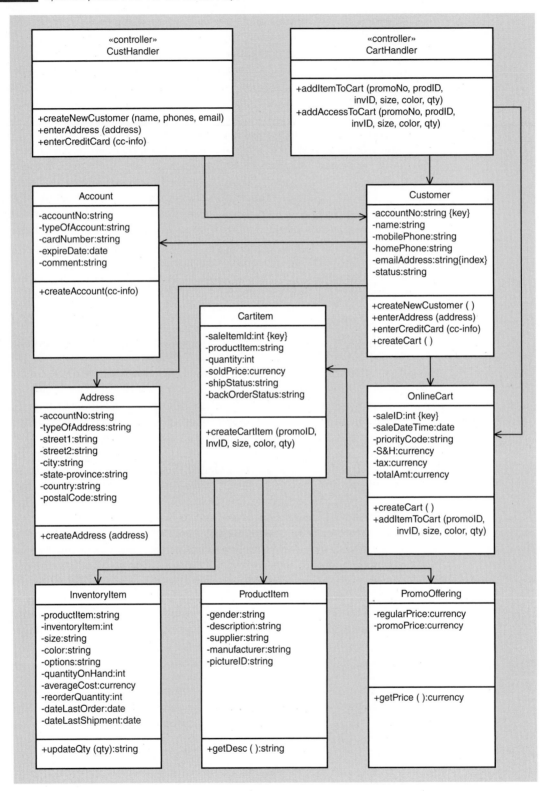

chapter. As you can see, this diagram provides excellent, thorough documentation of the design classes and serves as the blueprint for programming the system.

The two major additions to the domain layer classes are the two use case handlers. Additional navigation arrows have also been added to document which classes are visible from the controller classes. The other navigation arrows, which were defined during the first cut of the class diagram, have proved to be adequate for these two use cases. Additional use case development will enable us to add more navigation arrows.

Structuring the Major Components with Package Diagrams

As you learned previously, a package diagram in UML is simply a high-level diagram that allows designers to associate classes of related groups. The previous sections illustrated three-layer design, which includes the view layer, the domain layer, and the data access layer. In the interaction diagrams, the objects from each layer were shown together in the same diagram. However, designers sometimes need to document differences or similarities in the objects' relationships in these different layers—perhaps separating or grouping them based on a distributed processing environment. This information can be captured by showing each layer as a separate package. **Figure 11-19** illustrates how these layers might be documented.

The classes are placed inside the appropriate package based on the layer to which they belong. Classes are associated with different layers as they are developed in the interaction diagrams. To develop this package diagram, we simply extracted the information from design class diagrams and interaction diagrams for each use case. Figure 11-19 is only a partial package diagram

FIGURE **11-19**

Partial design of three-layer package diagram for RMO

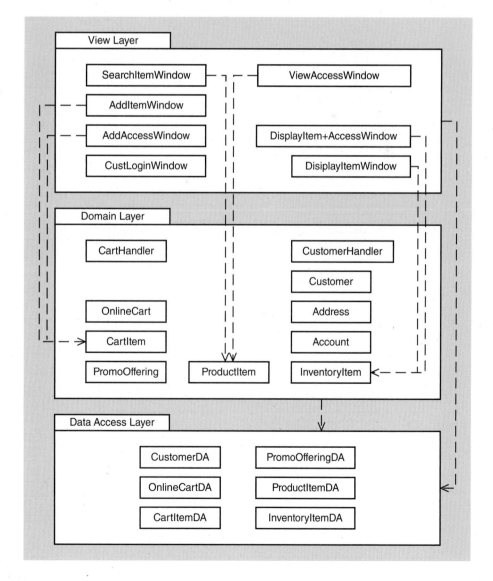

dependency relationship a relationship between packages, classes, or use cases in which a change in the independent item requires a change in the dependent item

because the packages contain only the classes from the use case interaction diagrams that were developed in this chapter.

The other symbol used on a package diagram is a dashed arrow, which represents a **dependency relationship**. The arrow's tail is connected to the package that is dependent, and the arrowhead is connected to the independent package. Dependency relationships are used in package diagrams, class diagrams, and even interaction diagrams. A good way to think about a dependency relationship is that if one element changes (the independent element), the other (dependent) element might also have to be changed. Dependency relationships can be between packages or between classes within packages. Figure 11-19 indicates that several classes in the view layer are dependent on classes in the domain layer. Thus, for example, if a change is made in the ProductItem class, the SearchItemWindow class should be evaluated to capture that change. However, the reverse isn't necessarily true. Changes to the view layer usually don't carry through to the domain layer.

Two examples of dependency relationships are given in Figure 11-19. The first, we have seen, is between classes. Another example is less detailed and indicates a dependency between packages. **Figure 11-20** indicates that the view layer and the domain layer depend on the data access layer. For some simple queries against the database, the view layer may directly access the data layer without requiring any involvement of the domain layer. These dependencies indicate that changes to the data structures, as reflected in the data access layer, usually require changes at the domain layer and the view layer.

FIGURE **11-20** *RMO subsystem packages*

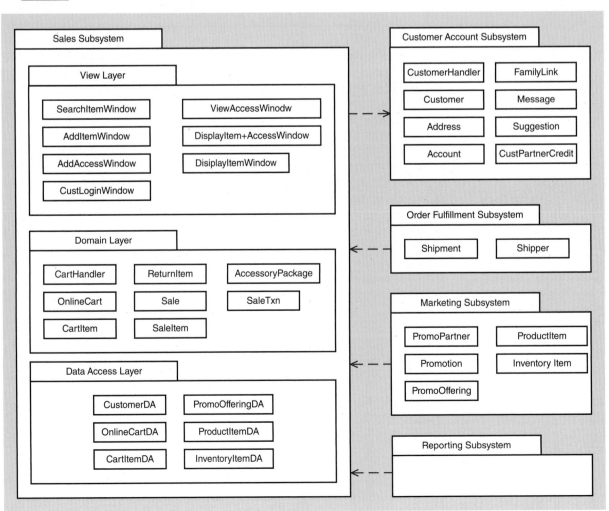

Package diagrams can also be nested to show different levels of packages. Figure 11-20 indicates that the packages as well as some of the classes contained within them are all part of the Order-entry subsystem. The RMO system can be divided into subsystems, and one way to document them is with package diagrams. A major benefit of this documentation is that different packages can be assigned to different teams of programmers to program the classes. The dependency arrows will help them recognize where communication among teams is needed to ensure an integrated system.

In summary, package diagrams show related components and dependencies. Generally, we use package diagrams to relate classes or other system components such as network nodes. The preceding figures show two uses of package diagrams: dividing a system into subsystems and showing the nesting within packages.

Implementation Issues for Three-Layer Design

Using design class diagrams, interaction diagrams, and package diagrams, programmers can begin to build the components of a system. Thus, implementation in this sense means constructing the system with a programming language such as Java, PHP, or such Visual Studio languages as VB or C#. Over the last few years, integrated development environment (IDE) tools have been developed to help programmers construct systems. Such tools as Jbuilder and Eclipse (for Java), Aptana (for PHP), Visual Studio (for Visual Basic), and C# and C++Builder (for C++) provide a high level of programming support, especially in building the view layer classes—the windows and window components of a system.

Unfortunately, these same tools have propagated some bad programming habits in some developers. The ease with which programmers can build GUI windows and automatically insert code has allowed them to put all the code in the windows. Each window component has several associated events where code can be inserted. Thus, some programmers find it easy to build a window with an IDE tool, let the tool automatically generate the class definition, and merely insert business logic code. No new classes need to be defined, and little other coding is required. Some of these tools also have database engines, so the entire system can be built with windows classes. However, taking such shortcuts exacts a price later.

The problem with this approach is the difficulty of maintaining the system. Code snippets scattered throughout the GUI classes are hard to find and maintain. Plus, when the user interface classes need to be upgraded, the programmer must also find and update the business logic. If a network-based system needs to be enhanced to include a Web front end, a programmer must rebuild nearly the entire system. Or if two user interfaces are desired, all the business logic is programmed twice. Finally, without the tool that generates the code, it is almost impossible to keep the system current. This problem is exacerbated by new releases of the IDE tools, which may not be compatible with earlier versions. Many programmers have had to completely rewrite the front end of a system because the new release of an IDE tool didn't generate code the same way the previous release did. Thus, we advise analysts and programmers to use good design principles in the development of new systems.

Based on the design principle "object responsibility," it is possible to define which program responsibilities belong to each layer. If you follow these guidelines when writing code, a system will be much easier to maintain throughout its lifetime. Let us summarize the primary responsibilities of each layer.

View layer classes should have programming logic to perform the following:

- Display electronic forms and reports.
- Capture such input events as clicks, rollovers, and key entries.
- Display data fields.
- Accept input data.
- Edit and validate input data.
- Forward input data to the domain layer classes.
- Start and shut down the system.

Domain layer classes should have responsibilities to perform the following:

- Create problem domain (persistent) classes.
- Process all business rules with appropriate logic.
- Prepare persistent classes for storage to the database.

Data access layer classes should have responsibilities to perform the following:

- Establish and maintain connections to the database.
- Contain all SQL statements.
- Process result sets (the results of SQL executions) into appropriate domain objects.
- Disconnect gracefully from the database.

Design Patterns

Systems that are based on good design principles aren't only easier to develop and put into operation the first time, they are also much easier to maintain. Such concepts as object responsibility, coupling, cohesion, protection from variations, and indirection were introduced in Chapter 10 and have been applied throughout the discussion in Chapter 11.

You are also familiar with the concepts of design patterns and with two specific patterns: three-layer design and use case controllers. Patterns exist at various levels of abstraction. At a concrete level, a pattern may be a class definition that is written in code to be used by any developer. At the most abstract level, a pattern might only be an approach to solving a problem. For example, the multilayer design pattern tends to be more abstract and recommends that it is better to separate system functions into three layers of classes; the GUI logic is placed in a set of view-layer classes that are separate and distinct from the domain layer and data access layer. Thus, multilayer design is an approach to building a system rather than a specific solution.

The use case controller pattern is more concrete. It defines a specific class or classes that act as the switchboard for all incoming messages from the environment. As with all patterns, there are multiple ways to implement the controller pattern. A single controller class can be defined to handle all messages from the view layer to the domain layer. Alternatively, a class can be defined for each use case or some combination of the two can be used. Regardless of the specific approach, the controller pattern does require a separate, specified class.

Adapter

We start with the adapter pattern because the concept is straightforward. The adapter pattern is also a good example of the design principles "protection from variations" and "indirection." An adapter pattern is roughly akin to an electrical adapter used for international travel. Thus, if you are traveling to England, you might decide to take your hair dryer with you. It has a switch for either 110 volts or 220 volts, so you think you can run it on either voltage. However, the plug on the end of the power cord has two flat prongs. Unfortunately, wall sockets in England have slots for three large prongs set at angles. You need something that can adapt the power cord's two prongs to the wall's three angled slots. **Figure 11-21** shows a typical electrical adapter you might use.

FIGURE **11-21**
Electrical adapter

The adapter design pattern works just like the electrical adapter; it plugs an external class into an existing system. The method signatures on the external class are different from the method names being called from within the system, so the adapter class is inserted to convert the method calls from within the system to the method names in the external class.

Figure 11-22 describes the details of the adapter design pattern. The sample diagram has four UML classes. The one labeled System represents the

FIGURE **11-22** *Adapter pattern template*

Name:	Adapter
Problem:	A class must be replaced, or is subject to being replaced, by another standard or purchased class. The replacing class already has a predefined set of method signatures that are different from the method signatures of the original class. How do you link in the new class with a minimum of impact so that you don't have to change the names throughout the system to the method names in the new class?
Solution:	Write a new class, the adapter class, which serves as a link between the original system and the class to be replaced. This class has method signatures that are the same as those of the original class (and the same as those expected by the system). Each method then calls the correct desired method in the replacement class with the method signature. In essence, it "adapts" the replacement class so that it looks like the original class.
Example:	There are several places in the RMO system where class libraries were purchased to provide special processing. These purchased libraries provide specialized services such as tax calculations and shipping and postage rates. From time to time, these service libraries are updated with new versions. Sometimes a service library is even replaced with one from an entirely different vendor. The RMO systems staff applies protection from variations and indirection design principles by placing an adapter in front of each replaceable class. 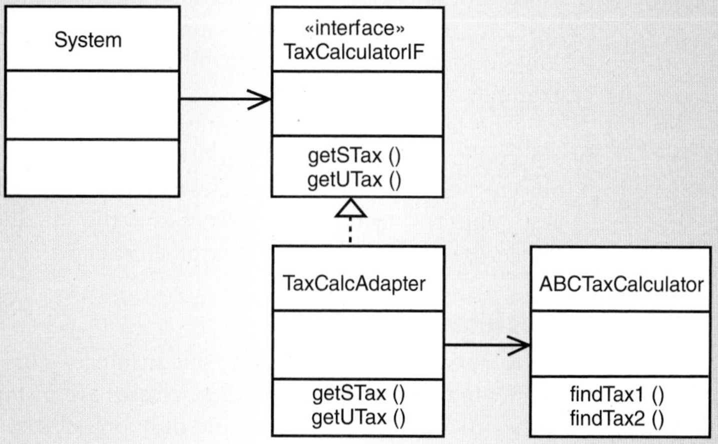
Benefits and consequences:	The adaptee class can be replaced as desired. Changes are confined to the adapter class and do not ripple through the system. Two classes are defined, an interface class and the adapter class. Passed parameters may add more complexity, and it is difficult to limit changes to the adapter class.

entire system. The classes within the system use such method names as getSTax() and getUTax() to access the tax routines. The TaxCalculator class has the method names findTax1() and findTax2(). The two UML classes in the middle represent the adapter. The top middle class symbol represents an interface class. An interface is useful to specify the method names; although not absolutely necessary, it is a simple way to specify and enforce the use of the correct method names. The adapter class then inherits those method names and provides the method logic for those methods. The body of each method simply extends a call to the final method name findTax1() or findTax2(). In other words, it "adapts," or translates, the method names from one to the other.

As you become familiar with this design pattern, you will find that it has a multitude of uses. It is a powerful and elegant solution to making a system more maintainable. Experienced developers use this pattern frequently—for foreign classes and for internally written classes that may need frequent upgrades. It is an excellent way to insulate the system from frequently changing classes.

Factory

In the discussions of detailed design, we have often expressed the need to have utility classes, which include the data access objects or controller classes. An adapter in an adapter pattern situation is also a utility class. What class should create these utility objects? In most situations, it doesn't make sense for domain classes to create them because it isn't a listed responsibility of domain classes. A popular solution in object-oriented programming is to have some classes that are factories. In other words, these classes instantiate objects from utility classes.

For example, an executing customer object may need to write some data. If the factory class is designed with static methods, which means they have global visibility, the customer object can just say to the factory, "Get me a reference to a data access object for the customer table." The factory will create a new data access object and return the reference. If a customer data access object already exists in memory, it simply returns the reference. The customer object doesn't have to be concerned about creating objects to access the database. It just uses whatever is passed to it. This reduces coupling, enhances cohesion, and assigns responsibilities to the right classes. **Figure 11-23** is an example of a factory class.

The factory class has private attributes to hold the references to the data objects that are created. When a request is made to get the reference to a data object, the method simply checks to see if the attribute is null. If so, it creates a new object, places its reference in the attribute, and returns the value. Otherwise, it just returns the parameter with the reference already in it.

Singleton

Some classes must have exactly one instance—for example, a factory class or the main window class. Because these classes are instantiated from only one place, it is easy to limit the logic to create only one object.

Other classes must have exactly one instance but can't be easily controlled by having only one place to invoke the constructor. Depending on the system's flow of logic, a particular class might get instantiated from multiple locations. However, only one instance needs to be created, so the first one that needs it creates it, and every other class uses the one that was initially created. Usually, these classes are service classes that manage a system resource, such as a database connection. In fact, the factory class that was just described is an excellent example. This common problem has a standard solution: the singleton pattern.

FIGURE **11-23** *Factory method pattern template*

Name:	Factory or Factory Method
Problem:	Who should be responsible for creating utility type objects that do not specifically belong to the problem domain classes? These utility objects may also be accessed from various places within the system, so a given object may need to be instantiated from several classes.
Solution:	Create an artifact that is a factory class. Its responsibility is only to instantiate utility classes. In many situations, only one instance of a particular utility class is allowed. Hence, all classes that need access to the class come through the factory. The factory ensures that only one instance is created.
Example:	Several places in the RMO system need to get data from an Order object and need to have a reference to an Order_DA [data access] object. The Order_DA object may or may not already have been instantiated. A data access factory is defined and an interface is created. The requesting object uses the methods defined in the interface to request the reference to the Order_DA object. It then can read the database of orders.

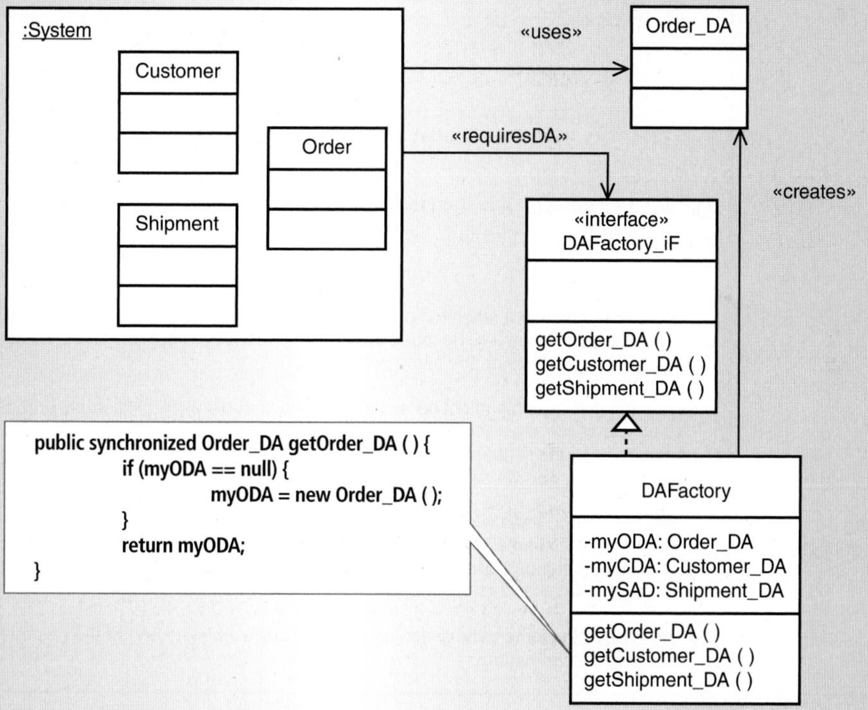

```
public synchronized Order_DA getOrder_DA ( ) {
        if (myODA == null) {
                myODA = new Order_DA ( );
        }
        return myODA;
}
```

Benefits and Consequences:	Higher cohesion of problem domain classes Less coupling between business logic layer and data layer Smaller, more maintainable classes

Figure 11-24 presents the template of the description for the singleton pattern. The singleton pattern provides a solution in which the class itself controls the creation of only one instance.

Notice that the singleton pattern has the same basic logic as the factory method pattern. The difference is that the singleton class has code that applies to itself as static methods. The approach of the singleton solution is that the

FIGURE **11-24** *Singleton pattern technique*

Name:	Singleton
Problem:	Only one instantiation of a class is allowed. The instantiation (new) can be called from several places in the system. The first reference should make a new instance, and later attempts should return a reference to the already instantiated object. How do you define a class so that only one instance is ever created?
Solution:	A singleton class has a static variable that refers to the one instance of itself. All constructors to the class are private and are accessed through a method or methods, such as getInstance(). The getInstance() method checks the variable; if it is null, the constructor is called. If it is not null, then only the reference to the object is returned.
Example:	In RMO's system, the connection to the database is made through a class called Connection. However, for efficiency, we want each desktop system to open and connect to the database only once, and to do so as late as possible. Only one instance of Connection—that is, only one connection to the database—is desired. The Connection class is coded as a singleton. The following coding example is similar to C# and Java: ```Class Connection
{
private static Connection conn = null;
public synchronized static getConnection ( )
{
if (conn == null) {
conn = new Connection ( );}
return conn;
}
}``` Another example of a singleton pattern is a utilities class that provides services for the system, such as a factory pattern. Because the services are for the entire system, it causes confusion if multiple classes provide the same services. An additional example might be a class that plays audio clips. Since only one audio clip should be played at one time, the audio clip manager will control that. However, for this to work, there must be only one instance of the audio clip manager.
Benefits and consequences:	There are other times when only one instance of an object is needed, but if it is instantiated from only one place, then a singleton may not be required. The singleton object controls itself and ensures that only one instance is created—no matter how many times it is called and wherever the call occurs in the system. The code to implement the singleton is very simple, which is one of the desirable characteristics of a good design pattern.

class has a static variable that refers to the created object. A method such as getConnection is defined and used to get the reference to the object. The first time the getConnection method is called, it instantiates an object and returns a reference to it. On later calls to the method, it simply returns a reference to the already instantiated object. The code is simple and elegant. The example doesn't show the constructor; however, to ensure that only one instance is created, all constructors are specified as private—not accessible—so no other class can accidentally invoke one.

In the singleton template, the pattern is represented by code. To specify this in your design, you should stereotype the class as a ≪singleton≫. Good programmers will recognize the stereotype and know exactly how to code the class.

Chapter Summary

Multilayer design of new systems isn't limited to architectural design. Detailed object-oriented design also identifies the various levels in a system. The identification of classes and their responsibilities follows the three-layer pattern explained in this chapter. The three layers are the view layer, the business (or logic) layer, and the data access layer.

Three-layer design is part of the overall movement in systems design based on design patterns. A design pattern is a standard solution or template that has proven to be effective to a particular requirement in systems design. The other pattern, introduced in Chapter 10, is a use case controller, which addresses the need to isolate the view layer from the business layer in a simple way that limits coupling between the two layers.

Detailed design is use case driven in that each use case is designed separately. This type of design is called *use case realization*. The two primary models used for detailed design are the design class diagram and the sequence diagram. Design class diagrams were discussed in Chapter 10.

Detailed design of use cases entails identifying problem domain classes that collaborate to carry out a use case. Each input message from an external actor triggers a set of internal messages. Using a sequence diagram or a communication diagram, the designer identifies and defines all these internal messages. In the first cut, only the problem domain classes and their internal messages are identified. Next, the solution is completed by adding the classes and messages for the view layer and the data access layer.

The final step is to convert each message, along with the passed parameters and return values, into method signatures located in the correct classes. This information is used to update the design class diagram. Changes are also made to the design class diagram to show required visibility between the classes in order to send messages in the sequence diagrams.

As classes are identified during the design process, they are added to the DCD. The DCD can also be partitioned into several layers or into subsystems. Package diagrams are used to partition the DCD into appropriate packages. Dependency between the classes and the packages is also added to the package diagram.

Popular design patterns include the adapter pattern, factory pattern, singleton pattern, and observer pattern. The adapter pattern implements the design principle "protection from variations" by allowing a changing piece of the system to simply plug into a more stable part of the system. When the pluggable piece of the system needs to change, it can just be unplugged and the updated component can be plugged in.

The factory and singleton patterns have much in common. Both return a reference to a specific object. Both allow only one instance of that object to exist in the system. The difference is that the factory pattern enforces a single occurrence for utility classes and the singleton only enforces a single occurrence for itself.

Key Terms

activation lifeline 335	persistent classes 345
communication diagrams 332	separation of responsibilities 345
dependency relationship 354	sequence diagrams 332
design patterns 330	use case realization 328

Review Questions

1. What is meant by the term *use case realization*?
2. What are the benefits of knowing and using design patterns?
3. What is the contribution to systems development by the Gang of Four?
4. What are the five components of a standard design pattern definition?
5. List five elements included in a sequence diagram.
6. How does a sequence diagram differ from an SSD?
7. What is the difference between designing with CRC cards and designing with sequence diagrams?
8. Explain the syntax of a message on a sequence diagram.
9. What is the purpose of the first-cut sequence diagram? What kinds of classes are included?
10. What is the purpose of the use case controller?
11. What is meant by an activation lifeline? How is it used on a sequence diagram?

12. Describe the three major steps in developing the set of messages for the first-cut sequence diagram.

13. What assumptions do developers usually make while doing the initial use case realization?

14. When doing multilayer design, what is the order in which layers should be designed? Why?

15. What is the "separation of responsibilities" principle?

16. Explain the two methods of accessing the database to create new objects in memory.

17. What symbols are used in a communication diagram, and what do they mean?

18. Explain the components of message syntax in a communication diagram. How does this syntax differ from that of a sequence diagram message?

19. Explain the method syntax on design classes.

20. What is meant by a dependency relationship? How is it indicated on a drawing?

21. List the major implementation responsibilities of each layer in a three-layer design.

22. What is the purpose of the adapter pattern?

23. What common element is found in the singleton pattern and the factory pattern? What is the basic difference between the two patterns?

Problems and Exercises

Problems 1 through 7 are based on the solutions you developed in Chapter 5 for problems 1 and 2, which involved a university library system. Alternatively, your instructor may provide you with a use case diagram and a class diagram.

1. **Figure 11-25** is an SSD for the use case *Check out books* in the university library system. Do the following:
   a. Develop a first-cut sequence diagram that only includes the actor and problem domain classes.
   b. Develop a design class diagram based on your solution. Be sure to include your controller class.

2. Using your solution to problem 1, do the following:
   a. Add the view layer classes and the data access classes to your diagram. You may do this with two separate diagrams to make them easier to work with and read.

   b. Develop a package diagram showing a three-layer solution with view layer, domain layer, and data access layer packages.

3. **Figure 11-26** is an activity diagram for the use case *Return books* in the university library system. Do the following:
   a. Develop a first-cut sequence diagram that only includes the actor and problem domain classes.
   b. Develop a design class diagram based on the domain class diagram.

4. Using your solution to problem 3, do the following:
   a. Add the view layer classes and the data access classes to your diagram.
   b. Develop a package diagram showing a three-layer solution with view layer, domain layer, and data access layer packages.

FIGURE **11-25**

System sequence diagram for the Check out books *use case*

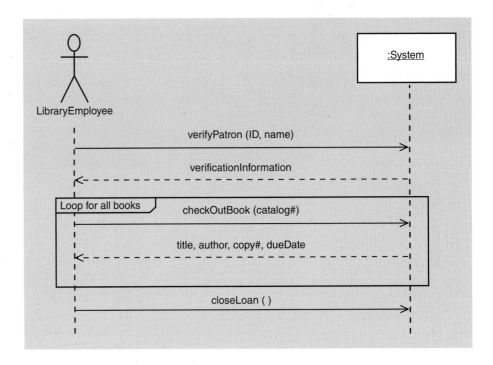

FIGURE **11-26**

Activity diagram for the Return books *use case*

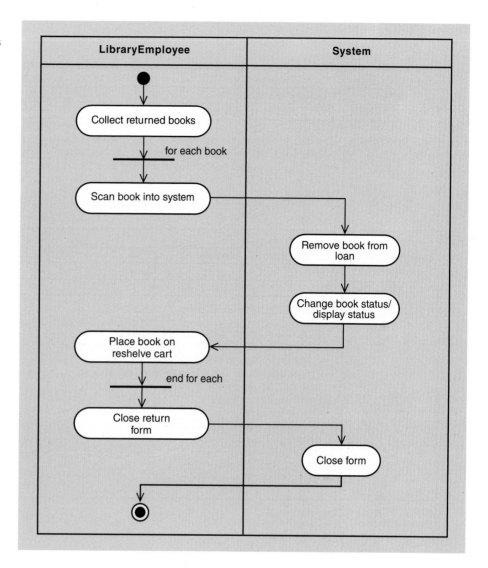

5. **Figure 11-27** is a fully developed use case description for the use case *Receive new book* in the university library system. Do the following:
   a. Develop a first-cut communication diagram that only includes the actor and problem domain classes.
   b. Develop a design class diagram based on the domain class diagram.

6. Using your solution to problem 5, do the following:
   a. Add the view layer classes and the data access classes to your diagram.
   b. Develop a package diagram showing a three-layer solution with view layer, domain layer, and data access layer packages.

7. Integrate the design class diagram solutions you developed for problems 1, 3, and 5 into a single design class diagram.

   Problems 8 through 14 are based on the solutions you developed for problems 3 and 4 in Chapter 5, which involved a dental clinic system. Alternatively, your instructor may provide you with a use case diagram and a class diagram.

8. **Figure 11-28** is an SSD for the use case *Record dental procedure* in the dental clinic system. Do the following:
   a. Develop a first-cut sequence diagram that only includes the actor and problem domain classes.
   b. Develop a design class diagram based on the domain class diagram.

9. Using your solution to problem 8, do the following:
   a. Add the view layer classes and the data access classes to your diagram.
   b. Develop a package diagram showing a three-layer solution with view layer, domain layer, and data access layer packages.

10. **Figure 11-29** is an activity diagram for the use case *Enter new patient information* in the dental clinic system. Do the following:
    a. Develop a first-cut sequence diagram that only includes the actor and problem domain classes.
    b. Develop a design class diagram based on the domain class diagram.

FIGURE **11-27** *Fully developed use case description for the* Receive new book *use case*

Use Case Name:	Receive new book	
Scenario:	Receive new book	
Triggering Event:	Newly purchased book arrives	
Brief Description:	The librarian decides on purchases of new books and places order (prior to this use case). Shipments of new books arrive. Each new book is assigned a library catalog number. Some books are simply additional copies of existing titles. Some books are new editions of existing titles. Some books are new titles and new physical books. The new book information is added to the system.	
Actors:	Library Employee	
Stakeholders:	Library Employee, Librarian	
Preconditions:	None	
Postconditions:	Book Title exists, Physical Book exists	
Flow of Activities:	Actor	System
	1. Collect new books from receipt of shipment. 2. For each book, research book category and catalog numbers. Assign tentative number. 3a. If new copy of existing title, enter book information and catalog number into system. 3b. If new edition of existing title, enter book information, edition information, and catalog number. 3c. If new title, assign general catalog number. Assign book copy number. 4. Mark book with number. 5. Place book on shelving cart. 6. Repeat for each book (back to step 2).	3a.1 Update catalog with new number.Verify that not duplicate. 3b.1 Update catalog with new number. Verify that not duplicate. 3c.1 Verify that catalog number not duplicate.
Exception Conditions:	Duplicate numbers require further research and reassignment of catalog numbers.	

FIGURE **11-28**

System sequence diagram for the Record dental procedure *use case*

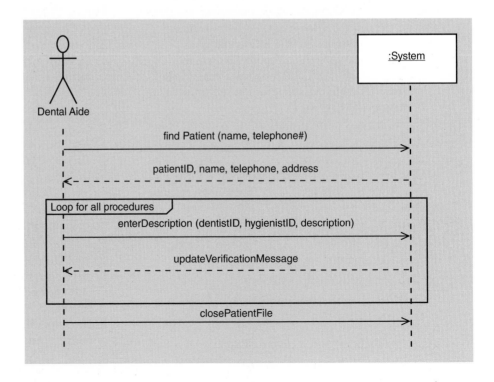

FIGURE **11-29**

Activity diagram for the Enter new patient information *use case*

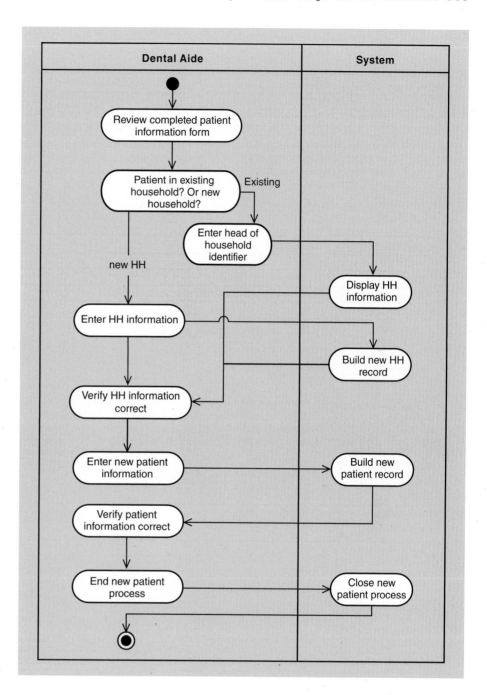

11. Using your solution to problem 10, do the following:
    a. Add the view layer classes and the data access classes to your diagram.
    b. Develop a package diagram showing a three-layer solution with view layer, domain layer, and data access layer packages.

12. **Figure 11-30** is a fully developed use case description for the use case *Print patient invoices* in the dental clinic system. Do the following:
    a. Develop a first-cut communication diagram that only includes the actor and problem domain classes.
    b. Develop a design class diagram based on the domain class diagram.

13. Using your solution to problem 12, do the following:
    a. Add the view layer classes and the data access classes to your diagram.
    b. Develop a package diagram showing a three-layer solution with view layer, domain layer, and data access layer packages.

14. Integrate the design class diagram solutions that you developed for problems 8, 10, and 12 into a single design class diagram.

15. In **Figure 11-31**, the package on the left contains the classes in a payroll system. The package on the right is a payroll tax subsystem. What technique would you use to integrate the payroll tax subsystem into the payroll system? Show how you would solve the problem by modifying the existing classes (in either figure). What new classes would you add? Use UML notation.

FIGURE **11-30** *Fully developed use case description for the* Print patient invoices *use case*

Use Case Name:	Print patient invoices	
Scenario:	Print patient invoices	
Triggering Event:	At the end of the month, invoices are printed	
Brief Description:	The billing clerk manually checks to see that all procedures have been collected. The clerk spot-checks, using the written records to make sure procedures have been entered by viewing them with the system. The clerk also makes sure all payments have been entered. Finally, he/she prints the invoice reports. An invoice is sent to each patient.	
Actors:	Billing Clerk	
Stakeholders:	Billing Clerk, Dentist	
Preconditions:	Patient Records must exist, Procedures must exist	
Postconditions:	Patient Records are updated with last billing date	
Flow of Activities:	Actor	System
	1. Collect all written notes about procedures completed this month. 2. View several patients to verify that procedure information has all been entered. 3. Review log of payments received and verify that payments have been entered. 4. Enter month-end date and request invoices. 5. Verify invoices are correct. 6. Close invoice print process.	2.1 Display patient information, including procedure records. 3.1 Display patient information, including account balance and last payment transactions. 4.1 Review every patient record. Find unpaid procedures. List on report as aged or current. Calculate and break down by copay and insurance pay.
Exception Conditions:	None	

FIGURE **11-31** *Payroll system packages and classes*

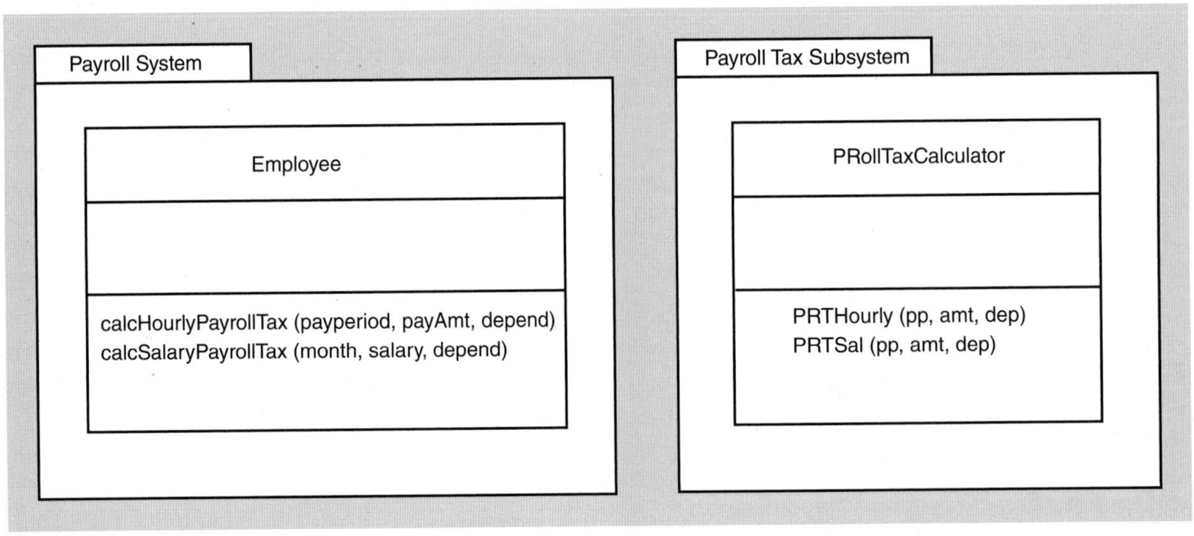

Payroll System

Employee

calcHourlyPayrollTax (payperiod, payAmt, depend)
calcSalaryPayrollTax (month, salary, depend)

Payroll Tax Subsystem

PRollTaxCalculator

PRTHourly (pp, amt, dep)
PRTSal (pp, amt, dep)

Case Study

MoveYourBooksNow.com Book Exchange

MoveYourBooksNow.com is a book exchange that does business entirely on the Internet. The company acts as a clearinghouse for buyers and sellers of used books.

To offer books for sale, a person must register with MoveYourBooksNow.com. The person must provide a current physical address and telephone number as well as a current e-mail address. The system maintains an open account for this person. Access to the system as a seller is through a secure, authenticated portal.

A seller can list books on the system through a special Internet form. Information required includes all the pertinent information about the book, its category, its general condition, and the asking price. A seller may list as many books as desired. The system maintains an index of all books in the system so buyers can use the search engine to search for books. The search engine allows searches by title, author, category, and keyword.

People who want to buy books come to the site and search for the books they want. When they decide to buy, they must open an account with a credit card to pay for the books. The system maintains all this information on secure servers.

When a request to purchase is made and the payment is sent, MoveYourBooksNow.com sends an e-mail notice to the seller of the book. It also marks the book as sold. The system maintains an open order until it receives notice that the book has been shipped. After the seller receives notice

that a listed book has been sold, the seller must notify the buyer via e-mail within 48 hours. Shipment of the order must be made within 24 hours of the seller sending the notification e-mail. The seller sends a notification to the buyer and MoveYourBooksNow.com when the shipment is made.

After receiving notice of shipment, MoveYourBooksNow.com maintains the order in shipped status. At the end of each month, a check is mailed to each seller for the book orders that have been in shipped status for 30 days. The 30-day waiting period allows the buyer to notify MoveYourBooksNow.com if the shipment doesn't arrive for some reason or if the book isn't in the same condition as advertised.

If they want, buyers can enter a service code for the seller. The service code is an indication of how well the seller is servicing book purchases. Some sellers are very active and use MoveYourBooksNow.com as a major outlet for selling books. Thus, a service code is an important indicator to potential buyers.

For this case, develop the following diagrams:

1. A domain model class diagram
2. A use case diagram
3. SSDs for two use cases, such as *Add a seller* and *Record a book order*
4. A first-cut sequence diagram for each of the above use cases
5. An integrated design class diagram that includes classes, methods, and navigation attributes

RUNNING CASE STUDIES

Community Board of Realtors

In Chapter 3, you identified use cases for the business events for the Community Board of Realtors. In Chapter 5, you elaborated on those use cases. In Chapter 4, you identified the classes associated with the business events. Using your solutions from those chapters, develop:

1. A first-cut DCD by using the problem domain classes that you identified in Chapter 4.
2. A first-cut communication diagram for the *Create new listing* use case (domain classes and controller class only).
3. A first-cut sequence diagram for the *Update agent information* use case (domain classes and controller class only).

4. A multilayer sequence diagram for the *Update agent information* use case that includes domain classes and data access classes.
5. A separate multilayer sequence diagram for the *Update agent information* use case that includes the domain classes and the view layer classes.
6. A final design class diagram that includes the classes from both use cases. Include elaborated attributes, navigation arrows, and all the method signatures from both use cases.

(continued on page 368)

(continued from page 367)

The Spring Breaks 'R' Us Travel Service

In Chapter 3, you identified use cases for the business events for the Spring Breaks 'R' Us Travel Service. In Chapter 5, you elaborated on those use cases. In Chapter 4, you identified the classes associated with the business events. Using your solutions from those chapters, develop:

1. A first-cut DCD by using the problem domain classes you identified in Chapter 4.
2. A first-cut communication diagram for the *Add a resort* use case (domain classes and controller class only).
3. A first-cut sequence diagram for the *Book a reservation* use case (domain classes and controller class only).

4. A multilayer sequence diagram for the *Book a reservation* use case that includes domain classes and data access classes.
5. A separate multilayer sequence diagram for the *Book a reservation* use case that includes the domain classes and the view layer classes.
6. A final design class diagram that includes the classes from both use cases. Include elaborated attributes, navigation arrows, and all the method signatures from both use cases.
7. A package diagram of the four subsystems (Resort relations, Student booking, Accounting and finance, and Social networking) that includes all the problem domain classes.

On the Spot Courier Services

In Chapter 10, you developed a first-cut design class diagram and CRC card solutions for two use cases: *Request a package pickup* and *Pickup a package*.

Let us extend your solution from that chapter by developing the following:

1. A first-cut sequence diagram for each use case (domain classes and controller classes only).
2. A multilayer sequence diagram for each use case that includes domain classes and data access classes.
3. A separate multilayer sequence diagram for each use case that includes the domain classes and the view layer classes. (We won't combine view and data access layers on the same drawing. It makes the drawing too complex.)
4. A final design class diagram that includes the classes from both use cases. Include elaborated attributes, navigation arrows, and all the method signatures from both use cases.

In Chapter 9, we defined four subsystems:

- Customer account (like customer account)
- Pickup request (like sales)
- Package delivery (like order fulfillment)
- Routing and scheduling

Even though these subsystems are somewhat arbitrary, we can treat each one as a separate package. Develop a package diagram for each of the four subsystems by assigning domain model classes to each package. A domain model class should belong to only one subsystem package. Normally, it is the subsystem that instantiates objects from that class. Also, show dependency relationships among the various packages and classes.

Sandia Medical Devices

Review your answers to the case-related questions in Chapter 10 and then do the following:

1. Develop a first-cut sequence diagram for the patient use case *View/respond to alert*.
2. Develop a multilayer sequence diagram that includes domain classes and data access classes.
3. Develop a separate multilayer sequence diagram that includes the domain classes and the view layer

classes. (We won't combine view and data access layers on the same drawing. It makes the drawing too complex.)
4. Update your DCD from Chapter 10 to include the methods you have identified. Also, include any changes you may have made to navigation visibility and attribute details.

Further Resources

Grady Booch, James Rumbaugh, and Ivar Jacobson, *The Unified Modeling Language User Guide*. Addison-Wesley, 1999.

Grady Booch, et al., *Object-Oriented Analysis and Design with Applications*, 3rd edition. Addison-Wesley, 2007.

Frank Buschmann, R. Meunier, H. Rohnert, P. Sommerlad, and M. Stal, *Pattern-Oriented Software Architecture: A System of Patterns*. John Wiley and Sons, 1996.

Alur Deepak, J. Crupi, and D. Malks, *Core J2EE Patterns: Best Practices and Design Strategies*. Sun Microsystems Press, 2001.

Hans-Erik Eriksson, Magnus Penker, Brian Lyons, and David Fado, *UML 2 Toolkit*. John Wiley and Sons, 2004.

Erich Gamma, R. Helm, R. Johnson, and J. Vlissides, *Elements of Reusable Object-Oriented Software*. Addison-Wesley, 1995.

Mark Grand, *Patterns in Java*, Volumes I and II. John Wiley and Sons, 1999.

Craig Larman, *Applying UML and Patterns: An Introduction to Object-Oriented Analysis and Design and the Unified Process*, 3rd edition. Prentice Hall, 2004.

David S. Linthicum, *Next Generation Application Integration: From Simple Information to Web Services*. Addison-Wesley, 2004.

James Rumbaugh, Ivar Jacobson, and Grady Booch, *The Unified Modeling Language Reference Manual*. Addison-Wesley, 1999.

12

Databases, Controls, and Security

Chapter Outline

- Databases and Database Management Systems
- Relational Databases
- Data Access Classes
- Distributed Database Architectures
- Database Design Timing and Risks
- Designing Integrity Controls
- Designing Security Controls

Learning Objectives

After reading this chapter, you should be able to:

- Design a relational database schema based on a class diagram
- Evaluate and improve the quality of a database schema
- Describe the different architectural models for distributed databases
- Determine when and how to design the database
- Explain the importance of integrity controls for inputs, outputs, data, and processing
- Discuss issues related to security that affect the design and operation of information systems

Downslope Ski Company: Designing a Secure Supplier System Interface

Downslope Ski Company is a medium-sized manufacturer of skis and snowboards. In the company's early years, ski manufacturing was simple and straightforward. However, manufacturing has become more complex in recent years with such newer technology as carbon-laced resins, other high-tech materials, and product features such as skis with integrated bindings. To ensure product quality, computers control such production processes as ingredient mixtures as well as furnace temperatures and curing times. These changes have gone hand in hand with increased attention to raw materials quality. The company has changed suppliers frequently to respond to global market changes and to ensure an adequate supply of high-quality raw materials.

As in most modern manufacturing companies, Downslope uses a just-in-time (JIT) manufacturing process, which means that it doesn't stockpile a large quantity of raw materials. It keeps about a five-day supply on hand and depends on its suppliers to restock materials at least weekly. To facilitate quick ordering and delivery, Downslope has a simple system used by its suppliers to check inventory levels and the production schedule. Suppliers access a password-protected Web page, view inventory and production schedule data, and enter data about planned shipments. The page incorporates JavaScript code that interacts with Downslope's Oracle Database management system.

Senior management had been considering implementing a more sophisticated version of the supplier system for some time, but this hadn't yet risen to the top of the priority list. The current system was designed under the assumption that supplier employees would use it interactively once or twice per week. However, suppliers have increasingly asked for a fully automated interface that would be more efficient and would support more frequent or continuous monitoring and inventory replenishment. Downslope's production and purchasing departments also have a list of enhancements that would streamline their operations.

The priority for a replacement system received a significant boost after a recent external audit of Downslope's financial records and information systems. The auditing firm was concerned about whether Downslope was adequately managing system-related risks and whether a major disruption or security breach would affect the company's financial health. The auditing firm hired an external security consultant to conduct penetration tests. Within a matter of minutes, the consultant bypassed username and password security to access the supplier Web page and then gained administrative access to the entire database. He also demonstrated how simple it would be for a hacker to destroy database contents by using an SQL injection attack or to download database content, including sensitive data on company finances and detailed payroll and benefit records for company employees. The auditing firm refused to certify Downslope's financial statements unless immediate short-term security measures were implemented and unless a replacement system was scheduled for implementation within six months.

Nathan Lopez, Downslope's system development project manager, was charged with developing a replacement system that better met supplier needs and addressed security-related issues to the satisfaction of the auditors. He met with Downslope senior managers to review the results of a quick study to determine the feasibility of various alternatives for upgrading the system and the extent of the security issues that needed to be addressed.

Nathan began the meeting by saying: "I met with our production and purchasing staff and with most of our suppliers to determine information and format requirements. As expected, there was little consistency in the suppliers' desired formats. I have been able to consolidate some of their needs into three basic formats. But we will need some format flexibility to enable us to add or change suppliers over time.

"Of course, our suppliers' need for greater access to our data and flexibility must be balanced with system and database security. To protect our data and systems, we need to ensure that only our suppliers can access the data, that the data is secured in transit, and that a security breach of a supplier system won't compromise our own systems. As the audit tests showed, there are weaknesses that must be addressed in the new system and across the existing infrastructure. We need more time to fully analyze the required changes, and we will want some outside assistance to design and test those changes."

The meeting lasted a long time, with considerable discussion about the benefits and dangers of the new system. Senior management decided that Nathan should study the situation for a couple of weeks more and develop a list of every possible breach of security, with potential solutions for each one. They also authorized him to contract with an outside security firm for immediate assistance. Only after the security issues were fully addressed would the project move forward.

Overview

Databases and database management systems are important components of a modern information system. Database management systems provide designers, programmers, and end users with sophisticated capabilities to store, retrieve, and manage data. Sharing and managing the vast amounts of data needed by a modern organization wouldn't be possible without a database management system.

In Chapter 4, you learned to construct a domain model class diagram, which is a conceptual model of the information manipulated and stored by an information system. To design and build the system, developers must transform conceptual models into more detailed design models. In this chapter, we transform the domain model class diagram into a detailed database model and implement that model by using a database management system.

Databases lie at the heart of modern organizations, supporting internal activities and interactions with suppliers, customers, and other external parties. Designers expend considerable effort to ensure that databases contain data that is correct and complete and that the database and the systems that interact with it are accessible yet secure. The trade-offs among data correctness, completeness, accessibility, and security are complex issues that must be addressed with a carefully designed system of controls and security measures. Although these measures are a critical issue in database design, they also extend to all other aspects of the system.

Databases and Database Management Systems

database (DB) an integrated collection of stored data that is centrally managed and controlled

database management system (DBMS) a system software component that manages and controls one or more databases

A **database (DB)** is an integrated collection of stored data that is centrally managed and controlled. A database typically stores information about dozens or hundreds of classes. A database is managed and controlled by a **database management system (DBMS)**. A DBMS is a system software component that is generally purchased and installed separately from other system software components (e.g., operating systems). Examples of modern database management systems include Microsoft SQL Server, Oracle, and MySQL.

Figure 12-1 illustrates the components of a typical database and its interaction with a DBMS, application programs, users, and administrators. The database consists of two related information stores: the physical data store and the schema. The **physical data store** contains the raw bits and bytes of data that are created and used by the information system (e.g., names, prices, and account balances). The **schema** contains descriptive information about the data stored in the physical data store, including:

physical data store database component that stores the raw bits and bytes of data

schema database component that contains descriptive information about the data stored in the physical data store

- Organization of individual stored data items into higher level groups, such as tables
- Associations among tables or classes (e.g., pointers from customer objects to related sale objects)
- Details of physical data store organization, including types, lengths, locations, and indexing of data items
- Access and content controls, including allowable values for specific data items, value dependencies among multiple data items, and lists of users allowed to read or update data items

A DBMS has four key components: an application program interface (API), a query interface, an administrative interface, and an underlying set of data access programs and subroutines. Application programs, users, and administrators never access the physical data store directly. Instead, they tell an appropriate DBMS interface what data they need to read or write, using names defined

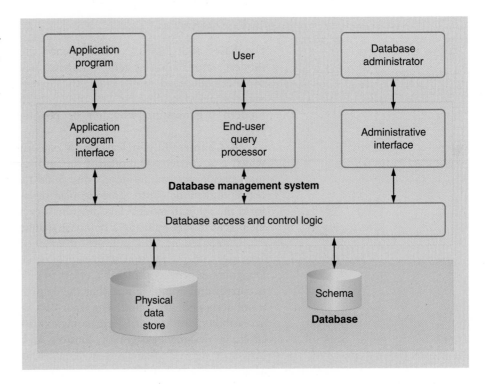

in the schema. The DBMS accesses the schema to verify that the requested data exist and that the requesting user has appropriate access privileges. If the request is valid, the DBMS extracts information about the physical organization of the requested data from the schema and uses that information to access the physical data store on behalf of the requesting program or user.

Databases and database management systems provide several important data access and management capabilities, including:

- Simultaneous access by many users and application programs
- Access to data without writing application programs (i.e., via a query language)
- Application of uniform and consistent access and content controls
- Integration of data stored on multiple servers distributed across multiple locations

For these and other reasons, databases and DBMSs are widely used in modern information systems.

DBMSs have evolved through a number of technology stages since their introduction in the 1960s. The most significant change has been the type of model used to represent and access the content of the physical data store. Early models, including the hierarchical and network models, have been replaced by the relational and object-oriented (OO) models. Most deployed databases and DBMSs are based on the relational model.

Relational Databases

relational database management system (RDBMS) a DBMS that organizes data in tables or relations

table a two-dimensional data structure of columns and rows

row one horizontal group of data attribute values in a table

A **relational database management system (RDBMS)** is a DBMS that organizes stored data into structures called **tables** or *relations*. Relational database tables are similar to conventional tables; that is, they are two-dimensional data structures of columns and rows. However, relational database terminology is somewhat different from conventional table and file terminology. A single row of a table is called a **row**, *tuple*, or *record*, and a column of a table is called an

FIGURE **12-2**
Partial display of a relational database table

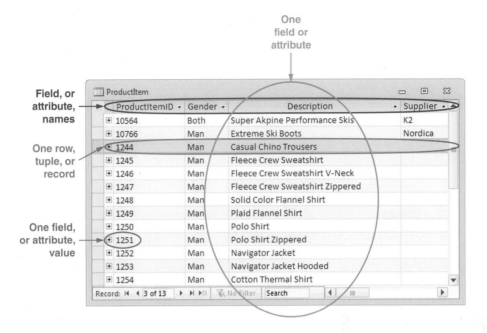

attribute *or* field. A single cell in a table is called an **attribute value**, *field value*, or *data element*.

Figure 12-2 shows the content of a table as displayed by the Microsoft Access relational DBMS. Note that the first row of the table contains a list of attribute names (column headings) and that the remaining rows contain a collection of attribute values, each of which describes a specific product. Each row contains the same attributes in the same order.

Each table in a relational database must have a unique key. A **key** is an attribute or set of attributes, the values of which occur only once in all the rows of the table. If only one attribute (or set of attributes) is unique, then that key is also called the table's **primary key**. If there are multiple unique attributes (or sets of attributes), then the database designer must choose one of the possible keys as the primary key.

Key attributes may be natural or invented. An example of a natural key attribute in chemistry is the atomic weight of an element in a table containing descriptive data about elements. Unfortunately, in business, few natural key attributes are useful for information processing, so most key attributes in a relational database are invented. Your wallet contains many examples of invented keys, including your Social Security number, driver's license number, and credit card numbers.

Primary keys are critical elements of relational database design because they are the basis for representing relationships among tables. Keys are the "glue" that binds rows of one table to rows of another table—in other words, keys relate tables to each other. For example, consider the class diagram fragment from the RMO example, which is shown in **Figure 12-3**, and the tables that are shown in **Figure 12-4**. The class diagram fragment shows an optional one-to-many association between the classes ProductItem and InventoryItem. The upper table in Figure 12-4 contains data representing the ProductItem class. The lower table contains data representing the InventoryItem class.

The association between the ProductItem and InventoryItem classes is represented by a common attribute value within their respective tables. The ProductID attribute (the primary key of the ProductItem table) is also stored within the InventoryItem table, where it is called a foreign key. A **foreign key** is an attribute that duplicates the primary key of a different (or foreign) table.

---

attribute   one vertical group of data attribute values in a table

attribute value   the value held in a single table cell

key   an attribute or set of attributes, the values of which occur only once in all the rows of the table

primary key   the key chosen by a database designer to represent relationships among rows in different tables

foreign key   an attribute that duplicates the primary key of a different (or foreign) table

FIGURE 12-3
Portion of the RMO class diagram

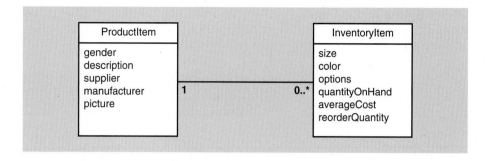

FIGURE 12-4
An association between rows in two tables represented with primary and foreign keys

ProductItem

ProductItemID	Gender	Description	Supplier
10564	Both	Super Akpine Performance Skis	K2
10766	Man	Extreme Ski Boots	Nordica
1244	Man	Casual Chino Trousers	
1245	Man	Fleece Crew Sweatshirt	
1246	Man	Fleece Crew Sweatshirt V-Neck	
1247	Man	Fleece Crew Sweatshirt Zippered	
1248	Man	Solid Color Flannel Shirt	
1249	Man	Plaid Flannel Shirt	
1250	Man	Polo Shirt	
1251	Man	Polo Shirt Zippered	
1252	Man	Navigator Jacket	
1253	Man	Navigator Jacket Hooded	
1254	Man	Cotton Thermal Shirt	

Record: 3 of 13 No Filter Search

InventoryItem

InventoryID	ProductID	Size	Color	Options	QuantityOnHand	Average Cost	RecorderQuantity
86779	1244	30/30	Khaki		45	$12.75	100
86780	1244	30/30	Slate		10	$12.75	100
86781	1244	30/30	LightTan		17	$12.75	100
86782	1244	30/31.	Khaki		22	$12.75	100
86783	1244	30/31	Slate		6	$12.75	100
86784	1244	30/31	LightTan		31	$12.75	100
86785	1244	30/32	Khaki		120	$12.75	100
86786	1244	30/32	Slate		28	$12.75	100
86787	1244	30/32	LightTan		21	$12.75	100
86788	1244	30/33	Khaki		7	$12.75	100
86789	1244	30/33	Slate		41	$12.75	100
86790	1244	30/34	LightTan		35	$12.75	50

Record: 13 of 13 No Filter Search

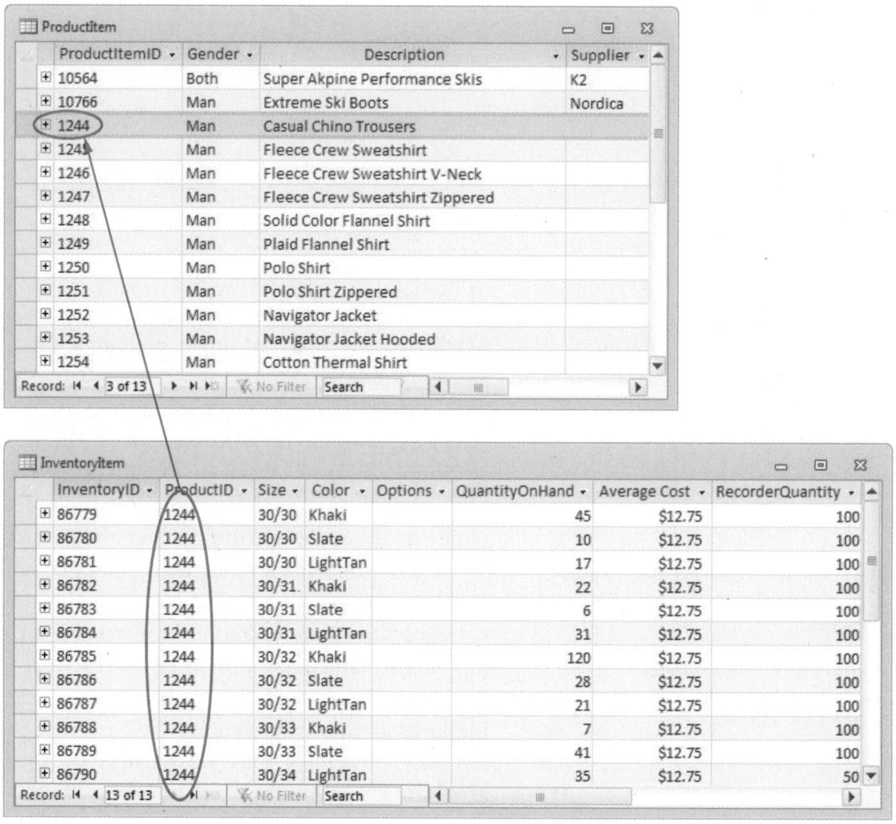

In Figure 12-4, the existence of the value 1244 as a foreign key within the InventoryItem table indicates that the values of Gender, Description and Supplier in the first row of the ProductItem table also describe inventory items 86779 through 86790.

Designing Relational Databases

The starting point for designing a relational database schema is either a class diagram or entity-relationship diagram (ERD). Starting with an ERD is the more traditional approach, but the widespread use of class diagrams for software design has made them more common than ERDs. For purposes of relational database design, there is no particular advantage to either diagram type. We will base our discussion and examples on class diagrams because they are more common today than ERDs.

As you have learned in earlier chapters, there are multiple types of class diagrams, including domain model and design class diagrams. For database design, the preferred starting point is the domain model class diagram because it omits many design details that aren't relevant to database design. To create a relational database schema from a domain model class diagram, follow these steps:

1. Create a table for each class.
2. Choose a primary key for each table (invent one, if necessary).
3. Add foreign keys to represent one-to-many associations.
4. Create new tables to represent many-to-many associations.
5. Represent classification hierarchies.
6. Define referential integrity constraints.
7. Evaluate schema quality and make necessary improvements.
8. Choose appropriate data types.
9. Incorporate integrity and security controls.

The following subsections discuss steps 1–8 in detail. Controls are discussed at the end of this chapter.

Representing Classes

The first step in creating a relational database schema is to create a table for each class on the class diagram. **Figure 12-5** shows a partial class diagram for the RMO customer support system, with 17 classes, including three specialized classes of Sale and two of OnlineCart. For the moment, we will treat each group of generalized and specialized classes as if it were a single class. The attributes of each table will be the same as those defined for the corresponding class in the class diagram. To avoid confusion, table and attribute names should match the names used in the class diagram, although abbreviations should be avoided. Initial table definitions for the classes in Figure 12-5 are shown in **Figure 12-6**.

Choosing Primary Keys

After creating tables for each class, the designer selects a primary key for each table. If a table already has an attribute or set of attributes that are guaranteed to be unique, then the designer can choose that attribute or set of attributes as the primary key. If the table contains no possible keys, then the designer must invent a new key attribute. Any name can be chosen for an invented key field, but the name should indicate that the attribute contains unique values. Typical names include Code, Number, and ID—possibly combined with the table name (for example, ProductCode and OrderID). **Figure 12-7** shows the class tables with primary key columns in boldface type.

Because key creation and management are critical functions in databases and information systems, many relational DBMSs automate key creation. DBMSs typically provide a special data type for invented keys (e.g., the AutoNumber type in Microsoft Access). The DBMS automatically assigns a key value to newly created rows and communicates that value to the application program for use in subsequent database operations. Embedding this capability in the DBMS frees the IS developer from designing and implementing customized key-creation software modules.

Invented keys that aren't assigned by the information system or DBMS must be given careful scrutiny to ascertain their uniqueness and usefulness over time. For example, employee databases in the United States commonly use Social Security numbers as keys. Because the U.S. government has a strong interest in guaranteeing the uniqueness of Social Security numbers, the assumption that they will always be unique seems safe. But will all employees who are stored in the database have a Social Security number? What if the company opens a manufacturing facility in Asia or South America?

Invented keys assigned by nongovernmental agencies deserve even more careful scrutiny. For example, FedEx, UPS, and most shipping companies assign a tracking number to each shipment they process. Tracking numbers are

FIGURE **12-5** *Subset of the RMO domain model class diagram*

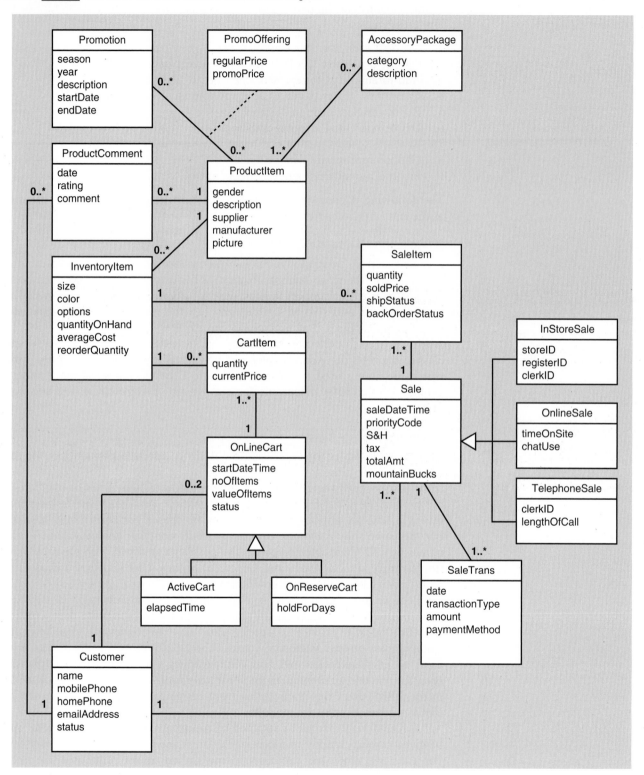

guaranteed to be unique at any given point in time, but are they guaranteed to be unique forever (that is, are they ever reused)? Could reuse of a tracking number cause a primary key duplication in your database? And what would happen if two different shippers assigned the same tracking number to two different shipments?

FIGURE **12-6**

Initial set of tables representing RMO classes

Table	Attributes
AccessoryPackage	Category, Description
CartItem	Quantity, CurrentPrice
Customer	Name, MobilePhone, HomePhone, EmailAddress, Status
InventoryItem	Size, Color, Options, QuantityOnHand, AverageCost, ReorderQuantity
OnlineCart	StartDateTime, NumberOfItems, ValueOfItems, Status, ElapsedTime, HoldForDays
ProductComment	Date, Rating, Comment
ProductItem	Gender, Description, Supplier, Manufacturer, Picture
PromoOffering	RegularPrice, PromoPrice
Promotion	Season, Year, Description, StartDate, EndDate
Sale	SaleDateTime, PriorityCode, ShippingAndHandling, Tax, TotalAmount, MountainBucks, StoreID, RegisterID, ClerkID, TimeOnSite, ChatUse, LengthOfCall
SaleItem	Quantity, SoldPrice, ShipStatus, BackOrderStatus
SaleTransaction	Date, TransactionType, Amount, PaymentMethod

FIGURE **12-7**

Class tables with primary keys identified in bold

Table	Attributes
AccessoryPackage	**AccessoryPackageID**, Category, Description
CartItem	**CartItemID**, Quantity, CurrentPrice
Customer	**AccountNumber**, Name, MobilePhone, HomePhone, EmailAddress, Status
InventoryItem	**InventoryItemID**, Size, Color, Options, QuantityOnHand, AverageCost, ReorderQuantity
OnlineCart	**OnlineCartID**, StartDateTime, NumberOfItems, ValueOfItems, Status, ElapsedTime, HoldForDays
ProductComment	**ProductCommentID**, Date, Rating, Comment
ProductItem	**ProductItemID**, Gender, Description, Supplier, Manufacturer, Picture
PromoOffering	**PromoOfferingID**, RegularPrice, PromoPrice
Promotion	**PromotionID**, Season, Year, Description, StartDate, EndDate
Sale	**SaleID**, SaleDateTime, PriorityCode, ShippingAndHandling, Tax, TotalAmount, MountainBucks, StoreID, RegisterID, ClerkID, TimeOnSite, ChatUse, LengthOfCall
SaleItem	**SaleItemID**, Quantity, SoldPrice, ShipStatus, BackOrderStatus
SaleTransaction	**SaleTransactionID**, Date, TransactionType, Amount, PaymentMethod

Representing Associations

Associations are represented within a relational database by foreign keys. Which foreign keys should be placed in which tables depends on the type of association being represented. The RMO class diagram in Figure 12-4 contains 10 one-to-many associations, two many-to-many associations, and two generalization/specialization association groups. We will deal with the generalization/specialization associations in a later step. The rules for representing one-to-many and many-to-many associations are as follows:

- One-to-many associations—Add the primary key attribute(s) of the "one" class to the table that represents the "many" class.
- Many-to-many associations—If no association class exists, create a new table to represent the association. Add the primary key attribute(s) of the associated classes to the table that represents the association.

One-to-Many Associations **Figure 12-8** shows the results of representing the 10 one-to-many associations within the tables from Figure 12-7. Each foreign key (shown in italics) represents a single association between the table containing the foreign key and the table that uses that same key as its primary key. For example, the attribute CustomerAccountNumber was added to the Sale table as a foreign key representing the one-to-many association between the Customer and Sale classes. The foreign key SaleID was added to the SaleTransaction table to represent the one-to-many association between Sale and SaleTransaction.

Many-to-Many Associations **Figure 12-9** expands the table definitions in Figure 12-8 by updating the PromoOffering table to represent the many-to-many association between Promotion and ProductItem. The primary key of the PromoOffering becomes the combination of PromotionID and ProductItemID. The old primary key PromoOfferingID is discarded. The two attributes that comprise the primary key are also foreign keys and are displayed in boldface and italics to indicate their dual status. PromotionID is a foreign key from the Promotion table, and ProductItemID is a foreign key from the ProductItem table.

Because there is no association class representing the many-to-many association between ProductItem and AccessoryPackage, a new table named AccessoryPackageContents is created. As with PromoOffering, it contains two foreign key columns that combine to form a primary key. The class diagram

FIGURE **12-8**

One-to-many associations represented by adding foreign key attributes (shown in italics)

Table	Attributes
AccessoryPackage	**AccessoryPackageID**, Category, Description
CartItem	**CartItemID**, *InventoryItemID*, *OnlineCartID*, Quantity, CurrentPrice
Customer	**AccountNumber**, Name, MobilePhone, HomePhone, EmailAddress, Status
InventoryItem	**InventoryItemID**, *ProductItemID*, Size, Color, Options, QuantityOnHand, AverageCost, ReorderQuantity
OnlineCart	**OnlineCartID**, *CustomerAccountNumber*, StartDateTime, NumberOfItems, ValueOfItems, Status, ElapsedTime, HoldForDays
ProductComment	**ProductCommentID**, *ProductItemID*, *CustomerAccountNumber*, Date, Rating, Comment
ProductItem	**ProductItemID**, Gender, Description, Supplier, Manufacturer, Picture
PromoOffering	**PromoOfferingID**, RegularPrice, PromoPrice
Promotion	**PromotionID**, Season, Year, Description, StartDate, EndDate
Sale	**SaleID**, *CustomerAccountNumber*, SaleDateTime, PriorityCode, ShippingAndHandling, Tax, TotalAmount, MountainBucks, StoreID, RegisterID, ClerkID, TimeOnSite, ChatUse, LengthOfCall
SaleItem	**SaleItemID**, *InventoryItemID*, *SaleID*, Quantity, SoldPrice, ShipStatus, BackOrderStatus
SaleTransaction	**SaleTransactionID**, *SaleID*, Date, TransactionType, Amount, PaymentMethod

FIGURE **12-9**

PromoOffering table modified to represent the many-to-many association between Product and Promotion

Table	Attributes
AccessoryPackage	**AccessoryPackageID**, Category, Description
AccessoryPackageContents	***AccessoryPackageID, InventoryItemID***
CartItem	***InventoryItemID, OnlineCartID,*** Quantity, CurrentPrice
Customer	**AccountNumber**, Name, MobilePhone, HomePhone, EmailAddress, Status
InventoryItem	**InventoryItemID**, *ProductItemID*, Size, Color, Options, QuantityOnHand, AverageCost, ReorderQuantity
OnlineCart	**OnlineCartID**, *CustomerAccountNumber*, StartDateTime, NumberOfItems, ValueOfItems, Status, ElapsedTime, HoldForDays
ProductComment	**ProductCommentID**, *ProductItemID*, *CustomerAccountNumber*, Date, Rating, Comment
ProductItem	**ProductItemID**, Gender, Description, Supplier, Manufacturer, Picture
PromoOffering	***PromotionID, ProductItemID***, RegularPrice, PromoPrice
Promotion	**PromotionID**, Season, Year, Description, StartDate, EndDate
Sale	**SaleID**, *CustomerAccountNumber*, SaleDateTime, PriorityCode, ShippingAndHandling, Tax, TotalAmount, MountainBucks, StoreID, RegisterID, ClerkID, TimeOnSite, ChatUse, LengthOfCall
SaleItem	***InventoryItemID, SaleID***, Quantity, SoldPrice, ShipStatus, BackOrderStatus
SaleTransaction	**SaleTransactionID**, *SaleID*, Date, TransactionType, Amount, PaymentMethod

contains other classes—including ProductComment, SaleItem, and CartItem— that are similar to PromoOffering. Though not shown the same way as PromoOffering on the class diagram, each represents a many-to-many association. For example, SaleItem represents a many-to-many association between Sale and InventoryItem and contains foreign keys for both associated classes.

Because the combination of foreign key values in SaleItem is always unique, the foreign key combination can serve as the table's primary key, and the invented key (SaleItemID) created earlier can be discarded. A similar situation exists for CartItem but not for ProductComment. With ProductComment, it is possible for a single customer to make multiple comments about the same product. If that were to happen, there would be two rows in the ProductComment table with the same values of CustomerAccountNumber and ProductItemID. Thus, the invented key ProductCommentID is retained and the two primary key attributes aren't part of the primary key.

Representing Classification Hierarchies

Classification hierarchies, such as the association among Sale, InStoreSale, TelephoneSale, and WebSale, are a special case in relational database design. Just as a specialized class inherits the data and methods of a generalized class, a table representing a specialized class inherits some or all of its data from the table representing its generalized class. This inheritance can be represented in multiple ways, including:

■ Combining all the tables into a single table containing the superset of all classes
■ Using separate tables to represent the child classes and using the primary key of the parent class table as the primary key of the child class tables

Either method is an acceptable approach to representing a classification hierarchy.

Figure 12-9 shows the definition of the Sale table under the first method. All the non-key attributes in the InStoreSale, TelephoneSale, and OnlineSale classes are stored in the Sale table. For any particular sale, some of the attribute values in each row will be empty or, in database terminology, NULL. For example, a row representing a telephone sale would have no values for the attributes StoreID, RegisterID, TimeOnSite, and ChatUse.

Figure 12-10 shows separate table definitions for specialized classes. The relationship among Sale, InStoreSale, TelephoneSale, and WebSale is represented by the foreign key SaleID in all three specialized class tables. In each case, the foreign key representing the inheritance association also serves as the primary key of the table representing the specialized class. A similar situation exists for OnlineCart, ActiveCart, and OnReserveCart.

Enforcing Referential Integrity

In general terms, referential integrity is a constraint on database content—for example, "A sale must be to a customer" and "A sale item must be something that we normally stock in inventory." For relational databases, the term **referential integrity** describes a consistent state among foreign key and

referential integrity   a consistent state among foreign key and primary key values

FIGURE **12-10**

Specialized classes of Sale and OnlineCart represented as separate tables

Table	Attributes
AccessoryPackage	**AccessoryPackageID**, Category, Description
AccessoryPackageContents	***AccessoryPackageID, InventoryItemID***
ActiveCart	***OnlineCartID***, ElapsedTime
CartItem	***InventoryItemID, OnlineCartID***, Quantity, CurrentPrice
Customer	**AccountNumber**, Name, MobilePhone, HomePhone, EmailAddress, Status
InStoreSale	***SaleID***, StoreID, RegisterID, ClerkID
InventoryItem	**InventoryItemID**, *ProductItemID*, Size, Color, Options, QuantityOnHand, AverageCost, ReorderQuantity
OnlineCart	**OnlineCartID**, *CustomerAccountNumber*, StartDateTime, NumberOfItems, ValueOfItems, Status, ElapsedTime, HoldForDays
OnlineSale	***SaleID***, TimeOnSite, ChatUse
OnReserveCart	***OnlineCartID***, HoldForDays
ProductComment	**ProductCommentID**, ProductItemID, CustomerAccountNumber, Date, Rating, Comment
ProductItem	**ProductItemID**, Gender, Description, Supplier, Manufacturer, Picture
PromoOffering	***PromotionID, ProductItemID***, RegularPrice, PromoPrice
Promotion	**PromotionID**, Season, Year, Description, StartDate, EndDate
Sale	**SaleID**, *CustomerAccountNumber*, SaleDateTime, PriorityCode, ShippingAndHandling, Tax, TotalAmount, MountainBucks
SaleItem	***InventoryItemID, SaleID***, Quantity, SoldPrice, ShipStatus, BackOrderStatus
SaleTransaction	**SaleTransactionID**, *SaleID*, Date, TransactionType, Amount, PaymentMethod
TelephoneSale	***SaleID***, ClerkID, LengthOfCall

primary key values. In most cases, a database designer wants to ensure that these references are consistent. That is, foreign key values that appear in one table must also appear as the primary key value of the related table.

The DBMS usually enforces referential integrity automatically once the schema designer identifies primary and foreign keys. For example, when a new row is added to a table containing a foreign key, the DBMS checks that the value also exists as a primary key value in the related table and rejects the new row if no such primary key value exists. The database designer "tells" the DBMS which columns are foreign keys and to which primary key columns they refer by creating a **referential integrity constraint**. For example, a referential integrity constraint for SaleID in the SaleItem table would be written in SQL as:

```
ADD CONSTRAINT FK_SaleItem_Sale FOREIGN KEY
SaleID REFERENCES Sale(ID)
```

Evaluating Schema Quality

Database design is the final step in a complex modeling process. There are multiple possibilities for database design errors, including errors in the domain class diagram, poor choice of primary keys, and errors in converting the class diagram into relational tables. Even if no obvious errors were made, some database designs are simply better than others.

Systems and programs are frequently revised, and new systems often replace old ones. But the databases embedded within information systems often survive several generations of programs. Because databases are difficult to change once they are populated with data, analysts take extra steps to ensure a high-quality database design.

A high-quality relational database schema has these features:

- Flexibility or ease of implementing future data model changes
- Lack of redundant data

A database schema is considered flexible and maintainable if changes to the database schema can be made with minimal disruption to existing data content and structure. For example, adding a new class to the schema shouldn't require redefining existing tables. Adding a new one-to-many association should only require adding new foreign keys and/or tables. Redundancy also plays a role in database longevity and usability. Excessive redundancy reduces schema flexibility and also reduces system performance.

Analysts use various formal and informal techniques to evaluate schema quality. A common informal technique is a design review by other technical staff or outside consultants. The most common formal technique is called normalization, which is described in detail in the next section.

Database Normalization

Normalization is a formal technique for evaluating and improving the quality of a relational database schema. It determines whether a database schema is flexible and whether it contains any of the "wrong" kinds of redundancy. It also defines specific methods to eliminate redundancy and improve flexibility. A complete discussion of normalization is well beyond the scope of this textbook. We will briefly define a few key concepts supplemented by short descriptions and examples.

First Normal Form

A table is in **first normal form (1NF)** if all rows contain the same number of columns. 1NF prohibits attributes such as Dependent, which is shown in **Figure 12-11**. Instead, the database designer must either provide a fixed number of columns in which dependent information is stored or must represent

referential integrity constraint a constraint, stored in the schema, that the DBMS uses to automatically enforce referential integrity

normalization a formal technique for evaluating and improving the quality of a relational database schema

first normal form (1NF) restriction that all rows of a table must contain the same number of columns

FIGURE **12-11** *Employee table with repeating attribute*

SSN	Name	Department	Salary	Dependent1	Dependent2	Dependent3 ...	DependentN
111-22-3333	Mary Smith	Accounting	40,000	John	Alice	Dave	
222-33-4444	Jose Pena	Marketing	50,000				
333-44-5555	Frank Collins	Production	35,000	Jan	Julia		

Dependent as a separate class with an association to the Employee class. The latter approach would result in at least one additional table for the dependent information, with an invented primary key and a corresponding foreign key in the employee table (for a one-to-many association) or in a table representing an associative entity (for a many-to-many association).

Functional Dependency

functional dependency a one-to-one association between the values of two attributes

A **functional dependency** is a one-to-one association between the values of two attributes. The association is formally stated as follows:

> *Attribute A* is functionally dependent on *attribute B* if for each value of *attribute B* there is only one corresponding value of *attribute A.*

The most precise way to determine whether functional dependency exists is to pick two attributes in a table and insert their names in the italic portions of the previous statement and ask whether the result is true. For example, consider the attributes ProductItemID and Description in the ProductItem table (see **Figure 12-12**). ProductItemID is an internally invented primary key that is guaranteed to be unique within the table. To determine whether Description is functionally dependent on ProductItemID, substitute Description for "attribute A" and ProductItemID for "attribute B" in the italicized portion of the functional dependency definition:

> *Description* is functionally dependent on *ProductItemID* if for each value of *ProductItemID* there is only one corresponding value of *Description.*

Now ask whether the statement is true for all rows that could possibly exist in the ProductItem table. If the statement is true, then Description is functionally dependent on ProductID. As long as the invented key ProductID is guaranteed to be unique within the ProductItem table, then the preceding statement is true. Therefore, Description is functionally dependent on ProductItemID.

FIGURE **12-12**
RMO ProductItem table

ProductItemID	Gender	Description	Supplier
10564	Both	Super Akpine Performance Skis	K2
10766	Man	Extreme Ski Boots	Nordica
1244	Man	Casual Chino Trousers	
1245	Man	Fleece Crew Sweatshirt	
1246	Man	Fleece Crew Sweatshirt V-Neck	
1247	Man	Fleece Crew Sweatshirt Zippered	
1248	Man	Solid Color Flannel Shirt	
1249	Man	Plaid Flannel Shirt	
1250	Man	Polo Shirt	
1251	Man	Polo Shirt Zippered	
1252	Man	Navigator Jacket	
1253	Man	Navigator Jacket Hooded	
1254	Man	Cotton Thermal Shirt	

Record: I◄ ◄ 3 of 13 ► ►I ►* | No Filter | Search

A less formal way to analyze functional dependency of Description on ProductItemID is to remember that the ProductItem table represents a specific product sold by RMO. If that product can have only a single description in the database, then Description is functionally dependent on the key of the table that represents products (ProductItemID). If it is possible for any product to have multiple descriptions, then the attribute Description isn't functionally dependent on ProductItemID.

Second Normal Form

second normal form (2NF) restriction that a table is in 1NF and that each non-key attribute is functionally dependent on the entire primary key

A table is in **second normal form (2NF)** if it is in 1NF and if each non-key attribute is functionally dependent on the entire primary key. A table violates 2NF when a non-key attribute is functionally dependent on only part of the primary key, which is only possible if the primary key contains multiple attributes. Thus, if the ProductItem table is in 1NF, it is also in 2NF because its primary key is a single column.

When a table's primary key consists of two or more attributes, the analyst must examine functional dependency of non-key attributes on each part of the primary key. For example, consider a modified version of the RMO PromoOffering table, as shown in **Figure 12-13**. Recall that this table represents a many-to-many association between Promotion and ProductItem. Thus, the table representing this association has a primary key consisting of the primary keys of Promotion (PromotionID) and ProductItem (ProductID).

If this table is in 2NF, then each non-key attribute must be functionally dependent on the combination of PromotionID and ProductItemID. The simplest way to test for 2NF is to test for functional dependency of non-key attributes on each subset of the primary key. Because the primary key contains two attributes, there are two statements that must be tested for each non-key attribute. For RegularPrice, these statements are:

RegularPrice is functionally dependent on PromotionID if for each value of PromotionID there is only one corresponding value of Regular Price.

RegularPrice is functionally dependent on ProductItemID if for each value of ProductItemID there is only one corresponding value of Regular Price.

If either statement is true, then a 2NF violation exists. In this example, the first statement is true, but the second is false. RegularPrice depends on ProductItemID regardless of what promotions, if any, a product participates in. Another way to think about this example is to think about the underlying association represented by the PromoOffering table. A product can be part of multiple promotions at the same time. Although a product's promotional price can be different in different promotions, its regular price is the same whether it participates in one promotion, three promotions, or none.

If a non-key attribute such as RegularPrice is functionally dependent on only part of the primary key, then you must remove the non-key attribute from its present table and place it in another table to satisfy the requirements of 2NF. Because the first functional dependency statement is true, RegularPrice belongs in a table that has ProductItemID as its primary key. If such a table doesn't already exist, it must be created. However, in this case, there is already a table with ProductItemID as its primary key: the ProductItem table. Thus, the

FIGURE **12-13**

Version of the RMO PromoOffering table that violates 2NF

column RegularPrice is removed from the PromoOffering table and added to the ProductItem table, thus ensuring that the PromoOffering table is in 2NF.

Third Normal Form

A table is in **third normal form (3NF)** if it is in 2NF and if no non-key attribute is functionally dependent on any other non-key attribute. To verify that a table is in 3NF, we must check the functional dependency of each non-key attribute on every other non-key attribute. This can be cumbersome for a large table because the number of pairs that must be checked grows quickly as the number of non-key attributes grows. The number of functional dependencies to be checked is $N \times (N - 1)$, where N is the number of non-key attributes. Note that functional dependency must be checked in both directions (i.e., A dependent on B, and B dependent on A).

A common example of a 3NF violation is an attribute that can be computed by a formula or algorithm that uses other stored values as inputs. Common examples of computable attributes include subtotals, totals, and taxes—for example, the attribute TotalAmount, which appears in the Sale table shown in **Figure 12-14**. Here is the formula for computing the TotalAmount:

$$TotalAmount = (\Sigma \ Quantity \times SoldPrice) + Shipping + Tax$$

Note that all the inputs to the formula aren't stored in the same table (see Figure 12-13). Violations of 3NF involving computable attributes can be localized to a single table or spread across multiple tables. In this case, Shipping and Tax are stored in the Sale table, whereas Quantity and SoldPrice are stored in related rows of the SaleItem table. An algorithm that computes TotalAmount for a particular sale needs to extract all matching rows in the SaleItem table by using the SaleID foreign key.

Computational dependencies are a form of redundancy because a change to the value of any input variable in the computation (e.g., Shipping) also changes the result of the computation (i.e., TotalAmount). The way to correct this type of 3NF violation is simple: Remove the computed attribute from the database. Eliminating the computed attribute from the database doesn't mean that its value is lost. For example, any program or method that needs TotalAmount can query the SaleItem table for matching values of Quantity and SoldPrice, sum the result of multiplying each Quantity and SoldPrice, and add Shipping and Tax.

Data Types

A **data type** defines the storage format and allowable content of a program variable, class attribute, or relational database attribute or column. **Primitive data types** are supported directly by computer hardware and programming languages and include integers, single characters, and real numbers (floating-point numbers). **Complex data types** are combinations of or extensions to primitive data types that are supported by programming languages, operating systems, and DBMSs. Examples include arrays and tables, strings (character arrays), dates, times, currency (money), audio streams, still images, motion video streams, and uniform resource locators (URLs or Web links).

FIGURE **12-14**

TotalAmount computed from attributes in two tables

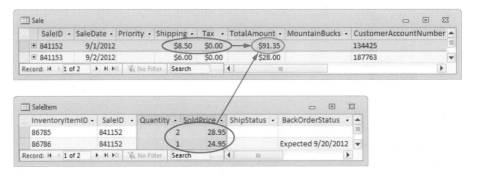

Type(s)	Description
datetimeoffset	Date, time, and time zone
int, small int, and bigint	Whole numeric values
float and real	Numeric values with fractional quantities
money	Currency values and related symbols (e.g., $ and €)
nchar and nvarchar	Fixed- and variable-length Unicode string
varbinary	Variable-length byte sequence up to 2 GB
xml	XML document up to 2 GB

A database designer must choose an appropriate data type for each attribute in a relational database schema. For many attributes, the choice of a data type is relatively straightforward. For example, designers can represent customer names and addresses as strings, inventory quantities as integers, and item prices as currency. RDBMSs support a variety of primitive and complex data types required by modern information systems. **Figure 12-15** contains a partial listing of some of the data types available in the Microsoft SQL Server RDBMS. The varbinary data type is typically used to store such data items as pictures, sound, and video encoded in such standardized formats as JPEG, MP3, and MP4.

Data Access Classes

In Chapter 10, you learned how to develop an OO design based on three-layer architecture. Under that architecture, data access classes implement the bridge between data stored in program objects and in a relational database. **Figure 12-16** illustrates the interaction among the RMO problem domain class Promotion, the data access class PromotionDA, and the relational database. The data access class has methods that add, update, find, and delete fields and rows in the table or tables that represent the class. Data access class methods encapsulate the logic needed to copy values from the problem domain class to the database and vice versa. Typically, that logic is a combination of program code in a language such as C++, C#, Java, Visual Basic, or PHP and embedded relational database commands in SQL.

The lower-left part of Figure 12-16 shows a fragment of Java code with an embedded SQL statement that implements the find method of PromotionDA. Similar code is needed for all other methods in the data access class.

Distributed Database Architectures

Because databases and DBMSs are such a crucial part of modern information systems, most organizations employ complex database architectures to improve reliability and performance. Although a detailed discussion of these architectures is beyond the scope of this textbook, a database design isn't complete unless these issues are addressed. In this section, we will briefly describe distributed database architectures and show how they might be applied for Ridgeline Mountain Outfitters.

Architectural approaches to support database services include:

single database server architecture one or more databases are hosted by a single DBMS running on a single server

replicated database server architecture complete database copies are hosted by cooperating DBMSs running on multiple servers

- **Single database server architecture**—One or more databases are hosted by a single DBMS running on a single server.
- **Replicated database server architecture**—Complete database copies are hosted by cooperating DBMSs running on multiple servers. The servers are

FIGURE **12-16** *Interaction among problem domain class, data access class, and the DBMS*

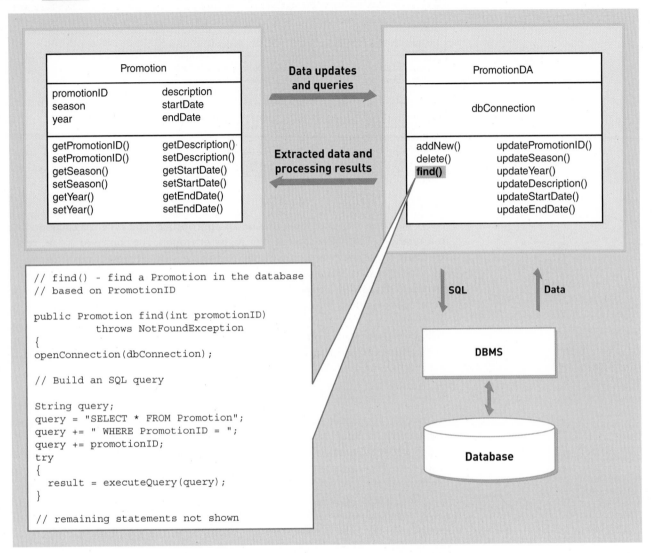

usually distributed across geographic locations. Application programs can access any server and usually make database updates to only one server. Servers periodically exchange update information to synchronize their database copies.

■ **Partitioned database server architecture**—Multiple distributed database servers are used and the database schema is partitioned, with some content on only one server and some content copied on all servers. Content that appears on multiple servers is periodically synchronized.

■ **Cloud-based database server architecture**—This architecture isn't really a separate architecture. Rather, it is a specific implementation of one or more of the other architectures by using the services of a cloud computing provider, such as Amazon or Google. The cloud provider hosts the database on multiple servers distributed across a predefined geographic area. Application programs access database services through the cloud provider. The cloud provider takes care of database synchronization and backup.

The primary advantage of single database server architecture is its simplicity. There is only one server to manage, and all clients are programmed to direct requests to that server. Disadvantages include susceptibility to server failure and possible overload of the network or server. All application programs that

partitioned database server architecture   multiple distributed database servers are used and the database schema is partitioned, with some content on only one server and some content copied on all servers

cloud-based database server architecture   use of a cloud computing service provider to provide some or all database services

depend on the server are disabled whenever the server is unavailable (such as during a crash or during hardware maintenance). Thus, single database server architecture is poorly suited to applications that must be available on a seven-day, 24-hour basis.

Replicated database servers make an information system more fault-tolerant. Applications can direct access requests to any available server, with preference to the nearest server. When a server is unavailable, clients can redirect their requests to another available server. In spite of their advantages, replicated database servers do have some drawbacks. When data is updated on one database copy, clients accessing that same data from another database copy receive an outdated response. To counteract this problem, each database copy must periodically be updated with changes from other database servers. This process is called **database synchronization**.

database synchronization updating one database copy with changes made to other database copies

The time delay between an update to a database copy and the propagation of that update to other database copies is an important database design decision. During the time between the original update and the update of database copies, application programs that access outdated copies aren't receiving responses that reflect current reality. Designers can address this problem by synchronizing more frequently or continuously. Synchronization then consumes a substantial amount of database server capacity, and a large amount of network capacity among the related database servers must be provided. The proper synchronization strategy is a complex trade-off among cost, hardware and network capacity, and the need of application programs and users for current data.

Designers can minimize the need for database synchronization by partitioning database contents among multiple database servers. **Figure 12-17** shows the division of a hypothetical database schema into two partitions. A different group of clients accesses each partition. It is seldom possible to partition a database schema into mutually exclusive subsets. Some portions of a database are typically needed by most or all users, and those portions must exist in each partition. For example, data in the region of overlap in Figure 12-17 should be

FIGURE **12-17**

Partitioning a database schema into client access subsets

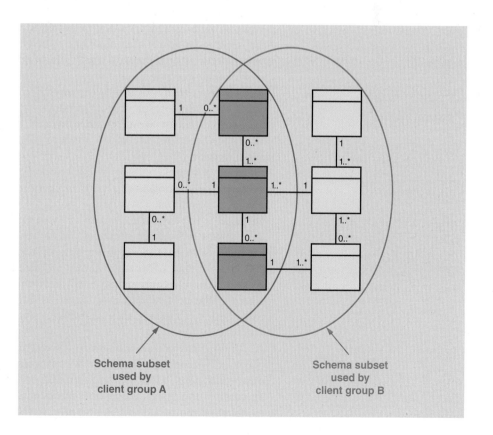

stored on each server with periodic synchronization. Thus, partitioning can reduce the problems associated with database synchronization, but it seldom eliminates them entirely.

Because advanced database architectures can be very complex and difficult to manage, many organizations outsource data services to a cloud computing vendor. Cloud computing vendors manage large server farms and deal with such issues as replication and synchronization every day. They have the requisite expertise to manage complex database architectures and can generally provide services more cheaply than user organizations. The primary drawbacks to using cloud providers are proprietary database interaction methods that vary among vendors and the need to maintain high-capacity Internet connections to support access to database services in the cloud.

RMO Distributed Database Architecture

The starting point for designing a distributed database architecture is information about the data needs of geographically dispersed users. Some of this information for RMO was gathered as an analysis activity and is summarized here:

■ Warehouse staff members (Portland, Salt Lake City, and Albuquerque) need to check inventory levels, query orders, record back orders and order fulfillment, and record order returns.

■ Phone-order staff members (Salt Lake City) need to check inventory levels; create, query, update, and delete orders; query customer account information; and query catalogs.

■ Customers using the online sales system and sales associates in retail stores need the same access capabilities as phone-order staff.

■ Marketing staff members (Park City) need to query and adjust orders, query and adjust customer accounts, and create and query product and promotion information.

RMO has already decided to manage its database by using the existing server cluster in the Park City data center. That same center will also host servers supporting the online sales, order fulfillment, and marketing systems. Thus, a high-capacity wide area network (WAN) will be required to connect the servers to local area networks (LANs) in the warehouses, phone-order centers, retail stores, headquarters, and data centers.

A single-server architecture is infeasible for the CSMS. There are many accesses from many locations at many different times. Inevitable database server downtime would result in lost productivity, sales, and reputation. In essence, the entire company would grind to a halt, and its future revenue stream would be jeopardized. As for many modern organizations, that risk is simply too much to bear.

A more complex alternative that addresses the risk is shown in **Figure 12-18**. Each remote location employs a combination of database partitioning and replication. A server at each warehouse stores a local copy of the order and inventory portions of the database. Servers in the phone-order center and retail stores host local copies of a larger subset of the database. Corporate headquarters relies on the central database server in the data center.

The primary advantages of this architecture are fault tolerance and reduced WAN capacity requirements. Each location could continue to operate independently if the central database server failed. However, as the remote locations continued to operate, their database contents would gradually drift out of synchronization.

The primary disadvantages to the distributed architecture are cost and complexity. The architecture saves WAN costs through reduced capacity requirements but adds costs for additional database servers. The cost of acquiring, operating, and maintaining the additional servers would probably be much higher than the cost of adding greater WAN capacity.

FIGURE **12-18**
Replicated and partitioned database server architecture for RMO

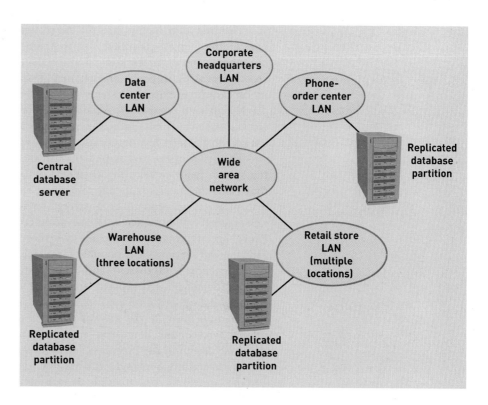

So, does the proposed architecture make sense for RMO? The answer depends on some data that hasn't yet been gathered and on answers to some questions about desired system performance, tolerances for downtime, cost and reliability of WAN connections, and the cost and availability of cloud services. RMO management must also determine its goals for system performance and reliability. The distributed architecture would provide higher performance and reliability but at substantially increased cost. Management must determine whether the extra cost is worth the expected benefits.

If management wants to reduce costs by using a cloud provider, it will save money on database servers and their management, but it will also require more expensive high-capacity or redundant network connections to each location. Even with such connections, the risk of database inaccessibility would be greater with a cloud provider than with locally replicated database servers.

Database Design Timing and Risks

Database design has profound impacts on other development activities. Key questions that many analysts raise are when to design the database and whether to design and build the entire database at once or spread those activities across multiple project iterations. Unfortunately, there are no simple answers to those questions.

The primary factors in deciding when and how to perform database design are:

■ Architecture—Decisions about DBMSs, database servers, and database distribution are tightly integrated with other architectural decisions, including network design, Web and component services, and security.

■ Existing databases—Most new or upgraded systems must interact with existing databases, with their pre-existing constraints. While adapting existing databases to new or updated systems, analysts must ensure their continued operation.

■ Domain model class diagram—Database design can't proceed until related parts of the class diagram have been developed.

The first two factors are the most significant; they make database design one of the highest-risk elements in many development projects. The risk is multidimensional, encompassing financial and operational risks through high costs, pervasive impacts on the new system, and possible disruption to or modification of existing systems and the business functions that rely on them. As described in Chapter 9, Agile development purposely addresses high risks in early iterations. Thus, in many projects, database design is performed in its entirety in the first few iterations.

The risks associated with database design are substantially reduced when the system being developed doesn't have to interact with existing databases. In that case, the most significant remaining risks are associated with architecture. The primary question to be answered is how the database architecture of the new system will interact with the architecture that supports existing systems. More specific questions include whether the new system will use existing DBMSs and servers—it is possible for a single DBMS or server to support multiple independent databases—and whether the existing servers and the network have sufficient unused capacity to support the new system. Those questions are typically asked and answered in an early project iteration or before the project is even approved. If the answer to both questions is yes, then the database-related risks of the new project are relatively low and the need to front-load database design in the project is substantially reduced.

When database-related risks are low, database design and deployment tasks can be spread across multiple iterations. Typically, the database schema is developed incrementally as related portions of the domain model class diagram are developed. As portions of the schema are developed, corresponding portions of the database are created and populated with test data. This enables development and testing of application software in general and data access classes in particular. As the database and application software move closer to completion, performance and stress testing can proceed in earnest.

Designing Integrity Controls

Controls are mechanisms and procedures that are built into a system to safeguard the system and the information within it. Here are a few scenarios that illustrate the need for controls:

- A furniture store sells merchandise on credit with internal financing. Salespeople sometimes sell furniture on credit to friends and relatives. How do we ensure that only authorized employees can extend credit and record payments and adjustments to credit accounts?
- A bookkeeper uses accounting software to generate electronic payments to suppliers. How does the system ensure that the payment is for goods or services that were actually received? How does the system ensure that no one can generate payments to a bogus supplier?
- An online retailer collects and stores credit card and other information about customers. How does the company ensure that customer data is protected and secure?

integrity control a control that rejects invalid data inputs, prevents unauthorized data outputs, and protects data and programs against accidental or malicious tampering

As shown in **Figure 12-19**, controls are incorporated into various parts of the system. Some of the controls—called **integrity controls**—must be integrated into the application programs that are being developed and the database that supports them. Other controls—usually called *security controls*—are part of the operating system and the network. Integrity controls ensure correct system function by rejecting invalid data inputs, preventing unauthorized data outputs, and protecting data and programs against accidental or malicious tampering. Security controls tend to be less application specific. The distinction between the two isn't precise because there is some overlap and because designers typically

FIGURE **12-19** *Security and integrity control locations*

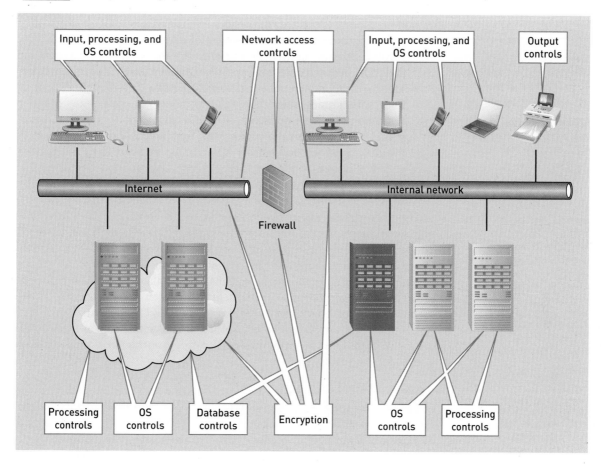

FIGURE **12-19** *Security and integrity control locations*

use both types. This section explains integrity controls. Later sections discuss security controls.

The primary objectives of integrity controls are to:

■ Ensure that only appropriate and correct business transactions occur.
■ Ensure that the transactions are recorded and processed correctly.
■ Protect and safeguard the assets of the organization (including hardware, software, and information).

As described in the following subsections, organizations incorporate many types of controls into their systems and databases to achieve these objectives. Each control type addresses such specific issues as accuracy of inputs and limiting access to authorized users. No control type is sufficient by itself to achieve all the objectives. Thus, a layered and multifaceted system of controls is required.

Input Controls

input control a control that prevents invalid or erroneous data from entering the system

Input controls prevent invalid or erroneous data from entering the system. Input controls can be applied to data entered by people or data transmitted from internal or external systems. Input controls can be implemented within application programs, the database schema, or both. Commonly used input control types include:

value limit control a control that checks numeric data input to ensure that the value is reasonable

■ **Value limit controls**—These check numeric data inputs to ensure that the amount entered is reasonable. For example, the amount of a sale or the amount of a commission usually falls within a certain range of values. A control might reject negative values or those that exceed a certain threshold, such as $10,000.

completeness control a control that ensures that all required data values describing an object or transaction are present

data validation control a control that ensures that numeric fields that contain codes or identifiers are correct

field combination control a control that reviews combinations of data inputs to ensure that the correct data are entered

access control a control that restricts which persons or programs can add, modify, or view information resources

transaction logging a technique by which any update to the database is logged with such audit information as user ID, date, time, input data, and type of update

complex update control a control that prevents errors that can occur when multiple programs try to update the same data at the same time or when recording a single transaction requires multiple related database updates

- **Completeness controls**—These ensure that all required data values describing an object or transaction are present. For example, when a shipping address is entered, the system might check whether enough information has been provided to ensure successful delivery.
- **Data validation controls**—These ensure that numeric fields containing codes or identifiers are correct. For example, a program for entering letter grades for a course might check that entered values match a set of predefined valid grades (such as A, B, C, D, and F) and reject any other entries.
- **Field combination controls**—These review various combinations of data inputs to ensure that the correct data are entered. For example, on an insurance policy, the application date must be prior to or the same as the effective date of policy coverage.

Access Controls

Access controls restrict which persons or programs can add, modify, or view information resources. Access controls are generally implemented through the operating system (as security controls) or through the DBMS to restrict access to individual attributes, tables, or entire databases. A DBMS stores access control information within the schema and applies controls each time data are read or written. Controls can be defined on such schema subsets as groups of related tables or objects, single tables or objects, or single attributes. For example, different controls might be applied to the name, Social Security number, and salary fields of an employee table. Also, controls on a single column might differ for read and write access.

Transaction Logging

Transaction logging is a technique by which any update to the database is logged with such audit information as user ID, date, time, input data, and type of update. Transaction logging provides a record of database changes that is stored in a separate location and can be checked independently of the database itself. The fundamental idea is to create an audit trail of all database updates and therefore track any errors or problems that occur. Most DBMSs include transaction logging as part of the DBMS software, although database designers or administrators can customize its application.

Transaction logging achieves two objectives. First, it helps discourage fraudulent transactions or malicious database changes. For example, if a person knows that his or her ID will be associated with every check request, that person isn't likely to request a bogus payment. Similarly, a disgruntled employee who might be tempted to delete important records knows that his or her actions are being logged.

Second, a logging system provides a recovery mechanism for erroneous transactions. A mid-level logging system maintains the set of all updates. The system can then recover from errors by "unapplying" the erroneous transactions. More sophisticated logging systems can provide "before" and "after" images of the attributes or rows that are changed by the transaction as well as the audit trail of all transactions. These sophisticated systems are typically used only for highly sensitive or critical data files, but they do represent an important control mechanism that is available when necessary.

Complex Update Controls

Complex update controls prevent errors that can occur when multiple programs try to update the same data at the same time or when recording a single transaction requires multiple related database updates. Because DBMSs support many application programs simultaneously, multiple programs may want to access and update a record or field at the same time. Update controls within a DBMS provide locking mechanisms to protect against multiple updates that might conflict with or overwrite each other. For example, while a sale is being processed, inventory levels of the products being purchased might be locked until the sale is completed, thus ensuring that the same inventory item isn't sold twice.

In addition, some transactions that are applied to the database have multiple parts, such as a financial transaction that must credit one account and debit a different account. Locking all updated rows until all changes are written to the database is a technique used to protect the data from partial updates of complex transactions.

Redundancy, Backup, and Recovery

Redundancy, backup, and recovery procedures are designed to protect software and data from hardware failure and from such catastrophes as fire and flood. Most operating systems and DBMSs incorporate support for all three. Many organizations that need continuous access to their data and systems employ redundant databases, servers, and sites. Each server or site hosts copies of the database and all application software. Updates made to one database copy are immediately or frequently synchronized with the other copy or copies to ensure consistency. If one site or server fails, the other(s) is (are) still accessible and the organization continues to function.

Backup procedures make partial or full copies of a database to removable storage media, such as magnetic tape, or to data storage devices or servers at another site. Unlike redundant sites or servers, backup copies stored off-site can't be accessed directly by application software. Instead, recovery procedures read the off-site copies and replicate their contents to a database server that can then provide access to programs and users. Backup and recovery operations can take from minutes to hours. Backups are typically scheduled during periods of low utilization. Recovery is performed when needed, and the database is unavailable until the recovery procedures are completed.

Output Controls

Systems outputs come in various forms, including output that is used by other systems, printed reports, and data displayed on traditional or mobile computer displays. **Output controls** ensure that output arrives at the proper destination and is accurate, current, and complete. It is especially important that outputs with sensitive information arrive at the proper destination and that they not be accessed by unauthorized persons. Common types of output controls include:

output control a control that ensures that output arrives at the proper destination and is accurate, current, and complete

- Physical access controls to printers—Printers and printed outputs should be located in secure areas accessible only by authorized personnel.
- Discarded output control—Although often ignored during system design, physical control of discarded printed outputs containing sensitive data is a must because "dumpster diving" is an effective way to access data without authorization. Sensitive printed documents should be segregated from other trash and shredded or burned.
- Access controls to programs that display or print—Program access controls restrict which users can access specific programs and program functions, usually via such a mechanism as a username and password. In some instances, a system designer might restrict program or function access by access device. This extra safeguard is used primarily for military or other systems that house workstations in secure areas and provide access to the system's information to anyone who has access to the area.
- Formatting and labeling of printed outputs—System developers ensure completeness and accuracy by printing control data on the output report. For example, every report should have a date and time stamp—for the time the report was printed and for the date(s) of the underlying data. To ensure that a document is complete, designers typically incorporate such formatting features as pagination in the "page__of__" format, control totals, and an "end of report" trailer.
- Labeling of electronic outputs—Electronic outputs typically include internal labels or tags that identify their source, content, and relevant dates. They may also include control totals or checksums that enable the recipient to determine whether content has been lost or altered.

FIGURE **12-20**
The fraud triangle

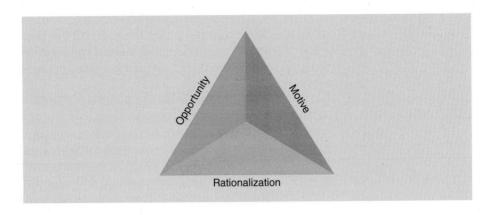

Integrity Controls to Prevent Fraud

System designers and administrators are rightly concerned with security breaches arising from outside the organization. But they often pay inadequate attention to an equally serious problem: the use of the system by authorized people to commit fraud. Obviously, integrity and security controls won't completely eliminate fraud. However, system developers should be aware of the fundamental elements that make fraud possible and incorporate controls to combat it.

In the 1950s, fraud researchers developed a widely used model called the **fraud triangle** (see **Figure 12-20**). Much as a fire needs fuel, heat, and oxygen to burn, fraud requires three elements:

fraud triangle model of fraud that states that opportunity, motivation, and rationalization must all exist for a fraud to occur

- Opportunity—the ability of a person to take actions that perpetrate a fraud. For example, unrestricted access to all functions of an accounts payable system enables an employee to generate false vendor payments.
- Motivation—a desire or need for the results of the fraud. Money is the usual motivation, although a desire for status or power as well as a need to be a "team player" may be contributing factors.
- Rationalization—an excuse for committing the fraud or an intention to "undo" the fraud in the future. For example, an employee might falsify financial reports to stave off bankruptcy, thus enabling fellow workers to keep their jobs. Or an employee might steal money to pay a gambling debt or medical bills, with the intention of repaying the money later.

System designers have little or no impact on motive and rationalization, but they can minimize or eliminate opportunity by designing and implementing effective controls. **Figure 12-21** contains several of the more important factors that increase the risk of fraud. This list isn't comprehensive, but it does provide a foundation on which developers can design a computer system that reduces the opportunity for fraud. As a system developer, you should include discussions with your users and within the project teams to ensure that adequate controls have been included to reduce fraud.

Designing Security Controls

security control a control that protects the assets of an organization from all threats, with a primary focus on external threats

Although the objective of **security controls** is to protect the assets of an organization from all threats, as indicated earlier, the primary focus is on external threats. In addition to the objectives enumerated earlier for integrity controls, security controls have two objectives:

- Maintain a stable, functioning operating environment for users and application systems (usually 24 hours a day, 7 days a week).
- Protect information and transactions during transmission across the Internet and other insecure environments.

FIGURE **12-21**
Fraud risks and prevention techniques

Factors affecting fraud risk	Risk-reduction techniques
Separation of duties	Design systems so those with asset custody have limited access to related records. Also, ensure that no one has sufficient system access to commit and cover up a fraud.
Records and audit trails	Record all transactions and changes in asset status. Log all changes to records and databases, and restrict log access to a few trusted persons.
Monitoring	Incorporate regular and systematic procedures to review records and logs for unusual transactions, accesses, and other patterns.
Asset control and reconciliation	Limit physical access to valuable assets, such as inventory, and periodically reconcile physical asset counts with related records.
Security	Design security features into individual systems and supporting infrastructure. Review and test security features frequently. Use outside consultants to conduct penetration testing attack and fraud vectors from external and internal sources.

The first objective—to maintain a stable operating environment—focuses on security measures to protect the organization's systems from external attacks from hackers, viruses, and worms as well as denial-of-service attacks. Most organizations have firewalls between their internal systems and the Internet (see Figure 12-19). Every time someone in an organization communicates through the Internet, there is the potential for a security violation and for undesirable access that could disrupt the internal systems. Thus, eliminating and controlling any undesirable access helps avoid disruption of the system.

The second objective—to protect transactions during transmission—focuses on the information that is sent or received via the Internet. Once a transaction is sent outside the organization, it could be intercepted, destroyed, or modified. Thus, security controls use techniques to protect data while they are in transit from the source to the destination.

The most common security control points are network and computer operating systems because they exercise direct control over such assets as files, application programs, and disk drives. All modern operating systems contain extensive security features that can identify users, restrict access to files and programs, and secure data transmission among distributed software components. Operating system security is the foundation of security for most information systems.

Access Controls

Access controls limit the ability of specific users to access such specific resources as servers, files, Web pages, application programs, and database tables. Operating systems, networking software, and DBMSs all provide access control systems, and all can be configured to share a common access control system. Most access control systems rely on these common principles and processes:

- **Authentication**—the process of identifying users who request access to sensitive resources. Users can be authenticated through usernames and passwords, smart cards, challenge questions and responses, or such biometric methods as fingerprint and retinal scans or voice recognition. **Multifactor authentication** uses multiple methods for increased reliability.

- **Access control list**—a list attached or linked to a specific resource that describes users or user groups and the nature of permitted access (e.g., read data, update data, and execute program). Users who don't appear in the access control list can't use the associated resource.

authentication the process of identifying users who request access to sensitive resources

multifactor authentication using multiple authentication methods for increased reliability

access control list a list attached or linked to a specific resource that describes users or user groups and the nature of permitted access

authorization the process of allowing or restricting a specific authenticated user's access to a specific resource based on an access control list

■ **Authorization**—the process of allowing or restricting a specific authenticated user's access to a specific resource based on an access control list.

To build an effective access control system, a designer must categorize system users and determine what type(s) of access each requires to every resource. **Figure 12-22** illustrates three user categories or types and the role of the access control system in allowing or restricting their access:

unauthorized user a person who isn't allowed access to any part or functions of the system

■ **Unauthorized users**—people who aren't allowed access to any part or functions of the system. Such users include employees who are prohibited from accessing the system, former employees who are no longer permitted to access the system, and such outsiders as hackers and intruders.

registered user a person who is authorized to access the system

■ **Registered users**—those who are authorized to access the system. Normally, various types of registered users are defined depending on what they are authorized to view and update. For example, some users may be allowed to view data but not update them, and other users can update only certain data fields. Some screens and functions of the new system may be hidden from other levels of registered users.

FIGURE **12-22** *Users and their access to computer systems*

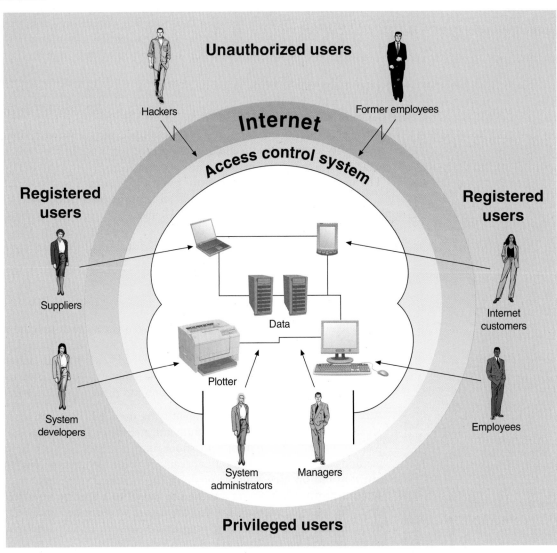

■ **Privileged users**—people who have access to the source code, executable program, and database structure of the system, including system programmers, application programmers, operators, and system administrators. These people may have differing levels of security access.

Data Encryption

No access control system is perfect, so designers must anticipate access control breaches and provide other measures to protect the confidentiality of data. Protective measures must also be applied to data that is stored or transmitted outside the organization's own network, such as transaction data sent by remote suppliers or customers and interactions between internal applications and cloud service providers. Common types of information that require additional protection include:

■ Financial information
■ Credit card numbers, bank account numbers, payroll information, healthcare information, and other personal data
■ Strategies and plans for products and other mission-critical data
■ Government and sensitive military information
■ Data stored on such portable devices as laptop computers and cell phones

The primary method of maintaining the security of data—data on internal systems and transmitted data—is by encrypting the data. **Encryption** is the process of altering data so unauthorized users can't view them. **Decryption** is the process of converting encrypted data back to their original state. Data stored in files or a database on hard drives or other storage devices can be encrypted to protect them against theft. Data sent across a network can be encrypted to prevent eavesdropping or theft during transmission. A thief or eavesdropper who steals or intercepts encrypted data receives a meaningless group of bits that is difficult or impossible to convert back into the original data.

An **encryption algorithm** is a complex mathematical transformation that encrypts or decrypts binary data. An **encryption key** is a binary input to the encryption algorithm—typically a long string of bits. The encryption algorithm varies the data transformation based on the encryption key so data can be decrypted only with the same key or a compatible decryption key. Many encryption algorithms are available, and a few—including Data Encryption Standard (DES) and several algorithms developed by RSA Security—are widely deployed governmental or Internet standards. An encryption algorithm must generate encrypted data that are difficult or impossible to decrypt without the encryption key. Decryption without the key becomes more difficult as key length is increased. Sender and receiver must use the same or compatible algorithms.

Figure 12-23 shows an example of **symmetric key encryption**, in which the same key encrypts and decrypts the data. A significant problem with

privileged user a person who has access to the source code, executable program, and database structure of the system

encryption the process of altering data so unauthorized users can't view them

decryption the process of converting encrypted data back to their original state

encryption algorithm a complex mathematical transformation that encrypts or decrypts binary data

encryption key a binary input to the encryption algorithm—typically a long string of bits

symmetric key encryption encryption method that uses the same key to encrypt and decrypt the data

FIGURE **12-23** *Symmetric key encryption*

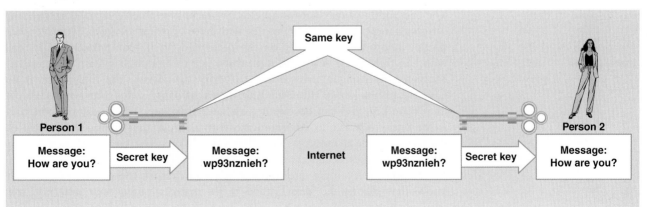

symmetric key encryption is that sender and receiver use the same key, which must be created and shared in a secure manner. Security is compromised if the key is transmitted over the same channel as messages encrypted with the key. Also, sharing a key among many users increases the possibility of key theft. Because of these risks, symmetric key encryption is primarily used with data stored in files and databases but not with data transmitted over networks.

Data stored by file and database servers can be encrypted with symmetric key encryption to protect against unauthorized access that bypasses the OS or DBMS to directly access the physical data store. Typically, the encryption key is stored on a different server that is queried via a secure logon when the server first boots up. Data stored on laptops and other portable devices are often encrypted to protect against unauthorized access due to loss, theft, or disposal.

remote wipe security measure that automatically deletes sensitive data from a portable device when unauthorized accesses are attempted

An additional security measure for portable devices is a technique commonly called **remote wipe**, which automatically deletes sensitive data from portable devices under certain conditions, such as repeated failure to enter a valid username and password or an attempt to access a database or file from an unauthorized application. Remote wipe is commonly used with apps on portable devices that sync sensitive data from a server. An authorized synchronization attempt triggers the remote wipe via command from the app or the server.

asymmetric key encryption encryption method that uses different keys to encrypt and decrypt the data

public key encryption a form of asymmetric key encryption that uses a public key for encryption and a private key for decryption

Asymmetric key encryption uses different but compatible keys to encrypt and decrypt data. **Public key encryption** is a form of asymmetric key encryption that uses a public key for encryption and a private key for decryption. The two keys are like a matched pair. Once information is encrypted with the public key, it can be decrypted only with the private key. It can't be decrypted with the same public key that encrypted it. Organizations that use this technique broadcast their public key so it is freely available to anybody who wants it. Then, when some entity—for example, someone who wants to order something from the vendor—wishes to transmit a secure message to a vendor, that customer reads the vendor's public key from a public source, such as a Web site. The customer encrypts the message with the public key and sends the message to the vendor. The vendor then decrypts the message with the private key. Because no one else has the private key, no one else can decrypt the message.

Some asymmetric encryption methods can encrypt and decrypt messages in both directions. That is, in addition to using the public key to encrypt a message that can be decrypted with the private key, an organization can also encrypt a message with the private key and decrypt it with the public key. Notice that both keys must still work as a pair, but the message can go forward or backward through the encryption/decryption pair. This second technique is the basis for digital signatures and certificates, which are explained in the next section. **Figure 12-24** illustrates an asymmetric key encryption transmittal.

Digital Signatures and Certificates

The encryption of messages is an effective technique to enable a secure exchange of information between two entities who have appropriate keys. However, how do you know that the entity on the other end of the communication is really who you think it is? A **digital signature** is a technique in which a document is encrypted by using a private key to verify who wrote the document. If you have the public key of an entity and then that entity sends you a message with its private key, you can decode it with the public key. You know that the party is the one you want to communicate with because that entity is the only one who can encode a message with that private key. The encoding of a message with a private key is called *digital signing*.

digital signature a technique in which a document is encrypted by using a private key to verify who wrote the document

Taking the example one step further, you can ask the question "How do I know that the public key I have is the correct public key and not some

FIGURE **12-24** *Asymmetric key encryption*

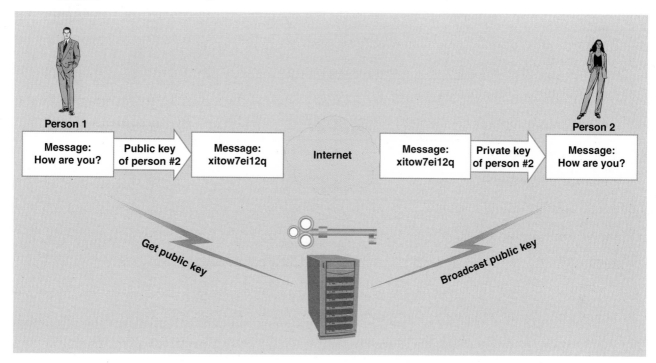

counterfeit key?" In other words, maybe someone is impersonating another entity and is passing out false public keys to be able to intercept encoded messages (such as financial transactions) and steal information. In essence, the problem is ensuring that the key that is purported to be the public key of some institution is in fact that institution's public key. The solution to that problem is a certificate.

A certificate—or **digital certificate**—is an institution's name and public key (plus other information, such as address, Web site URL, and validity date of the certificate), encrypted and certified by a third party. Many third parties, such as VeriSign and Equifax, are very well known and widely accepted **certifying authorities**. In fact, they are so well known that their public keys are built right into Netscape and Internet Explorer. As shown in **Figure 12-25**, you can know that the entities with whom you are communicating are in fact who they say they are and that you do have their correct public key.

An entity that wants a certificate with its name and public key goes to a certifying authority and buys a certificate. The certifying authority encrypts the data with its own private key (signs the data) and gives the data back to the original entity. Then, when someone, such as a customer, asks the entity for its public key, it sends the certificate. The customer receives the certificate and opens it with the certifying authority's public key. Again, the certifying authority is so well known that its public key is built into everyone's browser and is essentially impossible to counterfeit. The customer can now be sure that he or she is communicating with the original entity and can do so with encrypted messages by using the entity's public key.

A variation of this scenario occurs when the buyer and seller transmit their certificates to one another. Each participant can decrypt the certificate by using the certifying authority's public key to extract such information as name and address. However, to ensure that the public key contained within the certificate is valid, the certificates are transmitted to the certifying authority for

digital certificate an institution's name and public key (plus other information, such as address, Web site URL, and validity date of the certificate) encrypted and certified by a third party

certifying authority a widely accepted issuer of digital certificates

FIGURE **12-25** *Using a digital certificate*

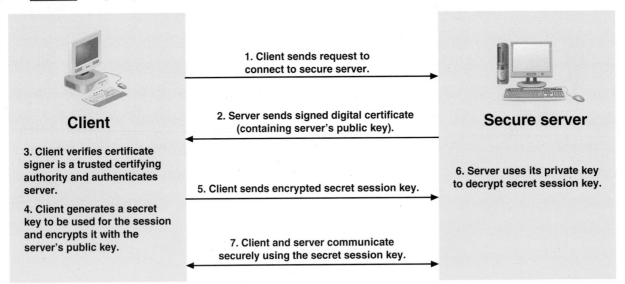

verification. The authority stores certificate data, including public keys, within its database and verifies transmitted certificates by matching their content against the database.

Secure Transactions

Secure Sockets Layer (SSL) a standard set of methods and protocols that address authentication, authorization, privacy, and integrity

Transport Layer Security (TLS) an Internet standard equivalent to SSL

Secure electronic transactions require a standard set of methods and protocols that address authentication, authorization, privacy, and integrity. Netscape originally developed the **Secure Sockets Layer (SSL)** to support secure transactions. SSL was later adopted as an Internet standard and renamed **Transport Layer Security (TLS)**, although the original name—SSL—is still widely used.

TLS is a protocol for a secure channel to send messages over the Internet. Sender and receiver first establish a connection by using ordinary Internet protocols and then ask each other to create a TLS connection. Sender and receiver then verify each other's identity by exchanging and verifying identity certificates as explained previously. At this point, either or both have exchanged public keys, so they can send secure messages. Because asymmetric encryption is quite slow and difficult, the two entities agree on a protocol and encryption method—usually a single-key encryption method. Of course, all the messages to establish a secure connection are sent by using the public key/private key combination. Once the encryption technique has been determined and the secret, single key has been transmitted, all subsequent transmission is done by using the secret, single key.

IP Security (IPSec) an Internet standard for secure transmission of low-level network packets

IP Security (IPSec) is a newer Internet standard for secure message transmission. IPSec is implemented at a lower layer of the network protocol stack, which enables it to operate with greater speed. IPSec can replace or complement SSL. The two protocols can be used at the same time to provide an extra measure of security. IPSec supports more secure encryption methods than SSL.

Secure Hypertext Transport Protocol (HTTPS) an Internet standard for securely transmitting Web pages

Secure Hypertext Transport Protocol (HTTPS or HTTP-S) is an Internet standard for securely transmitting Web pages. HTTPS supports several types of encryption, digital signing, and certificate exchange and verification. All modern Web browsers and servers support HTTPS. It is a complete approach to Web-based security, although security is enhanced when HTTPS documents are sent over secure TLS or IPSec channels.

Chapter Summary

Most modern information systems store data in a database and access and manage the data by using a DBMS. One of the key activities of system design is developing a relational or object database schema. A relational database is a collection of data stored in tables. A relational database schema is normally developed from a domain class diagram. Each class is represented as a separate table. One-to-many associations are represented by embedding foreign keys in class tables. Many-to-many associations are represented by creating additional tables containing foreign keys of the related classes.

Database design is usually performed in an early iteration of a development project to minimize project risk. Many new systems interact with existing databases, and the architectural decisions associated with databases are intertwined with other architectural issues. The combination of these factors increases the risk associated with database design tasks. When the risks are low, the database can be designed and implemented incrementally across multiple project iterations.

Because stored data is such an important organizational asset, database design incorporates controls to ensure the correctness, completeness, and security of data. Integrity and security controls extend beyond the database to all other system components, including application programs, operating systems, Web sites, and networks.

Key Terms

access control list   397

access controls   394

asymmetric key encryption   400

attribute   375

attribute value   375

authentication   397

authorization   398

certifying authorities   401

cloud-based database server
   architecture   388

completeness controls   394

complex data types   386

complex update controls   394

data type   386

data validation controls   394

database (DB)   373

database management system (DBMS)   373

database synchronization   389

decryption   399

digital certificate   401

digital signature   400

encryption   399

encryption algorithm   399

encryption key   399

field combination controls   394

first normal form (1NF)   383

foreign key   375

fraud triangle   396

functional dependency   384

input controls   393

integrity controls   392

IP Security (IPSec)   402

key   375

multifactor authentication   397

normalization   383

output controls   395

partitioned database server architecture   388

physical data store   373

primary key   375

primitive data types   386

privileged users   399

public key encryption   400

referential integrity   382

referential integrity constraint   383

registered users   398

relational database management system
   (RDBMS)   374

remote wipe   400

replicated database server architecture   387

row   374

schema   373

second normal form (2NF)   385

Secure Hypertext Transport Protocol
   (HTTPS or HTTP-S)   402

Secure Sockets Layer (SSL)   402

security controls   396

single database server architecture   387

symmetric key encryption   399

Review Questions

1. List the components of a DBMS and describe the function of each.

2. What is a database schema? What information does it contain?

3. Why are databases the preferred method of storing data used by an information system?

4. With respect to relational databases, briefly define the terms *row* and *attribute value*.

5. What is a primary key? Are duplicate primary key values allowed? Why or why not?

6. What is the difference between a natural key and an invented key? Which type is most commonly used in business information processing?

7. What is a foreign key? Why are foreign keys used or required in a relational database? Are duplicate foreign key values allowed? Why or why not?

8. Describe the steps used to transform a domain class diagram into a relational database schema.

9. What is referential integrity? Describe how it is enforced when a new foreign key value is created, when a row containing a primary key is deleted, and when a primary key value is changed.

10. What types of data (or attributes) should never be stored more than once in a relational database? What types of data (or attributes) usually must be stored more than once in a relational database?

11. What is relational database normalization? Why is a database schema in third normal form considered to be of higher quality than an unnormalized database schema?

12. Describe the process of relational database normalization. Which normal forms rely on the definition of functional dependency?

13. What is the difference between a primitive data type and a complex data type?

14. Briefly describe these distributed database architectures: replicated database servers, partitioned database servers, and cloud-based database servers. What are the comparative advantages of each?

15. What additional database management complexities are introduced when database contents are replicated in multiple locations?

16. Describe the risk factors associated with database design.

17. When should database design be performed? Can the database be designed iteratively or must the entire database be designed at once?

18. Explain four types of integrity controls for input forms. Which have you seen most frequently? Why are they important?

19. What are the objectives of integrity controls in information systems? In your own words, explain what each of the three objectives means. Give an example of each.

20. What are the four types of input controls used to reduce input errors? Describe how each works.

21. What is the basic purpose of transaction logging?

22. What are the two primary objectives of security controls?

23. Briefly define or describe authentication, access control lists, and authorization.

24. How does single-key (symmetric) encryption work? What are its strengths? What are its weaknesses?

25. How does public key (asymmetric) encryption work? What are its strengths? What are its weaknesses?

26. What is a digital certificate? What role do certifying authorities play in security systems?

27. What is a digital signature? What does it tell a user?

Problems and Exercises

1. The Universal Product Code (UPC) is a bar-coded number that uniquely identifies many products sold in the United States. For example, all printed copies of this textbook sold in the United States have the same UPC bar code on the back cover. Now consider how the design of the RMO database might change if all items sold by RMO were required by law to carry a permanently attached UPC (e.g., on a

label sewn into a garment or on a radio frequency ID tag attached to a product). How might the RMO relational database schema change under this requirement?

2. Assume that RMO will begin asking a random sample of customers who order by telephone about purchases made from competitors. RMO will give customers a 15 percent discount on their current order in exchange for answering a few questions. To store and use this information, RMO will add two new classes and three new associations to the class diagram. The new classes are Competitor and ProductCategory. Competitor has a one-to-many association with ProductCategory, and the existing Customer class also has a one-to-many association with ProductCategory. Competitor has a single attribute called Name. ProductCategory has four attributes: Description, DollarAmountPurchased, MonthPurchased, and YearPurchased. Revise the relational database schema shown in Figure 12-10 to include the new classes and associations. All tables must be in 3NF.

3. Assume that RMO will use a relational database, as shown in Figure 12-10. Assume further that a new catalog group located in Milan, Italy, will now create and maintain the product catalog. To minimize networking costs, the catalog group will have a dedicated database server attached to its LAN. Develop a plan to partition the RMO database. Which tables should be replicated on the catalog group's local database server? Update Figure 12-18 to show the new distributed database architecture.

4. Visit the Web site of an online catalog vendor similar to RMO (such as www.llbean.com) or an online vendor of computers and related merchandise (such as www.cdw.com). Browse the online catalog and note the various types of information contained there. Construct a list of complex data types that would be needed to store all the online catalog information.

5. This chapter described various situations that emphasized the need for controls. In the first scenario presented, a furniture store sells merchandise on credit. Based on the descriptions of controls given in this chapter, identify the various controls that should be implemented in the system to ensure that corrections to customer balances are made only by someone with the correct authorization. In the second scenario illustrating the need for controls, an accounts payable clerk uses the system to write checks to suppliers. Based on the information in this chapter, what kinds of controls would you implement to ensure that checks are written only to valid suppliers, that checks are written for the correct amount, and that all payouts have the required authorization? How would you design the controls if different payment amounts required different levels of authorization?

6. Look on the Web for an e-commerce site (such as Amazon.com or eBay). What kinds of security and controls are integrated into the system?

7. Examine the information system of a local business, such as a fast-food restaurant, doctor's office, video store, grocery store, etc. Evaluate the screens (and reports, if possible). What kind of integrity controls are in place? What kinds of improvements would you make?

8. Search the Web and find out what you can about Pretty Good Privacy. What is it? How does it work? Find what you can about a passphrase. What does it mean? Start your research at www.pgpi.org.

Case Study

Computer Publishing, Inc.

In only a decade, Computer Publishing, Inc. (CPI) grew from a small textbook publishing house into a large international company with significant market share in traditional textbooks, electronic books, and distance education courseware. CPI's processes for developing books and courseware were similar to those used by most other publishers, but those processes had proven cumbersome and slow in an era of rapid product cycles and multiple product formats.

Text and art were developed in a wide variety of electronic formats, and conversions among those formats were difficult and error-prone. Many editing steps were performed with traditional paper-and-pencil methods. Consistency errors within books and among books and related

products were common. Developing or revising a book and all its related products typically took a year or more.

CPI's president initiated a strategic project to re-engineer the way that CPI developed books and related products. CPI formed a strategic partnership with Davis Systems (DS) to develop software that would support the re-engineered processes. DS had significant experience developing software to support product development in the chemical and pharmaceutical industries by using the latest development tools and techniques, including object-oriented software and relational databases. CPI expected the new processes and software to reduce development time and cost. Both companies expected to license the software to other publishers within a few years.

(continued on page 406)

(continued from page 405)

A joint team specified the workflows and high-level requirements for the software. The team developed plans for a large database that would hold all book and courseware content through all stages of production. Authors, editors, and other production staff would interact with the database in a variety of ways, including traditional word-processing programs and Web-based interfaces. When required, format conversions would be handled seamlessly and without error. All content creation and modification would be electronic—no text or art would ever be created or edited on paper, except as a printed book ready for sale.

Software would track and manage content through every stage of production. Content common to multiple products would be stored in the database only once. Dependencies within and across products would be tracked in the database. Software would ensure that any content addition or change would be reflected in all dependent content and products, regardless of the final product form. For example, a sentence in Chapter 2 that refers to a figure in Chapter 1 would be updated automatically if the figure were renumbered. If a new figure were added to a book, it would be added automatically to the related courseware presentation slides. Related courseware and study material on the Web site would automatically reflect changes, such as a new answer to an end-of-chapter question.

1. Consider the contents of this textbook as a template for CPI's database content. Draw a class diagram that represents the book and its key content elements. Expand your diagram to include related product content, such as a set of PowerPoint slides, an electronic book formatted as a Web site or PDF file, and a Web-based test bank.
2. Develop a list of data types required to store the content of the book, slides, and Web sites. Are the relational DBMS data types listed in Figure 12-14 sufficient?
3. Authors and editors are often independent contractors, not publishing company employees. Consider the implications of this fact for controls and security. How would you enable authors and editors to interact with the database? How would you protect database content from hackers and other unauthorized accesses?

RUNNING CASE STUDIES

Community Board of Realtors

In Chapter 4, you developed a domain model class diagram. Using your previous solution or one provided to you by your instructor, update your domain model class diagram with any additional problem domain classes, new associations, or additional attributes that you have discovered as you worked through the previous chapters. Finalize this comprehensive domain model and then turn it in as part of your solution.

Using this comprehensive domain model class diagram, develop a relational database schema. In the schema, identify the foreign keys that are required.

Also, identify a key attribute for each table. You may need to add a key field if there isn't an attribute that could logically serve as the key. Remember that a candidate key for an association class is the combination of the keys of the connect classes. However, it may make sense to define a shorter, more concise key field.

Verify that each table is in first, second, and third normal form. Discuss any discrepancies you had to fix from your first solution. Discuss any tables that may not be in third normal form and why you are leaving it as not-normalized.

The Spring Breaks 'R' Us Travel Service

As with other social networking sites and systems, users of the Spring Breaks 'R' Us social networking system face such risks as identity theft, phishing attacks, and viruses. Review the following information related to social networking risks and security published by the United States Computer Emergency Readiness Team, including:

■ Socializing Securely: Using Social Networking Services (www.us-cert.gov/reading_room/safe_social_networking.pdf)
■ Cyber Security Tip ST06-003: Staying Safe on Social Network Sites (www.us-cert.gov/cas/tips/ST06-003.html)

(continued on page 407)

(continued from page 406)

■ Cyber Security Tip ST05-013: Guidelines for Publishing Information Online (www.us-cert.gov/cas/tips/ST05-013.html)

After reviewing this information, revisit the questions for this case in Chapter 6 for the Social Networking subsystem. Based on the contents of this chapter and the information contained in the readings, what specific controls and security measures should be incorporated into the Social Networking subsystem?

On the Spot Courier Services

In Chapter 4, you developed a domain model class diagram. Using your previous solution or one provided to you by your instructor, update your domain model class diagram with any additional problem domain classes, new associations, or additional attributes that you have discovered as you worked toward your solutions in the previous chapters. Finalize this comprehensive domain model and then turn it in as part of your solution.

Using this comprehensive domain model class diagram, develop a relational database schema. In the schema, identify the foreign keys that are required. Also, identify a key attribute for each table. You may need to add a key field if there isn't an attribute that could logically serve as the key. Remember that a candidate key for an association class is the combination of the keys of the connect classes. However, it may make sense to define a shorter, more concise key field.

Verify that each table is in first, second, and third normal form. Discuss any discrepancies you had to fix from your first solution. Discuss any tables that aren't in third normal form and why you are leaving them as not-normalized. (For example, in the United States, city and state are functionally dependent on zip code, but we leave all three fields in the same table. Why?)

In Chapter 6, you discussed hardware requirements, and in Chapter 10, you developed component and deployment diagrams. Based on your work in those chapters, take these steps:

1. For each user and each type of device, discuss what security precautions and techniques should be used to protect access to the device itself.
2. For each user and each type of device, discuss what security precautions and techniques should be used to protect access to the application programs, connect to the home system, and protect the data being transmitted to the foreign devices.
3. Discuss any security precautions and techniques you would recommend for the home office and the network servers.

Sandia Medical Systems Real-Time Glucose Monitoring

Part 1
Review the original system description in Chapter 2, the additional project information in Chapters 3, 4, and 8, and the domain class diagram shown in **Figure 12-26** to refamiliarize yourself with the proposed system. Assume that the type attribute of the AlertCondition class identifies one of three alert types:

1. Glucose levels that fall outside the specified range for 15 minutes (three consecutive readings)
2. Glucose levels that fall outside the specified range for 60 minutes (12 consecutive readings)
3. An average of glucose levels over a eight-hour period that falls outside a specified range

The specified range for an AlertCondition object is the set of values between and including lowerBound and upperBound. AlertCondition objects also include an effective time period specified by the attributes startHour and endHour, which enables physicians to set different alert parameters for sleeping and waking hours.

When an alert is triggered, an object of type Alert is created and associated with an alertCondition object. The dateTime attribute records when the Alert object was created, and the value(s) attribute record(s) the glucose levels (alert types 1 and 2) or average level (alert type 3) that fell outside the specified range. Each Alert object is indirectly related to a Patient object via the association between Alert and AlertCondition and the association between AlertCondition and Patient.

Develop a set of relational database tables based on the domain class diagram. Identify all primary and foreign keys, and ensure that the tables are in 3NF.

(continued on page 408)

(continued from page 407)

FIGURE **12-26**
Updated domain model class diagram for Sandia RTGM system

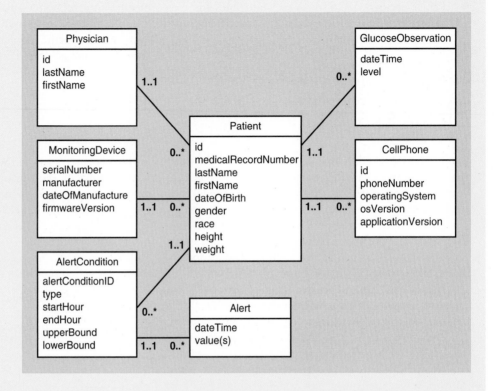

Part 2

Based on what you learned in this chapter about databases, controls, and system security, review your answers to the questions for this case in Chapter 6. Assume that the patient's cell phone and the centralized servers are different nodes in a replicated database architecture and are regularly synchronized. What changes, if any, should be made to your answers now that you have a deeper understanding of databases, controls, security, and related design issues?

Further Resources

W. Steve Albrecht, Chad O. Albrecht, Conan C. Albrecht, and Mark F. Zimbelman, *Fraud Examination*, 4th edition, Cengage Learning, 2012.

Alfred Basta and Melissa Zgola, *Database Security*, Cengage Learning, 2011.

Carlos Coronel, Steven Morris, and Peter Rob, *Database Systems: Design, Implementation, and Management*, 9th edition, Cengage Learning, 2010.

Michael E. Whitman and Herbert J. Mattord, *Principles of Information Security*, 4th edition, Cengage Learning, 2012.

13

Making the System Operational

Chapter Outline

- Testing
- Deployment Activities
- Planning and Managing Implementation, Testing, and Deployment
- Putting It All Together—RMO Revisited

Learning Objectives

After reading this chapter, you should be able to:

- Describe implementation and deployment activities
- Describe various types of software tests and explain how and why each is used
- Explain the importance of configuration management, change management, and source code control to the implementation, testing, and deployment of a system
- List various approaches to data conversion and system deployment and describe the advantages and disadvantages of each
- Describe training and user support requirements for new and operational systems

Tri-State Heating Oil: Juggling Priorities to Begin Operation

It was 8:30 on Monday morning, and Maria Grasso, Kim Song, Dave Williams, and Rajiv Gupta were about to begin the weekly project status meeting. Tri-State Heating Oil had started developing a new scheduling system for customer orders and service calls five months earlier. The target completion date was 10 weeks away, but the project was behind schedule. Early project iterations had accomplished far less than anticipated because key users had disagreed on what new system requirements to include and the system scope was larger than expected.

Maria began the meeting. "We've gained a day or two since our last meeting due to better-than-expected unit testing results," she said. "All the methods developed last week sailed through unit testing, so we won't need any time this week to fix errors in that code."

Kim frowned. "I wouldn't get too cocky just yet," she said. "All the nasty surprises in my last project came during integration testing. We're completing the user-interface classes this week, so we should be able to start integration testing with the business classes sometime next week."

Nodding enthusiastically, Dave said, "That's good! We have to finish testing those user-interface classes as quickly as possible because we're scheduled to start user training in three weeks. I need that time to develop the documentation and training materials and work out the final training schedule with the users."

Rajiv looked doubtful. "I'm not sure that we should be trying to meet our original training schedule with so much of the system still under development," he said. "What if integration testing shows major bugs that require more time to fix? And what about the unfinished business and database classes? Can we realistically start training with a system that's little more than a user interface with half a system behind it?"

"But we have to start training in three weeks," Dave replied. "We contracted for a dozen temporary workers so we could train our staff on the new system. Half of them are scheduled to start in two weeks, and the rest two weeks after that. It's too late to renegotiate their starting dates. We can extend the time they'll be here, but delaying their starting date means we'll be paying for people we aren't using."

Maria spoke up. "I think that Rajiv's concerns are valid," she said. "It's not realistic to start training in three weeks with so little of the system completed and tested. We're at least five weeks behind schedule, and there's no way we'll recapture more than four or five days of that during the next few weeks. I've already looked into rearranging some of the remaining coding to give priority to the work most critical to user training. There are a few batch processes that can be back-burnered for a while. Kim, can you rearrange your testing plans to handle all the interactive applications first?"

"I'll have to go back to my office and take another look at the dependencies among those programs," Kim replied. "Offhand, I'd say yes, but I need a few hours to make sure."

"Okay," Maria said. "Let's proceed under the assumption that we can rearrange coding and testing to complete a usable system for training in five weeks. I'll confirm that by e-mail later today, as soon as Kim gets back to me. I'll also schedule a meeting with the CIO to deliver the bad news about temporary staffing costs."

After a few moments of silence, Rajiv asked, "So, what else do we need to be thinking about?"

Well, let's see," Maria replied. "There's hardware delivery and setup, operating system and DBMS installation, importing data from the old database, the network upgrade, and stress testing for the distributed database accesses."

Rajiv smiled and said to Maria, "You must have been a juggler in your youth, which would have been good practice for keeping all these project pieces up in the air. Does management pay you by the ball?"

Maria chuckled. "I do think of myself as a juggler sometimes. And if management paid me by the ball, I could retire as soon as this project is finished!"

Overview

Developing any complex system is inherently difficult. For example, consider the complexity of manufacturing automobiles. Tens of thousands of parts must be fabricated or purchased. Laborers and machines must assemble those parts into small subcomponents, such as dashboard instruments, wiring harnesses, and brake assemblies, which are in turn assembled into larger subcomponents, such as instrument clusters, engines, and transmissions, which in turn must be constructed, tested, and passed on to subsequent assembly steps. The effort, timeliness, cost, and output quality of each step depend on all the preceding steps.

Implementing and deploying an information system is similar in many ways; it is a complex production and assembly process that must use resources efficiently,

FIGURE **13-1** *Activities of the implementation and deployment SDLC core processes*

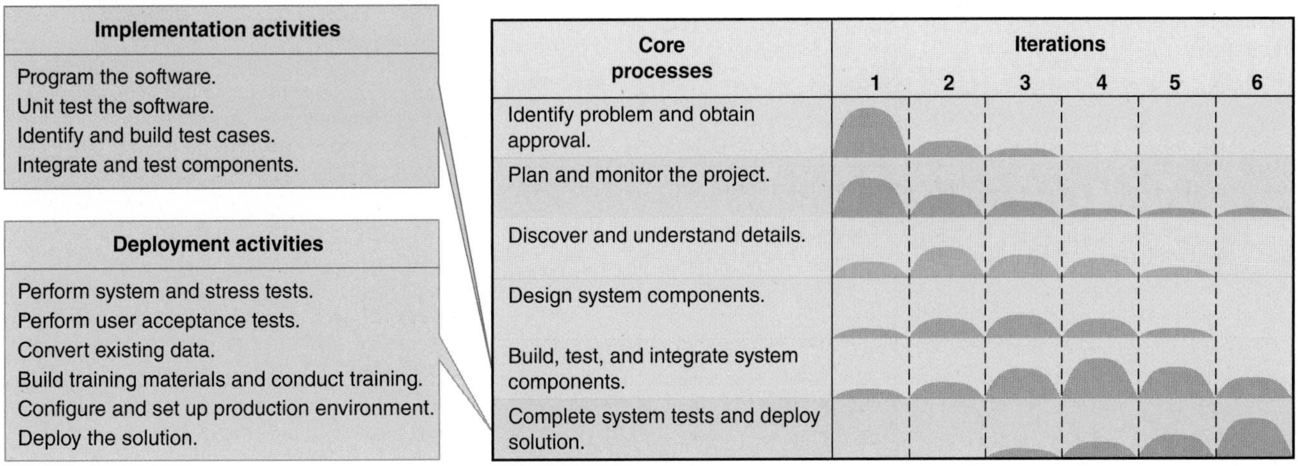

Implementation activities
Program the software.
Unit test the software.
Identify and build test cases.
Integrate and test components.

Deployment activities
Perform system and stress tests.
Perform user acceptance tests.
Convert existing data.
Build training materials and conduct training.
Configure and set up production environment.
Deploy the solution.

Core processes	Iterations					
	1	2	3	4	5	6
Identify problem and obtain approval.						
Plan and monitor the project.						
Discover and understand details.						
Design system components.						
Build, test, and integrate system components.						
Complete system tests and deploy solution.						

minimize construction time, and maximize product quality. But unlike automobile manufacturing, it isn't done once and then used to build thousands of similar units. Instead, implementation and deployment are unique to each project and must match that project's characteristics.

We have spent many chapters detailing the first four core processes of the system development life cycle (SDLC). Those core processes are the primary focus of this text, but additional processes and activities are needed to complete a system and put it to regular use. The core processes and activities covered in this chapter are summarized in **Figure 13-1**.

The fact that we are covering two core processes in a single chapter doesn't mean that they are simple or unimportant. Rather, they are complex processes that you will learn about in detail by completing other courses and reading other books as well as through on-the-job training and experience. Our purpose in covering them in this chapter is to round out our discussion of the SDLC and to show how all the core processes and activities relate to one another.

As you can see from Figure 13-1, program and testing are the primary implementation activities. You will learn about programming in other courses and from other textbooks, so we won't discuss how software components are constructed in this textbook. However, we will discuss testing activities in detail because they are such an integral part of multiple core processes, including project planning and monitoring, design, implementation, and deployment.

Testing

Testing activities are a key part of implementation and deployment activities, although different kinds of tests are used in each core process. Testing is the process of examining a component, subsystem, or system to determine its operational characteristics and whether it contains any defects. To conduct a test, developers must have well-defined standards for functional and nonfunctional requirements. From the requirements, test developers develop precise definitions of expected operational characteristics and what constitutes a defect. The developers can test software by reviewing its construction and composition or by designing and building the software, exercising its function, and examining the results. If the results indicate a shortcoming or defect, developers cycle back through earlier implementation or deployment activities until the shortcoming is remedied or the defect is eliminated.

Test types, their related core processes, and the defects they detect and operational characteristics they measure are summarized in **Figure 13-2**. Each type of testing is described in detail later in this section.

FIGURE **13-2**

Test types and corresponding operational characteristics and defects

Test type	Core process	Tested defects and operational characteristics
Unit testing	Implementation	Software component that doesn't correctly perform its function when tested in isolation—for example, a component for calculating sales tax that consistently computes sales tax incorrectly for one or more localities
Integration testing	Implementation	Software component that performs correctly in isolation but incorrectly when tested in combination with other components—for example, order entry and shipping cost calculation components that pass unit testing but fail when tested together due to conversion errors as data are passed from one component to the other
Usability testing	Implementation	Software that works but fails to satisfy one or more user requirements related to function or ease of use—for example, a user-interface component that forces a user to follow a needlessly complex procedure to complete a common and simple task
System and stress testing	Deployment	System or subsystem that doesn't correctly perform its function or fails to meet a nonfunctional requirement under normal operating conditions—for example, an order retrieval function that displays a result in two seconds when tested in isolation with a dummy database but requires 30 seconds when tested with other functions using a live database

test case a formal description of a starting state, one or more events to which the software must respond, and the expected response or ending state

test data a set of starting states and events used to test a module, group of modules, or entire system

unit test test of an individual method, class, or component before it is integrated with other software

An important part of developing tests is specifying test cases and data. A **test case** is a formal description of the following:

- A starting state
- One or more events to which the software must respond
- The expected response or ending state

The starting and ending states and the events are represented by a set of **test data**. For example, the starting state of a system may represent a particular set of data, such as the existence of a particular customer and order for that customer. The event may be represented by a set of input data items, such as a customer account number and order number used to query order status. The expected response may be a described behavior, such as the display of certain information, or a specific state of stored data, such as a canceled order.

Preparing test cases and data is a tedious and time-consuming process. At the component and method levels, every instruction must be executed at least once. Ensuring that all instructions are executed during testing is a complex problem. Fortunately, automated tools based on proven mathematical techniques are available to generate a complete set of test cases. Many test cases representing normal and exceptional processing situations should be prepared for each scenario.

Unit Testing

Unit testing is the process of testing individual methods, classes, or components before they are integrated with other software. The goal of unit testing is to identify and fix as many errors as possible before modules are combined into larger software units, such as programs, classes, and subsystems. Errors become much more difficult and expensive to locate and fix when many units are combined.

Few units are designed to operate in isolation. Instead, groups of units are designed to execute as an integrated whole. If a method is considered a unit,

FIGURE **13-3** *Sequence diagram for* Create new order

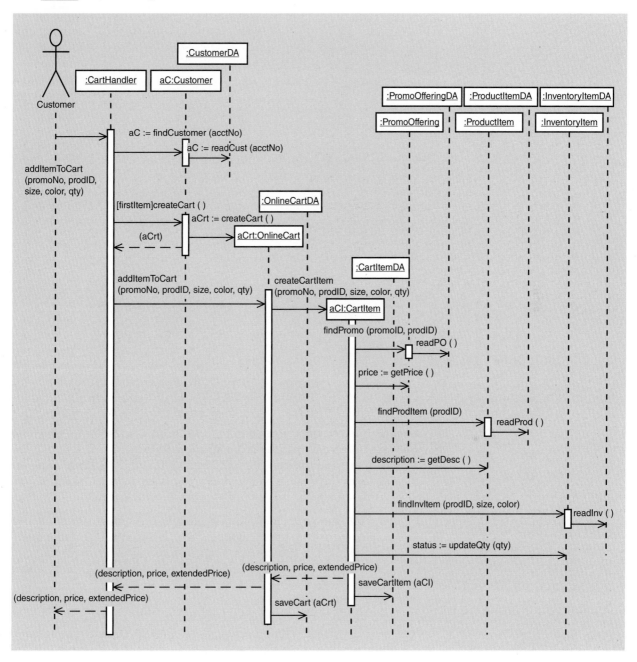

that method may be called by messages sent from methods in one or more classes and may, in turn, send messages to other methods in its own or other classes. These relationships are easily seen in a sequence diagram, such as in **Figure 13-3**, which duplicates Figure 11-13. For example, when the class CartItem receives a createCartItem() message, it performs internal processing and sends messages to six other methods—findPromo(), getPrice(), findProdItem(), getDescription(), findInvItem(), and updateQty()—in three other classes: PromoOffering, ProductItem, and InventoryItem.

If the createCartItem() method of CartItem is being tested in isolation, then two types of testing methods are required. The first method type is called a driver. A **driver** simulates the behavior of a method that sends a message to the

driver a method or class developed for unit testing that simulates the behavior of a method that sends a message to the method being tested

FIGURE **13-4**

Driver module to test createCartItem( )

```
main()
{
 // driver method to test CartItem::createCartItem()

 // declare input parameters and values

 int promoID = 23;
 int prodID = 1244;
 String size = "large";
 String color = "red";
 int quantity = 1;

 // perform test

 cartItem cartItem = new cartItem();
 cartItem.createcartItem(promoID,prodID,size,color,quantity);

 // display results

 System.out.println("price=" + cartItem.getPrice());
 System.out.println("description=" + cartItem.getDescription();
 System.out.println("status=" + cartItem.getStatus());
} // end main()
```

method being tested—in this example, the call by an OnlineCart object to createCartItem(). A driver module implements these functions:

- Sets the value of input parameters
- Calls the tested unit, passing it the input parameters
- Accepts return parameters from the tested unit and prints them, displays them, or tests their values against expected results and then prints or displays the results

Figure 13-4 shows a simple driver module for testing createCartItem(). A more complex driver module might use test data consisting of hundreds or thousands of test inputs and correct outputs stored in a file or database. The driver would loop through the test inputs and repeatedly call createCartItem(), check the return parameter against the expected value, and print or display warnings of any discrepancy. Using a driver allows a subordinate method to be tested before methods that call it have been written.

The second type of testing method used to perform unit tests is called a stub. A **stub** simulates the behavior of a method that hasn't yet been written. A unit test of createCartItem() would require three stub methods: getPrice(), getDescription(), and updateQty(). Stubs are relatively simple methods that usually have only a few executable statements. Each of the stubs used to test createCartItem() can be implemented as a statement that simply returns a constant, regardless of the parameters passed as input. **Figure 13-5** shows sample code for each of the three stub modules.

stub a method or class developed for unit testing that simulates the behavior of a method that hasn't yet been written

Integration Testing

integration test test of the behavior of a group of methods, classes, or components

An **integration test** evaluates the behavior of a group of methods, classes, or components. The purpose of an integration test is to identify errors that weren't or couldn't be detected by unit testing. Such errors may result from a number of problems, including:

- Interface incompatibility—For example, one method passes a parameter of the wrong data type to another method.

FIGURE 13-5
*Stub modules used by
createCartItem( )*

```
float getPrice()
{
 // stub method for CatalogProduct::getPrice()

 return(24.95);
} // end getPrice()

String getDescription()
{
 // stub method for Product::getDescription()

 return("mens khaki slacks");
} // end getDescription()

String updateQty(int decrement)
{
 // stub method for InventoryItem::updateQty()

 return("OK");
} // end updateQty()
```

- Parameter values—A method is passed or returns a value that was unexpected, such as a negative number for a price.
- Run-time exceptions—A method generates an error, such as "out of memory" or "file already in use," due to conflicting resource needs.
- Unexpected state interactions—The states of two or more objects interact to cause complex failures, as when an OnlineCart class method operates correctly for all possible Customer object states except one.

These four problems are some of the most common integration testing errors, but there are many other possible errors and causes.

Once an integration error has been detected, the responsibility for incorrect behavior must be traced to specific method(s). The person responsible for performing the integration test is generally also responsible for identifying the cause of the error. Once the error has been traced to a particular method, the programmer who wrote the method is asked to rewrite it to correct the error.

Integration testing of object-oriented software is very complex. Because an object-oriented program consists of a set of interacting objects that can be created or destroyed during execution, there is no clear hierarchical structure. As a result, object interactions and control flow are dynamic and complex.

Additional factors that complicate object-oriented integration testing include:

- Methods can be (and usually are) called by many other methods, and the calling methods may be distributed across many classes.
- Classes may inherit methods and state variables from other classes.
- The specific method to be called is dynamically determined at run time based on the number and type of message parameters.
- Objects can retain internal variable values (i.e., the object state) between calls. The response to two identical calls may be different due to state changes that result from the first call or occur between calls.

This combination of factors makes it difficult to determine an optimal testing order and to predict the behavior of a group of interacting methods and objects. Thus, developing and executing an integration testing plan for object-oriented software is an extremely complex process. Specific methods and techniques for dealing with that complexity are well beyond the scope of this textbook. See the

Further Resources section at the end of this chapter for object-oriented software testing references.

Usability Testing

A **usability test** is a test to determine whether a method, class, subsystem, or system meets user requirements. Because there are many types of requirements—functional and nonfunctional—many types of usability tests are performed at many different times.

The most common type of usability test evaluates functional requirements and the quality of a user interface. Users interact with a portion of the system to determine whether it functions as expected and whether the user interface is easy to use. Such tests are conducted frequently, as user interfaces are developed to provide rapid feedback to developers for improving the interface and correcting any errors in the underlying software components.

System, Performance, and Stress Testing

A **system test** is an integration test of the behavior of an entire system or independent subsystem. Integration testing is normally associated with the implementation core process, and system testing is normally associated with the deployment core process. The line separating integration testing from system testing is fuzzy, as is the line between implementation and deployment activities. The important differences are scope and timing. Integration tests are performed more frequently and on smaller component groups. System tests are performed less frequently on entire systems or subsystems.

For a system developed by using a traditional waterfall SDLC, system testing is concentrated near the end of the project. In a typical iterative project, some deployment activities, including system testing activities, are usually performed at the end of each iteration. In essence, the system is implemented and deployed incrementally.

System testing may also be performed more frequently. A **build and smoke test** is a system test that is typically performed daily or several times per week. The system is completely compiled and linked (built), and a battery of tests is executed to see whether anything malfunctions in an obvious way ("smokes").

Build and smoke tests are valuable because they provide rapid feedback on significant integration problems. Any problem that occurs during a build and smoke test must result from software modified or added since the previous test. Daily testing ensures that errors are found quickly and that they can be easily tracked to their sources. Less frequent testing provides rapidly diminishing benefits because more software has changed and errors are more difficult to track to their sources.

A **performance test**, also called a **stress test**, determines whether a system or subsystem can meet such time-based performance criteria as response time or throughput. **Response time** requirements specify desired or maximum allowable time limits for software responses to queries and updates. **Throughput** requirements specify the desired or minimum number of queries and transactions that must be processed per minute or hour.

Performance tests are complex because they can involve multiple programs, subsystems, computer systems, and network infrastructure. They require a large suite of test data to simulate system operation under normal or maximum load. Diagnosing and correcting performance test failures are also complex. Bottlenecks and underperforming components must first be identified. Corrective actions may include any combination of the following:

- Application software tuning or reimplementation
- Hardware, system software, or network reconfiguration
- Upgrade or replacement of underperforming components

User Acceptance Testing

user acceptance test a system test performed to determine whether the system fulfills user requirements

A **user acceptance test** is a system test to determine whether the system fulfills user requirements. Acceptance testing may be performed near the end of the project or it may be broken down into a series of tests conducted at the end of each iteration. Acceptance testing is a very formal activity in most development projects. Details of acceptance tests are sometimes included in the request for proposal (RFP) and procurement contract when a new system is built by or purchased from an external party. Customer payments to the developers are often tied to passing specific usability tests.

Deployment Activities

Once a new system has been developed and tested, it must be placed into operation. Deployment activities (see **Figure 13-6**) involve many conflicting constraints, including cost, the need to maintain positive customer relations, the need to support employees, logistical complexity, and overall risk to the organization. User acceptance and other test types were described in the previous section. Multiple types of tests are often performed concurrently because later project iterations typically include implementation and deployment activities. The following sections provide additional details about deployment activities other than testing.

Converting and Initializing Data

An operational system requires a fully populated database to support ongoing processing. For example, online order-entry and management functions of the RMO CSMS rely on stored information about products, promotions, customers, and previous orders. Developers must ensure that such information is present in the database at the moment the subsystem becomes operational.

Data needed at system startup can be obtained from these sources:

■ Files or databases of a system being replaced
■ Manual records
■ Files or databases from other systems in the organization
■ User feedback during normal system operation

Reusing Existing Databases

Most new information systems replace or augment an existing manual or automated system. In the simplest form of data conversion, the old system's database is used directly by the new system with little or no change to the database structure. Reusing an existing database is fairly common because of the difficulty and

FIGURE **13-6** *SDLC deployment activities*

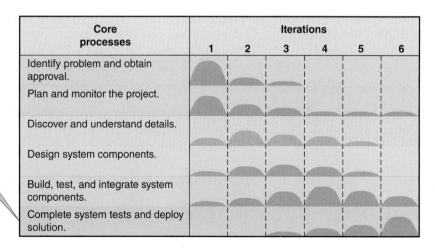

Deployment activities

Perform system and stress tests.
Perform user acceptance tests.
Convert existing data.
Build training materials and conduct training.
Configure and set up production environment.
Deploy the solution.

expense of creating new databases from scratch, especially when a single database often supports multiple information systems, as in today's enterprise resource planning (ERP) systems.

Although old databases are commonly reused in new or upgraded systems, some changes to database content are usually required. Typical changes include adding new tables, adding new attributes, and modifying existing tables or attributes. Modern database management systems (DBMSs) usually allow database administrators to modify the structure of a fully populated database. Such simple changes as adding new attributes or changing attribute types can be performed entirely by the DBMS.

Reloading Databases

More complex changes to database structure may require creating an entirely new database and copying and converting data from the old database to the new database. Whenever possible, utility programs supplied with the DBMS are used to copy and convert the data. In more complex conversions, implementation staff must develop programs to perform the conversion and transfer some or all of the data. The upper portion of **Figure 13-7** shows both approaches. In either case, the old database can be discarded once the conversion and transfer process is complete.

The middle of Figure 13-7 shows a more complex approach that uses an export utility, an import utility, and a temporary data store. This approach might be employed when the source and target databases employ different database technologies; no utility exists that can directly translate from one to the other, but a "neutral" format exists that can serve as a bridge.

Data from paper records can be entered by using the same programs being developed for the operational system. In that case, data-entry programs are

FIGURE **13-7** *Complex data-conversion example*

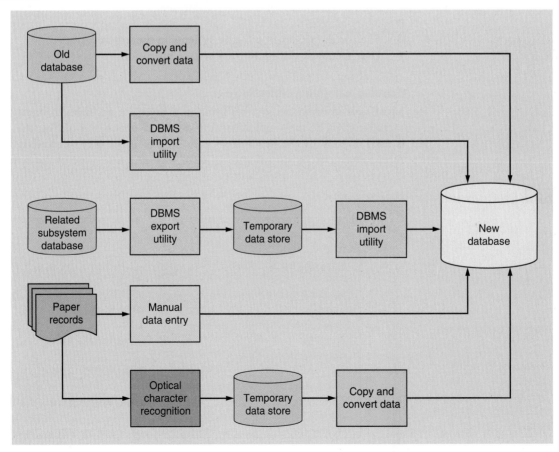

usually developed and tested as early as possible. Initial data entry can be structured as a user training exercise. For greater efficiency, data from paper records can also be scanned into an optical character recognition program and then entered into the database by using custom-developed conversion programs or a DBMS import utility.

In some cases, it may be possible to begin system operation with a partially or completely empty database. For example, a customer order-entry system need not have existing customer information loaded into the database. Customer information could be added the first time a customer places an order, based on a dialog between a telephone order-entry clerk and the customer. Adding data as they are encountered reduces the complexity of data conversion but at the expense of slower processing of initial transactions.

Training Users

Training two classes of users—end users and system operators—is an essential part of any system deployment project. End users are people who use the system from day to day to achieve the system's business purpose. System operators are people who perform administrative functions and routine maintenance to keep the system operating. **Figure 13-8** shows representative activities for each role. In smaller systems, a single person may fill both roles.

The nature of training varies with the target audience. Training for end users must emphasize hands-on use for specific business processes or functions, such as order entry, inventory control, or accounting. If the users aren't already familiar with those procedures, training must include them. Widely varying skill and experience levels call for at least some hands-on training, including practice exercises, questions and answers, and one-on-one tutorials. Self-paced training materials can fill some of this need, but complex systems also require some face-to-face training. If there is a large number of end users, group training sessions can be used, and a subset of well-qualified end users can be trained and then pass their knowledge on to other users.

System operator training can be much less formal when the operators aren't end users. Experienced computer operators and administrators can learn most or all they need to know by self-study. Thus, formal training sessions may not be required. Also, the relatively small number of system operators makes one-on-one training feasible, if it is necessary.

Determining the best time to begin formal training can be difficult. On one hand, users can be trained as parts of the system are developed and tested, which ensures that they hit the ground running. On the other hand, starting early can be frustrating to users and trainers because the system may not be stable or complete. End users can quickly become frustrated when using a buggy, crash-prone system with features and interfaces that are constantly changing.

In an ideal world, training doesn't begin until the interfaces are finalized and a test version has been installed and fully debugged. But the typical end-of-project

FIGURE **13-8**

Typical activities of end users and system operators

End-user activities	System operator activities
Creating records or transactions	Starting or stopping the system
Modifying database contents	Querying system status
Generating reports	Backing up data to archive
Querying database	Recovering data from archive
Importing or exporting data	Installing or upgrading software

crunch makes that approach a luxury that is often sacrificed. Instead, training materials are normally developed as soon as the interfaces are reasonably stable, and end-user training begins as soon as possible thereafter. It is much easier to provide training if system interfaces are completed in early project iterations.

Documentation and other training materials are usually developed before formal user training begins. Documentation can be loosely classified into two types:

- **System documentation**—descriptions of system requirements, architecture, and construction details
- **User documentation**—descriptions of how to interact with and use the system

Each documentation type is targeted to a different purpose and audience. The purpose of system documentation is to support development activities now and in the future. User documentation is created during implementation. The development team can't create user documentation earlier because many details of the user interface and system operation either haven't yet been determined or may change during development.

System Documentation

System documentation serves one primary purpose: providing information to developers and other technical personnel who will build, maintain, and upgrade the system. System documentation is generated throughout the SDLC by each core process and many development activities. System documentation developed during early project iterations guides activities in later iterations, and documentation developed throughout the SDLC guides future system maintenance and upgrades.

A system deployed for a customer is a collection of computing and network hardware, system software, and application software. Once the system has been developed, separate descriptions of it, such as written text and graphical models, are redundant with the system itself. In the early days of computing, there were few automated tools to support development of analysis and design models and even less support for automating the process of generating application software from those models. Developers in that era faced a recurring dilemma: how to minimize the duplicate effort of updating models and application software while ensuring that the system documentation was always "in sync" with the actual system. In the rush to complete and deploy systems and to maintain and upgrade them over time, system documentation updates were often neglected and documentation was frequently lost. As a result, systems were often scrapped "before their time" because it was cheaper to build a new system than to fix or upgrade a poorly documented existing system.

Modern application development tools and methods have largely solved the system documentation dilemma of earlier times. A modern integrated development environment provides automated tools to support all SDLC core processes. Requirements and design models, such as use case descriptions, class diagrams, and sequence diagrams, are developed by using the development tool and stored in a project library (see **Figure 13-9**). Changes to one model are automatically synchronized with related models. Application software is often generated in part or in its entirety directly from design models. When application software is altered at a later date, the development tools can "reverse engineer" appropriate changes to the models. Due to these capabilities, system documentation is always complete and in sync with the deployed system, thus simplifying future maintenance and upgrades.

User Documentation

User documentation provides ongoing support for end users of the system. It primarily describes routine operation of the system, including such functions as

system documentation descriptions of system requirements, architecture, and construction details, as used by maintenance personnel and future developers

user documentation descriptions of how to interact with and use the system, as used by end users and system operators

FIGURE **13-9** *System documentation stored within Microsoft Visual Studio*

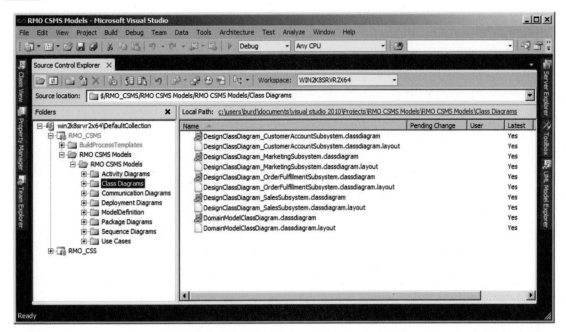

data entry, output generation, and periodic maintenance. Topics typically covered include:

- Software startup and shutdown
- Keystroke, mouse, or command sequences required to perform specific functions
- Program functions required to implement specific business procedures (e.g., the steps followed to enter a new customer order)
- Common errors and ways to correct them

For ease of use, user documentation typically includes a table of contents, a general description of the purpose and function of the program or system, a glossary, and an index.

User documentation for modern systems is almost always electronic and is usually an integral part of the application. Most modern operating systems provide standard facilities to support embedded documentation. **Figure 13-10** shows electronic user documentation of a typical Windows application. The table of contents can be displayed by clicking the book-shaped icon in the top toolbar, and the user can search for specific words or phrases by using the search tool. The center portion of the display shows individual pages of user documentation. The sample page includes embedded glossary definition hyperlinks (in green) and hyperlinks to other documentation pages (in blue).

Knowledge of how to use a system is as important an asset as the system itself. After initial training is completed, that practical knowledge is stored in the minds of end users. But experience such as that is difficult to maintain or effectively transfer to other users. Employee turnover, reassignment, and other factors make direct person-to-person transfer of operational knowledge difficult and uncertain. In contrast, written or electronic documentation is easier to access and far more permanent.

Developing good user documentation requires special skills and considerable time and resources. Writing clearly and concisely, developing effective presentation graphics, organizing information for easy learning and access, and communicating effectively with a nontechnical audience are skills for which there is high demand and limited supply. Development takes time, and high-quality results are achieved only with thorough review and testing. Unfortunately,

FIGURE **13-10** *Sample Windows Help and Support display*

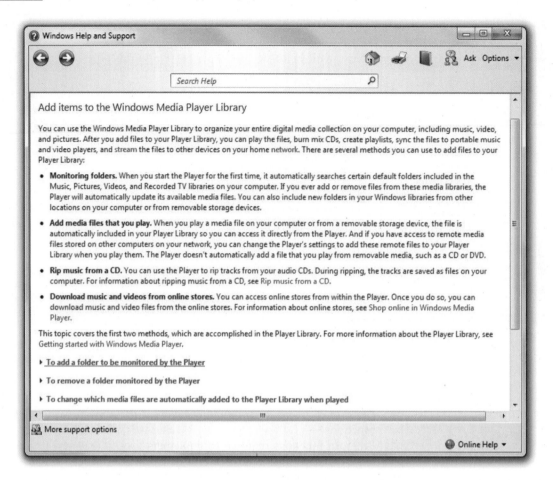

preparing user documentation is often left to technicians lacking in one or more necessary skills. Also, preparation time, review, and testing are often short-changed because of schedule overruns and the last-minute rush to tie up all the loose ends of implementation.

Configuring the Production Environment

Modern applications are built from software components based on interaction standards, such as Common Object Request Broker Architecture (CORBA), Simple Object Access Protocol (SOAP), and Java Platform Enterprise Edition (Java EE). Each standard defines specific ways in which components locate and communicate with one another. Each standard also defines a set of supporting system software to provide needed services, such as maintaining component directories, enforcing security requirements, and encoding and decoding messages across networks and other transport protocols. The exact system software, its hardware, and its configuration requirements vary substantially among the component interaction standards.

Figure 13-11 shows a typical support infrastructure for an application deployed using Microsoft .NET, a variant of SOAP. Application software components written in such programming languages as Visual Basic and C# are stored on one or more application servers. Other required services include a Web server for browser-based interfaces, a database server to manage the database, an Active Directory server to authenticate users and authorize access to information and software resources, a router and firewall, and a server to

FIGURE **13-11** *Infrastructure and clients of a typical .NET application*

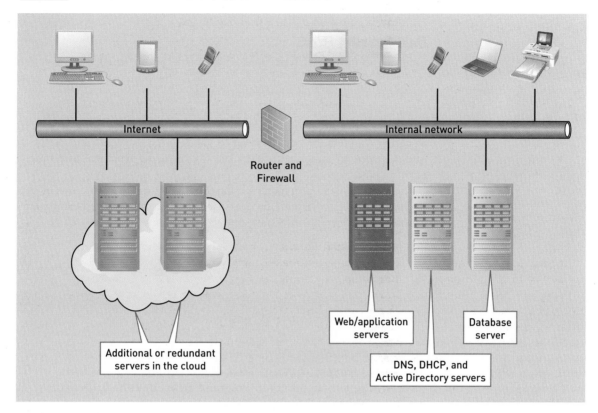

operate such low-level Internet services as domain naming (DNS) and Internet address allocation (DHCP).

Unless it already exists, all this hardware and system software infrastructure must be acquired, installed, and configured before application software can be installed and tested. In most cases, some or all of the infrastructure will already exist—to support existing information systems. In that case, developers work closely with personnel who administer the existing infrastructure to plan the support for the new system. In either case, this deployment activity typically starts early in the project so software components can be developed, tested, and deployed as they are developed in later project iterations.

Planning and Managing Implementation, Testing, and Deployment

The previous sections have discussed the implementation, testing, and deployment activities in isolation. In this section, we concentrate on issues that impact all those activities as well as other core processes, including project planning and monitoring, analysis, and design. In an iterative development project, activities from all core processes are integrated into each iteration and the system is analyzed, designed, implemented, and deployed incrementally. But how does the project manager decide which portions of the system will be worked on in early iterations and which in later iterations? And how does he or she manage the complexity of so many models, components, and tests?

Some of these issues were partly addressed in earlier chapters. But now that you understand implementation, testing, and deployment activities in depth, you can see that there are many interdependencies that must be accounted for. These interdependencies must be fully identified and considered when developing a workable iteration plan. Furthermore, automated tools must be utilized to manage each part of the development project and to ensure maximal coordination across iterations, core processes, and activities.

Development Order

One of the most basic decisions to be made about developing a system is the order in which software components will be built or acquired, tested, and deployed. Choosing which portions of the system to implement in which iterations is difficult, and developers must consider many factors, only some of which arise from the software requirements. Some of the other factors discussed in earlier chapters include the need to validate requirements and design decisions and the need to minimize project risk by resolving technical and other risks as early as possible.

A development order can be based directly on the structure of the system itself and its related issues, such as use cases, testing, and efficient use of development staff. Several orders are possible, including:

- Input, process, output
- Top-down
- Bottom-up

Each project must adapt one or a combination of these approaches to specific project requirements and constraints.

Input, Process, Output Development Order

input, process, output (IPO) development order a development order that implements input modules first, process modules next, and output modules last

The **input, process, output (IPO) development order** is based on data flow through a system or program. Programs or modules that obtain external input are developed first. Programs or modules that process the input (i.e., transform it into output) are developed next. Programs or modules that produce output are developed last. The key issue to analyze is dependency—that is, which classes and methods capture or generate data that are needed by other classes or methods? Dependency information is documented in package diagrams and may also be documented in a class diagram. Thus, either or both diagram types can guide implementation order decisions.

For example, the package diagram in **Figure 13-12** shows that the Customer account and Marketing subsystems don't depend on any of the other subsystems. The Sales subsystem depends on the Customer account and Marketing subsystems, and the Order fulfillment and Reporting subsystems depend on the Sales subsystem.

Data dependency among the packages (subsystems) implies data dependency among their embedded classes. Thus, the classes CustomerHandler, Customer, Address, Account, FamilyLink, Message, Suggestion, CustPartnerCredit, PromoPartner, Promotion, PromoOffering, ProductItem, and InventoryItem have no data dependency on the remaining RMO classes. Under the IPO development order, those three classes are implemented first.

The chief advantage of the IPO development order is that it simplifies testing. Because input programs and modules are developed first, they can be used to enter test data for process and output programs and modules. The IPO development order is also advantageous because important user interfaces (e.g., data-entry routines) are developed early. User interfaces are more likely to require change during development than during other portions of the system, so early development allows for early testing and user evaluation. If changes are needed, there is still plenty of time to make them. Early development of user interfaces also provides a head start for related activities, such as training users and writing documentation.

A disadvantage of the IPO development order is the late implementation of outputs. Output programs are useful for testing process-oriented modules and programs; analysts can find errors in processing by carefully examining printed

FIGURE **13-12** *Package diagram for the four RMO subsystems*

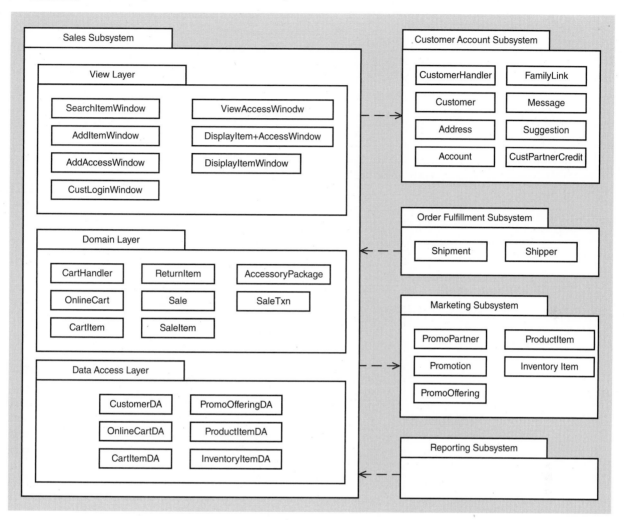

reports or displayed outputs. IPO development defers such testing until late in the development phase. However, analysts can usually generate alternate test outputs by using the query-processing or report-writing capabilities of a database management system (DBMS).

Top-Down and Bottom-Up Development Order

The terms *top-down* and *bottom-up* have their roots in traditional structured design and structured programming. A traditional structured design decomposes software into a series of modules or functions, which are hierarchically related to one another. As a visual analogy, consider a typical organization chart with the president or CEO at the top. In structured design, a single module (the president or CEO) controls the entire software program. Modules at the bottom perform low-level specialized tasks when directed to do so by a module at the next higher level. **Top-down development** begins with the CEO and works downward. **Bottom-up development** begins with the detailed modules at the lowest level and works upward to the CEO.

Top-down and bottom-up program development can also be applied to object-oriented designs and programs, although a visual analogy isn't obvious with object-oriented diagrams. The key issue is method dependency—that is, which methods call which other methods. Within an object-oriented subsystem or class, method dependency can be examined in terms of navigation visibility, as discussed in Chapters 10 and 11.

top-down development a development order that implements top-level modules first

bottom-up development a development order that implements low-level detailed modules first

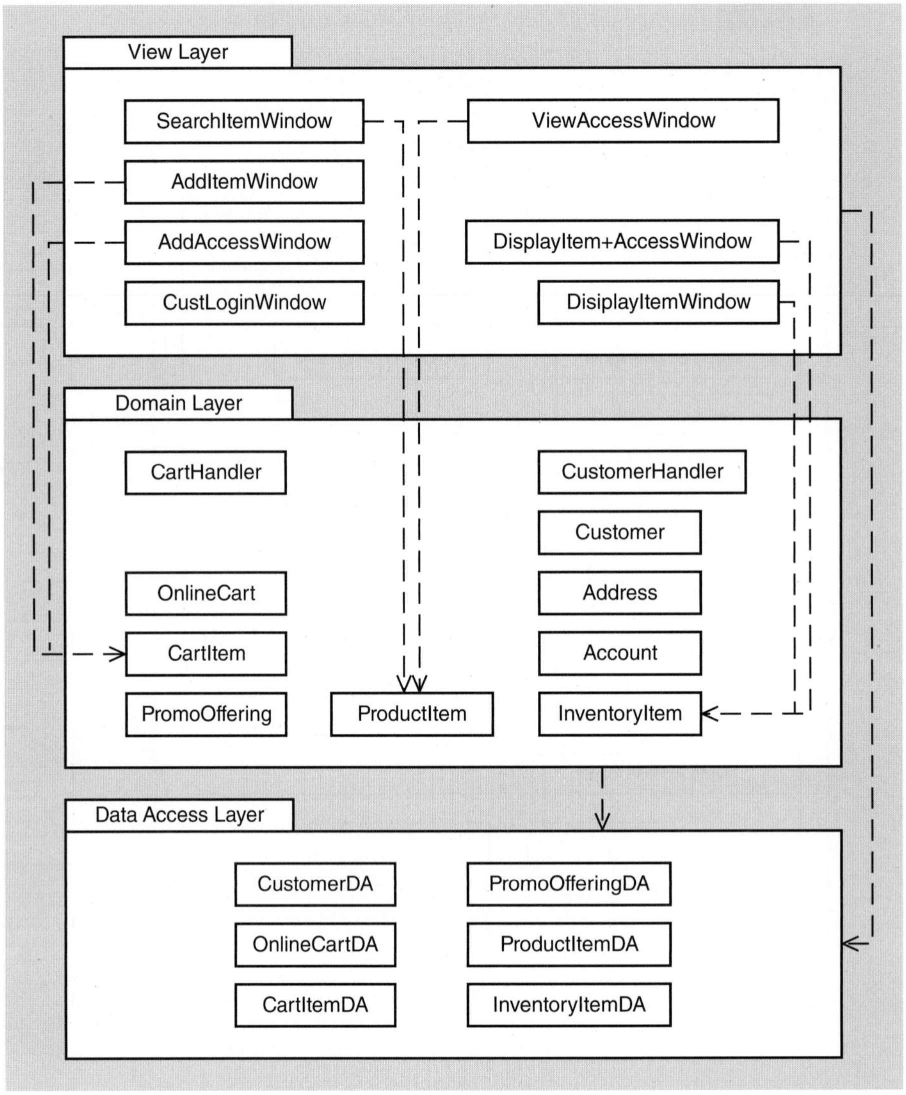

FIGURE **13-13**

Package diagram for a three-layer object-oriented design

For example, consider the three-layer design of part of the RMO Order-entry subsystem shown in **Figure 13-13**. The arrows connecting packages and classes show navigation visibility requirements. Methods in the view (user-interface) layer call methods in the domain layer, which in turn call methods in the data access layer. Top-down implementation would implement the view layer classes and methods first, the domain layer classes and methods next, and the data access layer classes and methods last. Bottom-up implementation would reverse the top-down implementation order.

Method dependency is also documented in a sequence diagram. For example, in Figure 13-3, method dependency is documented in the left-to-right flow of messages among objects. Rotating the figure 90 degrees clockwise creates a top-down and bottom-up visual analogy similar to an organizational chart. Top-down development would proceed through CartHandler, Customer, CustomerDA, OnlineCart, OnlineCartDA, CartItem, CartItemDA, and the set of classes: PromoOffering, PromoOfferingDA, ProductItem, ProductItemDA, InventoryItem, and InventoryItemDA. Bottom-up development would reverse this order, starting with InventoryItem and InventoryItemDA and ending with CartHandler.

The primary advantage of top-down development is that there is always a working version of a program. For example, top-down development in Figure 13-3 would begin with a partial or complete version of the CartHandler class and dummy (or stub) versions of the Customer and OnlineCart classes. This set of classes forms a complete program that can be built, deployed, and executed, although at this point, it wouldn't do very much when executed. As each method or class is implemented, stubs for the methods or classes on the next lower level are added. At every stage of development, the program can be executed and tested, and its behavior becomes more complex and realistic as development proceeds.

The primary disadvantage of top-down development order is that it doesn't use programming personnel very efficiently at the beginning of software development. Development has to proceed through two or three levels before a significant number of classes and methods can be developed simultaneously. In contrast, bottom-up development enables many programmers to be put to work immediately on classes that support a wide variety of use cases. Unfortunately, bottom-up development also requires writing a large number of driver methods to test bottom-level classes and methods, which adds additional complexity to the implementation and testing process. Also, the entire system isn't assembled until the topmost classes are written. Thus, testing for individual use cases and integration testing are delayed.

Other Development Order Considerations

IPO, top-down, and bottom-up development are only starting points for creating implementation and iteration plans. Other factors that must be considered include use case–driven development, user feedback, training, documentation, and testing. Use cases deserve special attention in determining development order because they are one of the primary bases for dividing a development project into iterations.

In most projects, developers choose a set of related use cases for a single iteration and complete analysis, design, implementation, and deployment activities. Developers choose which use cases to focus on first based on such factors as minimizing project risk, efficiently using nontechnical staff, or deploying some parts of the system earlier than others. For example, use cases with uncertain requirements or high technical risks are typically addressed in early iterations. Addressing uncertain requirements requires usability and other testing by nontechnical development staff, and those staff members may only be available at certain times in the project.

User feedback, training, and documentation all depend heavily on the user interfaces of the system. Early implementation of user interfaces enables user training and the development of user documentation to begin early in the development process. It also gathers early feedback on the quality and usability of the interface. Note the important role that this issue played in the opening case of this chapter.

Testing is also an important consideration when determining development order. As individual software components are constructed, they must be tested. Programmers must find and correct errors as soon as possible because they become much harder to find and more expensive to fix as components are integrated into larger units. It is important to identify portions of the software that are susceptible to errors and to identify portions of the software where errors can pose serious problems that affect the system as a whole. These portions of the software must be built and tested early, regardless of where they fit within the basic approaches of IPO, top-down, or bottom-up development.

Source Code Control

source code control system (SCCS) an automated tool for tracking source code files and controlling changes to those files

Development teams need tools to help coordinate their programming tasks. A **source code control system (SCCS)** is an automated tool for tracking source code files and controlling changes to those files. An SCCS stores project source code files in a repository, and it acts the way a librarian would—that is, implements check-in and checkout procedures, tracks which programmer has which files, and ensures that only authorized users have access to the repository.

FIGURE **13-14** *Project files managed by a source code control system*

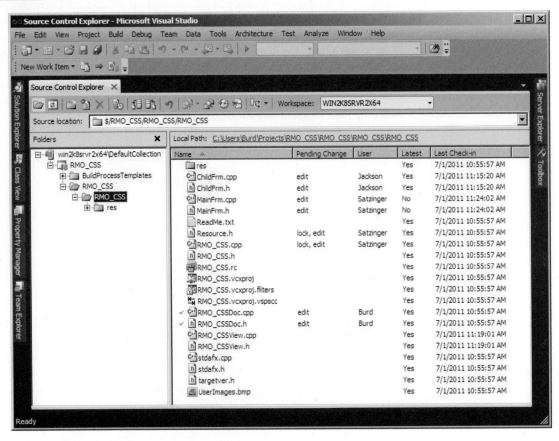

A programmer checks out a file in read-only mode when he or she wants to examine the code without making changes (e.g., to examine a module's interfaces to other modules). When a programmer needs to make changes to a file, he or she checks out the file in read/write mode. The SCCS allows only one programmer to check out a file in read/write mode. The file must be checked back in before another programmer can check it out in read/write mode.

Figure 13-14 shows the source code control display of Microsoft Visual Studio. Various source code files from the RMO CSS are shown in the display. Some files are currently checked out by programmers. For each file checked out in read/write mode, the program lists the programmer who checked it out, the date and time of checkout, and whether the copy currently stored in the central repository is the most current (latest) version.

An SCCS prevents multiple programmers from updating the same file at the same time, thus preventing inconsistent changes to the source code. Source code control is an absolute necessity when programs are developed by multiple programmers. It prevents inconsistent changes and automates coordination among programmers and teams. The repository also serves as a common facility for backup and recovery operations.

Packaging, Installing, and Deploying Components

As with the other disciplines discussed in this chapter, deployment activities are highly interdependent with activities of the other disciplines. In short, a system or subsystem can't be deployed until it has been implemented and tested. If a system or subsystem is large and complex, it is typically deployed in multiple stages or versions, thus necessitating some formal method of configuration and change management.

Important issues to consider when planning deployment include:

■ Incurring costs of operating both systems in parallel
■ Detecting and correcting errors in the new system
■ Potentially disrupting the company and its IS operations
■ Training personnel and familiarizing customers with new procedures

Different approaches to deployment represent different trade-offs among cost, complexity, and risk. The most commonly used deployment approaches are:

■ Direct deployment
■ Parallel deployment
■ Phased deployment

Each approach has different strengths and weaknesses, and no one approach is best for all systems. Each approach is discussed in detail here.

Direct Deployment

In a **direct deployment**, the new system is installed and quickly made operational, and any overlapping systems are then turned off. Direct deployment is also sometimes called **immediate cutover**. Both systems are concurrently operated for only a brief time (typically a few days or weeks) while the new system is being installed and tested. **Figure 13-15** shows a timeline for direct deployment.

The primary advantage of direct deployment is its simplicity. Because the old and new systems aren't operated in parallel, there are fewer logistical issues to manage and fewer resources required. The primary disadvantage of direct deployment is its risk. Because older systems aren't operated in parallel, there is no backup in the event that the new system fails. The magnitude of the risk depends on the nature of the system, the cost of workarounds in the event of a system failure, and the cost of system unavailability or less-than-optimal system function.

Parallel Deployment

In a **parallel deployment**, the old and new systems are operated for an extended period of time (typically weeks or months). **Figure 13-16** illustrates the timeline for parallel deployment. Ideally, the old system continues to operate until the new system has been thoroughly tested and determined to be error-free and ready to operate independently. As a practical matter, the time allocated for parallel operation is often determined in advance and limited to minimize the cost of dual operation.

The primary advantage of parallel deployment is relatively low operational risk. If both systems are operated completely (i.e., using all data and exercising all functions), the old system functions as a backup for the new system. Any failure in the new system can be mitigated by relying on the old system.

direct deployment or **immediate cutover**   a deployment method that installs a new system, quickly makes it operational, and immediately turns off any overlapping systems

parallel deployment   a deployment method that operates the old and the new systems for an extended time period

FIGURE

Direct deployment and cutover

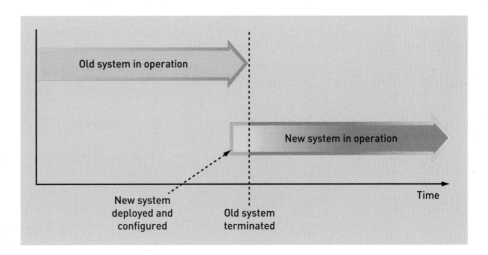

FIGURE **13-16**
Parallel deployment and operation

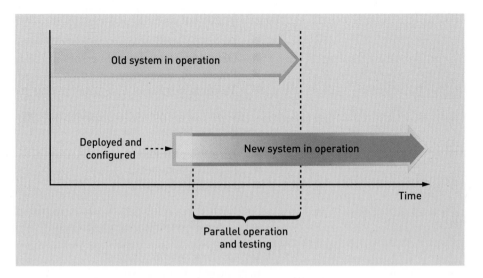

The primary disadvantage of parallel deployment is cost. During the period of parallel operation, the organization pays to operate both systems. Extra costs associated with operating two systems in parallel include:

■ Hiring temporary personnel or temporarily reassigning existing personnel
■ Acquiring additional computing and network capacity
■ Increasing managerial and logistical complexity

Parallel operation is generally best when the consequences of a system failure are severe. Parallel operation substantially reduces the risk of a system failure through redundant operation. The risk reduction is especially important for such "mission critical" applications as customer service, production control, basic accounting functions, and most forms of online transaction processing.

Full parallel operation may be impractical for any number of reasons, including:

■ Inputs to one system may be unusable by the other, and it may not be possible to use both types of inputs.
■ The new system may use the same equipment as the old system (e.g., computers, I/O devices, and networks), and capacity may be insufficient to operate both systems.
■ Staffing levels may be insufficient to operate or manage both systems at the same time.

When full parallel operation isn't possible or feasible, a partial parallel operation may be employed instead. Possible modes of partial parallel operation include:

■ Processing only a subset of input data in one of the two systems. The subset could be determined by transaction type, geography, or sampling (e.g., every 10th transaction).
■ Performing only a subset of processing functions (e.g., updating account history but not printing monthly bills)
■ Performing a combination of data and processing function subsets

Partial parallel operation always entails the risk that significant errors or problems will go undetected. For example, parallel operation with partial input increases the risk that errors associated with untested inputs won't be discovered.

Phased Deployment

phased deployment a deployment method that installs a new system and makes it operational in a series of steps or phases

In a **phased deployment**, the system is deployed in a series of steps or phases. Each phase adds components or functions to the operational system. During each phase, the system is tested to ensure that it is ready for the next phase. Phased deployment can be combined with parallel deployment, particularly when the new system will take over the operation of multiple existing systems.

FIGURE **13-17** *Phased deployment with direct cutover and parallel operation*

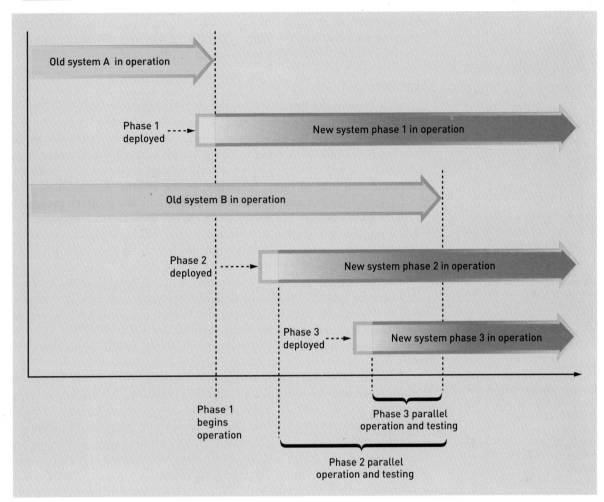

Figure 13-17 shows a phased deployment with direct and parallel deployment of individual phases. The new system replaces two existing systems. The deployment is divided into three phases. The first phase is a direct replacement of one of the existing systems. The second and third phases are different parts of a parallel deployment that replace the other existing system.

The primary advantage of phased deployment is reduced risk because failure of a single phase is less problematic than failure of an entire system. The primary disadvantage of phased deployment is increased complexity. Dividing the deployment into phases creates more activities and milestones, thus making the entire process more complex. However, each phase contains a smaller and more manageable set of activities. If the entire system is simply too big or complex to install at one time, the reduced risks of phased deployment outweigh the increased complexity inherent in managing and coordinating multiple phases.

Change and Version Control

Though not formal activities of the implementation or deployment core processes, change and version control are key parts of managing software development, testing, and deployment. Medium- and large-scale systems are complex and constantly changing. Changes occur rapidly during implementation and more slowly during deployment and after the system is in use. System complexity and rapid change create a host of management problems, particularly for testing and postdeployment support.

Change and version control tools and processes handle the complexity associated with testing and supporting a system through multiple versions. Tools and processes are typically incorporated into implementation activities from the beginning and continue throughout the life of a system. Most organizations use a common set of tools and procedures for all their systems.

Versioning

Complex systems are developed, installed, and maintained in a series of versions to simplify testing, deployment, and support. It isn't unusual to have multiple versions of a system deployed to end users and yet more versions in different stages of development. A system version created during development is called a *test version*. A test version contains a well-defined set of features and represents a concrete step toward final completion of the system. Test versions provide a static system snapshot and a checkpoint to evaluate the project's progress.

An **alpha version** is a test version that is incomplete but ready for some level of rigorous integration or usability testing. Multiple alpha versions may be built depending on the size and complexity of the system. The lifetime of an alpha version is typically short—days or weeks.

A **beta version** is a test version that is stable enough to be tested by end users over an extended period of time. A beta version is produced after one or more alpha versions have been tested and known problems have been corrected. End users test beta versions by using them to do real work. Thus, beta versions must be more complete and less prone to disastrous failures than alpha versions. Beta versions are typically tested over a period of weeks or months.

A system version created for long-term release to users is called a **production version, release version,** or **production release**. A production version is considered a final product, although software systems are rarely "finished" in the usual sense of that term. Minor production releases (sometimes called **maintenance releases**) provide bug fixes and small changes to existing features. Major production releases add significant new functionality and may be the result of rewriting an older release from the ground up.

Keeping track of versions is complex. Each version needs to be uniquely identified for developers, testers, and users. In applications designed to run under Windows, users typically view the version information by choosing the About item from the standard Help menu (see **Figure 13-18**). Users seeking support or reporting errors in a beta or production version use this feature to report the system version to testers or support personnel.

alpha version   a test version that is incomplete but ready for some level of rigorous integration or usability testing

beta version   a test version that is stable enough to be tested by end users over an extended period of time

production release, release version, or **production release**   a system version that is formally distributed to users or made operational for long-term use

maintenance release   a system update that provides bug fixes and small changes to existing features

FIGURE **13-18**

About box of a typical Windows application

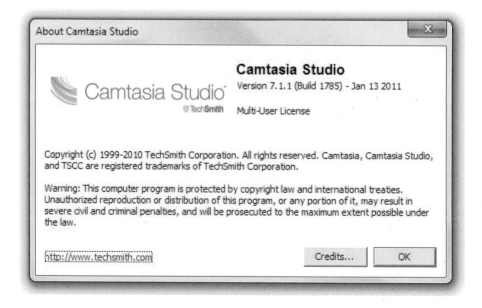

Controlling multiple versions of the same system requires sophisticated version control software, which is often built into development tools or can be obtained through a separate source code and version control system, as described later in this chapter. Programmers and support personnel can extract the current version or any previous version for execution, testing, or modification. Modifications are saved under a new version number to protect the accuracy of the historical snapshot.

Beta and production versions must be stored as long as they are installed on any servers or user machines. Stored versions are used to evaluate future bug reports. For example, when a user reports a bug in version 1.0, support personnel extract that release from the archive and attempt to replicate the user's error. Feedback provided to the user is specific to version 1.0, even if the most recent production release is a higher-numbered version.

Submitting Error Reports and Change Requests

To manage the risks associated with change, most organizations adopt formal control procedures for all systems under development and in operation. Formal controls are designed to ensure that potential changes are adequately described, considered, and planned before being implemented and deployed. Typical change control procedures include these:

- Standard reporting methods
- Review of requests by a project manager or change control committee
- For operational systems, extensive planning for design and implementation

Figure 13-19 shows a sample error (bug) report that has been completed by a tester or system developer. In this case, error reporting is integrated into the

FIGURE **13-19** *Sample error report in Microsoft Visual Studio*

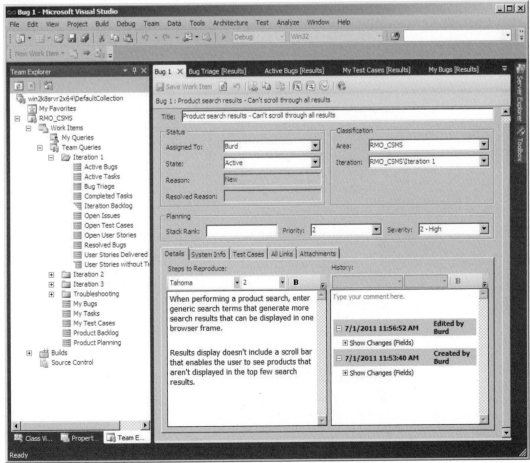

application development tool and source code control system, which enables the project manager to centrally manage all reports, assign reports to specific developers, and track each report to its resolution.

Similar tools can be used to report and manage errors and requests for new features in operational systems. In the case of new features, the request is usually submitted to a change control committee that reviews the change request to assess the impact on existing computer hardware and software, system performance and availability, security, and operating budget. Approved changes are added to the list of pending changes for budgeting, scheduling, planning, and implementation.

Implementing a Change

Change implementation follows a miniature version of the SDLC. Most of the SDLC activities are performed, although they may be reduced in scope or sometimes completely eliminated. In essence, a change for a maintenance release is an incremental development project in which the user and technical requirements are fully known in advance. Analysis activities are typically skimmed or skipped, design activities are substantially reduced in scope, and the entire project is typically completed in one or two short iterations.

Planning for a change includes these activities:

■ Identify what parts of the system must be changed.
■ Secure resources (such as personnel) to implement the change.
■ Schedule design and implementation activities.
■ Develop test criteria and a testing plan for the changed system.

production system the version of the system used from day to day

test system a copy of the production system that is modified to test changes

Whenever possible, changes are implemented and tested on a copy of the operational system. The **production system** is the version of the system used day to day. The **test system** is a copy of the production system that is modified to test changes. The test system may be developed and tested on separate hardware or on a redundant system. The test system becomes the operational system only after complete and successful testing.

Putting It All Together—RMO Revisited

In a medium-sized or large-scale development project, managers usually feel overwhelmed by the sheer number of activities to be performed, their interdependencies, and the risks involved. In this section, we give you a glimpse of the interplay among those issues by showing how Barbara Halifax's team developed an iteration plan for RMO's Customer Support System (CSS). But keep in mind that no single example can adequately prepare you to tackle iteration planning for a complex project. That is why iteration planning and other project planning tasks are typically performed by developers with years of experience.

Before reading the rest of this section, you may want to review earlier descriptions of the RMO case in Chapters 2, 3, 4, 6, and 9. Some basic parameters for the project are already described, including subsystem boundaries, project length, and number of iterations.

Chapter 9 describes Barbara's early planning decisions. In this section, she expands on those decisions, makes some changes to her earlier decisions, makes additional key decisions, and develops the revised iteration plan shown in **Figure 13-20**. The sections that follow describe key issues and decisions that underlie that plan.

Upgrade or Replace?

Upgrading the current CSS "in place" was ruled out early in project planning for these reasons:

■ The current infrastructure is near capacity.
■ RMO expects to save money by having an external vendor host the CSMS.

FIGURE **13-20** *Revised CSMS iteration plan*

Iteration	Description
1	Define business models and development/deployment environment. Define essential use cases and rough class diagram. Storyboard sales processing. Finalize deployment environment. Select and acquire network components, system software, hardware, and development tools. Create a CSS database copy with minimal data content as a starting point for CSMS database. Construct a simple prototype for adding a customer order (no database updates) and perform usability testing.
2	Define class, use case, sequence diagrams, and programs, concentrating on the key use cases (*Search for item, Fill shopping cart, Check out shopping cart, Look up customer,* and *Create customer account*). Deploy infrastructure components, including operating systems, Web/application servers, and DBMS by the middle of the iteration. Update database schema based on newly defined or revised classes and associations. Perform usability, unit, and integration testing to validate database design, customer/sales function set, and user interfaces.
3	Loop through iteration 2 use cases again and make all changes determined at the end of the previous iteration. Expand requirements and design to cover additional sales use cases and essential customer account and order-fulfillment use cases. Perform usability, unit, and integration testing.
4	Loop through iteration 3 use cases again and make all changes determined at the end of the previous iteration. Expand requirements and design to cover remaining Marketing subsystem use cases for products and promotions. Develop customer-oriented online help for all functions implemented in previous iterations. Prepare training materials and conduct training for phone and retail stores sales personnel. Finalize the new database and prepare it for data migration. Develop data migration (import) procedures. Test and refine data migration procedures by importing all data from the CSS database.
5	Loop through iteration 4 use cases again and make all changes determined at the end of the previous iteration. Continue training for phone and retail stores sales personnel. Conduct usability tests with a large number of actual or simulated customers. Make any needed changes to user interfaces, including online help. Conduct performance and stress testing and make any needed changes. Create a copy of the CSMS deployment environment at the Park City data center for use as a test system for version 2.0 development. Conduct use acceptance testing. Import all CSS database changes since the last import. Place version 1.0 into production.
6	Monitor system performance and user comments. Develop a change list and classify them as "ASAP" or "version 2.0." Implement ASAP changes. Expand requirements and design to cover essential use cases from the Reporting subsystem and those related to social networking. Migrate database updates from CSMS to CSS database twice per day. If no problems are encountered with CSMS, discontinue data migration and old system operation at the end of this iteration.
7	Loop through iteration 6 use cases again and make all changes determined at the end of the previous iteration. Expand requirements and design to cover all remaining use cases. Update database design as needed to support version 2.0 use cases. Program iteration 7 and use cases and conduct unit and integration testing.
8	Develop customer-oriented online help for all functions implemented in iterations 6 and 7. Prepare training materials and conduct training for sales, marketing, and management personnel. Conduct usability tests with a large number of actual or simulated customers. Make any needed changes to user interfaces, including online help. Update the production database with any structural changes in the test database.
9	Continue training for sales, marketing, and management personnel. Conduct performance and stress testing and make any needed changes. Conduct use acceptance testing. Place version 2.0 into production.

- Existing CSS programs and Web interfaces are a hodgepodge developed over 15 years.
- Current system software is several versions out of date.
- Infrastructure that supports the current CSS can be repurposed to expand SCM capacity.

In short, it would be too complex to upgrade the current CSS without disrupting operations, and the risks of upgrading old infrastructure and application software are simply too great. By building and deploying an entirely new system, RMO will make a clean break from the existing CSS and its supporting infrastructure. A new hosted infrastructure will be developed for the CSMS. After the first deployment phase, the existing CSS infrastructure will be updated to match the hosted environment and serve as a test environment for later development and deployment activities.

Phased Deployment to Minimize Risk

The schedule described in Chapter 9 didn't call for phased deployment, but neither did it directly consider such deployment issues as database development, data migration, and training. To minimize deployment risks, the CSMS will be deployed in two versions. Version 1.0 will reimplement most of the existing CSS use cases with minimal changes. Version 2.0 will incorporate bug fixes and incremental improvements to version 1.0 and will add additional functionality not present in the CSS, including social networking, feedback/recommendations, business partners, and Mountain Bucks.

The two-phase deployment minimizes project risk by dividing a single large deployment into two smaller deployments. Another key risk mitigation feature is maintaining the current CSS and its database as a backup for at least one iteration after version 1.0 deployment. If a serious problem arises with version 1.0, RMO can revert to the current CSS simply by redirecting Web site accesses back to its internal servers.

Database Development and Data Conversion

Many of the classes in the CSMS class diagram are already represented in the existing CSS database. However, there are some new classes and associations and some changes to existing classes. Thus, there is some degree of compatibility between the old and new databases but not enough to enable an upgraded version of the current database to directly interface with both systems. Thus, a new CSMS database will need to be built, and data will need to be migrated from the CSS database prior to deploying version 1.0.

Database development and migration prior to version 1.0 deployment will occur over multiple iterations. The iteration plan calls for creating a copy of the CSS database early in the project and making incremental changes to it. All data in the production CSS database will be migrated to the CSMS database near the end of the fourth iteration. If problems are encountered, they will be resolved and the migration will be repeated as early as possible during the fifth iteration. Migrating much of the data during the fourth iteration will enable fifth-iteration testing of user interfaces with real data from real customers and products and system and stress testing with a "production sized" database.

At the end of the fifth iteration, all CSS database changes since the last full migration will be copied to the CSMS database. Copying only the changes will enable migration within a matter of hours. The CSS system will be offline during the migration. Cutover to the CSMS will occur as soon as the migration is completed. To minimize risk, additional data conversion routines will copy new data from the CSMS database back to the CSS database twice per day during the fifth iteration. If disaster strikes, the CSS can be restarted with a current and complete database. If CSMS version 1.0 passes all user acceptance tests during the fifth iteration, the CSS will be turned off and data migration will cease.

Development Order

The IPO development order is the primary basis for the development plan. By starting with a copy of the CSS database, a set of test data will exist from the first iteration, thus enabling the highest risk use cases to be tackled first. These involve the entire Sales subsystem and customer-facing portions of the Order-fulfillment subsystem. The risks arise from new technology, uncertainty about requirements, and the operational importance of sales and order fulfillment to RMO. By tackling those use cases first, Barbara allowed her development staff plenty of time to resolve uncertainties and test related software. Note that significant testing of these functions began in iteration 2 and continued through most of the project.

Documentation and Training

Training activities were spread throughout later project iterations for both production versions. Initial training exercises covered the highest-risk portion of the system prior to deployment. They also enabled developers to do integration and performance testing on the sales-related use cases long before deployment. Additional training continued as new functions were added to the system, providing a gradual ramping up of user skills and developer workload.

Chapter Summary

Implementation and deployment are complex processes because they consist of so many interdependent activities. Testing is a key activity of implementation and deployment. Software components must be constructed in an order that minimizes the use of development resources and maximizes the ability to test the system and correct errors. Unfortunately, those two goals often conflict. Thus, a program development plan is a trade-off among available resources, available time, and the desire to detect and correct errors prior to system deployment.

Configuration and change management activities track changes to models and software through multiple system versions, which enables developers to test and deploy a system in stages. Versioning also improves postdeployment support by enabling developers to track problem support to specific system versions. Source code control systems enable development teams to coordinate their work.

Key Terms

alpha version   432

beta version   432

bottom-up development   425

build and smoke test   416

direct deployment   429

driver   413

immediate cutover   429

input, process, output (IPO)
    development order   424

integration test   414

maintenance release   432

parallel deployment   429

performance test   416

phased deployment   430

production system   434

production version, release version, or
    production release   432

response time   416

source code control system
    (SCCS)   427

stub   414

system documentation   420

system test   416

test case   412

test data   412

test system   434

throughput   416

top-down development   425

unit testing   412

usability test   416

user acceptance test   417

user documentation   420

Review Questions

1. List and briefly describe each activity of the SDLC core processes *Build, test, and integrate system components* and *Complete system tests and deploy solution*.

2. Define the terms *unit test, integration test, system test*, and *user acceptance test*. During which SDLC activity (or activities) is each test type performed?

3. What is a test case? What are the characteristics of a good test case?

4. What is a driver? What is a stub? With what type of test is each most closely associated?

5. List possible sources of data used to initialize a new system database. Briefly describe the tools and methods used to load initial data into the database.

6. How do user documentation and training activities differ between end users and system operators?

7. List and briefly describe the three basic approaches to program development order. What are the advantages and disadvantages of each?

8. How can the concepts of top-down and bottom-up development order be applied to object-oriented software?

9. What is a source code control system? Why is such a system necessary when multiple programmers build a program or system?

10. Briefly describe direct, parallel, and phased deployments. What are the advantages and disadvantages of each deployment approach?

11. Define the terms *alpha version*, *beta version*, and *production version*. Are there well-defined criteria for deciding when an alpha version becomes a beta version or a beta version becomes a production version?

Problems and Exercises

1. Describe the process of testing software developed with the IPO (input, process, output), top-down, and bottom-up development orders. Which development order results in the fewest resources required for testing? What types of errors are likely to be discovered earliest under each development order? Which development order is best, as measured by the combination of required testing resources and ability to capture important errors early in the testing process?

2. Assume that you and three of your classmates are charged with developing the first prototype to implement the RMO use case *Create/update customer account*. Create a development and testing plan to write and test the classes and methods. Assume that you have two weeks to complete all tasks.

3. Talk with a computer center or IS manager about the testing process used with a recently deployed system or subsystem. What types of tests were performed? How were test cases and test data generated? What types of teams developed and implemented the tests?

4. Consider the issue of documenting a system by using only electronic models developed with an integrated development tool, such as Microsoft Visual Studio or Oracle JDeveloper. The advantages are obvious (e.g., the analyst modifies the models to reflect new requirements and automatically generates an updated system). Are there any

disadvantages? (Hint: The system might be maintained for a decade or more.)

5. Talk with an end user at your school or work about the documentation and training provided with a recently installed or distributed business application. What types of training and documentation were provided? Did the user consider the training to be sufficient? Does the user consider the documentation to be useful and complete?

6. Assume you are in charge of implementation and deployment of a new system that is replacing a critical existing system that is used 24 hours a day. To minimize risk, you plan to phase in deployment of new subsystems over a period of six weeks and operate both systems in parallel for at least three weeks beyond the last new subsystem deployment. Because there aren't enough personnel to operate both systems, you plan to hire up to 30 temporary workers during the parallel operation period. How should you use the temporary workers? In answering that question be sure to consider these issues:

   i. Some current personnel will be trained before subsystem deployments, and those employees will train other employees.

   ii. Employees newly trained on the system will probably not reach their former levels of efficiency for many weeks.

Case Study

HudsonBanc Billing System Upgrade

Two regional banks with similar geographic territories merged to form HudsonBanc. Both banks had credit card operations and operated billing systems that had been internally developed and upgraded over three decades. The systems performed similar functions, and both operated primarily in batch mode on mainframe computers. Merging the two billing systems was identified as a high-priority cost-saving measure.

HudsonBanc initiated a project to investigate how to merge the two billing systems. Upgrading either system was quickly ruled out because the existing technology was considered old and the costs of upgrading the system were estimated to be too high. HudsonBanc decided that a new component-based, Web-oriented system should be built or purchased. Management preferred the purchase option because it was assumed that a purchased system could be brought online more quickly and cheaply. An RFP (request for proposal) was prepared, many responses were received, and after months of business modeling and requirements activities, a vendor was chosen.

Hardware for the new system was installed in early January. Software was installed the following week, and a random sample of 10 percent of the customer accounts was copied to the new system. The new system was operated in parallel with the old systems for two months. To save costs involved with complete duplication, the new system computed but didn't actually print billing statements. Payments were entered into both systems and used to update parallel customer account databases. Duplicate account records were checked manually to ensure that they were the same.

After the second test billing cycle, the new system was declared ready for operation. All customer accounts were migrated to the new system in mid-April. The old systems were turned off on May 1, and the new system took over operation. Problems occurred almost immediately.

The system was unable to handle the greatly increased volume of transactions. Data entry and customer Web access slowed to a crawl, and payments were soon backed up by several weeks. The system wasn't handling certain types of transactions correctly (e.g., charge corrections and credits for overpayment). Manual inspection of the recently migrated account records showed errors in approximately 50,000 accounts.

It took almost six weeks to adjust the incorrect accounts and update functions to handle all transaction types correctly. On June 20, the company attempted to print billing statements for the 50,000 corrected customer accounts. The system refused to print any information for transactions more than 30 days old. A panicked consultation with the vendor concluded that fixing the 30-day restriction would require more than a month of work and testing. It was also concluded that manual entry of account adjustments followed by billing within 30 days was the fastest and least risky way to solve the immediate problem.

Clearing the backlog took two months. During that time, many incorrect bills were mailed. Customer support telephone lines were continually overloaded. Twenty-five people were reassigned from other operational areas, and additional phone lines were added to provide sufficient customer support capacity. System development personnel were reassigned to IS operations for up to three months to assist in clearing the billing backlog. Federal and state regulatory authorities stepped in to investigate the problems. HudsonBanc agreed to allow customers to spread payments for late bills over three months without interest charges. Setting up the payment arrangements further aggravated the backlog and staffing problems.

1. What type of installation did HudsonBanc use for its new system? Was it an appropriate choice?
2. How could the operational problems have been avoided?

RUNNING CASE STUDIES

Community Board of Realtors

Assume that the Multiple Listing Service that is under development will replace an existing system developed many years ago. The database requirements and design for the old and new systems are very similar. Unfortunately, the existing system stores its data in a Microsoft Access database, which provides little support for simultaneous access and updates by multiple users. An important reason for replacing the current system is to upgrade to a DBMS that can easily support many simultaneous accesses.

(continued on page 440)

(continued from page 439)

The current plan is to use Microsoft SQL Server as the new DBMS and to migrate all data from the existing Microsoft Access database immediately prior to full deployment. Perform these tasks to prepare for this migration:

1. Investigate data migration from Microsoft Access to SQL Server. What tools are available to assist in or perform the migration? If there are multiple possible tools, which should you use and why?

2. Develop plans to test the migration tools/strategy in advance of full deployment. When should the test be performed, and how will you determine whether the test has been "passed"?

The Spring Breaks 'R' Us Travel Service

Review the case-related questions and tasks as well as your responses from Chapters 8 and 9. As described in previous chapters, assume the new system will upgrade an existing system and add new social networking functions to it. Specifically, review your answer to question 2 in Chapter 9 in light of the more detailed understanding of the risks, costs, and benefits of various implementation orders and deployment approaches that you gained by reading this chapter.

1. For each subsystem—Resort relations, Student booking, Accounting and finance, and Social networking—specify which other subsystem(s) it depends on for input data?

2. Can the four subsystems be developed and deployed independently? If so, in which order should they be developed and deployed? If not, explain why not and describe how you would develop and deploy the system.

On the Spot Courier Services

In Chapter 8, we identified these four subsystems:

- Customer account subsystem (such as customer account)
- Pickup request subsystem (such as sales)
- Package delivery subsystem (such as order fulfillment)
- Routing and scheduling subsystem

In Chapter 8, you also decided on a development order for these four subsystems, assuming a single two-person team. In Chapter 9, you created individual subsystem iteration schedules and a combined project schedule. In Chapter 6, you identified equipment that would be needed for the system.

Your assignment for this chapter is to develop a test plan for each subsystem and for the overall project as well as to develop a conversion/deployment schedule.

1. For your test plan, do the following:

   a. Develop an iteration test plan (i.e., one that applies to and can be used within a subsystem iteration mini-project). Discuss which types of testing (as identified in this chapter) you would include and why. Estimate how much time will be needed for each type of test. Discuss what types of testing might be combined or scheduled with an overlap.

   b. Develop a total project test plan to integrate all the subsystems. Discuss which types of testing you would include and why. (Don't put them on a schedule yet.)

2. Develop a conversion/deployment plan. Discuss these:

   a. Data conversion: Which parts of the data must be saved from the old spreadsheet/manual system? Which parts of the data can just be discarded (i.e., not moved to the new system)? Discuss specific tables that you identified in Chapter 12.

   b. Deployment: Based on your decisions about which subsystems should be deployed first (Chapter 8), your overall testing plan, and your data conversion decisions, develop an overall schedule for testing and deployment of the new system. How would you characterize your solution: direct, parallel, or phased conversion? Support your answer by discussing the logic behind your decisions.

3. Revisit your solution in Chapter 6 regarding the types of equipment that will be needed. Include in your discussion your current recommendation for hosting the system. Add to your deployment schedule the activities to purchase equipment and set up the hosting environment.

(continued on page 441)

(continued from page 440)

Sandia Medical Devices

Refer to the case information provided at the end of Chapters 8 and 9 and the domain class diagram at the end of Chapter 11. Review and update your results from performing the tasks at the end of Chapter 9 based on the information provided in this chapter. Then, answer these questions:

1. What integration and system tests are required, and when should they be incorporated into the iteration schedule?

2. What are the documentation and user training requirements for the system, and when should they be incorporated into the iteration schedule?

3. Assume that after deployment and a three-month testing and evaluation period, updates to the first Android-based system (client and server) will be implemented and another client-side version will be implemented for the iPhone. Develop an iteration plan for implementing and deploying the second version of the system.

Further Resources

Robert V. Binder, *Testing Object-Oriented Systems: Models, Patterns, and Tools.* Addison-Wesley, 2000.

Mark Fewster and Dorothy Graham, *Software Test Automation.* Addison-Wesley, 1999.

Jerry Gao, H.-S. Jacob Tsao, and Ye Wu, *Testing and Quality Assurance for Component-Based Software*, Artech House Publishers, 2003.

William Horton, *Designing and Writing Online Documentation: Hypermedia for Self-Supporting Products* (2nd edition). John Wiley & Sons, 1994.

William Horton, *Designing Web-Based Training: How to Teach Anyone Anything Anywhere Anytime.* John Wiley & Sons, 2000.

William Horton, *e-Learning by Design.* Pfeiffer, 2011.

International Association of Information Technology Trainers (ITrain) Web site, http://itrain.org.

David Yardley, *Successful IT Project Delivery.* Addison-Wesley, 2002.

14

Current Trends in System Development

Chapter Outline

- Trends in System Development Methodologies
- Trends in Technology Infrastructure
- Trends in Application Software Availability
- The Web as an Application Platform

Learning Objectives

After reading this chapter, you should be able to:

- Describe the elements of the Unified Process (UP)
- Compare and contrast the features of Extreme Programming and Scrum development
- Describe the major trends in connectivity, Internet, and telephone technologies
- List and describe the various methods of deploying application software
- List and describe the various elements that enhance Web applications
- Describe the various approaches to developing Rich Internet Applications

OPENING CASE

Valley Regional Hospital: Measuring a Project's Progress

Claire Haskell, the vice president of technology at Valley Regional Hospital (VRH), listened quietly to Henry Williams's progress report on the new patient records system. Henry was the project leader for the team that was developing the system. Also in the meeting were the project's sponsor, Charlie Montgomery, who was the director of patient information and records, and Jason Smith, the director of software development. Months before, Jason and Henry had asked Claire to try a new development approach called Extreme Programming (XP) for this recently approved project. They had already spoken with Charlie, and he had agreed to try the XP development method. Claire approved the project and their request to try the new approach even though she knew very little about it.

During his presentation, Henry kept talking about how wonderfully the team was working together and how much fun they were having. Although she was glad that the team was functioning well, Claire wanted more specifics. She wanted to know whether the new system was on schedule and within budget. After about 20 minutes of listening patiently, she couldn't wait any longer.

"I need to see a schedule," she told Henry, "and I need a report on the team's progress."

Henry projected a schedule on the screen, but that did little to help; it had no familiar milestones, such as analysis, design, and programming. Instead, she saw other terms: iteration, user stories, and refactoring.

At this point, Claire became worried. She turned to Charlie and said pointedly: "Exactly how is the project progressing from your viewpoint?"

His answer surprised her.

"The records administrators and I are extremely pleased with the demos we are seeing," he said. "We are also satisfied with the quality of the system we saw during our acceptance testing. From what we have seen so far, the system seems to be exactly what we need. But as far as the schedule is concerned, I'm not certain whether the entire system will be delivered on time. I think it will, but I'm not involved in the day-to-day development."

Claire felt a little better. At least the system was doing what it needed to do. But she still wanted reassurance from the project leader. "Henry, are we going to hit the completion date?" she asked. "The system needs to be ready on time."

"We are progressing on schedule so far and everything looks fine," Henry responded. "No, I can't show you a traditional schedule—one with major milestones. But here is a short-term schedule for the next two months of work."

Claire wasn't satisfied. She asked Henry to stay and talk with her privately after the meeting ended, at which point she became agitated.

"Henry, we need more accountability for this project," she said. "The only solution I see is to meet with you frequently to monitor its progress. I want a rough schedule for the rest of the project on my desk on Monday morning. That gives you three days to develop one. Then, I want you to meet with me every Monday from here on out so we can be sure we are on track and hit the delivery date."

Although he wasn't pleased with Claire's suggestion, Henry reluctantly agreed.

Overview

So far, this book has focused on teaching you the processes and skills associated with a system development project. You have learned the "soft" skills associated with managing projects, interacting in teams, gathering information, and making presentations. You have also learned the "hard" skills—those associated with problem solving, building requirements models, and designing new systems. You have learned many important concepts about projects, iterative development, and the SDLC. In short, you have developed a solid working knowledge of system development and obtained a bag of tools to get you started developing information systems for businesses and other organizations.

The approach presented in this textbook isn't the only method for developing systems. As you move forward with your career in information systems, you will find that the industry is wide and varied. There are companies using methodologies and techniques that have been around for 30 years or longer. There are other groups that are trying various approaches in an attempt to improve the speed and efficiency of the development process as well as the quality of the end result. Many organizations have a mix of older, well-established methods for some projects and newer techniques and methods for other projects. No matter which type of organization you work for, the methods and techniques you have learned from this textbook are fundamental and will serve you well.

We begin this chapter with a review of three of the more current development methodologies, along with their associated practices and techniques. These are the Unified Process (UP), Scrum, and Extreme Programming (XP). As indicated earlier in this text, most new approaches, including these three, are based on the Agile philosophy and an iterative life cycle. In this chapter, you will learn about the details of each of these three approaches.

Following the discussion of these three methodologies, we will look at some of the new technology trends in the technology and software application industry. These trends are a major driving force in requiring a more agile and iterative approach to application software development. Software development is very different today than it was even a few years ago, and there are two major reasons for this. One is our ability to connect and network with other like-minded developers throughout the world thanks to the Internet. The other is the proliferation of so many types of devices that support computer applications.

We will look at three trends that are currently impacting the world of software development.

The first trend is the extremely wide variety of consumer devices that have computing capabilities: cameras, ebook readers, smartphones, automobile GPS devices, gaming equipment, tablet and laptop computers, and even household appliances. Although the operating system software and communication software for these devices isn't a focus of information system development, many of these devices support custom applications and browser applications, and those are within the realm of information systems. We will briefly look at the proliferation of computing devices.

A second trend is the approach to the distribution of application software, especially consumer applications. For example, the open-source movement has provided a plethora of components, tools, and applications that can be used by developers and end users. We will look at two major movements in this arena: software as a service (SAAS) and open source.

The third trend is the movement toward Web-based applications. With the widespread availability of Internet access and the incredible amount of information and services provided through the Web, browser-based applications have become the largest source of new application development in the world. Although this is true for business applications and consumer-oriented applications, the most dramatic growth by far is in the development of new services for the consumer. We will review some of the more important aspects of application software development for the Web.

Trends in System Development Methodologies

In Chapter 8, you were introduced to Agile development. The Agile philosophy has proven to be an effective way to approach software development in today's fast-paced, continually changing landscape of computer applications. However, the Agile philosophy only proposes principles; it isn't meant to be a complete methodology, with practices and action steps. In this section, we present three methodologies that incorporate Agile principles but are also complete methodologies, with specific techniques and practices.

These three methodologies—UP, XP, and Scrum—are among the most popular approaches to application software development, but they aren't always found in their purest forms. Frequently, organizations either mix and match techniques from the three or only adopt a specific set of practices. However, adoption of these methodologies continues to expand throughout all types of organizations that develop software applications.

The Unified Process

The Unified Process (UP) is an object-oriented system development methodology originally offered by Rational Software, which is now part of IBM. Developed by Grady Booch, James Rumbaugh, and Ivar Jacobson—the three pioneers behind the success of the Unified Modeling Language (UML)—the UP defines a complete methodology that uses UML for system models and describes a new, adaptive system development life cycle. In the UP, the term *development process* is synonymous with development methodology.

The UP is now widely recognized as a highly influential innovation in software development methodologies for object-oriented development using an adaptive approach. The original version of UP defined an elaborate set of activities and deliverables for every step of the development process. More recent versions are streamlined, with fewer activities and deliverables, simplifying the methodology. The methodology used in this textbook is an adaptation of UP principles.

As discussed previously, adaptive methodologies—including the UP—are all based on an iterative approach to development. You learned in Chapter 1 that each iteration is like a mini-project, in which requirements are defined based on analysis tasks, system components are designed, and those components are then implemented—at least partially—through programming and testing. However, one of the big questions in adaptive development is what the focus of each iteration should be. In other words, do iterations early in the project have the same objectives and focus as those done later? The UP answers this question by dividing a project into four major phases.

UP Phases

A phase in the UP can be thought of as a goal or major emphasis for a particular portion of the project. The four phases of the UP life cycle are: inception, elaboration, construction, and transition, as shown in **Figure 14-1**.

Each phase of the UP life cycle describes the emphasis or objectives of the project team members and their activities at that point in time. Thus, the four phases provide a general framework for planning and tracking the project over time. Within each phase, several iterations are planned to give the team enough flexibility to adjust to problems or changing conditions. The emphases or objectives of the project team in each of the four phases are described briefly in **Figure 14-2**.

FIGURE **14-1**

The Unified Process system development life cycle

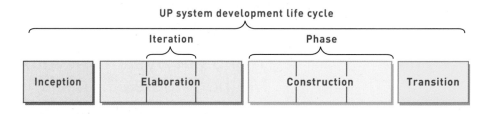

UP system development life cycle

Iteration   Phase

| Inception | Elaboration | Construction | Transition |

FIGURE **14-2**

UP phases and objectives

UP phase	Objective
Inception	Develop an approximate vision of the system, make the business case, define the scope, and produce rough estimates for cost and schedule.
Elaboration	Define the vision, identify and describe all requirements, finalize the scope, design and implement the core architecture and functions, resolve high risks, and produce realistic estimates for cost and schedule.
Construction	Iteratively implement the remaining lower-risk, predictable, and easier elements and prepare for deployment.
Transition	Complete the beta test and deployment so users have a working system and are ready to benefit as expected.

Inception Phase As in any project-planning phase, the inception phase consists of the project manager developing and refining a vision for the new system in order to show how it will improve operations and solve existing problems. Essentially, the project manager makes the business case for the new system, proving that the new system's benefits will outweigh the cost of development. The scope of the system must also be defined so it is clear what the project will accomplish. Defining the scope includes identifying many of the key requirements for the system.

The inception phase is usually completed in one iteration, and as with any iteration, parts of the actual system may be designed, implemented, and tested. As software is developed, team members must confirm that the system vision still matches user expectations or that the technology will work as planned. Sometimes, prototypes are discarded after proving that point.

Elaboration Phase The elaboration phase usually involves several iterations, and early iterations typically complete the identification and definition of all the system requirements. Because the UP is an adaptive approach to development, the requirements are expected to evolve and change after work starts on the project.

The elaboration phase's iterations also complete the analysis, design, and implementation of the system's core architecture. Usually, the aspects of the system that pose the greatest risk are identified and implemented first. Until developers know exactly how the highest-risk aspects of the project will work out, they can't determine the amount of effort required to complete the project. By the end of the elaboration phase, the project manager should have more realistic estimates for the project's cost and schedule, and the business case for the project can be confirmed. Remember that the design, implementation, and testing of key parts of the system are completed during the elaboration phase. One other major objective of the elaboration phase is to do the necessary research and fact-finding so all the user requirements are identified. During the elaboration phase, a high percentage of time is spent on understanding and analysis.

Construction Phase The construction phase involves several iterations that continue the design and implementation of the system. The core architecture and highest-risk aspects of the system are already complete. Now the focus of the work turns to the routine and predictable parts of the system—for example, detailing the system controls, such as data validation, fine-tuning the user interface design, finishing routine data maintenance functions, and completing the help and user preference functions. The team also begins to plan for deployment of the system.

Transition Phase During the transition phase, one or more final iterations involve the final user acceptance and beta tests, and the system is made ready for operation. After the system is in operation, it will need to be supported and maintained.

UP Disciplines

As we mentioned earlier, the four UP phases define the project sequentially by indicating the emphasis of the project team at any point in time. To make iterative development manageable, the UP defines disciplines to use within each iteration. A **UP discipline** is a set of functionally related activities that contributes to one aspect of the development project. UP disciplines include business modeling, requirements, design, implementation, testing, deployment, configuration and change management, project management, and environment. Each iteration usually involves activities from all disciplines.

UP discipline a set of functionally related activities that combine to enable the development process in a UP project

FIGURE **14-3**

UP disciplines used in varying amounts in each iteration

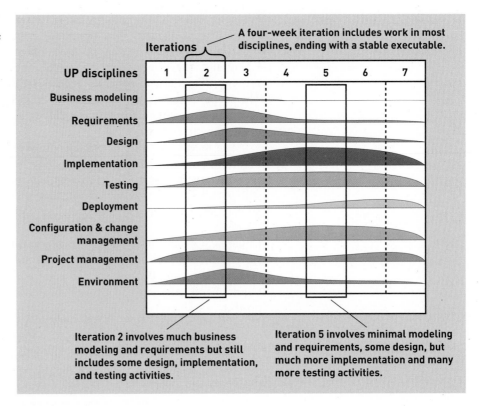

Figure 14-3 shows how the UP disciplines are involved in each iteration, which is typically planned to last four weeks. The size of the shaded area under the curve for each discipline indicates the relative amount of work included from each discipline during the iteration. The amount and nature of the work differs from iteration to iteration. For example, in iteration 2, much of the effort focuses on business modeling and requirements definition, with much less effort focused on implementation and deployment. In iteration 5, very little effort is focused on modeling and requirements and much more effort focused on implementation, testing, and deployment. But most iterations involve some work in all disciplines.

Figure 14-4 shows the entire UP life cycle: phases, iterations, and disciplines. It includes all the key UP life cycle features and is useful for understanding how a typical information system development project is managed.

The previous figures illustrate how the phases include activities from each discipline. But what about the detailed activities that occur within each discipline? The disciplines can be divided into two main categories: system development activities and project management activities. The six main UP development disciplines are:

- Business modeling
- Requirements
- Design
- Implementation
- Testing
- Deployment

For each iteration, the project team must understand the business environment (business modeling), define the requirements that that portion of the system must satisfy (requirements), design a solution for that portion of the system that satisfies the requirements (design), write and integrate the computer code

FIGURE 14-4

UP life cycle with phases, iterations, and disciplines

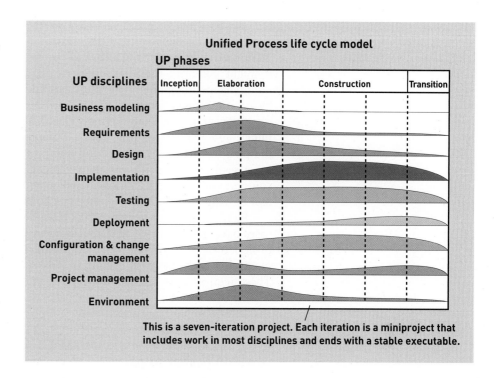

This is a seven-iteration project. Each iteration is a miniproject that includes work in most disciplines and ends with a stable executable.

that makes that portion of the system work (implementation), thoroughly test that portion of the system (testing), and then, in some cases, put the part of the system that is completed and tested into operation for users (deployment).

Three additional support disciplines are necessary for planning and controlling the project:

- Configuration and change management
- Project management
- Environment

Configuration and change management involves setting up processes to support the coding activities. This includes such guidelines as when and how to release code as well as when and how to manage releases and versions. Project management refers to the tasks that were discussed in Chapter 9, such as planning the iterations, assigning work, and verifying that work has been completed. The environment discipline involves those tasks required to establish the working environment, including the tools to be used by the team. It also includes those guidelines about how to work together in an iterative Agile project.

All nine UP disciplines are employed throughout the lifetime of a project but to different degrees. For example, in the inception phase, there is one iteration. During the inception phase iteration, the project manager might complete a model showing some aspect of the system environment (the business modeling discipline). The scope of the system is delineated by defining many of the key system requirements and listing use cases (the requirements discipline). To prove technological feasibility, some technical aspect of the system might be designed (the design discipline), programmed (the implementation discipline), and tested to make sure it will work as planned (the testing discipline). In addition, the project manager makes plans for handling changes to the project (the configuration and change management discipline), working on a schedule and cost/benefit analysis (the project management discipline), and tailoring the UP phases, iterations, deliverables, and tools to match the needs of the project (the environment discipline).

The elaboration phase includes several iterations. In the first iteration, the team works on the details of the domain classes and use cases addressed in the iteration (the business modeling and requirements disciplines). At the same time, it might complete the description of all use cases to finalize the scope (the requirements discipline). The use cases addressed in the iteration are designed by creating design class diagrams and interaction diagrams (the design discipline), programmed using Java or Visual Basic .NET (the implementation discipline), and fully tested (the testing discipline). The project manager works on the plan for the next iteration and continues to refine the schedule and feasibility assessments (the project management discipline), and all team members continue to receive training on the UP activities they are completing and the system development tools they are using (the environment discipline).

By the time the project progresses to the construction phase, most of the use cases have been designed and implemented in their initial form. The focus of the project turns to satisfying other technical, performance, and reliability requirements for each use case, finalizing the design, and implementing the design. These requirements are usually routine and lower risk, but they are key to the success of the system. The effort focuses on designing system controls and security and on implementing and testing these aspects.

As a system development methodology, the Unified Process must be tailored to the development team and the specific project. Choices must be made about which deliverables to produce and the level of formality, or ceremony, to be used. Sometimes, a project requires formal reporting and controls. Other times, it can be less formal. The UP should always be tailored to the project, although the UP does tend toward more ceremony than the next two methodologies.

Extreme Programming

Extreme Programming (XP) is an adaptive, agile development methodology that was created in the mid-1990s. The word *extreme* sometimes makes people think that this methodology is completely new and that developers who embrace XP are radicals. However, XP is really an attempt to take the best practices of software development and extend them "to the extreme." Extreme programming has these characteristics:

■ Takes proven industry best practices and focuses on them intensely
■ Combines those best practices (in their most intense forms) in a new way to produce a result that is greater than the sum of the parts

Figure 14-5 lists the core values and practices of XP. In the following sections, we first present the four core values of XP and then we explain its 12 primary practices. Finally, we describe the basic structure of an XP project and the way XP is used to develop software.

FIGURE **14-5**
XP core values and practices

XP core values	XP practices
• Communication • Simplicity • Feedback • Courage	• Planning • Testing • Pair programming • Simple designs • Refactoring the code • Owning the code collectively • Continuous integration • On-site customer • System metaphor • Small releases • Forty-hour week • Coding standards

XP Core Values

The four core values of XP—communication, simplicity, feedback, and courage—drive its practices and project activities. You will recognize the first three as best practices for any development project. You will also notice that the fourth is a desired value for any project, even though it might not be stated explicitly. Here are brief descriptions of the four core values of XP:

- Communication—One of the major causes of project failure is a lack of open communication among the right players at the right time and at the right level. Effective communication involves not only documentation but also verbal discussion. The practices and methods of XP are designed to ensure that open, frequent communication occurs.
- Simplicity—Even though developers have always advocated keeping solutions simple, they don't always follow their own advice. XP includes techniques to reinforce this principle and make it a standard way of developing systems.
- Feedback—As with simplicity, getting frequent, meaningful feedback is recognized as a best practice of software development. Feedback on functionality and requirements should come from the users, feedback on designs and code should come from other developers, and feedback on satisfying a business need should come from the client. XP integrates feedback into every aspect of development.
- Courage—Developers always need courage to face the harsh choice of doing things right or throwing away bad code and starting over. But all too frequently, they haven't had the courage to stand up to a too-tight schedule, resulting in bad mistakes. XP practices are designed to give developers the courage to "do it right."

XP Practices

XP's 12 practices embody the basic values just presented. These practices are consistent with the Agile principles explained earlier in this chapter.

Planning Some people describe XP as glorified hacking or as the old "code and fix" methodology that was used in the 1960s. That isn't true; XP does include planning. However, as an adaptive technique, it recognizes that you can't know everything at the start. As indicated earlier, XP embraces change. XP planning focuses on making a rough plan quickly and then refining it as things become clearer. This reflects the Agile development philosophical dictum that change is more important than detailed plans. It is also consistent with the idea that individuals—and their abilities—are more important than an elaborate process.

The basis of an XP plan is a set of stories that users develop. A story describes what the system needs to do. XP doesn't use the term *use case*, but a user story and a use case express a similar idea. Planning involves two aspects: business issues and technical issues. In XP, the business issues are decided by the users and clients, whereas technical issues are decided by the development team. The plan, especially in the early stages of the project, consists of the list of stories (from the users) and the estimates of effort, risk, and work dependencies for each story (from the development team). As in Agile development, the idea is to heavily involve the users in the project rather than have them to simply sign off on specifications.

Testing Every new piece of software requires testing, and every methodology includes testing. XP intensifies testing by requiring that the tests for each story be written first—before the solution is programmed. There are two major types of tests: unit tests, which test the correctness of a small piece of code, and acceptance tests, which test the business function. The developers write the unit tests, and the users write the acceptance tests. Before any code can be integrated into the library of the growing system, it must pass the tests. By having the tests written first, XP automates their use and executes them frequently. Over time, a library of required tests is created, so when requirements change and the code needs to be updated, the tests can be rerun quickly and automatically.

Pair Programming More than any other, this practice is one for which XP is famous. Instead of simply requiring one programmer to watch another's work, **pair programming** divides up the coding work. First, one programmer might focus more on design and double-checking the algorithms while the other writes the code. Then, they switch roles; thus, over time, they both think about design, coding, and testing. XP relies on comprehensive and continual code reviews. Interestingly, research has shown that pair programming is more efficient than programming alone. It takes longer to write the initial code, but the long-term quality is higher. Errors are caught quickly and early, two people become familiar with every part of the system, all design decisions are developed by two brains, and fewer "quick and dirty" shortcuts are taken. The quality of the code is always higher in a pair-programming environment.

pair programming XP practice in which two programmers work together on designing, coding, and testing software

Simple Designs Opponents say that XP neglects design, but that isn't true. XP conforms to the principles of Agile Modeling, as described in Chapter 8, by avoiding the "Big Design Up Front" approach. Instead, it views design as so important that it should be done continually, although in small chunks. As with everything else, the design must be verified immediately by reviewing it along with coding and testing.

So, what is a simple design? It is one that accomplishes the desired result with as few classes and methods as possible and that doesn't duplicate code. Accomplishing all that is often a major challenge.

Refactoring the Code **Refactoring** is the technique of improving the code without changing what it does. XP programmers continually refactor their code. Before and after adding any new functions, XP programmers review their code to see whether there is a simpler design or a simpler method of achieving the same result. Refactoring produces high-quality, robust code.

refactoring revising, reorganizing, and rebuilding part of a system so it is of higher quality

Owning the Code Collectively In XP, everyone is responsible for the code. No one person can say "This is my code." Someone can say "I wrote it," but everyone owns it. Collective ownership allows anyone to modify any piece of code. However, because unit tests are run before and after every change, if programmers see something that needs fixing, they can run the unit tests to make sure the change didn't break something. This practice embodies the team concept that developers are building a system together.

Continuous Integration This practice embodies XP's idea of "growing" the software. Small pieces of code—which have passed the unit tests—are integrated into the system daily or even more often. Continuous integration highlights errors rapidly and keeps the project moving ahead. The traditional approach of integrating large chunks of code late in the project often resulted in tremendous amounts of rework and time lost while developers tried to determine just what went wrong. XP's practice of continuous integration prevents that.

On-Site Customer As with all adaptive approaches, XP projects require continual involvement of users who can make business decisions about functionality and scope. Based on the core value of communication, this practice keeps the project moving ahead rapidly. If the customer isn't ready to commit resources to the project, the project won't be very successful.

System Metaphor This practice is XP's unique and interesting approach to defining an architectural vision. It answers the questions "How does the system work?" and "What are its major components?" And it does it by having the developers identify a metaphor for the system. For example, Big Three automaker Chrysler's payroll system was built as a production-line metaphor, with its system components using production-line terms. Everyone at Chrysler understood a production line, so a payroll transaction was treated the same way; developers started with a basic transaction and then applied various processes to complete it. Of course, a system metaphor should be easily understood or well known to the members of the development team. It can guide members toward a vision and help them understand the system.

Small Releases A release is a point at which the new system can be turned over to users for acceptance testing and even for productive use. Consistent with the entire philosophy of growing the software, small and frequent releases provide upgraded solutions to the users and keep them involved in the project. Frequent releases also facilitate other practices, such as immediate feedback and continual integration.

Forty-Hour Week and Coding Standards These final two practices set the tone for how the developers should work. The exact number of hours a developer works isn't the issue. The issue is that the project shouldn't be a death march that burns out every member of the team. Neither should the project be a haphazard coding exercise. Developers should follow standards for coding and documentation. XP uses just the engineering principles that are appropriate for an adaptive process based on empirical controls.

XP Project Activities

Figure 14-6 shows an overview of the XP system development approach. It is divided into three levels: system (the outer ring), release (the middle ring), and

FIGURE **14-6**

XP development approach

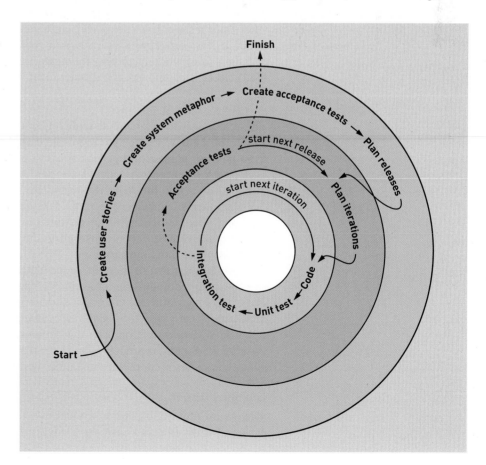

iteration (the inner ring). System-level activities occur once during each development project. A system is delivered to users in multiple stages called *releases*. Each release is a fully functional system that performs a subset of the full system requirements. A release is developed and tested within a period of no more than a few weeks or months. The activities in the middle ring cycle multiple times— once for each release. Releases are divided into multiple iterations. During each iteration, developers code and test a specific functional subset of a release. Iterations are coded and tested in a few days or weeks. There are multiple iterations within each release, so the iteration ring (inner) cycles multiple times.

The first XP development activity is creating user stories, which are similar to use cases in object-oriented analysis. A team of developers and users quickly documents all the user stories the system will support. Developers then create a class diagram to represent objects of interest within the user stories.

Developers and users then create a set of acceptance tests for each user story. Releases that pass the acceptance tests are considered finished. The final system-level activity is to create a development plan for a series of releases. The first release supports a subset of the user stories, and subsequent releases add support for additional stories. Each release is delivered to users and performs real work, thus providing an additional level of testing and feedback.

The first release-level activity is planning a series of iterations. Each iteration focuses on a small (possibly just one) system function or user story. The iterations' small size allows developers to code and test them within a few days. A typical release is developed by using from a few to a few dozen iterations.

After the iteration plan is complete, work begins on the first iteration-level activity. Code units are divided among multiple programming teams, and each team develops and tests its own code. XP recommends a test-first approach to coding. Test code is written before system code. As code modules pass unit testing, they are combined into larger units for integration testing. When an iteration passes integration testing, work begins on the next iteration.

When all iterations of a release have been completed, the release undergoes acceptance testing. If a release fails acceptance testing, the team returns it to the iteration level for repair. Releases that pass acceptance testing are delivered to end users, and work begins on the next release. When acceptance testing of the final release is completed, the development project is finished.

Scrum

Those of you who are familiar with rugby are aware that when a team gets possession of the ball, it attempts to go the entire distance in one continuous play— from point of possession to the score. The team works together, passing the ball back and forth; even when tackled, it can maintain possession and keep the ball in play. Originally, this "rugby" approach was applied to product development.

One interesting element in rugby is a scrum, which is used to get a ball back into play after a penalty. The defining characteristics of a scrum are that it begins quickly, is a very intense effort, involves the entire team, and usually only lasts for a short duration.

Combining some of these principles of rugby with the Agile philosophy gave rise to a methodology—the objective of which is to be quick, agile and intense and to go the entire distance. This methodology is referred to as the Scrum approach. Over time, the techniques have been refined to fit into a powerful adaptive software development methodology. **Figure 14-7** illustrates an overview of the Scrum approach. There are three important Scrum areas to understand: the philosophy, the organization, and the practices.

Scrum Philosophy

The Scrum philosophy is based on the Agile Development principles described earlier. Scrum is responsive to a highly changing, dynamic environment in

FIGURE **14-7** *Scrum software development process*

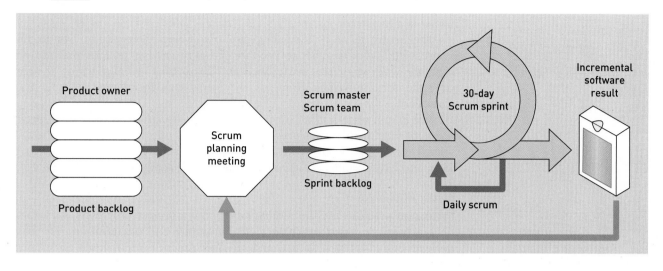

which users might not know exactly what is needed and might also change priorities frequently. In this type of environment, changes are so numerous that projects can bog down and never reach completion. Scrum excels in this type of situation.

Scrum focuses primarily on the team level. It is a type of social engineering that emphasizes individuals more than processes and describes how teams of developers can work together to build software in a series of short mini-projects. Key to this philosophy is the complete control a team exerts over its own organization and its work processes. Software is developed incrementally, and controls are imposed empirically—by focusing on things that can be accomplished.

The basic control mechanism for a Scrum project is a list of all the things the system should include and address. This list—called the **product backlog**—includes user functions (such as use cases), features (such as security), and technology (such as platforms). The product backlog list is continually being prioritized, and only a few of the high-priority items are worked on at a time, according to the current needs of the project and its sponsor.

Scrum Organization

The three main organizational elements that affect a Scrum project are the **product owner**, the **Scrum master**, and the Scrum team or teams.

The product owner is the client, but he or she has additional responsibilities. Remember that in Agile development, the user and client are closely involved in the project. In Scrum, the product owner maintains the product backlog list. For any function to be included in the final system, it must first be placed on the product backlog. Because the product owner maintains that list, any request must first be approved and agreed to by the product owner. In traditional development projects, the project team initiates the interviews and other activities to identify and define requirements. In a Scrum project, the primary client controls the requirements. This forces the client and user to be intimately involved in the project. Nothing can be accomplished until the product owner creates the backlog.

The Scrum master enforces Scrum practices and helps the team complete its work. A Scrum master is comparable to a project manager in other approaches. However, because the team is self-organizing and no overall project schedule exists, the Scrum master's duties are slightly different. He or she is the focal point for communication and progress reporting—just as in a traditional project. But the Scrum master doesn't set the schedule or assign

product backlog   a prioritized list of user requirements used to choose work to be done in a Scrum project

product owner   the client stakeholder for whom the system is being built

Scrum master   the person in charge of a Scrum project—similar to a project manager

tasks. The team does. One of the primary duties of the Scrum master is to remove impediments so the team can do its work. In other words, the Scrum master is a facilitator.

The Scrum team is a small group of developers—typically five to nine people—who work together to produce the software. For projects that are very large, the work should be partitioned and delegated to smaller teams. If necessary, the Scrum masters from all the teams can coordinate multiple team activities.

The Scrum team sets its own goal for what it can accomplish in a specific period of time. It then organizes itself and parcels out the work to members. In a small team, it is much easier to sit around a table, decide what needs to be done, and have members of the team volunteer or accept pieces of work.

Scrum Practices

The Scrum practices are the mechanics of how a project progresses. Of course, the practices are based on the Scrum philosophy and organization. The basic work process is called a **sprint**, and all other practices are focused on supporting a sprint.

sprint a time-controlled mini-project that implements a specific portion of a system

A Scrum sprint is a firm 30-day period called a *time box*, with a specific goal or deliverable. At the beginning of a sprint, the team gathers for a one-day planning session. In this session, the team decides on the major goal for the sprint. The goal draws from several items on the prioritized product backlog list. The team decides how many of the highest-priority items it can accomplish within the 30-day sprint. Sometimes, lower-priority items can be included for very little additional effort and can be added to the deliverables for the sprint.

After the team has agreed on a goal and has selected items from the backlog list, it begins work. The scope of that sprint is then frozen, and no one can change it—neither the product owner nor any other users. If users do find new functions they want to add, they put them on the product backlog list for the next sprint. If team members determine that they can't accomplish everything in their goal, they can reduce the scope for that sprint. However, the 30-day period is kept constant.

Every day during the sprint, the Scrum master holds a daily Scrum, which is a meeting of all members of the team. The objective is to report progress. The meeting is limited to 15 minutes or some other short time period. Members of the team answer only three questions:

■ What have you done since the last daily Scrum (during the last 24 hours)?
■ What will you do by the next daily Scrum?
■ What kept you or is keeping you from completing your work?

The purpose of this meeting is simply to report issues, not to solve them. Individual team members collaborate and resolve problems after the meeting as part of the normal workday. One of the major responsibilities of the Scrum master is to note the impediments and see that they are removed. A good Scrum master clears impediments rapidly. The Scrum master also protects the team from any intrusions. The team members are then free to accomplish their work. Team members do talk with users to obtain requirements, and users are involved in the sprint's work. However, users can't change the items being worked on from the backlog list or change the intended scope of any item without putting it on the backlog list.

At the end of each sprint, the agreed-on deliverable is produced. A final half-day review meeting is scheduled to recap progress and identify changes that need to be made for the following sprints. By time-boxing these activities—the planning, the sprint, the daily Scrum, and the Scrum review—the process becomes a well-defined template to which the team easily conforms, which contributes to the success of Scrum projects.

Trends in Technology Infrastructure

The technology infrastructure refers to the computing devices along with the operating system and communication software that allow those devices to function. The rapid pace of change in microprocessors for the past 30 years has allowed manufacturers to bring a continual stream of faster, more powerful computing devices to market. At first, there was the tremendous growth in the speed and computing capabilities of personal computers. The same increased capacity also occurred in reasonably priced server computers. No longer were large, expensive mainframe computers necessary to support high-speed and high-capacity computers. More recently, as the increase in capacity of personal and server computers has become more moderate, the most rapid changes are occurring in such handheld mobile devices as tablets and smartphones.

It is always difficult to predict where the next area of rapid growth will occur, and we won't try to do so in this book. However, these advances in equipment have led to more advanced and sophisticated application software. And not only have the software applications become more complex, but they also have had to be developed more rapidly to keep pace. Developers must be quick and agile just to keep up. This trend will continue in the future and will undoubtedly open up new opportunities for providing software applications in many different forms and on many different devices.

We will discuss the technology infrastructure in the following three areas: client and end-user devices, Internet and telephone communications, and back-end computing infrastructure for hosting applications.

Client Computing Devices

This is a broad category of devices that users interact with to communicate, play games, retrieve information, and perform everyday tasks. The most common devices are desktop, laptop, and notebook computers. However, the rapid rise of small portable devices, such as tablets and ebook readers, has broadened the utility of screen-based user computing. The exponential growth of Web technology has made these devices even more useful—for businesses and for individuals. And not only has Web technology and availability expanded exponentially (with wired and wireless LANs), it has been coupled with wireless telephone communications and is now available everywhere and anytime. The combination of these elements—portable, mobile devices and the wide availability of the Internet—has opened up a whole new world of opportunity and challenge for today's developers.

The capabilities of these devices continues to expand. Cross-capability also continues to drive new devices with expanded functions. For example, smartphones are no longer just telephone devices; they can also be used to play games, take pictures, take video, send and receive e-mail (with graphic attachments), browse the Internet, and watch TV. In the United States, the number of smartphones surpasses the number of PCs. Similarly, today's automobiles provide not just transportation but maps and directions; they also take pictures and videos of the view in front of and behind the vehicle. And book-reading devices no longer just provide downloaded books; they can now be used to review and purchase online books, music, and video as well as watch movies and TV. Many of these new devices are telephone and Wi-Fi enabled; thus, connectivity is always available by utilizing the most rapid methods. This trend toward wider connectivity, enhanced functionality, and increased mobility will continue into the foreseeable future. **Figure 14-8** shows the iPhone and several of the applications that are available for it.

Software for mobile devices is also changing. Three types of applications are currently found on stationary and mobile devices. The first type is a stand-alone application—for example, a word-processing application or an image-editing application—that executes on a single computer without requiring connectivity

FIGURE **14-8**
iPhone with several applications

Oleksiy Mark/Shutterstock.com

device-top application a computer application that is built to execute on a local computer (such as a desktop) without requiring a client/server connection

free-standing Internet application a client/server application that is self-contained and doesn't require the use of a Web browser

browser-based application a client/server application that executes locally within the control of a Web browser

to any other device. We refer to this type as a **device-top application**. The second type is a device-top application that is freestanding. **Free-standing Internet applications** are indeed device-top applications, but they access the Internet for certain functions—for example, iTunes. iTunes allows you to create playlists and to play music stored on your computer. However, when the computer has access to the Internet, iTunes also connects to Apple's iTunes site to provide broader and richer multimedia capabilities. The third type of application is browser-based software. A **browser-based application** is an Internet application that must have a browser, such as Internet Explorer or Firefox, to execute.

Before 2010, stand-alone device-top applications for mobile devices dominated the software market. Since then, new free-standing or browser-based Internet applications have outpaced stand-alone applications. Later sections of this chapter will further discuss software specifics for Internet-enabled applications.

These new devices produce all types of challenges for software developers—operating system developers and information system developers. For example, many of these devices require 24/7 availability and sometimes even 24/7 connectivity. They require high-definition multimedia capabilities, with user interfaces that are rich, engaging, and intuitive. The small screen size of many of these devices brings extra challenges regarding how to best present information that is elaborate and readable. Users now expect almost instant availability and access to information.

Although the news media focuses primarily on consumer devices for the general public, organizations and businesses continue to move rapidly toward the adoption of these new technologies. Product-based industries continue to seek new ways to handle, store, and distribute their products more efficiently and effectively. Service-based industries utilize stationary and mobile devices to provide enhanced levels of information and service for their clients. Opportunities abound for new applications to be developed and deployed on these devices.

Internet and Telephone Communications

The Internet is an incredibly complex system of interconnected computers and smaller networks. Sometimes, it is called the network of networks because it consists of millions of private, public, business, academic, and government networks. What ties all these individual computers and networks together is the **Internet backbone**, which consists of the primary data paths from large, strategically located networks and routers.

Internet backbone the primary data routes between large, strategically interconnected networks and routers on the Internet

The communication infrastructure also continues to change rapidly. Historically, the technologies of the Internet, the telephone, and television have grown up separately. Recently, we have seen the migration and merging of these three technologies' capabilities—from Internet to telephony, from telephony to Internet, from broadcast TV to Internet TV, and so forth. One can use the Internet not only to have audio and video communication sessions but to also place telephone calls. One can also access the Internet through wireless telephone connections. And one can watch TV and movies on either Internet-enabled or telephone-enabled devices.

Long-distance telephone communication and long-distance Internet communication have very similar requirements and capability. Because of that, some of the largest voice carriers are also the owners of the largest Internet backbone. In fact, Verizon become the world's most connected Internet backbone in 2010. Verizon is also the first company with plans to increase backbone speeds to 100 Gbits per second. In recent years, the growth of purely Internet traffic has moderated slightly. However, this has been more than offset by increases in telephone traffic, which includes synchronous phone calls, text, messaging, and multimedia.

last mile the final leg that delivers connectivity from the Internet network to the customer

The **last mile** has typically been and continues to be the most challenging element in telephone communications and Internet access. The term refers to that part of the communication link from the last backbone node to the local user. There are various methods used to implement this final link. Most common is by using telephone lines or cable TV (CATV) lines (either copper or optical fiber). Telephone lines have always been bidirectional. Now CATV has added bidirectional communication capability. More recently, due to the rapid growth of mobile devices, wireless solutions are becoming more prevalent. With the advent of 4G networks, the communication speeds and bandwidth are increasing dramatically. The advent of 4G networks will accelerate the penetration of mobile devices, including smartphones.

Back-End Computing

Back-end computing refers to the server computers that provide the content—dynamic and static—for all applications that access servers through the Internet. These applications include Internet-enabled free-standing applications as well as browser-based applications. Obviously, any type of client-server application requires server support. As more and more client computers try to access a particular application, the workloads on the servers becomes incredibly heavy. In Chapter 6, we discussed the elements of designing the network, including some of the alternatives: cloud computing, colocation, virtual servers, and virtual private networks. Those alternatives are only available because of the expansion of back-end computing capabilities.

Several factors are driving the need for large back-end computing services. The trend toward continuous connectivity to the Internet is one of them.

Another is the type of applications that are now desired by businesses and consumers. Today, many applications in the world of commerce keep a history of all the transactions that occur. For example, your phone company keeps track of every call made on your phone, with a tremendous amount of detail: date, time, to whom, how long, and so forth. The amount of data storage needed to maintain this amount of data is tremendous. Whether good or bad, organizations are keeping records of all types of activities, including purchases, credit card transactions, phone calls, Web sites visited, and even mouse clicks. Massive amounts of data storage are required to maintain this history (and then data mining it).

Another interesting trend is the consolidation of processing in central locations. When desktop and laptop computers first became relatively inexpensive, many applications and much business processing were done either on client computers or on localized networks. However, as larger and larger amounts of data are being captured, analyzed, and shared, it is more productive to use centralized servers, maintaining a single centralized copy and distributing only the results of data mining or data processing. Examples of the centralized processing and storage include such things as Google apps, Office 365, and Apple's cloud for consumers. (You can archive your music on Apple's "cloud" and not have to store it locally.)

All these factors have provided the impetus for organizations to consolidate computers in colocation facilities and server farms or purchase hosting services from companies that provide virtual servers or cloud computing. The two biggest activities on the Web today are searching and social networking. Companies—such as Google, Yahoo!, Microsoft, Facebook, and Twitter—that provide those services have tremendous server farms consisting of tens of thousands of server computers. Other large companies that sell hosting services also have server farms in the thousands of computers. **Figure 14-9** shows a typical server farm within a data center.

Content delivery networks (CDN) are also a rapidly increasing component of back-end computing. Because the Internet allows worldwide connectivity to any Web site, if the Web site is hosted at one location, many clients would have long transmission distances, with corresponding delays. To ameliorate this problem, many Web sites distribute their hosts at several locations around the world.

FIGURE

Server farm within a data center

Eimantas Buzas/Shutterstock.com

CDN providers, such as Akamai, Limelight, and EdgeCast, are companies that host and deliver this content from locations that are physically located closer to large markets of clients. CDN works especially well for such static content as images, audio, and video.

Trends in Application Software Availability

Partly due to the changes in the technology infrastructure described earlier, people and organizations are finding new ways to deploy and provide applications. Historically, when a large or small organization needed a software application to support some organizational procedure, it either developed that software itself or, if the problem was general enough, purchased software and modified it to fit the in-house procedure. One of the major trends in today's information systems environment is that new methods are available for obtaining software functionality. Let us discuss several of the more prevalent ones.

Software as a Service (SAAS)

A service is something that we purchase that does something for us. For example, we consider our utilities to be services. We don't have to have our own power generator to get electricity. We just buy what we need as we need it. Another example is service on our vehicles. When something breaks on our cars, we go to auto mechanics and ask them to fix it. We don't need to have our own repair shop. We purchase only the service.

software as a service (SAAS) a software delivery model similar to a utility, in which the application and its associated data are accessed via the Internet without locally installed programs

Software as a service (SAAS) follows that same basic idea. If an organization requires some services—for example, bookkeeping and accounting functions—it can either build or buy an accounting software system. Alternatively, it could find a firm that provides accounting services and buy only the accounting services it needs. As with any other utility type of service, it would purchase and pay only for those services it requires. It doesn't have to purchase—or install or maintain—the software system. Sometimes, this is referred to as **on-demand software**.

on-demand software another term for SAAS

Although the impetus for SAAS originally involved business software, more general-purpose consumer functions are also being provided as a service—for example, the editing and manipulation of photos and other graphics. You can purchase iPhoto or Adobe Photoshop and install it on your personal computer if you want to manipulate your own photos. Alternatively, there are now many Web-based photo services that provide many of these same functions. In many of those instances, not only does the user not have to buy the application software, but the service is also financed by advertising and is provided free to users.

SAAS software is usually hosted on a server farm, and the functionality is distributed over the Internet or a VPN. SAAS can be divided into two categories: client-data-oriented services and tools services. Each of these can also be further divided into business or organizational services and end-user services.

Client-data-oriented services are those services that maintain information about and data for each client. Each client has an account and requires authentication before it can use the service. If there are multiple employees that use the system, each must have access to a log-on capability. Common business-oriented SAAS services include accounting, customer relationship management (CRM), human resource management (HRM), content management (CM), supply chain management (SCM), and enterprise resource planning (ERP). Common end-user SAAS services include blog hosting, content management hosting, and photograph hosting. In all these cases, the client allows the SAAS provider to host and maintain the client's data. Depending on the sensitivity and importance of the data, this can require a high level of trust on the part of the client toward the SAAS provider.

FIGURE **14-10**

Comparison of owning software versus SAAS

Costs	Purchasing/owning software	SAAS
Software license	substantial	not required
Development or customizing	substantial	not required
Implementation and installation	substantial	not required
Usage fees	not required	as consumed or used
Configuration	not required	one time initialization
IT support staff	substantial	not required
Application support staff	substantial	not required
Training of users	required	required
Servers, networks, data storage	substantial	as consumed or used
Internet usage	required (possibly)	required

The major impetus for using an SAAS provider is the reduced cost of the service. For many end-user services, the cost is funded by advertising; therefore, the service is free to the user. For businesses, the cost is reduced substantially because none of the overhead of owning the software is required. **Figure 14-10** itemizes the primary differences between SAAS and owning one's own software. Although the figure doesn't specify dollar amounts, it should be evident that, overall, using SAAS is less costly—in the number of costs and the amount of each one.

SAAS tools often don't save the user's data. They just provide computer tool capability. Examples of SAAS that are tools include computer-aided design (CAD) tools and semiconductor design tools for businesses. There are also Web sites that generate such things as passwords, md5 encryption, or public key and private key combinations. In these instances, the users save their own data back on their local workstation. There is no need for extensive data storage capability by the service provider.

Application software that will be used in an SAAS environment must be developed with that end use in mind. Business functions must be developed to be able to handle all the varieties that occur across different companies. The software must be option-driven so a given business or user can configure the system to perform the functions as needed. Developing SAAS applications is much more complex; in fact, the software is often adjusted over time as new users require new functionality.

Other critically important issues with SAAS applications have to do with the security of the data. First, each client's data must be secure from outside intrusion. The data center where the equipment is maintained must also be physically secure, with adequate backup and recovery processes in place. Second, there must be clear and robust isolation of client data among the multiple client databases. Each client's data must be secure and incapable of being viewed by any other client. Just like providers of cloud computing and colocation services, SAAS providers must maintain high levels of security.

Open-Source Software

open-source software a method of developing, delivering, and licensing software that makes the application source code freely available to any interested developer or client

Open-source software (OSS) is one of the truly remarkable phenomena in the recent growth of the software industry. In the very earliest days of computer programming, application software was developed for a specific business or organization in order to satisfy a specific need. These software applications were at first developed internally. However, because of the commonality of business needs within specific industries, software development companies soon became prevalent. A software firm that specialized in a specific industry could consolidate its knowledge, expertise, and central pool of program source code to provide application software more efficiently and effectively. This phenomenon started in such narrow industry markets as life insurance, health insurance, and banking. (These early systems were often priced in the millions of dollars due to the limited customer base for these specific products.) The next step was the formation of software companies that developed applications for the general market. Because the market for these products is so large, these applications are usually sold for reasonable prices—the $30 to $1,000 range—even though they cost millions of dollars to develop. Such products include Microsoft Office, Adobe Photoshop, Intuit QuickBooks, and TurboTax.

At some point, developers began using a different model to provide applications—methods with various names, such as shareware, freeware, and free software. In 1998, the term *open-source software* came into existence with the formation of the Open Source Initiative (OSI), which is the organization that currently defines the terms and conditions of open-source definitions. The definition of open-source software includes a method for distributing application software and a method for licensing the software. Open-source software is distributed in source code form. It may also be distributed as a binary executable, but "open" means the source code is freely distributed. An open-source license gives the licensee the right to copy, modify, and redistribute the source code. Redistribution can usually be as a modified application or part of a larger, more elaborate application. Usually, credit must be given to the creator of the original source code.

The OSI has set specific requirements that have become the industry standard for the definition and use of open-source software. According to the OSI, simply giving away the source code doesn't make an application open source. It must conform to several criteria, including such items as:

- Source code—The source code must be open for distribution.
- Freely redistributable—Recipients of the source code may also distribute it.
- Derived works—The source code can be modified and distributed in its modified form.
- Distribution of license—The open license should apply to all derived or modified software.
- No discrimination—The license can't restrict who can receive or use the software.
- No related restrictions—The license can't restrict other software that may be distributed in conjunction with the open-source software.

It is estimated that the value of open-source applications distributed each year approximates $60 billion. There are over 180,000 open-source projects in the world today. There are also over 1,400 different unique versions of open-source licenses. Obviously, this has become a major part of the software application industry, and it must be an important consideration in the development and use of application software within any organization or even for an individual developer. Many businesses and organizations use open-source software as part of their normal business operations and as internal business tools. Specifically, considerable software development is done by using open-source software development tools. **Figure 14-11** illustrates a few of the many types of open-source software.

The business model for open-source software is also an interesting phenomenon. Because the software isn't sold or licensed for a fee, there has to be some other method for generating revenue. There are all types of organizations that develop open-source software—from individual programmers to large

FIGURE **14-11**
Several open-source software applications

Category	Name	Description
Business	Open Office	Word processing, spreadsheets, presentations, drawing, and simple database functions
	Open Project	Project management
Databases	MySQL	Database management system
	PostgreSQL	Database management system
Development	Eclipse	Java-based IDE with toolkit
	NetBeans	Java-based IDE with toolkit
Graphics applications	GIMP	Graphics manipulation
	KTooN	Vector animation toolkit
Security and privacy	GNU Privacy Guard	Encryption tool
	ClamWin	Antivirus program
Web development	Aptana Studio	Comprehensive Web-development IDE for Web languages
	SeaMonkey	Browser, e-mail, newsgroup, HTML-authoring tool
Servers/Internet	Unix	Operating system; several versions
	Apache	Web server and other similar projects
Internet applications	WordPress	Blog system
	phpBB	Bulletin board system
	Joomla	Content management system

foundations or organizations. The methods for funding these organizations also vary widely. The most obvious method, of course, is to request donations—either from individuals or other businesses. The developing organization will often have additional products or services that are sold for a price. Such things as installation services, code modification services, training, or technical support can also provide revenue. Another option is to have multitier or proprietary add-ons to the source code. Often, a base system is free, but a professional version with additional capabilities is sold for a price or for a subscription fee. Finally, some organizations have internal open-source groups that are funded by the organization itself. Universities have often begun research projects that became well-accepted open-source applications or system software. In addition to universities, some businesses believe that participation in the open-source community makes good business sense and will fund projects and applications for open-source distribution.

Perhaps the most interesting aspect of open-source software is how the development work gets done. For small open-source projects, an individual often develops the system based on his or her own knowledge and skills. Similarly, some projects have two or three developers who have worked together on other projects and who pool their skills and resources to develop the system. However, for large projects—for example, Apache, PostgreSQL, or

FIGURE **14-12** *Open-source software types of stakeholders*

Group	Description	Responsibilities
Directors	A core group that has control of the software system and organizes the work to be done. Sometimes, an organization can afford to have full-time paid directors.	Establish the vision and strategy. Approve major upgrades/enhancements. Collaborate on design of new features. Set schedules. Set the organizational structure. Choose team leaders.
Team leaders	Experienced developers who take responsibility for major portions of the system; often involved in software vision decisions.	Collaborate on designing new features. Organize the work for their portion. Verify the work quality. Maintain control lists of work in progress. Manage the tasks and developer assignments. Approve and manage developers.
Developers/ participants	Usually volunteer developers who accept assignments and do the actual programming.	Write code. Conduct tests. Write documentation and training.
User community	Users who have downloaded and installed the open-source software for use in their organization. Typically, they are technically experienced.	Install the software application. Join the community of users. Post comments in forums and discussions. Log identified errors. Suggest enhancements.
End users	The people who use the application to do their work. Typically, they are not technically experienced.	Use the application software. Refer to the documentation and community discussion boards to get answers. Identify errors. Suggest enhancements.

WordPress—a large group of developers is involved. Endeavors such as these usually have several ongoing projects and function much like a major organization. One major difference is that the organization that develops the system may not have an office or central location. Meetings and communication are often done entirely through online tools and online documentation. **Figure 14-12** lists some of the groups that may be involved in a large open-source project. Remember that for most large projects, the participants reside in locations throughout the world. Communication and coordination are done entirely through e-mail, online meetings, conversations, bulletin boards, discussion groups, and tracking logs.

The Web as an Application Platform

As the use of the Internet in mobile and computing devices becomes more widespread, a fundamental shift in our society—the way we work and the way we interact—is occurring. For many people—either at work or at home—the Web browser is the computer program they use the most; in fact, for some people, it is the only application they use. For many others, Web-based applications are used in almost all aspects of their employment. This trend toward immediate connectivity has also caused a fundamental shift in software development and deployment. The Web has become the primary environment for the deployment of new software applications and systems. The majority of new applications are being written for the Internet—either as browser-based applications or as free-standing Internet applications.

Development of this type follows the same pattern that has been discussed throughout this book. In this section, we focus on browser-based applications as a new type of development platform for application software.

FIGURE **14-13**
Evolution of the types of Web pages

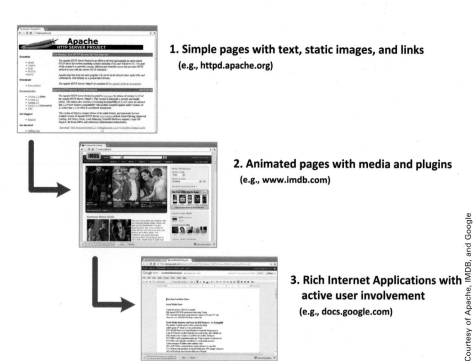

1. **Simple pages with text, static images, and links**
   (e.g., httpd.apache.org)

2. **Animated pages with media and plugins**
   (e.g., www.imdb.com)

3. **Rich Internet Applications with active user involvement**
   (e.g., docs.google.com)

Three major types of Web applications can be identified, as shown in **Figure 14-13**. The first type—Web pages—became prevalent in the mid- to late-1990s. These pages weren't really applications. They were information documents with static text, static images, and hyperlinks to navigate from page to page throughout the site. Their intent was to provide information and content via a hypertext type of document. Forms were also supported to enter information to be sent back to the site and captured by a program running on the server.

The second type of Web site began to appear early in the 2000s. These sites were more interactive, with animated graphics and plug-in modules that allowed richer content, and the functionality began to approach desktop applications. During this period, many new tools were invented to make Web sites much more dynamic, including such things as Java applets, Active X controls, and JavaScript scripting language. Other browser plug-ins—such as Flash, QuickTime, and Shockwave—allowed Web pages to function as rich applications that delivered all types of multimedia. The addition of efficient back-end databases permitted more dynamic information to be added to these pages. This was the era of attractive and engaging Web sites.

Most recently, we have seen the rapid growth of Web sites that are not only dynamic but are also truly interactive—much like desktop applications. In 2004, the term **Web 2.0** was first used to characterize this Web trend. It hasn't been precisely defined, but it generally refers to these highly interactive Web sites. Another term for this type of Web site, which we will discuss later in this section, is Rich Internet Application (RIA). The tools that allow these types of Web sites include more advanced JavaScript capabilities, such as Ajax, Java applets, widgets, plug-ins, and other components that execute within a browser on the user's Internet device. These not only allow interaction with the Web site via hotlinks, but they also support interaction and processing on the Web page itself. Thus, these pages function like a desktop software application.

In addition to interactivity, a major element of this recent phase is the ability to communicate and collaborate by using Web technology within the browsers. Such things as chatting, conferencing, sharing documents, and sharing photos and other personal relics have added a social collaboration element to

Web 2.0 a loosely defined, nonstandard term used to refer to Web sites that permit user-generated content and user interaction, such as social networking sites

Web sites. Increasingly, the users themselves are using these tools to set up their own Web sites as blogs, wikis, or accounts on such sites as Facebook and Twitter. In fact, today the largest use of the Internet is for social networking. Thus, whereas most Web sites 10 years ago were created by existing organizations and businesses, there are many more Web sites today created by individuals for personal reasons or to start a small business. Anyone can learn the opinions of others, express their own opinions, set up a blog, or even begin an online service or business. Web-based tools are available for all types of users to do almost anything on the Web.

The Internet is so all-encompassing today that it is next to impossible to categorize all the tools, techniques, methods, approaches, and capabilities that are found there. And, of course, it is an evolving landscape. Some of the tools and techniques that are popular today will continue to grow; others will fade away. In the following sections, we discuss three areas of Web application development. However, it should be recognized that these three topics aren't generalized classifications and especially not mutually exclusive topics.

Add-ons and Application Program Interfaces (APIs)

APIs are a powerful technique in the growth of capabilities for Web-based systems. In Chapter 10, an API was defined as the set of public methods that are exposed to external systems. In other words, they are the method names that any external component utilizes to plug into the system. Most OSS applications are also distributed with API documentation. However, many proprietary systems also publish an API so other developers can provide additional tools for users of these systems. Let us define a few of these terms and look at some examples.

Types of Web Software Components

plug-ins a software component that adds specific capabilities to a larger software application

Plug-ins are software components that add specific capabilities to a larger software application. They are found at many different levels of Web applications. Some work directly with the browser—for example, ones that allow browsers to play video, scan for malware, or add developer tools. In addition, many Web applications have plug-ins—for example, the OSS application WordPress. WordPress is a blogging software application that users can download and host on a server. It has many third-party functions (i.e., written by non-WordPress employees) that can be downloaded and installed. These allow WordPress to do such things as caching, keeping statistics, adding more menus, and locking pages. Plug-ins can be written by third parties because WordPress was built with specific access points and has a well-defined API that developers can use to integrate their plug-ins with the system. **Figure 14-14** lists a few popular plug-ins for Wordpress blogging software out of the hundreds that are available.

widget a type of plug-in that focuses on enhancing the user interface with additional capability

A **widget** is a type of plug-in, but it usually has a user interface component. In other words, it is a plug-in that can be placed on a Web page and is visible to the user. There are two kinds of widgets: browser widgets and application widgets. An example of a browser widget would be a time-and-temperature widget or a stock market widget that is displayed in one corner of the browser. It is always there—no matter what Web page is being viewed at the time. An example of an application widget would be something that enables the user to write blogs by using rich text formatting. A different application widget might maintain and display statistics about the blog, such as user comments or visits to the page. A **gadget** is another term that is used to describe a widget. Gadgets are most frequently used on a desktop, although some Web sites refer to their widgets as gadgets. **Figure 14-15** illustrates some Google gadgets that are available for Web pages for Google accounts.

gadget another term for a widget—often used for widgets that reside on a desktop

theme a type of add-on to an application that allows the look and feel, such as colors and layout, to be changed

A **theme** is a type of add-on that focuses on the look and feel of either the browser or the Web application. The development and use of a theme depends

FIGURE **14-14**

Sample plug-ins available on a WordPress blog

Plug-in name	Plug-in description
Ads Manager Plug-in	Quickly and easily inserts any ad code unit to your posts from Forum topics.
Akismet	Protects your blog from comment and trackback spam by accessing the Akismet database.
Artiss YouTube Embed	Embeds YouTube videos in your blog.
Awesome Flickr Gallery	Creates and customizes a gallery of your Flickr photos on your blog.
Fast Secure Contact Form	Easily configures and adds contact forms to your blog to allow users to send e-mails to the site administrator
Google Analytics Popular Posts	Uses Google Analytics API to fetch data from your analytics account and post it on your blog.
Social Sharing Toolkit	Enables sharing of your blog content via popular social networks.
What Others Are Saying	Uses the RSS field in your Blogroll to display the most recent post from sites that you link to.
WP to Twitter	Posts a Twitter status update from your blog.
WP Super Cache	Generates a static html file from your dynamic WordPress blog for faster service.

FIGURE **14-15** *Sample Google Web page gadgets*

Courtesy of Google

on the availability of an API that defines how to integrate CSS files and images into the application. Themes are a powerful technique that can drastically change the entire layout and look of a Web page. An example that shows dramatic differences of exactly the same content but with different themes is found at *www.csszengarden.com*.

A **toolbar** is a type of add-on that can provide multiple capabilities to the browser. It functions a little like a menu in that it provides a selection of functions or hotlinks for the user. Browser toolbars can be supplied by the browser provider; third-party toolbars are also available. These often consist of links that go to specific services provided by the third party. Toolbars may also be added to specific pages as part of the installation of a plug-in. If the plug-in has various functions that it can perform, it can add a menu or a toolbar to allow the user to access those functions.

Web mini-apps—sometimes referred to as *Web apps*—come in many varieties. Some of them are stand-alone applications that don't require a browser—for example, those that are available for such smartphones as iPhone and Android. (This is probably the largest number of apps available today.) Many of these function as desktop or device-top applications and don't require a browser to execute. Others are complete Web sites and execute their code within the browser. Still other mini-apps are plug-ins that are attached to a parent Web site and can only be accessed through the parent. Perhaps the most common example of these are Facebook apps. Facebook provides an extensive API definition that allows third-party developers to create many different types of games, personal apps, and even commercial apps.

Development

Historically, most software was developed by organizations—either regular businesses for their internal use or software development firms. However, open-source software, Web mini-apps, and add-ons are being developed by entrepreneurial independent programmers. Literally tens of thousands of programmers have joined the ranks of independent entrepreneurs. The locus of software development has moved out of business and industry and into the consumer arena. Everyone wants "cool stuff" for their phones, tablets, and gaming devices, which has nothing to do with such business applications as accounting or inventory management.

Another reason a cottage industry has developed is the ease of entry into software development. In the past, software development required large mainframes or large servers that cost hundreds of thousands if not millions of dollars. Only those corporate employees with access to these expensive resources could be programmers or developers. Today, anyone with a laptop and access to the Internet can obtain open-source software and begin developing applications.

In earlier chapters, we discussed the various stakeholders in the development of a software system. These include the client, the user, the architect, the analyst, the designer, and the programmer. When an entrepreneur develops software, he or she plays all these roles. Problems can occur if the developer's view of the problem to be solved is too limited. However, in some situations, the scope of the application is often small enough that even that isn't a major problem.

Mashup Applications

The open-source perspective also permeates the development and use of Web applications. Today, thousands of Web sites provide APIs to access the various services provided on the Web site. Through the use of these open APIs, the services on one Web site can be added to the total presentation of a different Web site. The idea of a **mashup** is to "mash" the services of two or more Web sites together to provide a new service or new way of viewing information. Mashups are an important trend in software, particularly social software and Web 2.0.

toolbar a type of add-on usually comprised of iconic menu items that access the capabilities of the application or plug-ins in a user-friendly fashion

Web mini-app a software application that provides a complete set of functions but that must be executed within the confines of another application

mashup a type of Web site that combines the functionality of several other Web sites through the use of predefined APIs

Obviously, the key to mashups is the availability of open APIs provided by various Web services. You will remember that an API is defined as a set of method calls to a class or a component. For Web-based APIs, those calls are expressed as particular URLs, which return to the originating source Web site. In other words, a mashup is a combination of data or services from multiple, physically separated Web sites.

One way to categorize these APIs is by whether they provide data or a service through this URL access.

Data-type APIs often provide indexes of documents, images, videos, or items for sale. Another example of a data-type API is a news aggregator for news feeds or podcasts. Service-type APIs do such things as convert data from one form to another—for example, a language translator or a URL shortener. Another example of a service-type API is a security Web site that performs authentication or encryption. Communication services, such as instant messaging or e-mail, can also be embedded within a parent Web site through a service-type API. Of course, some APIs provide both. For example, the Google Maps API not only provides the raw geolocator information but will render it in a user-friendly viewer.

The more popular open APIs used in mashups include Google Maps API, YouTube API, Twitter API, Flickr API, Facebook API, eBay API, and Google Search API. **Figure 14-16** is a snapshot of a mashup that lets you build your own dashboard page. It uses APIs from many Web sites.

FIGURE **14-16** *Mashup of several APIs to create a dashboard page*

Courtesy of Pageflakes

FIGURE **14-17** *Comparison of results from Google and Yahoo! search engines*

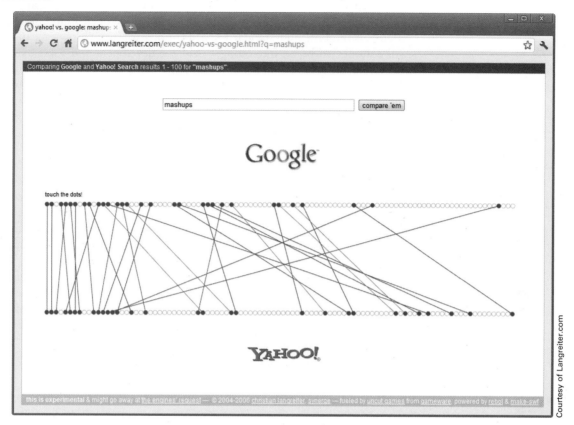

Courtesy of Langreiter.com

From a developer's viewpoint, open APIs are a simple yet powerful tool to enhance and extend a Web site. In fact, some APIs are so simple that the end users are able to add functionality to their Web site themselves. It isn't uncommon to have personal blogs or Web sites with search boxes, maps, or videos embedded in the Web site. This trend will probably continue as more open APIs are made available. **Figure 14-17** shows a Web site that compares the capability of Google and Yahoo!'s search engines. In this case, two simple APIs are used and then the results are displayed in an interesting graphical format.

Rich Internet Applications (RIA)

Another very powerful trend in the use of the Web as an application platform is the development of **Rich Internet Applications (RIAs)**. An RIA is a Web application that is built to have the same rich functionality and responsiveness as a desktop application. Because it is Internet connected, it also frequently delivers rich multimedia. Desktop applications can be very versatile and extremely responsive—their functions constructed to do exactly what the user requires—with a very efficient and rapid response. In addition, the layout and presentation of the desktop can be tailored for optimal user friendliness.

Historically, Web applications have had more limited capabilities. As explained previously, the early vision of the Web (and hence the development of Web browsers) was as a means of presenting information. Recently, however, as more tools have become available, Web applications have permitted a much richer user experience, with many services and capabilities. Today, Web applications are approaching desktop applications in versatility and usability.

The key to the development of RIAs is the addition of computing on the local client computer—within the browser itself. The delay going to and from the server prohibited a rapid and rich user experience in purely URL-driven

Rich Internet Applications (RIAs)
a type of Web site that provides active user interaction as well as delivers rich multimedia

Web sites. RIAs require enhanced computing capability locally and limited round trips to the server. We will now describe five of the major approaches to developing RIAs. As is frequently the case, there are open-source solutions, proprietary solutions, and standard specifications for new browser capabilities. It will continue to be an interesting world for developers and users as these various platforms compete and migrate to new levels. It is difficult to illustrate in a single image all the power and capability of an RIA Web site. **Figures 14-18** and **14-19** illustrate two Web sites that include interesting animation and allow dynamic user interaction.

JavaScript and Ajax Libraries

JavaScript has become the de facto standard for adding desktop-like computing within browsers. It is a powerful, object-based scripting language that can detect and trigger actions based on keystrokes within browsers. It is also used to access and manipulate all the components on the Web page as part of the document object model (DOM). (This is a hierarchical tree dynamically built by the browser of every element on a given Web page.) Detecting keystrokes and manipulating the DOM are two critical factors that allow JavaScript to support almost any type of desktop-like behavior that is limited to the local client.

In the late-1990s, Microsoft developed a set of JavaScript methods and techniques to allow data access to the server in order to dynamically update and manipulate the Web page without having to refresh the entire Web page. This set of methods and techniques, termed Ajax (an acronym for

FIGURE **14-18** *RIA from Nike Plus*

Courtesy of nikerunning.nike.com

FIGURE **14-19** *RIA from Mini Cooper*

Courtesy of Miniusa.com

Asynchronous JavaScript with XML), allows the Web application to communicate with the server in the background (asynchronously) to request data, wait for it, and then process it when it arrives. This added capability permitted JavaScript to extend its desktop-manipulation capabilities to include dynamic data updating.

The ability of JavaScript to support rich Web apps has been enhanced by the proliferation of libraries of JavaScript tools. The most popular libraries include Dojo, jQuery, MooTools, and YUI (Yahoo! User Interface). With the use of one of these libraries, Web site developers have a whole set of functions—such as user interaction, grids, graphs, Ajax, widgets, forms, and pop-ups—that add the richness of a desktop application to a Web page.

ICEFaces and JavaFX

ICEFaces is an open-source framework that provides Java language APIs to build and deploy server-based RIAs. The underlying framework that enables ICEFaces is Ajax; hence, it depends on the JavaScript language but doesn't require programmers to interact with JavaScript. ICEFaces applications are JavaServer Faces (JSF) applications and are included in the Java Enterprise Edition toolset. The purpose of ICEFaces is to enable Java developers to utilize their skill set with the Java EE development model and to protect them from doing low-level JavaScript and Ajax programming.

Another, more recent Java-based toolset is JavaFX, which was introduced in 2007. As a relatively new language and platform, it is in the early stages of providing comprehensive support for RIAs. However, JavaFX consists entirely of Java constructs, and it executes as part of the Java Virtual Machine (JVM) runtime environment. Because it runs under the Java Runtime Environment (JRE), it can be used to build RIAs for most devices that can run the JRE. For JavaFX 1.3, this includes desktops, browsers, and mobile phones. Another advantage of using the JRE to produce RIAs with JavaFX is that there is more consistency across browsers and platforms than with languages, such as JavaScript, that depend on the implementation provided by each browser.

Adobe Flash Platform

The Adobe Flash platform was introduced in 1996 and is best known as a multimedia platform for adding animation and interactivity to Web pages. Flash's power is in its ability to animate many elements on a Web page. It can capture user input from all computer components, including a mouse, a keyboard, a microphone, or a camera. It also supports bidirectional streaming of audio and video. The Adobe Flash player is available on desktops, within browsers, and on some mobile phones (but not the iPhone). With this rich history and background, Adobe has extended the Flash platform to enable the development and deployment of RIAs. One major advantage of Flash is that the Flash player has a 95–99 percent penetration rate on computers. It also has a very rich set of multimedia functions that are supported by a deep set of development tools. However, Flash is a proprietary product owned and licensed by Adobe and hence must be licensed from Adobe, although some components have been released as open-source products.

Microsoft Silverlight

Microsoft Silverlight is an application framework, similar to Adobe Flash, that is part of Microsoft's Windows Communication Foundation (WCF) .NET RIA Services to support the development and deployment of RIAs. Silverlight was introduced in 2007 and went through several rapid iterations before reaching a rich set of components to enhance the user experience with Silverlight 5, which was released in the fall of 2011. Silverlight requires a plug-in to execute, and plug-ins for all the major browsers are available for download. As of 2011, it had a penetration rate of approximately 75 percent; thus, it has grown very rapidly, although not as rapidly as JavaScript or Adobe.

A Silverlight application is developed as part of a Web Services application within the WCF .NET RIA services. The Web Services app provides the functionality on the server side, and the Silverlight app provides the functionality on the client side. User interfaces are defined by using Extensible Application Markup Language (XAML), which can be used to define graphics and animations that execute within the browser plug-in. Data access is also done asynchronously; thus, pages can dynamically send and retrieve data to and from the server without reloading pages.

HTML5

The World Wide Web Consortium (W3C) is the body that defines standards for HTML and XHTML. As these standards are defined and agreed upon, all Web browsers are expected to conform to them. Of course, in the real world, standards aren't always conformed to, and even when a standard is agreed upon, differences in implementation frequently produce different results. The Web Hypertext Application Technology Working Group (WHATWG), a working group of the W3C, began developing **HTML5** in 2004, and in 2008, it published the first working draft of the specification. As of 2011, the specification is still in the Last Call stage of the Working Draft, with a target date for Recommendation of 2014. Even though it appears that the specification won't

HTML5 the new HTML specification that standardizes RIA specifications for built-in browser delivery

be fully completed for years to come, in reality, many of RIA capabilities identified in the specification have already been implemented in today's browsers. In fact, some proponents of HTML5 see it as the wave of the future, replacing other RIA approaches.

The advantage of the HTML5 specifications for RIA Web pages is that rich presentation of audio, video, graphics, and animation is built into the browsers, without the need for specialized plug-ins or languages. For example, the new HTML5 has such tags as <video>, <audio>, and <canvas> as well as other vector graphics manipulation tags to provide rich display of data and images. The specification also includes APIs that can be used for manipulating elements on a Web page. Such APIs as drag-and-drop, file handling, geolocation, and SQL database access are included in the specification.

Chapter Summary

One of the most active trends in software development is adaptive development methodologies. The world is changing in many ways, with new consumer devices, new services, and new technology. To keep pace with these changes, the way software is developed has also changed.

The most formal adaptive process is the Unified Process (UP). It was one of the first methodologies to be formalized, with specific definitions for iterations and processes. Other more radical adaptive methodologies are now being promoted and used. Two of the more popular ones are Extreme Programming and Scrum.

Extreme Programming (XP) and Scrum are methodologies that embody Agile principles. Two core elements of XP are that system tests are written first and that programmers work in pairs to design, code, and test the software. Thus, when a function is completed, it has not only been designed and coded, it has been reviewed and tested.

The Scrum approach defines a specific goal that can be completed within four weeks. During that four-week sprint, the project team is protected from all outside distractions so it can complete the defined goal. A product backlog of all outstanding requests is maintained by the client, and changes to the work the team is doing are only allowed between sprints.

Major trends in technology, mobile computing devices, software availability, and Web access are the driving forces behind the need to develop application software more rapidly.

Client computing devices include all those devices that allow users to communicate, play games, retrieve information, and perform other life-related tasks. The most common client computing devices are desktop, laptop, and notebook computers. However, the number of computers will soon be surpassed by the number of mobile devices, such as smartphones and tablet computing devices. These devices have become so pervasive because of the expanded availability of Internet access

through wireless and telephone connections. Finally, the availability of so many connected computing devices has necessitated the growth of large-scale data centers with very large server farms. These server farms, which sometimes consist of thousands of computers, are needed to support the Internet activity of popular Web sites and applications.

Historically, most software was developed within large organizations and was private to that organization. However, there is a trend toward sharing software applications among many users and organizations. One such method is called Software as a Service (SAAS), in which third-party companies have software that can be used by many companies. Instead of delivering and installing the applications to the purchasing organization, only the use of the software is sold. The idea is to sell a service, such as accounting, the same way a utility is sold to individual households.

Another trend is providing software applications free of charge as open-source software. Open-source software is usually developed by individuals and teams of people that are distributed throughout the world and work together on an application in a loosely configured project. The business model for this type of development is to not generate revenue from the basic application but sell add-on services and capabilities.

A final—and extremely important—trend is the use of the Web as an application platform. The largest percentage of new application software being developed is for Web applications. Coupled with the open-source software trend, many Web-based software applications provide APIs so Web-based applications can share functionality and even be combined to provide new uses of Web-based software. Most Web sites are also including sophisticated local computing capabilities to provide a Rich Internet Application (RIA). RIAs allow a Web-based application to function much like a desktop application, with multimedia and active user interactions.

Key Terms

browser-based application   458

device-top application   458

free-standing Internet applications   458

gadget   467

HTML5   474

Internet backbone   459

last mile   459

mashup   469

on-demand software   461

open-source software   462

pair programming   452

plug-ins   467

product backlog   455

product owner   455

refactoring   452

Rich Internet Applications (RIAs)   471

Scrum master   455

software as a service (SAAS)   461

sprint   456

theme   467

toolbar   469

UP discipline   447

Web 2.0   466

Web mini-apps   469

widget   467

Review Questions

1. What are the four UP phases, and what is the objective of each?
2. What are the six UP development disciplines?
3. What are the three UP support disciplines?
4. List the basic principles of Agile Modeling.
5. Why is the word *extreme* included as part of Extreme Programming?
6. List the core values of XP.
7. List the XP practices.
8. What is the product backlog used for in a Scrum project?
9. Explain how a Scrum sprint works.
10. Explain the difference between device-top applications, free-standing applications, and browser-based applications.
11. What is the "last mile" problem?
12. Explain what SAAS is and why it is an economically attractive alternative.
13. What are the six licensing criteria that are usually associated with open-source software?
14. List eight or 10 popular open-source applications. You may include some not mentioned in the text.
15. Explain the evolution of Web pages.
16. Explain how an API works.
17. What is the difference between widgets and plug-ins and themes?
18. What is meant by Web 2.0?
19. What is a mashup and why is it an important contribution to Web 2.0?
20. What is an RIA and why is it an important contribution to Web 2.0?
21. What is the difference between JavaScript and JavaFX?
22. What is HTML5?

Problems and Exercises

1. The Unified Process (UP) was first developed by a company called Rational, which is now owned by IBM. On the IBM Web site, find any information about UP tools available through IBM/Rational. Briefly describe the suite of tools available. Also, look on the IBM Web site and other Web sites (such as the Agile Modeling Web site) for opinions on the relationships and commonality between the UP and Agile Modeling. Report your findings.

2. Consider XP's team-based programming approach in general and its principle of allowing any programmer to modify any code at any time in

particular. No other development approach or programming management technique follows this particular principle. Why not? In other words, what are the possible negative implications of this principle? How does XP minimize these negative implications?

3. Visit the Web sites of the Agile Alliance (www.agilealliance.org) and Agile Modeling (www.agilemodeling.com). Find some articles on project management in an Agile environment. Summarize key points that you think make project management more difficult in this environment than in a traditional, predictive project. Do the same for key points that make project management easier for an Agile project.

4. Visit the Web site of the World Wide Web Consortium (www.w3.org) and review recent developments related to the HTML5 standard.

What are some of the basic components of the standard? What are the major additions to HTML4?

5. Find a company in your community that uses Scrum or XP (or variations thereof) as its development methodology. Learn how the company has applied the methodology and how it applies the principles and practices. Also, research what development tools it uses and how well the methodology is supported.

6. Find someone in your community who is working on a software development project that is using Agile principles. How was the team trained to use Agile Development? How was this approach adopted in the organization? What is the general feeling about its success? What aspects does this developer like? Which aspects does he or she find frustrating or difficult to use?

Case Study

Midwestern Power Services

Midwestern Power Services (MPS) provides natural gas and electricity to customers in four Midwestern states. Like most power utilities, MPS has seen significant changes in federal and state regulations in the last several years. Federal deregulation opened the floodgates of change, but there was little guidance from the federal government on how that would shape the industry's future. State legislatures also significantly changed their laws and regulations. The industry went through tremendous upheaval, with significant problems created by power shortages at several California power companies. Today, regulations such as the Sarbanes-Oxley Act are changing the landscape again. These new regulations seriously affect all areas of business, including accounting, record keeping, power purchases, distribution agreements, and customer consumption and billing.

New and proposed regulations seek to increase controls and expand competition for electricity and natural gas. The final form these regulations will take is unknown, and the exact details will probably vary from state to state.

MPS needed to rapidly prepare its systems for these new regulations. Three systems are most directly affected: one for purchasing wholesale natural gas, one for purchasing wholesale electricity, and one for billing customers for combined gas and electric services. The billing system isn't currently structured to separate supply and distribution charges, and it has no direct ties to the natural gas and electricity purchasing systems. MPS's general ledger accounting system is also affected because it is used to account for MPS's own electricity-generating operations.

MPS plans to restructure its accounting, purchasing, and billing systems in the following ways:

■ Customer billing statements will clearly distinguish between charges for supply and distribution of gas and electricity. The wholesale suppliers of each power commodity will determine prices for supply. Revenues will be allocated to appropriate companies—for example, distribution charges to MPS and supply charges to wholesale providers.

■ MPS will create a new payment system for wholesale suppliers to capture per-customer revenues and to generate payments from MPS to wholesale suppliers. Daily payments will be made electronically based on actual payments by customers.

■ MPS will restructure its own electricity-generating operations into a separate profit center—similar to other wholesale power providers. Revenues from customers who choose MPS as their electricity supplier will be matched to generation costs.

MPS's current systems were developed internally many years ago. The general ledger accounting and natural gas purchasing systems are mainframe based. They were developed in the mid-1990s, and incremental changes have been made ever since. All the programs are written in Visual Basic; DB2 (a relational DBMS) is used for data storage and management. There are approximately 50,000 lines of Visual Basic code.

The billing system was rewritten from the ground up in the mid-1990s and has been slightly modified since that time. The system runs on a cluster of servers that use the UNIX operating system. The latest version of

(continued on page 478)

(continued from page 477)

Oracle (a relational DBMS) is used for data storage and management. Most of the programs are written in C++, although some are written in C and others use Oracle Forms. There are approximately 80,000 lines of C and C++ code.

MPS has a network that is used primarily to support terminal-to-host communications, Internet access, and printer and file sharing for personal computers. The billing system relies on the network for communication among servers in the cluster. The servers that support the accounting and purchasing systems are connected to the network, although that connection is primarily used to back up data files and software to a remote location.

MPS is currently in the early stages of planning the system upgrades. It hasn't yet committed to specific technologies or development approaches. It also hasn't yet decided whether to upgrade individual systems or replace them entirely. The target date for completing all system modifications is three years from now, but the company is actively seeking ways to shorten that schedule.

1. What would you recommend as an approach to upgrading the three listed applications—a single total project or three individual projects? Explain your decision.

2. Describe the pros and cons of the UP approach versus XP and Scrum development approaches to upgrading the existing systems or developing new ones. Do the pros and cons change if the systems are replaced instead of upgraded? Do the pros and cons vary by system? If so, should different development approaches be used for each system?

3. Assume that MPS has had very little experience with developing projects by using adaptive techniques. Do you think it would be viable for them to attempt an adaptive approach for these three systems? What would be your recommendation for each? Which method would you recommend and why?

4. Assuming MPS decided to use one of the three methodologies discussed in the chapter, make a list of potential problems and the steps they should take to avoid those problems. List any activities they should consider to ensure success.

5. Assuming that MPS decided to develop and deploy each system individually, what would you recommend for a development approach? Would you recommend any SAAS solutions? Would you recommend using MS Visual Studio and IIS or a Unix environment by using some open-source applications? Explain your decision.

RUNNING CASE STUDIES

Community Board of Realtors

The Community Board of Realtors Multiple Listing Service is a small system with limited requirements. In Chapter 9, you identified a complete list of use cases and divided the system into two subsystems. Using the results from your earlier work, please do the following:

1. Based on Figure 14-6—the XP methodology—divide your use cases into releases and iterations within each release. Develop a project iteration plan that includes the necessary activities at each level (system, release, iteration) for integration testing and acceptance testing. Compare your answer to this question to the project iteration plan you developed for Chapter 9.

2. Discuss the requirements of this system for mobility devices. What use cases would be best utilized on a mobile device? What use cases would be best with a desktop user interface?

3. Would this application be suited as an SAAS application? What would be critical factors for a company to consider if it wanted to offer an SAAS version of this application?

4. Can you identify any use cases that would best be implemented as a mashup Web application? Discuss which ones might fit this requirement and why?

The Spring Breaks 'R' Us Travel Service

Recall from Chapter 2 that SBRU's initial system included four major subsystems: resort relations, student booking, accounting and finance, and social networking.

In Chapter 9, you developed a comprehensive list of use cases for each of the four subsystems. You also developed a project plan by using the adaptive SDLC and

(continued on page 479)

(continued from page 478)

other principles that you learned in this textbook. For this chapter, you will develop a project plan by using XP. The approach for ordering the use cases will be based on user interface issues.

1. For each subsystem, build a table with the following columns:
   a. Use case name—as defined in Chapter 9
   b. Primary use device—whether this use case will be used primarily on a mobile device or a desktop device or equally prevalent on each
   c. Development platform—whether this use case best fits as a desktop app, a free-standing Internet app, or a browser-based app

d. RIA—whether this use case should be an RIA (for browser-based applications)
2. Based on your table from question 1, organize the use cases by using two criteria:
   a. The logical ordering based on functional similarity (i.e., as you did in Chapter 9)
   b. Group them together as much as possible based on primary use device, development platform, and RIA characteristics. In other words, you are trying to help the team so everyone is building similar types of use cases together.
3. Assuming two separate XP programmer teams, build a project iteration plan as you did in Chapter 9.

On the Spot Courier Services

As you read this chapter, you probably noted that the development methodology used in this textbook has many things in common with the Unified Process, Scrum, and Extreme Programming. In fact, our objective in this textbook is to teach you the principles common to all these methodologies without forcing you to accept only one.

Given these four methodologies—Satzinger-Jackson-Bird (SJB), UP, Scrum, XP—and what you now know about Agile and iterative development, do the following for the development of the On the Spot system:

1. Choose a single methodology. Why did you choose that one?

2. Mix and match practices from each methodology. Discuss which ideas you like from each methodology.

Given the trends in new technology, software availability, and the Web as an application platform, answer these questions:

1. What kind of equipment would be best and most stable for the truck drivers to use?
2. What would you recommend as the development approach and platform: a custom application using Visual Studio and .NET, a custom application using Java or similar language, a Web application using ASP.NET, a Web application using PHP and JavaScript, or some other combination. Discuss your recommendation.

Sandia Medical Devices

Based on the discussion of hardware, Internet, and software technology trends in this chapter, it should be clear to you that the Real-Time Glucose Monitoring (RTGM) system is an interesting combination of older and newer technology. Except for the interface to software and data on mobile phones, the server-side portions of the system are a relatively traditional business-oriented application that can be implemented by using old-fashioned technology. What makes the RTGM system "new" are its client-side functions, including the automated collection of glucose levels, the regular transmission of that data to servers, the integration of communication between patients and health-care providers, and the integration

of those functions within software installed on a portable device that can be carried in a user's pocket.

With that in mind, answer the following questions. You may need to do some additional research to fully address them.

1. The chapter classified apps on portable devices as device-top, free-standing Internet, and browser-based. Which type is most appropriate for the client-side portions of the RTGM system? Be sure to consider such issues as client-server communication requirements and frequency, user interface quality, and portability across devices and operation systems.

(continued on page 480)

(continued from page 479)

2. Which (if any) social networking capabilities might make a useful addition to the RTGM system? Be sure to consider the HIPAA requirements described for this case at the end of Chapter 6.

3. When recorded glucose levels generate high-priority alerts, physicians or other health-care providers initiate direct contact with the patient. An ordinary phone call over the cellular phone network is one way to support direct contact. Because any client-side device used with the RTGM system must be fully Internet-capable, an Internet telephony application, such as Skype, is another possible way of supporting synchronous voice or video communication with the patient. Should Skype or a similar Internet telephony application be used with the RTGM system? Why or why not? If such an application is used, should it support video? Why or why not?

Data mining is an increasingly important technique for medical research. The ability to scan medical records of large numbers of patients over extended time periods enables researchers to better evaluate the effectiveness of drugs and therapies, more accurately connect disease risk levels to specific patient characteristics, and identify patterns of transmission or occurrence, progression, and treatment response for rare diseases and conditions. What types of medical research might be enabled or better supported by the data collected by the RTGM system? Would your answer change if the database were extended to include additional information that might be gathered from the patient's mobile phone (e.g., location information when each glucose level was captured, size and content of the patient's contact list, call history, and the volume of text messages and Internet browsing activity)?

Further Resources

Agile Alliance, www.agilealliance.org.

Scott W. Ambler, *Agile Modeling: Effective Practices for eXtreme Programming and the Unified Process*. John Wiley and Sons, 2002.

Scott Ambler, John Nalbone, and Michael J. Vizdos, *The Enterprise Unified Process: Extending the Rational Unified Process*, Prentice Hall, 2005.

Ken Auer and Roy Miller, *Extreme Programming Applied: Playing to Win*. Addison-Wesley, 2002.

Kent Beck, *Extreme Programming Explained: Embrace Change*. Addison-Wesley, 1999.

Mike Cohn, Succeeding with Agile: Software Development Using Scrum, Addison-Wesley, 2010.

Philippe Kruchten, *The Rational Unified Process: An Introduction*. Addison-Wesley, 2004.

Craig Larman, *Agile and Iterative Development: A Manager's Guide*. Addison-Wesley, 2004.

"Manifesto for Agile Software Development," Agile Alliance, www.agilemanifesto.org.

Pete McBreen, *Questioning Extreme Programming*. Addison-Wesley, 2003.

Andrew Pham and Phuong-Van Pham, *Scrum in Action*, Course Technology, 2011.

Ken Schwaber and Mike Beedle, *Agile Software Development with Scrum*. Prentice Hall, 2002.

INDEX